Vice President of Editorial	Melissa Wagner
Director of Acquisitions	Mikaela Felty
Director of Book Development	Jocelynn Hessler
Production Manager	Mike Brehm
Cover Design	Jose Rosado

Published in the United States of America by
IGI Global Scientific Publishing
701 East Chocolate Avenue
Hershey, PA, 17033, USA
Tel: 717-533-8845 | Fax: 717-533-7115
Website: https://www.igi-global.com E-mail: cust@igi-global.com

Library of Congress Cataloging-in-Publication Data

LCCN: 2025027499 (CIP Data Pending)
ISBN13: 9798337339627
Isbn13Softcover: 9798337339634
EISBN13: 9798337339641

British Cataloguing in Publication Data
A Cataloguing in Publication record for this book is available from the British Library.

Chalcogenide-Based Materials for Optoelectronics, Energy, and Sustainability

Karthik Kannan
https://orcid.org/0000-0001-5438-2460
National Chung Cheng University, Taiwan & Karpagam Academy of Higher Education, India

Vinaya Tari
https://orcid.org/0000-0002-8352-252X
Universitas Airlangga, Indonesia

IGI Global
Scientific Publishing
Publishing Tomorrow's Research Today

Table of Contents

Preface xvii

Acknowledgment xxiii

Chapter 1
Introduction to Chalcogenide-Based Nanocomposites Advances in Optoelectronics Energy Systems and Sustainability 1
Periasamy Palanisamy, Nehru Institute of Engineering and Technology, Anna University, Comibatore, India
S. Satheeskumar, K.S. Rangasamy College of Technology, India

Chapter 2
Synthesis, Functional Properties, and Characterization of Chalcogenide-Based Nanocomposites 39
Chinnaiah Karuppaiya, Kalasalingam Academy of Research and Education, India
Gurushankar Krishnamoorthy, Krishna Chaitanya Institute of Technology and Sciences, India
Paramanantham Muniyandi, V.S.S. Government Arts College, India

Chapter 3
Green Synthesized Metal-Chalcogenides for Sustainable Development 59
Vinaya Satyawan Tari, Faculty of Science and Technology, Universitas Airlangga, Surabaya, Indonesia

Chapter 4
Transition Metal Chalcogenides, Bimetallic Chalcogenides, Perovskites, and Nanocomposites: Advanced Materials for Optoelectronic and Thermoelectric Applications 81
Hemalatha Parangusan, Mannar Thirumalai Naicker College, India
G. Anandha Babu, Bannari Amman Institute of Technology, India
K. Karuppa Samy, Khalifa University, UAE
Deepalekshmi Ponnamma, Qatar University, Qatar
Akram Alfantazi, Khalifa University, UAE

Chapter 5
Rhenium and Tantalum-Based Transition Metal Chalcogenides: Next-Generation Materials for Optoelectronics 115
Sudipta Koley, Department of Physics, Amity Institute of Applied Sciences, Amity University, Kolkata, India

Chapter 6
Tailored Chalcogenide Materials for Eco-Efficient Electronic and Energy Systems 141
R. Naveen Kumar, Chandigarh Group of Colleges Jhanjeri, Mohali, India
Rubi Sarkar, Brainware University, Kolkata, India
Tanu Chandel, Chandigarh Group of Colleges Jhanjeri, Mohali, India

Chapter 7
Chalcogenide Nanocomposites Enabling Materials for Future Energy Solutions 161
Aman Kumar, Department of Physics, Keral Verma Subharti College of Science, Swami Vivekanand Subharti University, India

Chapter 8
Sustainability of Financial Energy Literacy: A Literature Review 185
José G. Vargas-Hernandez, Financial Literacy Research Center, Universidad Cristóbal Colón, Mexico & Tecnológico Nacional de México, ITS Fresnillo, Mexico
Teresa I. Salazar-Echeagary, Universidad Autònoma de Sinaloa, Mexico
Julieta E. Salazar-Echeagary, Universidad Autónoma de Sinaloa, Mexico

Chapter 9
Advancements in Chalcogenide Nanocomposites of Next-Generation Materials for Dye-Sensitized Solar Cells: Emerging Trends and Future Solar Innovations 211
Hari Shankar Biswas, Surendranath College, India
Amit Kr. Kundu, Sripat Singh College, India
Dilip K. Maiti, University of Calcutta, India
Prasenjit Mandal, Santipur College, India

Chapter 10
Chalcogenide-Based Nanocomposites as Counter Electrodes for Quantum Dot-Sensitized Solar Cells 249
S. Shanmugan, Koneru Lakshmaiah Education Foundation, India

Chapter 11
Cost-Effective Chalcogenide-Based Nanocomposite Electrodes for High-Performance Electrochemical and Photoelectrochemical Water Splitting 277
P. Selvakumar, Department of Science and Humanities, Nehru Institute of Technology, Coimbatore, India
T. C. Manjunath, Rajarajeswari College of Engineering, India
Santosh Kumar Nathsharma, Stewart Science College, India

Chapter 12
Chalcogenide-Based Nanocomposites for High-Performance Oxygen Evolution and Reduction Reactions 303
G. Anandha Babu, Bannari Amman Institute of Technology, India
Hemalatha Parangusan, Mannar Thirumalai Naicker College, India
R. Sangeetha, Mannar Thirumalai Naicker College, India
S. Jayanthi, The Standard Fireworks Rajaratnam College for Women, India
G. Dhanalakshmi, Mannar Thirumalai Naicker College, India

Chapter 13
Novel Chalcogenide-Based Nanocomposites as Electrocatalysts for Nitrogen Reduction Reactions 329
P. Selvakumar, Department of Science and Humanities, Nehru Institute of Technology, Coimbatore, India
Santosh Kumar Nathsharma, Stewart Science College, India
T. C. Manjunath, Rajarajeswari College of Engineering, India

Chapter 14
Chalcogenide-Based Nanocomposites for CO2 Transformation Into Valuable Chemicals and Clean Energy Generation 359
Krishnamoorthy Gurushankar, Krishna Chaitanya Institute of Technology and Sciences, India
Karthik Kannan, National Chung Cheng University, Taiwan & Karpagam Academy of Higher Education, India

Chapter 15
Chalcogenide-Based Nanocomposites for the Removal of Emerging Organic Micropollutants in Wastewater 381
Suganthi Nachimuthu, Government Arts College, Bharathidasan University, Karur, India
Kuppusamy Pushpanathan, Government Arts College, Bharathidasan University, Karur, India

Chapter 16
Agricultural Applications of Chalcogenide-Based Materials: Toward Sustainable, Smart, and Nutrient-Efficient Farming 409
Md. Shoeab Akhter, International University of Business, Agriculture, and Technology, Bangladesh
M. Shohidullah Miah, International University of Business, Agriculture, and Technology, Bangladesh
Habibur Rahman, International University of Business, Agriculture, and Technology, Bangladesh
S. M. Maksudur Rahman, International University of Business, Agriculture, and Technology, Bangladesh
A. J. M. Sirajul Karim, International University of Business, Agriculture, and Technology, Bangladesh
Sakibul Islam Ratul, Daffodil International University, Dhaka, Bangladesh
Mohammed Ataur Rahman, International University of Business, Agriculture, and Technology, Bangladesh

Chapter 17
Enhancing Pharmaceutics With AI Through Predictive Modeling of Crystal Structures and Atom Properties for Improved Solubility and Bioavailability .. 439
Sekar Kidambi Raju, School of Computing, SASTRA University, Thanjavur, India
Subhash Kannan, K. Ramakrishnan College of Engineering, Samayapuram, India
Ganesh Karthikeyan Varadarajan, School of Computing, SASTRA University, Thanjavur, India
Raj Anand Sundaramoorthy, School of Computing, SASTRA University, Thanjavur, India
Marwa M. Eid, Jadara University Research Center, Jadara University, Jordan
El-Sayed M. El-Kenawy, Applied Science Research Center, Applied Science Private University, Amman, Jordan

Compilation of References 473

About the Contributors 559

Index 569

Detailed Table of Contents

Preface xvii

Acknowledgment xxiii

Chapter 1
Introduction to Chalcogenide-Based Nanocomposites Advances in Optoelectronics Energy Systems and Sustainability 1
Periasamy Palanisamy, Nehru Institute of Engineering and Technology, Anna University, Comibatore, India
S. Satheeskumar, K.S. Rangasamy College of Technology, India

Chalcogenide nanocomposites are nanosized materials wherein chalcogen elements (sulfur, S; selenium, Se, and tellurium, Te) are combined with a wide range of metals and/or semiconducting materials, have attracted researchers' interest due to their peculiar optoelectronic, energy storage and environmental sanitization characteristics. This chapter describes recent progress on the synthesis, properties, and applications of nanocomposites based on chalcogenides in the field of optoelectronics, energy, and sustainability. Potential applications include solar cells, light-emitting devices, photodetectors, energy storage systems (such as batteries and supercapacitors) and environmental sustainability-related fields such as photocatalysis and water purification. This chapter describes new trends and perspectives in the fabrication of chalcogenide nanocomposites, which are proposed as advanced materials for next-generation energy technology, including eco-friendly applications

Chapter 2
Synthesis, Functional Properties, and Characterization of Chalcogenide-Based Nanocomposites 39
Chinnaiah Karuppaiya, Kalasalingam Academy of Research and Education, India
Gurushankar Krishnamoorthy, Krishna Chaitanya Institute of Technology and Sciences, India
Paramanantham Muniyandi, V.S.S. Government Arts College, India

Chalcogenides exist in the form of S, Se, and Te and their nanocomposite developed with different combinations of structures and composition gaining great importance in the field of optoelectrics and energy-related applications. Hence, to identify the synthesis routes for the preparation of chalcogenide based nanocomposites from various literatures are important for developing novel chalcogenide nanocomposites, which were identified. Newly proposed chalcogenide nanocomposites with different forms such as quantum dots, core/shell, heterostructures, and nanostructures especially in the optoelectronics and energy-field were compiled, and their properties were discussed. Efficient analytical techniques used to characterize the synthesized S, Se, and Te nanocomposites were addressed.

Chapter 3
Green Synthesized Metal-Chalcogenides for Sustainable Development 59
Vinaya Satyawan Tari, Faculty of Science and Technology, Universitas Airlangga, Surabaya, Indonesia

The growing demand for eco-friendly and energy-efficient materials has motivated researchers and Scientists in the green synthesis of chalcogenide compound materials based on sulfur (S), selenium (Se), and tellurium (Te). These are critical in applications spanning optoelectronics, photovoltaics, catalysis, and energy storage. However, methods involved in conventional synthesis often carry toxic solvents, high temperatures, and hazardous precursors, posing environmental and safety concerns. This chapter presents progress in the field of the green synthesis of metal chalcogenides in their nano-phased forms with their biomedical and environmental applications, which leads to sustainable development. This chapter will help bridge the knowledge gap between industrial and academic scientists.

Chapter 4
Transition Metal Chalcogenides, Bimetallic Chalcogenides, Perovskites, and Nanocomposites: Advanced Materials for Optoelectronic and Thermoelectric Applications 81
Hemalatha Parangusan, Mannar Thirumalai Naicker College, India
G. Anandha Babu, Bannari Amman Institute of Technology, India
K. Karuppa Samy, Khalifa University, UAE
Deepalekshmi Ponnamma, Qatar University, Qatar
Akram Alfantazi, Khalifa University, UAE

The accelerating advancement of modern technologies underscores an urgent need for the design and development of next-generation materials that exhibit multifunctionality, property tunability, and superior performance across a spectrum of applications. This chapter investigates into the state-of-the-art progress in four essential material systems such as transition metal chalcogenides (TMCs), bimetallic chalcogenides (BMCs), perovskites, and nanocomposites, which have emerged as frontrunners in the realms of optoelectronics and thermoelectrics. Transition metal chalcogenides, offer a versatile platform for applications ranging from photodetection and photovoltaics to thermoelectric energy conversion. The integration of bimetallic species within chalcogenide matrices introduces synergistic enhancements. The incorporation of these materials into nanocomposite architectures further enables tailored interfacial engineering

Chapter 5
Rhenium and Tantalum-Based Transition Metal Chalcogenides: Next-Generation Materials for Optoelectronics.. 115
Sudipta Koley, Department of Physics, Amity Institute of Applied Sciences, Amity University, Kolkata, India

Transition metal Dichalcogenides, have emerged as a novel class of two-dimensional layered materials with exceptional potential for optoelectronic applications. Due to their unique electronic band structures, strong light-matter interactions, and tunability, they bridge the gap between conventional semiconductors and low-dimensional nanomaterials. From first-principles density functional theory, the electronic band structure, and optical absorption spectra, of these chalcogenides in mono and few-layer configurations will be studied. Theoretical simulations reveal strong in-plane anisotropic dielectric functions showing their relevance in advanced light detection systems. Tantalum-based dichalcogenides, showcase a rich polymorphic phase encompassing 1T, 1T', and 2H phase. This chapter aims to provide a comprehensive theoretical foundation for researchers exploring novel transition metal dichalcogenide systems for optoelectronics and to highlight the potential of Re- and Ta-based dichalcogenides as advanced functional materials for future electronic and photonic applications.

Chapter 6
Tailored Chalcogenide Materials for Eco-Efficient Electronic and Energy Systems 141
R. Naveen Kumar, Chandigarh Group of Colleges Jhanjeri, Mohali, India
Rubi Sarkar, Brainware University, Kolkata, India
Tanu Chandel, Chandigarh Group of Colleges Jhanjeri, Mohali, India

Abstract: Chalcogenide-based materials, which are primarily made up of sulfur, selenium, and tellurium, have become an essential category of compounds for advanced optoelectronic and energy applications due to their distinctive electronic, optical, and thermal characteristics. These materials hold particular importance in the fields of photodetectors, solar cells, thermo electrics, and non-volatile memory devices. Their adjustable band gap, elevated refractive index, and capacity to create glassy phases render them suitable for sustainable energy solutions. This chapter delves into the synthesis, characterization, and application of chalcogenides, emphasizing recent developments and their contribution to facilitating environmentally friendly technologies. Through comparative analysis and experimental findings, we demonstrate how advancements in material engineering improve performance in energy harvesting and storage. We conclude by underscoring future opportunities and challenges in scaling these materials for commercial use.

Chapter 7
Chalcogenide Nanocomposites Enabling Materials for Future Energy Solutions 161
Aman Kumar, Department of Physics, Keral Verma Subharti College of Science, Swami Vivekanand Subharti University, India

The many functions of chalcogenide nanocomposites in cutting-edge energy systems are critically examined in this talk, with a focus on synthetic processes, environmental effects, and performance improvement techniques. Sustainable synthesis methods—such as hydrothermal, solvothermal, and UV-assisted techniques—are given special consideration because they have many benefits, including reduced ecological impact, scalability, and cost effectiveness. Important drawbacks are methodically assessed, such as inadequate cycle stability and subpar energy conversion or storage efficiency. The suggested remediation techniques include compositional engineering to increase electrical conductivity, electrode architectural modification to promote ion/electron transport, and thorough explanation of physicochemical degradation routes under operating settings.In order to solve these issues, the discussion emphasizes the critical role that multidisciplinary research—which integrates materials science, electrochemistry, nanotechnology, and environmental engineering—plays.

Chapter 8
Sustainability of Financial Energy Literacy: A Literature Review................... 185
José G. Vargas-Hernandez, Financial Literacy Research Center,Universidad Cristóbal Colón, Mexico & Tecnológico Nacional de México, ITS Fresnillo, Mexico
Teresa I. Salazar-Echeagary, Universidad Autònoma de Sinaloa, Mexico
Julieta E. Salazar-Echeagary, Universidad Autónoma de Sinaloa, Mexico

This study aims to analyze the implications of the financial energy literacy (FEL) with sustainability as interdisciplinary concepts. The holistic approach of sustainability of financial energy literacy (SFEL) is critical in an environment where energy costs are rising and environmental sustainability is becoming increasingly urgent. The method employed is the meta-analytical and meta-cognitive analysis based on a conceptual, theoretical and empirical literature review published in the last five years. It is concluded that financial energy literacy is a crucial element for achieving sustainable development goals, as it empowers individuals to make informed decisions that benefit both their financial well-being and the environment of individuals, families, organizations and communities.

Chapter 9
Advancements in Chalcogenide Nanocomposites of Next-Generation Materials for Dye-Sensitized Solar Cells: Emerging Trends and Future Solar Innovations 211
Hari Shankar Biswas, Surendranath College, India
Amit Kr. Kundu, Sripat Singh College, India
Dilip K. Maiti, University of Calcutta, India
Prasenjit Mandal, Santipur College, India

Dye-sensitized solar cells (DSSCs) are a promising photovoltaic technology known for low cost, flexibility, and eco-friendliness. Chalcogenide-based nanocomposites, with their unique electronic, optical, and catalytic features, have attracted significant interest for boosting DSSC efficiency. This review explores next-generation chalcogenide nanocomposites—sulfides, selenides, tellurides, and hybrids—used in key DSSC components like photoanodes, counter electrodes, and electrolytes. It highlights synthesis methods, structural traits, and the functional improvements these nanomaterials offer. Challenges, recent progress, and future perspectives for integrating chalcogenide nanocomposites into commercially viable DSSCs are also discussed.

Chapter 10
Chalcogenide-Based Nanocomposites as Counter Electrodes for Quantum Dot-Sensitized Solar Cells 249
S. Shanmugan, Koneru Lakshmaiah Education Foundation, India

This study explores a sustainable and cost-effective method to improve dye-sensitized solar cells (DSSCs) by developing nickel sulfide (NiS) nanoparticle composites combined with activated carbon (AC) derived from sugarcane waste. The NiS/AC nanocomposites were prepared through a facile hydrothermal process assisted by ultrasonication and characterized using XRD, TEM, UV-Vis, Raman spectroscopy, and XPS to analyze their structure and optical features. Notably, the composite with 20% AC showed enhanced optical transmittance and a lowered bandgap from 1.82 to 1.56 eV, facilitating better visible light absorption. Electrochemical tests revealed that NiS/AC-20 electrodes had improved triiodide reduction kinetics compared to those with lower AC content. The optimized NiS/AC counter electrode exhibited a power conversion efficiency of 8.54%, supported by a short-circuit current density of 19.5 mA cm^{-2} and an open-circuit voltage near 0.759 V.

Chapter 11
Cost-Effective Chalcogenide-Based Nanocomposite Electrodes for High-Performance Electrochemical and Photoelectrochemical Water Splitting 277
P. Selvakumar, Department of Science and Humanities, Nehru Institute of Technology, Coimbatore, India
T. C. Manjunath, Rajarajeswari College of Engineering, India
Santosh Kumar Nathsharma, Stewart Science College, India

Water splitting, also known process with immense potential in the global shift toward sustainable energy systems. This process involves breaking down water (H_2O) into its, wind, or hydropower. As the world seeks alternatives to fossil fuels to mitigate climate change and reduce greenhouse gas emissions, hydrogen has emerged as a promising energy carrier due to its high energy content and environmentally friendly combustion, which produces only water as a by-product.demonstrating the decomposition of water using direct current electricity. Since then, the field has evolved significantly, driven by advancements in electrochemical systems, materials science, and catalysis. materials science, and catalysis. Modern water splitting can be broadly categorized into three primary methods: electrolysis, photoelectrochemical (PEC) water splitting, and thermochemical water splitting. Among these, electrolysis is the most mature and widely researched technique, involving the application of electrical energy to drive the non-spontaneous reaction.

Chapter 12
Chalcogenide-Based Nanocomposites for High-Performance Oxygen Evolution and Reduction Reactions 303
G. Anandha Babu, Bannari Amman Institute of Technology, India
Hemalatha Parangusan, Mannar Thirumalai Naicker College, India
R. Sangeetha, Mannar Thirumalai Naicker College, India
S. Jayanthi, The Standard Fireworks Rajaratnam College for Women, India
G. Dhanalakshmi, Mannar Thirumalai Naicker College, India

The intensifying global demand for clean and sustainable energy solutions has accelerated the quest for high-performance electrocatalysts capable of efficiently mediating key electrochemical reactions—most notably, the oxygen evolution reaction (OER) and the oxygen reduction reaction (ORR). To overcome intrinsic drawbacks of pristine chalcogenide materials—such as limited electroactive surface area and inherently poor electrical conductivity—research efforts have increasingly focused on engineering chalcogenide-based nanocomposites via strategic incorporation with conductive matrices, heterostructured architectures, and multifunctional co-catalysts. Transition metal chalcogenides (TMCs), in particular, offer significant potential due to their versatile redox behavior, variable oxidation states, and electronically tunable layered structures. This chapter offers an in-depth exploration of recent advances in chalcogenide-based nanocomposites, which have emerged as highly promising candidates for next-generation OER and ORR electrocatalysis.

Chapter 13
Novel Chalcogenide-Based Nanocomposites as Electrocatalysts for Nitrogen Reduction Reactions 329
P. Selvakumar, Department of Science and Humanities, Nehru Institute of Technology, Coimbatore, India
Santosh Kumar Nathsharma, Stewart Science College, India
T. C. Manjunath, Rajarajeswari College of Engineering, India

Despite this abundance, atmospheric nitrogen is chemically inert due to the presence of a strong triple bond (N≡N) that has a high bond dissociation energy of about 941 kJ/mol. This inertness renders N_2 unavailable for direct assimilation by most living organisms, thereby necessitating its conversion into reactive nitrogen species such as ammonia (NH_3), nitrate (NO_3^-), or nitrite (NO_2^-), which are essential for the biosynthesis of amino acids, nucleotides, and other vital biomolecules. The process by which inert nitrogen gas is converted into ammonia is termed nitrogen fixation, and among the strategies employed to achieve this transformation, Nitrogen Reduction Reactions (NRR) have emerged as a focal point of research, especially in the context of sustainable and environmentally friendly chemistry.Traditionally, the conversion of nitrogen gas into ammonia on an industrial scale has been accomplished through the Haber–Bosch process, developed in the early 20th century.

Chapter 14
Chalcogenide-Based Nanocomposites for CO2 Transformation Into Valuable Chemicals and Clean Energy Generation 359
Krishnamoorthy Gurushankar, Krishna Chaitanya Institute of Technology and Sciences, India
Karthik Kannan, National Chung Cheng University, Taiwan & Karpagam Academy of Higher Education, India

Optimizing the interaction between the nanocomposites and CO2, researchers aim to develop sustainable methods for producing essential chemicals while simultaneously addressing environmental concerns related to carbon emissions. This innovative strategy not only contributes to resource recovery but also supports the transition towards a more sustainable energy. Therefore, in this chapter, we explore the concept of chalcogenide nanocomposites for CO2 transformation into valuable chemicals and clean energy generation. In addition, by addressing a significant gap in renewable energy research, this chapter offers comprehensive guidance for researchers working on the development of chalcogenide-based materials specifically designed for CO2 transformation into valuable chemicals and clean energy generation.

Chapter 15
Chalcogenide-Based Nanocomposites for the Removal of Emerging Organic Micropollutants in Wastewater 381
Suganthi Nachimuthu, Government Arts College, Bharathidasan University, Karur, India
Kuppusamy Pushpanathan, Government Arts College, Bharathidasan University, Karur, India

Nowadays, a fast growing in population creates organic and inorganic waste in the form of extensive effluence and which causes water pollution. Industries, agriculture, domestic processes, etc instigated a huge pollution to the water atmospheres everywhere in the world via various types of chemical elements and substances known as "emerging pollutants". Among them, an organic form is high attention as a result of their comparatively excessive intensities in water. The structural and behavioural diversity of these emerging organic pollutants creates their potential dangers to living organisms. This chapter deals about the chalcogenide-based nanocomposites are utilized like catalysts for removing environmental organic micropollutants in wastewater treatment. Owing to their low bandgaps, higher charge carrier movement with high visible-light absorbing properties, chalcogenide-based nanocomposites are broadly used as photocatalysts. High efficacy, low-cost, precluding long-lasting pollutants, as well as direct utilization of solar energy are an advantage of the photocatalytic approach.

Chapter 16
Agricultural Applications of Chalcogenide-Based Materials: Toward Sustainable, Smart, and Nutrient-Efficient Farming.. 409
Md. Shoeab Akhter, International University of Business, Agriculture, and Technology, Bangladesh
M. Shohidullah Miah, International University of Business, Agriculture, and Technology, Bangladesh
Habibur Rahman, International University of Business, Agriculture, and Technology, Bangladesh
S. M. Maksudur Rahman, International University of Business, Agriculture, and Technology, Bangladesh
A. J. M. Sirajul Karim, International University of Business, Agriculture, and Technology, Bangladesh
Sakibul Islam Ratul, Daffodil International University, Dhaka, Bangladesh
Mohammed Ataur Rahman, International University of Business, Agriculture, and Technology, Bangladesh

Food chain has been altered through farming nanotechnology, which energizes such elements as Sulfur (S) and Selenium (Se), and several industries, such as crop farming, aquaculture, tree forestry, animal and poultry farms, and plant nutrients, have utilized this technology. Sulphur nanofertilizers reduce the stress level on the environment, increase the fertility of the soil, and optimize nutrient release. Selenium agents impede the occurrence of micronutrient insufficiency in human beings and animals, and Chalcogenide concentration minimizes the intake of synthetic additives and optimizes the conversion rate of feeds. The use of sulfur fertilizer helps activate sprouting and improves tolerance to diseases. Intelligent sensors with chalcogenides and optoelectronic agricultural plans can provide real-time information on the moisture of the soil, soil nutrients, and stress. These products increase nutrition, demand fewer resources, and provide adaptable agriculture approaches in an evolving climate.

Chapter 17
Enhancing Pharmaceutics With AI Through Predictive Modeling of Crystal Structures and Atom Properties for Improved Solubility and Bioavailability .. 439
Sekar Kidambi Raju, School of Computing, SASTRA University, Thanjavur, India
Subhash Kannan, K. Ramakrishnan College of Engineering, Samayapuram, India
Ganesh Karthikeyan Varadarajan, School of Computing, SASTRA University, Thanjavur, India
Raj Anand Sundaramoorthy, School of Computing, SASTRA University, Thanjavur, India
Marwa M. Eid, Jadara University Research Center, Jadara University, Jordan
El-Sayed M. El-Kenawy, Applied Science Research Center, Applied Science Private University, Amman, Jordan

The research study explores how sophisticated machine learning algorithms, i.e., our suggested model SENET, coupled with SVM and ADAM, could enhance drug solubility and bioavailability through the prediction of crystallographic structures and atom property classification. The current research aims to revolutionize pharmaceutical drug development processes by embracing AI-based preclinical Testing. We investigate the sophisticated field of crystallographic structures and dissect atomistic data in depth to reveal determining factors influencing crystal forms. The aim of the research is to enhance drug solubility and bioavailability by conducting this study, which will significantly transform pharmaceutical practice. The study shows that sophisticated machine learning methods can potentially forecast drug behavior in complicated situations. This study discloses the potential for groundbreaking drug discovery by speeding up the progress in pharmaceutical science, reducing the development cost, and optimizing patient benefit through enhanced drug delivery and action.

Compilation of References 473

About the Contributors 559

Index 569

Preface

The rapid advancement of materials science and nanotechnology has brought about a transformative era in energy, electronics, and environmental sustainability. Among the materials driving this innovation, chalcogenide-based nanocomposites have emerged as a promising class of materials with unique and versatile properties. Their structural diversity, tunable electronic behavior, and adaptability across multiple domains make them ideal candidates for next-generation technologies.

This edited volume, *Chalcogenide-Based Nanocomposites: Next-Generation Materials for Optoelectronics, Energy, and Sustainability*, was conceived with the aim of offering a comprehensive reference that bridges the gap between fundamental research and applied technology in this exciting and rapidly growing field. The book explores the underlying physics, chemistry, engineering, and electrochemical science of chalcogenide-based materials and showcases their multifaceted roles in optoelectronics, energy conversion and storage, and environmental sustainability.

The book brings together a wide spectrum of chalcogenide-based systems, including transition metal chalcogenides, bimetallic chalcogenides, perovskites, and hybrid nanocomposites. Their applications are equally diverse: from solar cells, photodetectors, and LEDs to neuromorphic devices, batteries, fuel cells, supercapacitors, and water-splitting catalysts. In environmental contexts, their unique properties -particularly at the nanoscale offer promising solutions for wastewater treatment and pollution control through enhanced photocatalytic activity.

Our goal is to provide readers with both foundational understanding and cutting-edge insights into the synthesis, properties, and practical applications of these materials. The chapters delve into critical mechanisms, interfacial chemistry, and the challenges and opportunities related to scalability and industrial implementation. Furthermore, we draw comparisons with other leading materials such as MXenes and graphene-based composites, highlighting the synergistic advantages that chalcogenides can offer.

A key strength of this work lies in its collaborative nature. Contributions from leading researchers across the globe bring diverse perspectives, experimental approaches, and theoretical insights to the forefront. As editors, we are grateful for their expertise, dedication, and forward-looking contributions that elevate this book beyond a mere collection of chapters to a truly interdisciplinary and impactful resource.

This book is intended for a broad audience: researchers, scientists, and engineers in academia and industry; graduate students and educators in materials science, chemistry, physics, and environmental engineering; as well as policy analysts and entrepreneurs exploring sustainable technologies. Whether your interest lies in the discovery of new materials, the development of scalable energy systems, or the advancement of sustainable environmental technologies, this book offers a rich and authoritative foundation.

We hope this volume will serve as a catalyst for further innovation and collaboration in the field of chalcogenide-based nanomaterials. By uniting theory, experimentation, and application, it aspires to accelerate breakthroughs that will shape the future of energy, electronics, and environmental sustainability.

ORGANIZATION OF THE BOOK

In curating this volume, our goal as editors has been to assemble a cohesive and forward-looking reference that captures the state-of-the-art in chalcogenide-based nanocomposite materials across multiple fields. Each chapter contributes a unique perspective and builds on the foundational theme of developing next-generation materials for optoelectronics, energy, and sustainability.

Chapter 1, *Introduction to Chalcogenide-Based Nanocomposites: Advances in Optoelectronics, Energy Systems, and Sustainability*, sets the stage by introducing the fundamental concepts, recent developments, and broad applicability of chalcogenide nanocomposites. It outlines the pivotal role of sulfur, selenium, and tellurium-based materials in fields ranging from solar energy and LEDs to water purification and environmental remediation. This chapter establishes the contextual foundation for the rest of the book, highlighting the relevance of these materials in eco-friendly energy technologies.

Chapter 2, *Synthesis, Functional Properties, and Characterization of Chalcogenide-Based Nanocomposites*, explores the diverse synthesis routes and structures from quantum dots to heterostructures—and the corresponding functional properties that make chalcogenide nanocomposites particularly promising. It also addresses the advanced characterization techniques required to probe their structure–property relationships, which are critical for tailoring materials for specific applications in energy and optoelectronics.

Chapter 3, *Green Synthesized Metal-Chalcogenides for Sustainable Development*, focuses on eco-conscious methods of synthesizing chalcogenide nanomaterials.

Emphasizing green chemistry principles, this chapter surveys recent advances in low-toxicity, energy-efficient synthesis methods. The chapter also connects green synthesis to biomedical and environmental applications, bridging industrial interest with academic research in the pursuit of sustainability.

Chapter 4, *Transition Metal Chalcogenides, Bimetallic Chalcogenides, Perovskites, and Nanocomposites: Advanced Materials for Optoelectronic and Thermoelectric Applications*, provides an in-depth analysis of key material classes at the forefront of energy conversion and optoelectronic technologies. It highlights how structural and compositional engineering especially in bimetallic systems and perovskites can unlock multifunctional performance characteristics, enabling advancements in photodetectors, thermoelectric, and solar cells through strategic interfacial design.

Chapter 5, *Rhenium and Tantalum-based Transition Metal Chalcogenides: Next-Generation Materials for Optoelectronics*, delves into the theoretical and computational modeling of Re- and Ta-based dichalcogenides. These materials, known for their polymorphic phases and tunable band structures, exhibit strong light-matter interactions ideal for optoelectronic applications. This chapter provides a solid theoretical framework for researchers exploring novel transition metal dichalcogenides, supported by density functional theory insights.

Chapter 6, *Tailored Chalcogenide Materials for Eco-Efficient Electronic and Energy Systems*, investigates how chalcogenides' tunable properties can be leveraged for sustainable electronic and energy systems. Focusing on photodetectors, non-volatile memory devices, and solar technologies, the chapter emphasizes material customization through bandgap engineering and phase manipulation. It also critically assesses performance metrics in relation to environmental compatibility and future scalability.

Chapter 7, *Chalcogenide Nanocomposites: Enabling Materials for Future Energy Solutions*, provides a critical assessment of the synthetic strategies and application-specific engineering approaches that improve the performance of chalcogenide nanocomposites in real-world energy systems. The discussion covers eco-friendly synthesis techniques, performance bottlenecks in energy storage and conversion, and degradation mechanisms under operational conditions. This chapter highlights the importance of interdisciplinary collaboration in overcoming these complex challenges.

Chapter 8, *Sustainability of Financial Energy Literacy: A Literature Review*, offers a novel and interdisciplinary exploration of sustainability from a socio-economic lens. Through a rigorous literature review, it introduces the concept of Sustainable Financial Energy Literacy (SFEL) and its significance in achieving global sustainability goals. By empowering individuals and communities to make informed energy and financial decisions, this chapter positions SFEL as a critical but often overlooked dimension in the sustainable development dialogue.

Chapter 9, *Advancements in Chalcogenide Nanocomposites of Next-Generation Materials for Dye-Sensitized Solar Cells*, explores the growing role of chalcogenide-based nanomaterials in dye-sensitized solar cells (DSSCs)—a cost-effective and flexible class of photovoltaic technology. This chapter reviews the integration of sulfides, selenides, tellurides, and their hybrids into critical DSSC components such as photoanodes, counter electrodes, and electrolytes. Highlighting emerging trends in synthesis and performance enhancement, it also provides insight into the challenges and opportunities for scaling these materials in future solar energy innovations.

Chapter 10, *Chalcogenide-Based Nanocomposites as Counter Electrodes for Quantum Dot-Sensitized Solar Cells*, presents a compelling study on sustainable electrode design using nickel sulfide (NiS) and activated carbon derived from agricultural waste. Through a simple hydrothermal-ultrasonic synthesis route, the resulting NiS/AC nanocomposites show enhanced optical properties and electrochemical activity. With a reported power conversion efficiency of 8.54%, this work demonstrates how chalcogenide nanocomposites can serve as low-cost, high-performance alternatives to conventional counter electrodes in solar cell architectures.

Chapter 11, *Cost-Effective Chalcogenide-Based Nanocomposite Electrodes for High-Performance Electrochemical and Photoelectrochemical Water Splitting*, addresses the global shift toward green hydrogen production. It outlines the foundational principles and technological progress in electrolysis and photoelectrochemical (PEC) water splitting, with a specific focus on chalcogenide-based materials. The chapter emphasizes how nanocomposites can enhance catalytic activity and efficiency in water splitting applications, thereby contributing to the hydrogen economy and carbon-neutral energy systems.

Chapter 12, *Chalcogenide-Based Nanocomposites for High-Performance Oxygen Evolution and Reduction Reactions*, investigates the design of electrocatalysts to drive key reactions in energy devices such as fuel cells and metal–air batteries. It focuses on the engineering of transition metal chalcogenides (TMCs) into nanocomposite forms that overcome limitations like low conductivity and poor surface activity. The chapter reviews advanced structural strategies such as heterostructure and incorporation with conductive matrices that significantly improve OER and ORR performance in renewable energy systems.

Chapter 13, *Novel Chalcogenide-Based Nanocomposites as Electrocatalysts for Nitrogen Reduction Reactions*, explores the challenge of converting atmospheric nitrogen (N_2) into ammonia (NH_3) through sustainable electrocatalysis. Traditional methods like the Haber–Bosch process are energy-intensive and environmentally taxing. This chapter highlights how chalcogenide-based nanocomposites can serve as promising electrocatalysts for the nitrogen reduction reaction (NRR), offering a green and efficient alternative for ammonia production, which is crucial for agriculture and chemical manufacturing.

Chapter 14, *Chalcogenide-Based Nanocomposites for CO_2 Transformation into Valuable Chemicals and Clean Energy Generation*, examines the urgent need for carbon capture and utilization (CCU) strategies. It presents chalcogenide nanocomposites as efficient catalytic platforms for converting carbon dioxide into fuels and valuable chemical products. The chapter bridges materials science with environmental sustainability, offering pathways for mitigating greenhouse gas emissions while generating clean energy and value-added chemicals.

Chapter 15, *Chalcogenide-Based Nanocomposites for the Removal of Emerging Organic Micropollutants in Wastewater*, focuses on the environmental remediation potential of chalcogenide photocatalysts. As organic micropollutants increasingly contaminate water supplies, the photocatalytic properties of chalcogenide nanocomposites driven by visible-light activation, low bandgaps, and high surface reactivity are critical in breaking down persistent contaminants. The chapter emphasizes their practical advantages, such as low cost, solar-driven activity, and minimal environmental impact, in wastewater treatment applications.

Chapter 16, *Agricultural Applications of Chalcogenide-Based Materials: Toward Sustainable, Smart, and Nutrient-Efficient Farming*, presents an emerging frontier where chalcogenide materials intersect with agricultural innovation. From sulfur and selenium-based nanofertilizers to smart sensors embedded with chalcogenide components, this chapter discusses their potential to improve soil health, plant nutrition, and crop yields while reducing environmental stress. The integration of optoelectronic technologies in precision agriculture signals a shift toward more resilient and resource-efficient farming systems.

Chapter 17, *Enhancing Pharmaceutics with AI Through Predictive Modeling of Crystal Structures and Atom Properties for Improved Solubility and Bioavailability*, concludes the volume with an interdisciplinary perspective on pharmaceuticals. This chapter showcases how artificial intelligence, including models like SENET, SVM, and ADAM, can predict crystallographic structures and atom-level properties, thereby enhancing drug solubility and bioavailability. While not directly focused on chalcogenides, this work highlights the broader integration of AI in materials and chemical science, reinforcing the importance of computational tools in next-generation materials research.

CONCLUSION

As editors of *Chalcogenide-Based Nanocomposites: Next-Generation Materials for Optoelectronics, Energy, and Sustainability*, it is our hope that this comprehensive reference will serve not only as a source of current knowledge but also as a catalyst for future innovation in this rapidly advancing field. The chapters compiled in this volume reflect a diverse and interdisciplinary spectrum of research that spans fundamental science, emerging technologies, and real-world applications.

Chalcogenide-based nanocomposites have demonstrated remarkable potential in a wide array of domains from solar energy harvesting and electrocatalysis to water purification, agriculture, and even pharmaceutical innovation. What unites these varied applications is the inherent versatility, tunability, and sustainability of chalcogenide materials, especially when engineered at the nanoscale and integrated into hybrid or composite architectures. Through advanced synthesis methods, structural manipulation, and novel functionalization strategies, researchers are unlocking new frontiers in energy efficiency, environmental remediation, and material design.

This book brings together leading voices from across disciplines and geographies to provide a timely and in-depth exploration of the state of the art. Each chapter contributes to a holistic understanding of how chalcogenide-based nanocomposites can meet the growing demands of energy, environmental, and technological sustainability in the 21st century.

We envision this book as a valuable resource for researchers, industry professionals, educators, and policy-makers alike those who are not only investigating the fundamental science but also translating discoveries into scalable solutions for global challenges. As the urgency for clean energy and sustainable materials continues to rise, we believe the insights and innovations shared in these pages will contribute meaningfully to the shaping of a more resilient and sustainable future.

Acknowledgment

The editors extend their deepest gratitude to everyone who contributed to the success of this project, with special appreciation to the authors and reviewers who took part in the review process. This book would not have been possible without their invaluable support and commitment.

First and foremost, we sincerely thank all the authors for their meaningful contributions. We truly value the time, effort, and expertise they invested in developing their chapters and enriching the overall quality of this book.

In addition, the editors wish to recognize the vital role of the reviewers in refining the clarity, coherence, and scholarly quality of the chapters. We are especially thankful to those contributors who graciously served both as authors and reviewers, and we deeply appreciate their dedication in fulfilling these dual responsibilities.

Editor Karthik Kannan acknowledges that this work was financially supported by the Advanced Institute of Manufacturing with High-tech Innovations (AIM-HI), through the Featured Areas Research Center Program under the Higher Education Sprout Project, funded by the Ministry of Education (MOE), Taiwan.

Editor Karthik Kannan wishes to express his sincere gratitude to Professor Yong-Song Chen, Department Chair and Associate Dean of the College of Engineering, Department of Mechanical Engineering, National Chung Cheng University, Taiwan, for his continuous encouragement and support. He also extends heartfelt thanks to the Advanced Institute of Manufacturing with High-Tech Innovations (AIM-HI), Department of Mechanical Engineering, National Chung Cheng University, Taiwan, for their unwavering support and collaboration throughout this project.

Dr. Karthik Kannan also extends heartfelt thanks to the Department of Physics, Center for Energy and Environment, Karpagam Academy of Higher Education, Coimbatore, Tamil Nadu, India, for their unwavering support and collaboration throughout this project.

Editor Vinaya Tari would like to convey her deepest appreciation to Professor Dwi Kusuma Wahyuni, Department of Biology, Faculty of Science and Technology, Universitas Airlangga, Surabaya, East Java, Indonesia, for her steadfast guidance, encouragement, and invaluable support during the course of this work.

Editor Vinaya Tari is thankful to the administration of Universitas Airlangga, Surabaya, for continuous support. Last but not least, both editors extend their sincere gratitude to their parents and to the universe for its generous blessings.

Karthik Kannan

National Chung Cheng University, Taiwan & Karpagam Academy of Higher Education, India

Vinaya Tari

Universitas Airlangga, Indonesia

Chapter 1

Introduction to Chalcogenide-Based Nanocomposites Advances in Optoelectronics Energy Systems and Sustainability

Periasamy Palanisamy
https://orcid.org/0000-0003-1069-5822
Nehru Institute of Engineering and Technology, Anna University, Comibatore, India

S. Satheeskumar
K.S. Rangasamy College of Technology, India

ABSTRACT

Chalcogenide nanocomposites are nanosized materials wherein chalcogen elements (sulfur, S; selenium, Se, and tellurium, Te) are combined with a wide range of metals and/or semiconducting materials, have attracted researchers' interest due to their peculiar optoelectronic, energy storage and environmental sanitization characteristics. This chapter describes recent progress on the synthesis, properties, and applications of nanocomposites based on chalcogenides in the field of optoelectronics, energy, and sustainability. Potential applications include solar cells, light-emitting devices, photodetectors, energy storage systems (such as batteries and supercapacitors)

DOI: 10.4018/979-8-3373-3962-7.ch001

and environmental sustainability-related fields such as photocatalysis and water purification. This chapter describes new trends and perspectives in the fabrication of chalcogenide nanocomposites, which are proposed as advanced materials for next-generation energy technology, including eco-friendly applications

INTRODUCTION

The chalcogenides are compounds containing elements from the chalcogen group of the periodic table (S, Se, Te). They are well known for their extraordinary properties, which derive from the strong interactions between the chalcogen atoms, metals and semiconductors (Bouroushian, 2010). The as-prepared materials have diverse electronic, optical, and thermal properties, and thereby can be used as versatile materials in optoelectronics, energy storage, and environmental protection(Ha et al., 2023). Chalcogenides, such as cadmium selenide (CdSe) and copper indium gallium selenide (CIGS), have been used as solar cells and light-emitting devices (LEDs) in the past. Nevertheless, thanks to these properties, many applications emerged after the combination with nanomaterials, which is the fabric to produce chalcogenide composite material. Composite materials of chalcogenides were prepared with other nanomaterials, such as graphene or carbon nanotubes, to improve their efficacy, (Stamford & Azapagic, 2019).

Another aspect of the relevance of chalcogenide-based nanocomposites is their capability to be tailored at the nanoscale, where quantum confinement and high surface-area phenomena play a significant role, leading to extraordinary properties compared to the respective bulk materials. These progresses in the nanocomposites led to better optical absorption, charge transport characteristics, as well as mechanical properties and opened up new opportunities in the newly arising areas such as flexible electronics, renewable energy and sustainable technology(Voigt et al., 2022). Chalcogenide-based nanocomposites are composites that involve chalcogenide compounds with other nanostructured materials. The composites offer the combined benefits of both components, enhancing performance in various applications. Those materials are usually modified by adding chalcogenide with other parts like metals, carbon-based nanomaterials and metal oxides, (Singh et al., 2015).

Figure 1. Typical Chalcogenide compounds

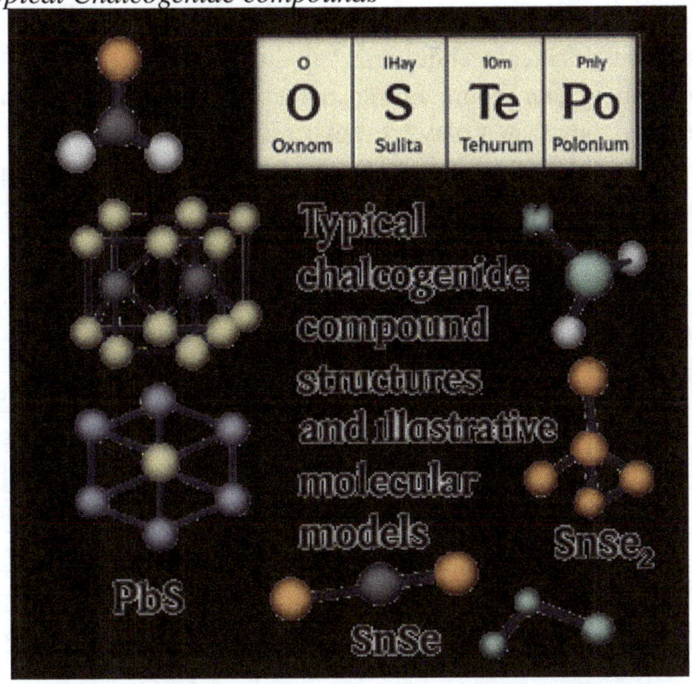

A typical class of nanocomposites derived from chalcogenide is metal chalcogenide nanocomposites, where chalcogenide elements, such as selenium or sulfur, are mixed with metals in the form of cadmium, copper or tin (Fig.1). Examples of these are CdSe (cadmium selenide), PbS (lead sulfide), and SnS (tin sulfide), with all having semiconducting properties which can be employed in photovoltaic cells, light-emitting diodes (LEDs) and photodetectors(Sivasamy et al., 2024a; Kannan K. & Tari V. 2025). Another class comprises chalcogenide-carbon nanocomposites, i.e., composite materials formed by chalcogenides and carbon materials such as graphene or carbon nanotubes. These composites inherit excellent conductivity and mechanical strength from carbon and unique optical and electronic performance from chalcogenides(A. K. Singh & Jen, 2021). Chalcogenide-metal oxide nanocomposites represent another critical class, since the hybridization of chalcogenides with materials such as TiO_2 or ZnO may significantly enhance their photocatalytic activity, rendering them suitable for environmental applications (Table.1) (photocatalysis, water purification, CO_2 reduction), (Ganguly et al., 2020)

In this chapter, we will examine recent developments in chalcogenide-based nanocomposites, primarily from the perspectives of optoelectronics and energy storage, with a more specific focus on sustainability. The growing demand for sustainable energy, the need for energy conversion devices to be more efficient, and the development of environmentally friendly technology have led to chalcogenide nanocomposites being significant in the context of global challenges. Chalcogenide nanocomposites have critical technological applications in various fields, including optoelectronics, energy, environmental sustainability, and more. In optoelectronics, chalcogenide nanocomposites play a crucial role in the development of efficient and cost-effective solar cells, photodetectors, LEDs, and lasers, (Njema & Kibet, 2025; Gurushankar et al., 2024).

Table 1. Applications of chalcogenides in various fields

Field	Application Examples	Notable Chalcogenides
Electronics	Transistors, memory devices	MoS_2, WSe_2, GeTe, Sb_2Te_3
Optoelectronics	IR sensors, fiber optics, photonics	As_2S_3, CdTe, chalcogenide glass
Energy	Solar cells, thermoelectrics, batteries	CIGS, CdTe, MoS_2, SnSe
Data Storage	Phase-change memory, optical discs	GeTe, Sb_2Te_3
Sensors / Biomedicine	Biosensing, imaging, drug delivery	Se, Te nanoparticles, quantum dots
Superconductivity	Magnetic levitation, MRI, power	$NbSe_2$, $YBa_2Cu_3O_7$
Catalysis	Petrochemical conversion, synthesis	MoS_2, WS_2

A direct or afterglow effect, including a glow center on the wavelength and/or energy, can be dependent on the type of material used. Concerning the application of these devices, materials such as CdTe, CIGS, and CdSe, their band gap energies are at the basis of these descriptions of the direct interband level1 and the strong absorption of light. In the field of energy, chalcogenide NCs also demonstrate their importance in energy storage devices such as batteries and supercapacitors with excellent capacity, rate capability, and cycling stability. Their applications in fuel cells and hydrogen storage devices are also significant in promoting the development of a hydrogen economy based on efficient and clean energy conversion technologies. Chalcogenides also provide an essential breakthrough in solutions for environmental sustainability(Mandal et al., 2025). Their application for photocatalysis and water treatment, environmental remediation, is garnering more and more attention, especially in environmentally friendly energy areas of photocatalytic and CO_2 reduction. Through the photocatalytic nature of chalcogenide materials, these systems can potentially have essential implications in clean energy generation and environmental remediation of fossil fuels(Wang et al., 2022).

FUNDAMENTAL PROPERTIES OF CHALCOGENIDE-BASED NANOCOMPOSITES

The properties of Chalcogenide nanocomposites are unique, and this uniqueness makes them highly versatile in various applications (Fig. 2). They exhibit beneficial optical properties, including tunable band gaps and photoluminescence. They also show different electrical conductivities and possess charge carrier mobility, which is high, hence being very suitable in semiconductors and sensors(Rana et al., 2018a). Nanostructured surfaces with controlled crystallinity are also observed in these materials, which enhances their functionality in energy storage and catalysis. Moreover, they are thermally stable, with variable thermal conductivity, meaning they can be used in high temperatures, and therefore their potential continues to grow in the sphere of electronics and energy systems(Said et al., 2024).

Figure 2. Properties of chalcogenide nanomaterials

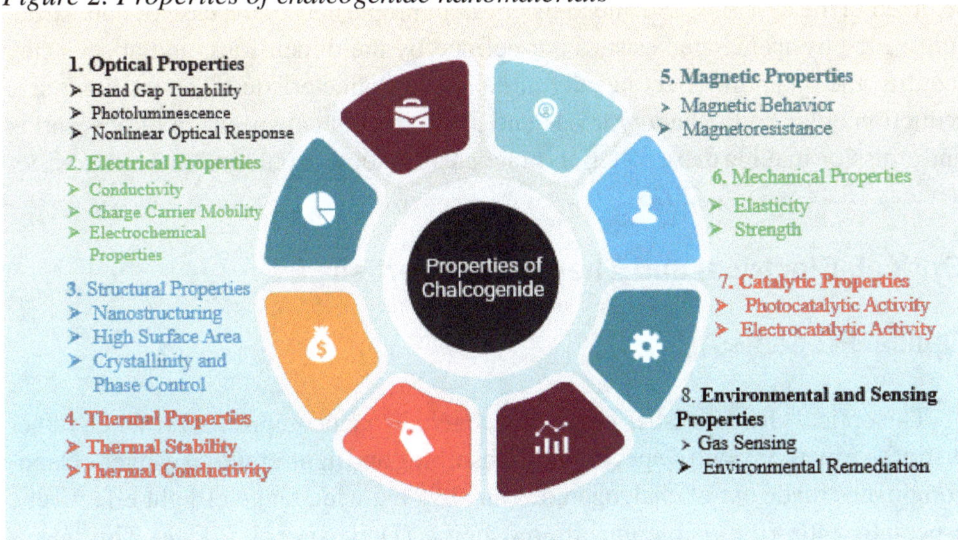

Atomic Structure and Composition of Chalcogenides

The atomic structure and composition of chalcogenide nanocomposites are crucial in determining their fundamental properties. The chalcogenides are usually made by mixing chalcogen elements (sulfur, selenium or tellurium) with metallic atoms or semiconductor materials. Their atomic structure and crystal structure determine essential functions which include light absorption, electron mobility and strength.

Regarding a nuclear structure, a chalcogenide-based nanocomposite is an assembly of crystals or an amorphous structure of the chalcogenide, depending on the process and synthesis conditions(Phogat et al., 2024). As an example, cadmium selenide (CdSe), a widespread semiconductor, uses the zinc blende lattice type, and both cadmium atoms and selenium atoms have a tetrahedral structure where other atoms (four atoms in three dimensions) are bound. The copper indium gallium selenide (CIGS), a well-known material in the solar cell sector, possesses a chalcopyrite structure, and it significantly influences the physical and optical features of this material(Rani et al., 2015).

The bond between metal and chalcogenoid elements in these chalcogenide materials is covalent, although it can have an ionic nature when the electronegativity of the metal and chalcogen is different. To illustrate this, in the case of tin sulfide (SnS), which is a material with some semiconducting capability, this characteristic can be attributed to the nature of tin and sulfur, which is ionic in nature(Ates et al., 2021). Conversely, the direct band gaps and high photovoltaic efficiency are realized in the case of CdSe due to its covalent properties. In the case of nanostructures, the physical characteristics are defined by the dimensions, and at such tiny lengths, the quantum effect heavily alters these characteristics. The resultant size reduction enlarges the energy levels and alters the optical spectrum of absorption and emission making these materials practical in optoelectronic and energy harvester devices(Hossain et al., 2023).

Optical, Electrical and Thermal Properties

Optical Properties

The optical characteristics of the chalcogenide nanocomposites present a rather distinctive material with respect to light-absorbing and light-emitting gadgets. Nanocomposites made out of chalcogenide can absorb a wide range of light effectively, especially within the visible and infrared region, (Dutkova et al., 2024a). This makes them suitable candidates for solar cells, photodetectors, and LEDs. Because of their composition and structure, chalcogenides are significantly tunable in terms of the band gap at which they absorb light in films and crystals. To illustrate, the CdSe quantum dots can be used due to the presence of a direct band gap that enables a high level of light absorption and photoluminescence. As the size diminishes, the band gap increases because this causes a blue shift of the absorption spectrum. Its size-dependent impairment of optical properties, known as the so-called quantum size effect, enables its tunability to diverse types of optoelectronic applications not only in solar energy conversion, but also in biomedical imaging. Moreover, their chalcogenide QDs exhibit an extremely high fluorescence quantum yield (photoluminescence QY) and tunable fluorescence, which is highly important in display devices, biosensors, and luminescent devices. They can be easily tuned in with their size following the rule of quantum effect that makes them highly appealing in all applications in optoelectronics, (Dutkova et al., 2024b; Hamam et al., 2025).

Electrical Characteristics:

The chalcogenide nanocomposites possess semiconductor properties and are therefore most suitable for various electronic and optoelectronic applications. If one takes as the building blocks the semiconductor nanostructures CdSe, PbS and SnSe to construct photodetectors, transistors and solar cells(Abd El-Ghany, 2025). Such materials can be synthesized with high mobility of charge carriers, low recombination of electrons and holes, and significant photovoltaic effects. The electrical conductivity of the chalcogenide-based nanocomposite can also be manipulated by doping with other metals or semiconductors, which can play an effective role in creating composites with improved carrier mobility and charge transport ability., (Abd El-Ghany, 2025; Sarkar et al., 2023). An example is CIGS (Copper Indium Gallium Selenide) thin films, which are typically used in solar cells, having high electron mobility and a considerable hole diffusion length; thus, this material is suitable for high-efficiency photovoltaic devices. Moreover, the components of chalcogenide quantum dots constituent transistors and displaying enormous on/off ratio and low threshold voltage are, thus, suitable for low power and bendable electronics, (Gaman et al., 1999).

Thermal Properties

The thermal performance of chalcogenide-based nanocomposites is another key consideration in many applications involving energy storage and conversion, (Zhao et al., 2008a). Many chalcogenides, such as TPV materials tin selenide (SnSe), have a relatively low thermal conductivity, a desirable property for thermoelectrics. Thermoelectric materials are used for converting waste heat to electric power, and their low thermal conductivity ensures that the heat gradient necessary for an efficient energy conversion is preserved, (Flasck & Rockstad, 1973). Meanwhile, for practical application in high-temperature environments, such as solar energy apparatus and energy storage devices, the thermal stability of those materials is of great importance. Several chalcogenide alloys, such as SnS and SnSe, have also been reported as promising candidates as the active layer in high-performance thermoelectric devices, wherein they work efficiently via the mechanisms of low thermal conductivity and high Seebeck coefficients, (Shi et al., 2019; Zhao et al., 2008b)

Characterization of Chalcogenide Nanocomposites

Characterization of chalcogenide nanocomposites is highly valuable in understanding the structure, properties and performance of the nanocomposites. Various advanced experiments are employed to study the cationic structure, optical performance, electrical conductivity, and morphological properties.

X-ray Diffraction (XRD)

A typical method for crystallographic order and phase identification of chalcogenide nanocomposites is X-ray diffraction. Based on the diffraction patterns, we can determine the crystal structure, lattice parameters, and crystallinity of the material. This method is also significant to characterize thin films and nanostructures deposited on the optoelectronic gadgets(Hairan & Azizi, 2023).

Electron Microscopy

SEM and TEM are complementary methods for studying the morphologies and dimensions of the nanostructured materials. The imaging SEM allows for the examination of the sample surface at the nanoparticle level, including size, distribution, and morphology. TEM, however, provides atomic-level details of internal structure. The methods are indispensable for analyzing the nanostructuring of chalcogenide-based nanocomposites.

UV-Vis Spectroscopy

UV-Vis spectroscopy is used to investigate the optical absorption and band gap of chalcogenide nanocomposites. This approach enables the measurement of absorption spectra in both the UV and visible ranges, facilitating an understanding of the material's electronic structure and optical properties. The blue shift of the peak positions could be utilized to study the quantum size effects in nanoclusters.

Fourier Transform Infrared Spectroscopy (FTIR)

FTIR spectroscopy is used to study the chemical bonding and functional group structure in chalcogenide-based nanocomposites. In the case of a chalcogenide, this is particularly useful for investigating bonding interactions between the chalcogenide and other components (e.g., carbon nanotubes or metal oxides), thereby determining the electronic properties and stability of the composite.

SYNTHESIS OF CHALCOGENIDE NANOCOMPOSITES

The preparation and construction of chalcogenide nanocomposites are vital for the fine-tuning of performances and functions of materials for various applications in optoelectronics, energy storage, and environmental remediation, (Martin & Nolas, 2014; Ridhova et al., 2021). Novel, low-cost, and scalable synthetic procedures are required to manufacture these materials for use in industrial applications at a commercial scale. In this part, we look into the commonly employed techniques for the synthesis of chalcogenide nanocomposites, such as CVD, PVD, sol-gel methods, hydrothermal and solvothermal processes, and template-mediated processes. Issues of scalability and quality control associated with the synthesis techniques will also be addressed, (Ridhova et al., 2021).

Vapor Deposition

Chemical Vapour Deposition (CVD) and Physical Vapour Deposition (PVD) are two of the most popular methods for fabricating thin films and nanostructures of chalcogenides. The two approaches provide accurate control of the composition, morphology, and thickness of the material requirements when considering the material for optoelectronic devices, solar cells and photodetectors. These methods are broadly used because they are capable of providing high-quality, even films of excellent adhesion to the substrate, (Abrutis et al., 2008).

Figure 3. Various preparation techniques for chalcogenide nanoparticles

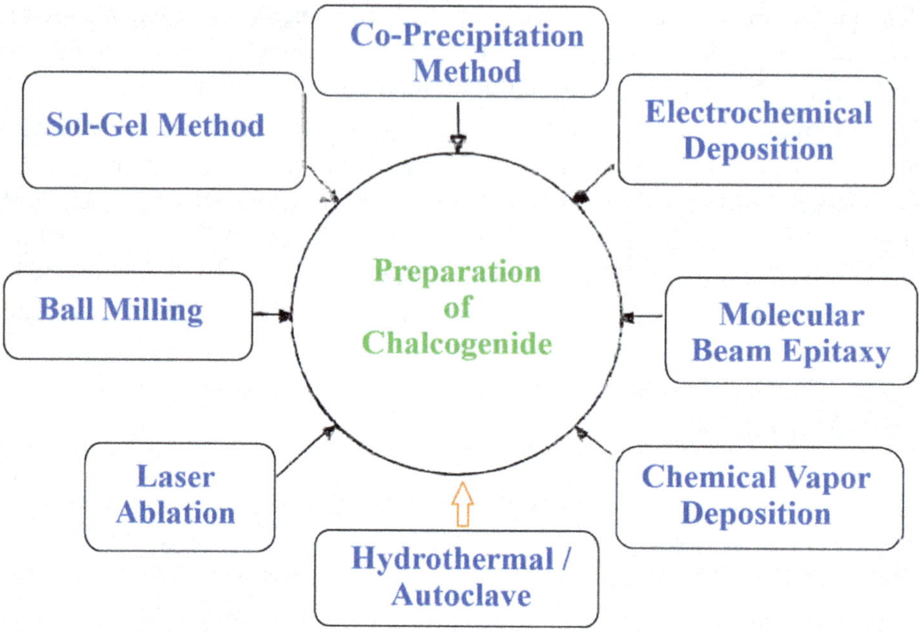

Chemical Vapor Deposition (CVD)

CVD is a method of depositing material onto a substrate from a vapor phase of a precursor in a high-temperature reactor. The chalcogenide-containing precursor gases (e.g., CdSe, CdTe, and CIGS) decompose in a hot substrate, leaving a deposited material. The major benefits of CVD are the capability to deposit very high purity films with a controlled stoichiometry and very good crystal quality. Low-temperature CVD (LTCVD) has recently been advanced to deposit chalcogenide thin films on flexible substrates, which is necessary for making flexible electronics and large-area applications. For instance, CVD-fabricated CIGS solar cells also have a high yield and their large-scale manufacture is one of the steps in reducing the cost of solar energy generation, (Abrutis et al., 2008).

Physical Vapor Deposition (PVD)

In contrast to PVD, CVD tends to rely on chemical reactions to deposit its material from a gas onto a solid. This approach is especially applicable for the production of metal-chalcogenide films that can be used in photovoltaic and optoelectronic devices. One of the most widely used PVD methods, which constitutes a process for depositing chalcogenide thin films, for example CdTe or ZnO, is sputtering. The most significant advantage of PVD is the precise control over the film thickness; therefore, it is widely employed to grow multi-layer films, such as in solar cells and LED applications. Advantages of PVD include: Scalable film deposition, Deposition at low temperature, minimizing thermal degradation, Homogeneous coatings across the substrate, Layer-by-layer deposition, allowing complex, multi-layer coatings, (Ramachandran et al., 2023).

Sol-Gel Method

The sol-gel process is a wet chemical approach to preparing chalcogenide nanocomposites, providing a low-cost, flexible and plausible method for fabricating their thin films, nanoparticles and nanorods. The procedure starts with the sol-gel reaction, where a sol, a colloidal solution of alkoxide or chalcogenide salts, is transformed into a gel through chemical reactions (hydrolysis and condensation). The most attractive feature of the sol-gel process is its flexibility in manipulating the morphology and composition of the material at the nanoscale. Meanwhile, the optical efficiency of CdSe nanoparticles fabricated by the sol-gel method is relatively high, as the CdSe particles are on the nanometer scale and have a modified surface. Furthermore, the sol-gel process can be utilized to prepare chalcogenide films at low temperatures, which is favorable for flexible electronics and large-area use. Nowadays, the attention has shifted to the combination of chalcogenides with graphene or metal oxide nanoparticles through the sol-gel method to enhance electronic conductivity and stability of the composites, namely for photocatalysis and energy storage applications, (Arachchige & Brock, 2007).

Hydrothermal and Solvothermal Methods

The hydrothermal/solvothermal process, which involves high temperatures and pressures, converts precursors dissolved in solvents into nanostructured materials. These approaches are universally employed for the preparation of chalcogenide nano-materials, in particular, nanoparticles, nanowires and nanorods with good crystallinity and regular shapes, (Rajamathi & Seshadri, 2002). Where the chalcogenide material synthesized is the II-VI chalcogenide, the hydrothermal method can employ water as the solvent for synthesis, and the solvothermal method can employ solvents other than water, as ethylene glycol or other alcohols large enough to act as a chalcogenide growth template for the chalcogenide material being synthesized. The great benefit of these procedures is that the size, morphology, and composition of the nanostructures can be tailored by changing the synthesis parameters (temperature, pressure, reaction time, among others). For instance, hydrothermally synthesized CdSe nanorods have demonstrated excellent optical properties, rendering CdSe-containing materials beneficial for use in solar cells and light-emitting diodes (LEDs). For example, tin selenide (SnSe) nanowires synthesized from the solvothermal method have shown an excellent thermoelectric behavior that is essential for energy harvesting applications, (Rajamathi & Seshadri, 2002)

Template-Assisted Synthesis

Template-assisted synthesis is also used, which is generally based on template material such as porous substrates, polymers or anodized alumina, for the control of the size and shape of the newly formed resultant nanocomposites(Kaur et al., 2022). Therefore, the template serves as a mold for the growth of chalcogenide nanomaterials in the form of nanowires, nanorods, or nanotubes. This is especially useful in the fabrication of high aspect ratio nanostructures necessary for a variety of nanodevices such as sensors, catalysts and optoelectronic devices. Template-based routes have been designed to fabricate ordered arrays of nanostructures, with the potential to enhance charge transport and electronic behavior in energy storage media. The most well-known application of TAS is the fabrication of chalcogenide semiconductors (e.g., CdS, CdSe, ZnS) nanowires or nanorods, which are extensively used for photovoltaic cells and photodetectors(Kaur et al., 2022).

Se/Te-based Chalcogenide Nanocomposites

Selenium based Nanocomposites

Selenium-containing chalcogenide nanocomposites exhibit notable photoelectrical, semiconductor, and biocompatible properties that make them crucial in optoelectronics, energy, and biomedicine. These materials demonstrate tunable bandgaps, efficient light-to-electric energy conversion, and exceptional infrared transparency, making them valuable for devices such as photocells, photodetectors, and IR optical fibers. Selenium nanomaterials also possess catalytic and antimicrobial characteristics, enabling applications in biosensing, antimicrobial surfaces, and environmental remediation.

Material Properties and Device Applications

- **Physical and Optical Properties**: Amorphous Se-based chalcogenide glasses exhibit infrared transmission between 2–15 μm, tunable bandgaps (size-dependent), photoconductivity, and photoluminescence, all essential for photonics, memory, and sensor devices.

- **Nano-thermodynamic Effects**: Selenium nanoparticle properties vary with composition and nano-structuring, especially when alloyed with tellurium, enhancing performance in photovoltaics and photodetector applications.
- **Biomedical and Catalytic Uses**: SeNPs show high biocompatibility, oxidative properties, and surface-functionalization potential, leading to development in biosensing, drug delivery, and selective antimicrobial activity.

Synthesis Strategies. Green synthesis and bio-mediated routes are increasingly popular, using plant extracts, microorganisms, or biomimetic conditions to yield environmentally benign selenium nanocomposites with tunable surface properties and high production efficiency. These methods are essential for producing stable, nontoxic nanomaterials suitable for biomedical, energy, and environmental sustainability applications

Tellurium-Based Nanocomposites

Tellurium-based chalcogenide nanomaterials are newly prominent for their unique anisotropic structure, p-type semiconductor behavior, enhanced charge transport, and synergistic catalytic features. These features allow the construction of high-performance optoelectronic and energy devices but require careful handling due to tellurium's rarity and environmental instability in some forms.

Material Properties and Device Applications

Electronic and Optoelectronic Properties. Tellurium shows strong spin-orbit coupling, excellent hole mobility, and can be fabricated into van der Waals crystals for advanced electronics, transistors, and photodetectors. Alloying Te with polymers (e.g., polyaniline/Te nanocomposites) significantly enhances thermoelectric performance through improved carrier transport and reduced thermal conductivity.

- **Nanostructure Engineering**: Te readily forms nanowires, nanorods, and nanoheterostructures due to its anisotropic growth, which can be leveraged for synergistic charge transfer and light absorption in photovoltaic and photonic devices.
- **Photodetection and Sensing**: Amorphous Te-Se alloys and nanocomposites display enhanced gain and sensitivity for IR photodetectors, extending their use to low-light imaging and communications technology.

Synthesis Approaches. Modern fabrication routes include solution-phase reduction, template-assisted synthesis, green chemical techniques, and in situ assembly within polymers or with other nanostructures, aiming for precise control over shape, heterostructure, and surface chemistry for specific device needs. The trend to combine Te with Se or polymers creates composite systems with new functionalities and greater environmental tolerance.

Selenium-Tellurium Alloy and Hybrid Systems. Combining selenium and tellurium yields alloyed nanocomposites with enhanced thermal, optical, and electronic properties compared to either element alone. These alloys are of particular interest for next-generation solar cells, thermoelectric converters, and photodetectors.

- **Alloy Effects**: Se-Te alloys display size- and shape-dependent thermal conductivity, band structure tuning, and mechanical flexibility, optimal for flexible photovoltaics and photonics.
- **Photodetector and Thermoelectric Devices**: Recent research reports amorphous Te-Se alloys with outstanding IR photodetection capability and high power factors in thermoelectric applications, addressing performance and integration challenges for future devices.
- **Sustainable Synthesis and Applications**: Green and scalable synthesis techniques for Se-Te nanocomposites are under development, focusing on eco-friendly routes for energy and environmental systems.

Synthesis and Scalability Challenges of Chalcogenide Nanocomposites

Although considerable progress has been made in chalcogenide nanocomposite synthesis, there are still many challenges to be addressed in terms of their scalability, compositional uniformity and morphology manipulation.

- Up-scalability: Several synthetic routes (CVD or sol-gel) struggle when scaling up from laboratory to industry. For instance, CVD generally demands high temperature and the real-time control of the precursory gases, both of which face significant challenges for the scale-up deposition on large-area substrates. Additionally, it is generally challenging to achieve uniformity when transitioning from small to large volumes using a sol-gel process.
- Material Uniformity – How to obtain homogeneous spreading of the materials at the nanoscale is another challenge. Chalcogenide nanocomposites are known to be sensitive to process variables, and even slight variations in temperature, precursor, or time cause changes in the material properties. This variation can have an interchanging influence on the performance of the end-

device, especially for applications that require accurate optical and electrical behavior.

- Morphology Configuration: Control of the nanostructure morphology is essential for the governing properties of chalcogenide nanocomposites in optoelectronics as well as energy storage devices. Although some approaches permit control of size and shape at the nanometric level, crystallinity, size uniformity, and specific surface area are often compromised, particularly in mass production. Providing a material with consistently sized nanostructures that retain their optimal properties is therefore a significant challenge in industrial applications.
- Toxicity and the environment concern: Toxic materials used, such as Cadmium in CdSe, or PbS, will bring about environmental issues and health hazards for large-scale production. Attention is now focused on the pursuit of environmentally friendly, safe chalcogenides that will replace hazardous chalcogenides.

OPTOELECTRONIC APPLICATIONS OF CHALCOGENIDE-BASED NANOCOMPOSITES

In recent years, chalcogenide-based nanocomposites have received significant interest in optoelectronics due to their excellent optical properties, such as high absorption coefficient, tunable band gap, and high photoluminescence. They find applications in photodetectors, photovoltaics, light-emitting diodes and lasers. Introducing nanostructure to the chalcogenide-based composites has additionally boosted the performance of the chalcogenide-based nanocomposite in charge transport, energy conversion efficiency and optical emission(Xia et al., 2016; Yang & Lin, 2025). This part presents different optoelectronic applications of chalcogenide-based nanocomposites in photodetectors, solar cells, LEDs, lasers, and photonics (Figure 4).

Figure 4. Optoelectronic applications of chalcogenide

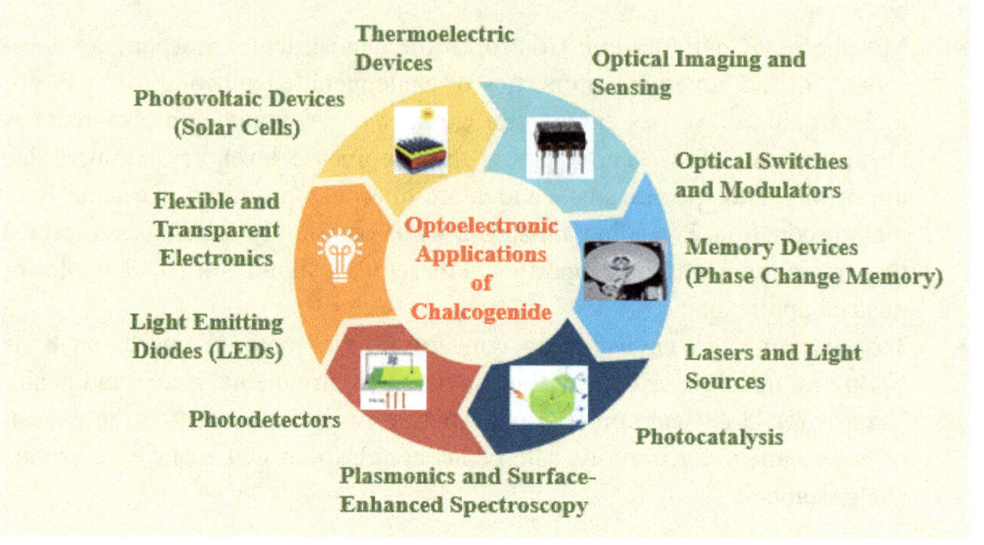

Photodetectors and Photovoltaic Devices of Chalcogenides

Photodetectors and photovoltaic devices play a key role in optoelectronics, and nanocomposites based on chalcogenides have contributed significantly to the development in these fields. Photodetectors and photovoltaic devices are known for imaging systems, night vision equipment and remote sensing(Xia et al., 2016).

Chalcogenides in Photodetectors

Nanocomposites of chalcogenides, such as cadmium selenide (CdSe), lead sulfide (PbS), and copper indium gallium selenide (CIGS), have proven helpful in photodetectors due to their direct band gaps, which are ideal for light absorption and charge carrier generation. As an example, CdSe quantum dots are a very light-sensitive material, and the size-dependent optics is adaptable to make a sensitive photodetector with very low dark current(J. Singh et al., 2025). Their optical absorption and charge collection efficiency are also increased since they adopt a nanostructured morphology. Therefore, it is expected that these materials will result in high-performance imaging and sensing applications. In addition, carbonaceous composite material nanohybrids with chalcogenide, such as graphene or exclusive carbon nanotubes, are also explored for enhanced charge delivery and increased mechanical strength of photodetectors. As an illustration, the flexibility of graphene-CdSe nanocomposites exhibits good electron mobility and response time and can be utilized as a prospective system of high-speed photodetectors and infrared sensors(J. Singh et al., 2025b).

Chalcogenides in Photovoltaic Devices

CdTe and CIGS chalcogenide materials are the two most common ones as thin-film solar cells used in this group due to their cost-effectiveness and performance. However, CIGS is among the most studied materials for harvesting solar energy because it has high absorption efficiency, as well as a tunable band gap that can be well matched to the solar spectrum (Bidikoudi & Stathatos, 2024; J. Singh et al., 2025b). The efficiency was enhanced, and the cost of deposit was reduced due to the recent advances in nanosized structures. Multijunction solar cells have also been produced this way, with the chalcogenide layers stacked one on top of the other to capitalize more efficiently on various components of the solar spectrum and consequently achieve a higher overall efficiency. Other nanostructured materials could also be combined with chalcogenides in this way, including quantum dots or graphene, to enhance light absorption and charge separation, resulting in more efficient and stable solar devices (Bidikoudi & Stathatos, 2024).

Application in LEDs and Lasers

The applications of chalcogenide-based nanocomposites also have enormous significance in LEDs and lasers, which are crucial in the design of efficient lighting and communication tools.

Chalcogenides in LEDs

Chalcogenide QDs Although several chalcogenide QDs have been attracting attention in recent years, including CdSe, PbS and ZnS, they cannot be easily tuned in wavelength because of their broader emission. It renders the chalcogenides applicable in full-colour visible screens and efficient lighting(Mal et al., 2016). Chalcogenide nanocrystals are typically functionalized on their surface to enhance optical performance and stability. As an example, CdSe quantum dots can be surface-passivated with a significant impact on their quantum efficiency and heat stability, which further applies them to the commercial manufacturers of LEDs, (Saini et al., 2020). The ability to tune the optical properties of chalcogenide nanocomposites using size and shape alone can tune highly efficient LEDs having a specifically designed color emission that can be utilized in displays, lighting and backlight, (Mal et al., 2016).

Chalcogenides in Lasers

Chalcogenide nanocomposites are also used in the synthesis of laser devices, such as mid-IR lasers. CdSe, PbS and PbSe are common laser materials due to their high photoluminescence quantum yields and efficient emission of light with a specific wavelength. Lasers play a critical part in telecommunication, medical diagnostics and sensor systems. Quantum dots in chalcogenides have become at the forefront of research with laser diodes (LDs) on such structures(Ates et al., 2021b). An example would be in the case of PbSe quantum dots, which are applied to near-infrared lasers, which are incorporated in optical communications and in remote sensing. The greatest attribute of chalcogenide QDs in laser applications is that they have high power and a narrow linewidth of the emission, such that chalcogenide QDs can be considered a viable option in the next generation of laser technology, (Ates et al., 2021b; Rana et al., 2018b).

Chalcogenide Nanocomposites for Solar Energy Devices: Material and Efficiency Enhancement

The high efficiency of the solar cells is due to the role that the chalcogenide nanocomposites play. The CdTe, CIGS and PbS are considered to have been the most frequently used materials in thin film solar cells because of their efficiency of conversion of sunlight to electricity(Bidikoudi & Stathatos, 2024b). Due to low cost, high absorption coefficient and high photovoltaic efficiency, these materials have become the subject of exhaustive research (Figure 5).

Adoption of nanostructured chalcogenides has significantly enhanced the low intensities of solar cells. Nano-patterning or nanostructuring of the chalcogenide layers is referred to as one of the primary methods of achieving higher efficiency of the solar cells through a greater surface area for light absorption. More than 20 per cent cell conversion efficiency is already replacing nanocrystalline thin films of CdTe and CIGS, (Saadat & Amiri, 2024; Sivasamy et al., 2024b). To this end, there are the quantum dot-based solar cells, which are today more commonly referred to as quantum-dot solar cells (QDSCs), where coatings of quantum dot whose optical absorption and charge transfer can be tuned are added to the chalcogenide solar cells to enhance the optical absorption and charge transfer. The suggested use of this technology is that it will be used to improve the effectiveness of solar cells way beyond the traditional Shockley-Queisser limit, (Ates et al., 2021b).

Figure 5. Chalcogenide nanocomposites for solar energy devices

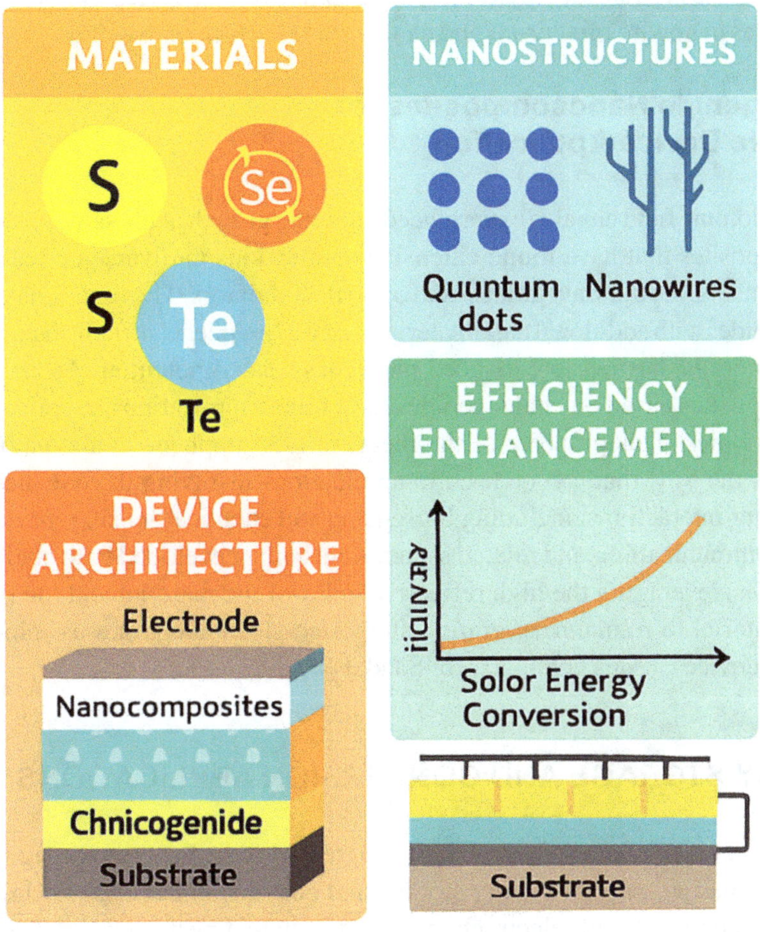

Nanostructuring for Better Optoelectronic Performance

There is a possibility to improve the performance of the chalcogenide-based nanocomposites in optoelectronic applications through nanostructuring. Having reduced the material to the nanoscale, there is also an opportunity to leverage quantum effects to increase the optical absorption and emission efficiency of the corresponding material. To give an example, the possibility of controlling optical characteristics of the chalcogenide quantum dots by precisely engineering the size and shape of nanostructures through the Quantum size effect (QSE). The result is narrower emission spectra, higher photoluminescence efficiencies and improved charge carrier dynamics. Chalcogenide nanocomposites are particularly excellent materials in the high surface area application (photovoltaic devices, photo-detectors, catalysis) as nanowires, nanorods and nanocapsules, (Bidikoudi & Stathatos, 2024b).

Chalcogenide Nanocomposites for Photonic Device Application

The photonic field has also experienced booming growth of the same chalcogenide nanocomposites that have found extensive applications in diverse optical devices, such as sensors, lasers, waveguides, and optical modulators. The functionalization of chalcogenides with additional nanomaterials, such as graphene, metallic nanoparticles, or 2D materials, has led to enhanced photonics-related potential. As an example, nanocomposites based on chalcogenides have found applications in optical sensing and plasmonics. This property of the metal sulfide-graphene oxide nanocomposites forms the SPR that can be used in applications involving the enhancement of light-matter interactions, including biosensing and chemical sensing. Furthermore, optical communication- and related systems are being developed with chalcogenide waveguides, leveraging the high refractive index of the material and the flexibility of the material to miniaturize grating-supported slot waveguides as miniaturized photonic devices, (Ates et al., 2021b; Saadat & Amiri, 2024).

ENERGY STORAGE AND CONVERSION APPLICATIONS

Chalcogenide nanocomposites have progressed from being real magical nanostructures (particularly for energy storage and conversion devices), owing to their unprecedented electrical, electrochemical, and optical performance. These materials are being thoroughly investigated for their potential application in batteries, supercapacitors, fuel cells, hydrogen storage and thermoelectric devices (Figure 5). By virtue of the extraordinary ionic and electronic conduction at the nanometer size, chalcogenides are promising for promoting the performance of these energy systems with better efficiency, capacity, stability, and sustainability, (Wani et al., 2025). This part discusses the application of chalcogenide-based nanocomposites in essential areas of energy storage and conversion, (Yuan et al., 2017).

Figure 6. Energy storage and conversion applications

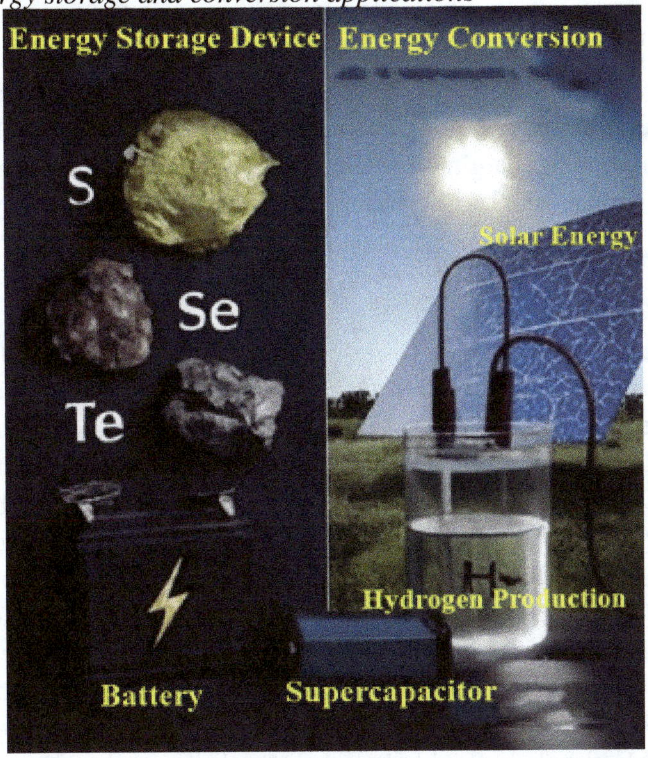

Chalcogenides as Cathodes for LIBs and SIBs

One of the most widely deployed energy storage technologies is the lithium-ion (Li-ion) and sodium-ion (Na-ion) batteries, which span from portable electronics to the electric vehicle (EV) market. Their operation is closely connected to the choice of electrode materials, and the chalcogenide nanocomposites offer a promising method of obtaining both the capacity and the cycle life on them, (Palchoudhury et al., 2023).

Lithium-Ion Batteries

Cathodes in Li-ion batteries are of lithium cobalt oxide ($LiCoO_2$) or lithium iron phosphate ($LiFePO_4$), and the anodes are either graphite or silicon. However, these materials have low electrical capacity, low cycle life, and rate performance. To overcome these issues, owing to their high capacity and high-rate performance, chalcogenide-based nanocomposites, including tin selenide (SnSe) and tin sulfide (SnS), have been contemplated. SnSe and SnS are of specific interest because they boast high theoretical capacities and high volumetric energy storage, which are needed to increase the energy density of Li-ion batteries. These materials have been observed to exhibit high charge-discharge efficiency, low polarization, and good rate capacity. Therefore, it is estimated that they can be introduced as a next-generation anode material. Moreover, their nanostructures may enable a greater acceptance of volumetric variation of charging-discharging reactions to support a greater cycling output, (Palchoudhury et al., 2023; Yuan et al., 2017)

Sodium-Ion Batteries

Despite the wide use of Li-ion batteries, the scarcity of lithium and its price have drawn significant interest in sodium-ion batteries (NIBs). The resource of sodium is much more plentiful and comparatively cheaper than lithium, and it can be comprehensively deployed in large-scale energy storage systems and grid technology in the form of NIBs. The Chalcogenides, e.g. $Na_2FeP_2O_7$, Na_2S, etc., may be considered to be the prospective electrode materials in the NIBs, and they can yield both high energy density and high stability. The sodium chalcogenides, (Na_2S) and (Na_2Se), have exhibited great specific capacity and electrochemical stability and are possible anode materials of sodium ion batteries. Research has been conducted to optimize the electronic conductivity and structural stability of materials, to enhance the performance of these new candidates as NIBs. And, in such a way, compared to the regular materials, the chalcogenide-based electrodes will allow Na-ion batteries to conduct better charge-discharge cycles and greater efficiency, (Yuan et al., 2017).

Supercapacitors and Chalcogenide Nanocomposites

Supercapacitor, more often known as ultracapacitor, is a kind of energy storage device that supports electrostatic and a high quantity of charge/discharge cycles. Batteries do not possess the same power density and energy density as supercapacitors. Electrode materials, Such as Chalcogenide nanocomposites, have a high surface area, electrical conductivity, and ion storage capacity, and can be explored to enhance the energy density of supercapacitors further.

Chalcogenide Nanocomposites in Supercapacitors

The MnO_2-CdSe, NiS-CdTe, and TiO_2-SnSe-based nanocomposites have proved to be unique in the world of supercapacitors because they can hold a significant amount of charge density at the electrode/electrolyte interface. These nanocomposites have been observed to increase the specific capacitance and rate performance of a supercapacitor, enhancing the transfer of charge and also providing a greater active surface. As an illustration of the above discussion, the results of combining environmentally friendly and plentiful MCs with carbon-based materials (including graphene or carbon nanotubes) in a cross with supercapacitors are significantly enhanced through improved electrical conductivity and mechanical characteristics. Such energetically dense, cyclic-stable materials, as well as those with long cycle life, are the candidate materials in the high-performance energy storage applications within electric vehicles, renewable energy systems, and portable electronics, (Ahmad et al., 2024).

Chalcogenides in Fuel Cells and Hydrogen Storage

Nanocomposites containing chalcogenides are also being investigated for applications in fuel cells, hydrogen storage, and the key functions that underpin the future hydrogen economy. Fuel cells are devices that convert chemical energy directly into electricity, and a hydrogen fuel cell is an attractive source of clean energy, generating electricity from water as the only product, (Karol et al., 2024).

Chalcogenides in Fuel Cells

Solid oxide fuel cells (SOFCs) operating at high temperatures leverage the high ionic conductivity and stability of chalcogenide materials, such as TiS_2 and MoS_2. These structures are essential in the efforts to increase the overall performance and longevity of fuel cells. In particular, some classes of transition metal chalcogenides (such as MoS_2) have been observed to exhibit catalytic behavior that assists the electrochemical reactions in fuel cells. In addition, chalcogenide nanocomposites may also serve as catalysts in hydrogen production via water electrolysis, supporting a sustainable hydrogen economy. The large surface area of nanostructured chalcogenides makes them suitable for acting as an efficient cathode to enhance electrochemical reactions for hydrogen production, which in turn decreases the overall energy consumption of this process, (Karol et al., 2024b).

Hydrogen Storage

Chalcogenide nanocomposites also hold great promise for hydrogen storage, a decisive factor in the hydrogen economy. The hydrogen storage medium should have high volumetric energy and show reversible hydrogen uptake and discharge behavior. Given their relatively simple elements, these materials have been investigated for use as solid-state hydrogen storage materials, as they are capable of absorbing significant quantities of hydrogen at modest pressures and temperatures. The recent progress on the chalcogenide-based nanocomposites has also witnessed the emergence of light-weight and high-capacity materials for hydrogen storage, which is a prerequisite for fuel cells/hydrogen-powered vehicles(Eisa et al., 2023; Karol et al., 2024a).

Thermoelectric Energy Harvesting Applications

Thermoelectric materials convert heat into electricity, providing a clean method for harvesting waste heat from car exhaust and industrial processes to power electrical devices. Chalcogenide-based compounds, such as tin selenide (SnSe), lead telluride (PbTe), and bismuth telluride (Bi_2Te_3), are widely studied for their thermoelectric behavior. Materials based on chalcogens, for example, SnSe, have exhibited an excellent thermoelectric performance because of appropriate properties of low thermal conductivity and high seebeck coefficient, which are essential for the conversion of heat to electricity(Shahzad et al., 2024). Growth of chalcogenide structured materials at the nano level, which has additional advantage in tuning up the thermoelectric performance (reduction in thermal conductivity with simultaneous high electrical conductivity) for effective energy harvesting, has been progressed. For instance, SnSe nanowires demonstrate promising thermoelectric performance, which makes them promise for energy harvesting. The further optimization of the figure of merit (ZT), which represents a primary performance measure for thermoelectric materials, is still in focus as a prerequisite for the successful application of these materials for large-scale energy recovery, (Karol et al., 2024a).

Issues on Storage and Conversion of Energy based on Chalcogenide Nanocomposites

Although chalcogenide nanocomposites show great promise in energy storage and conversion devices, they also face challenges such as scalability, material stability, and cost.

- Scalability: A lot of synthesis techniques of the chalcogenide nanocomposites (such as CVD or sol-gel procedures) are not straightforward to scale up from laboratory to industrial scales. It is still a big challenge to realise the uniformity and homogeneity of the properties of the materials at the scale level, especially for the electrode materials and supercapacitors.
- Material Stability: Chalcogenides, in particular CdTe and PbS, may tend to be unstable in extreme environments (e.g. high-temperature, aggressive electrolytes). It is critical to develop stable and robust materials to guarantee the long-term stability of energy storage and conversion devices.

- Cost: Though many chalcogenide-based nanocomposites exhibit superb performance, the cost could be a serious concern, especially for the applications of large-scale energy storage. The toxic nature of some materials, such as cadmium and lead, as well as concerns over environmental impact and recycling, also make it even more difficult and costly to bring them to market.
- Energy conversion efficiency: Despite their high efficiencies of energy conversion, energy is lost during charge/discharge cycles in batteries and energy harvesting. Reducing energy losses and enhancing the performance of these materials is a continued bottleneck.

ENVIRONMENTAL SUSTAINABILITY AND GREEN TECHNOLOGIES

Chalcogenide nanocomposites might be of importance to environmental and green technology programs. Such structures exhibit excellent photocatalytic, electrochemical and optical characteristics and have been demonstrated to have an enormous utility in ecological assessment, energy capture, water treatment and mitigation of carbon gases. Better still, their ability to transform harmful pollutants into harmless by-products, create alternative and green sources of energy, and promote eco-friendly processes and technologies, is in tandem with the global trends towards more environmentally friendly and sustainable systems, (Tari et al., 2024). This chapter outlines the various environmental usages of chalcogenide nanocomposites, especially in the area of remediation, transformation of sunlight into energy, purification of water and the aspect of green synthetic techniques, (Munonde et al., 2024; Sivasamy et al., 2024c).

Utility in Environment Processing Applications of Chalcogenide Nanocomposites

One of the moral imperative challenges of the modern world is the restoration of the contaminated ecosystems, particularly in conditions where the increase in industrial waste, heavy metals, and toxic organic pollutants is considered. These environmental issues have been addressed through the great potential of chalcogenide nanocomposites, which offer high surface area, reactivity, and catalytic properties.

Photocatalytic Remediation

Hybrid composites of chalcogenide nanoparticle structures comprise cadmium sulfide (CdS), titanium dioxide (TiO_2) and zinc sulfide (ZnS) and turn out to be highly excellent photocatalysts in the degradation of organic pollutants under the influence of solar light. The materials can withstand heat and continually facilitate the formation of a pair of electrons and holes through catalysis in the sunlight. This process results in the activation of oxidation or reduction of poisonous substances within the environment. As an example, nanocomposites prepared on CdS have been well-studied in terms of photocatalytic wastewater treatment of organic pollutants (dyes, pesticides and drugs). Nano-structured Chalcogenide photocatalysts have displayed high activity, owing to the increase in surface area and active reaction sites. Tuning is also possible with chalcogenide nanocomposite band gaps, which makes them useful in terms of solar energy absorption in solar-driven cleansing of the environment, (Karol et al., 2024b; Shahzad et al., 2024; Sivasamy et al., 2024c).

Heavy Metal Eliminant

Removal of heavy metals, such as lead, cadmium, and arsenic, in polluted water is also a significant application of chalcogenide nanocomposites, (Tari et al., 2025). Chalcogenide nanocomposites promise to be effective adsorbents of such metals because surface groups of functional groups can be present to interact with metal ions. An eco-friendly example is that the presence of the CdSe and the ZnS nanocomposites helped to remove the presence of heavy metal ions in aqueous solutions at a very cheap cost. These can be used in water treatment, pollution prevention methods, and decontamination systems, and in this case, their adsorptive powers and the ability to be reused as a material are a prospective aspect in the remediation of a polluted environment, (Munonde et al., 2024).

Photocatalysis and Solar Energy for Sustainable Energy Production

Photocatalysis is regarded as one of the most hopeful clean energy technologies to produce renewable energy and clean our environment, which can convert sunlight to chemical energy for degrading pollutants or producing clean fuels. The chalcogenide nanocomposites, therefore, can be utilized as active materials for solar energy conversion as they exhibit strong light-harvesting ability, efficient photogenerated charge separation and high catalytic activity(Karol et al., 2024b; Shahzad et al., 2024).

Photocatalytic Hydrogen Production

Photocatalysis as a route for producing hydrogen has been the most promising application of chalcogenide nanocomposites in solar energy harvesting. Water splitting for the production of hydrogen fuel using solar energy is considered a sustainable solution for the world's energy needs. Chalcogenide-type photocatalysts (such as CdS, CdSe, and Cu2ZnSnS4 [CZTS]) have shown remarkable performance in hydrogen evolution in the visible region. For instance, CdS nanocomposites are often paired with Pt nanoparticles to enhance the photocatalytic efficiency of hydrogen evolution. Due to their tunable band gap and large surface area, chalcogenide materials are promising for light absorption and charge separation, making them potential materials for photo-solar hydrogen production, (Ates et al., 2021b; Karol et al., 2024b).

Carbon Dioxide Reduction

Apart from hydrogen generation, efforts have also been made to explore the applications of chalcogenide nanocomposites for the reduction of carbon dioxide (CO_2) (i.e., the conversion of CO_2 into valuable chemicals such as methane or methanol) and reducing the impacts of greenhouse gas emissions. CdSe, ZnS and Cu_2S have already been shown to be efficient photocatalysts for the reduction of CO_2 under solar radiation and thus to contribute to one of the major global problems, i.e., carbon dioxide emission to the atmosphere, providing renewable fuels, (Ates et al., 2021b).

Water Purification and Desalination Technologies Based on Chalcogenides

Dealing with the world's current water shortage is one of the most serious challenges we face today, and economically viable water purification and desalination technologies are becoming increasingly important. Among them, chalcogenide-derived nanocomposites have exhibited promising applications in pollutant degradation and seawater desalination through photocatalysis, adsorption, and membrane filtration, (Shahzad et al., 2024; Wani et al., 2025).

Water Purification Process

Applications of chalcogenide nanocomposites, such as TiO2-CdS and ZnO-CdTe, have been used for water purification via photocatalysis, where solar irradiation is exploited to break down organic contaminants and disinfect them by yielding highly reactive hydroxyl radicals (OH·). These nanocomposite materials have been validated to be highly effective for degrading pollutants such as pesticides, pharmaceuticals and industrial dyes in water, (Eisa et al., 2023; Karol et al., 2024b).

Desalination

Desalination is another area where chalcogenide-based nanocomposites have potential – specifically, the removal of salt and other dissolved solids contained in seawater. There is also potential in Chalcogenide-based materials due to their high surface area and selectivity in membrane filtration, which can remove salts and other impurities from water. Moreover, the reverse osmosis membranes employing chalcogenide-based nanocomposites can be utilized, which is a crucial step towards improving the merits of desalination and the cost, (Abu Elella et al., 2024)

Eco-Friendly Synthesis Protocols of Chalcogenide Nanocomposites

The sustainability of chalcogenide nanocomposites is also a factor that influences their implementation in green technologies. The traditional pathways involved in the manufacturing process of the preparation involve the chemical vapour deposition (CVD) processes, the high temperature-reaction mechanism; these are usually set alongside the application of toxic pollutants and also energy-demanding processes. Therefore, the alternative of reducing the environmental load of these materials is addressed through the green synthesis methods that are becoming invested, (Karol et al., 2024b).

Green Synthesis Strategies New interest in developing solvent-free and low-temperature synthesis routes to chalcogenide nanocomposites, e.g. hydrothermal synthesis, sol-gel approaches, and biomimetics. As an example, the use of extracts of plants to reduce the CdS and ZnS nanostructures has been recently described as a potential, (Abu Elella et al., 2024) Environmentally friendly method, since toxic solvents are not needed, and the amount of energy involved is minimal. Moreover, recyclable templates and biodegradable precursors are under exploration, which encourages the development of more environmentally friendly synthesis methods for chalcogenide nanocomposites. The green techniques are earth-friendly, economical and scalable and can reduce the impact of such materials on the environment during the industrial process.

The Role of Chalcogenide Nanocomposites in Carbon Emission Mitigation

With the issue of climate change a central challenge facing the global community, the fight to reduce carbon emissions is more pressing than ever. There are different pathways through which chalcogenide nanocomposites help with carbon savings, as follows:

- Photocatalytic CO_2 reduction: Chalcogenide nanocomposites like CdS or ZnS are emerging in the conversion of CO_2 into useful fuels (methane, methanol) and lead to the reduction of carbon footprints of industrial processes.
- Energy Saving Solar Cells: These chalcogenides help solar cells reduce their reliance on fossil fuels, while promoting clean energy. As solar power becomes more prevalent, chalcogenides help reduce the greenhouse gases emitted during energy production.
- Pollution Remediation: Photocatalytic decomposition/degradation of pollutants, such as industrial effluent, heavy metals, and organic dyes, can help to remediate water contamination and restrict the release of hazards to the environment, thus reducing carbon emissions associated with pollution.
- Thermoelectric Materials: The chalcogenide thermoelectric materials, including SnSe and PbTe, can be utilized in waste heat hydroelectricity material, harvesting the waste heat of industry and reducing industrial carbon emissions.

Due to improvements in energy efficiency, the opportunity for renewable energy supply, and reduced environmental contamination, engineered chalcogenide nanocomposites play an essential role in addressing the global challenge of decreasing carbon emissions and the need for a sustainable and low-carbon economy.

CHALLENGES AND FUTURE OUTLOOK

Although chalcogenide-based nanocomposites have exhibited tremendous potential and rapid growth, numerous insurmountable challenges remain that need to be addressed for their practical application in commercial and industrial fields. These obstacles may hinder chalcogenide-based nanocomposites from entering the mainstream market due to (a) fabrication limitations, (b) social and environmental constraints and (c) demand for additional improvements in the properties of the material. This article summarizes the main challenges in the synthesis, performance optimization, and upscaling of chalcogenide nanocomposites, along with the future perspectives of these materials to be envisioned through the prism of global sustainability.

The development of chalcogenide-based nanocomposites for these applications is associated with various technological problems, with fabrication and scale-up issues being among the major ones. Although laboratory-scale methods, including chemical vapor deposition (CVD), sol-gel reactions, and hydrothermal techniques, are capable of synthesizing high-quality chalcogenide nanocomposites, such processes are generally complex and costly to scale for practical commercial production.

However, some issues need to be agreed upon when scaling up and commercializing these materials:

- **Material Uniformity and Consistency**: During the scaling from small-scale laboratory synthesis to full-scale production, it is challenging to maintain size, composition and morphology uniformity of nanostructures. The nanostructuring of chalcogenides is very sensitive to process variation in temperature, precursor concentration and pressure. The difference in those parameters during large-scale synthesis could cause some irregularities in the material, and that could harm the optoelectronic or energy storage device performance.
- **Cost of Production**: Most of the chalcogenide materials used in nanocomposites, for example, Cu, Cd, Pb, are pretty expensive and, in some cases, even toxic. As such, for chalcogenide nanocomposites to be competitive with other commercial materials, cost-effective large-scale production methods need to be improved. However, synthesizing pure chalcogenide materials using nontoxic solvents and precursors is still a challenge.

Scalable Film Deposition Techniques Film deposition techniques for solar cells or photodetectors are usually limited for large-scale applications due to the following: Methods such as chemical and physical vapor deposition (CVD, PVD), though of proven utility at a laboratory level, present some difficulties in scaling up for standard commercial production. Novel, large-area low-cost deposition processes must be developed for these to be mass-produced efficiently and cost-effectively.

CONCLUSION AND FUTURE DIRECTION

In this chapter, we highlight chalcogenide-based nanocomposites for applications in optoelectronics, energy storage, and sustainability. These materials, synthesized by blending chalcogenides based on sulfur, selenium, and tellurium with other nanomaterials, offer a wide variety of unique properties, including tunable optical absorption, high charge carrier mobility, and a high surface area, rendering them excellent candidates for several state-of-the-art technological applications. Solar energy, energy storage, environmental remediation and smart sensors are just a few of the many potential applications of chalcogenide-based nanocomposites. The work concludes with a brief presentation of key results, a few words on the implications for further research, and a reflection on how chalcogenide nanocomposites can generate sustainable energy solutions. The chalcogenide-based composites studies, based on nanocomposite materials, have led to several significant results that demonstrate their prospective technological outlook:

- Optoelectronic Applications: Chalcogenide nanocomposites, such as CdSe, PbS and chalcopyrite-based CIGS, have shown great potential in photovoltaic devices, photodetectors, LEDs and lasers. The possibility of tuning their optical properties by modulating their size and shape, along with the application of these materials due to their strong absorption coefficient and superb charge transport properties, has led to their use in next-generation optoelectronic devices.
- Energy in Storage and Conversion: Chalcogenide materials have been proven to be up-and-coming materials in batteries, supercapacitors, fuel cells, and thermoelectric devices. Due to their large capacity, ionic conductivity, and good electrochemical stability, MOF runners are excellent candidates for enhancing the performance and efficiency of energy storage systems, such as those for renewable energy integration and sustainable energy storage solutions.
- Environ-me-grail sustainability: The application of chalcogenide nanocomposites in environmental remediation through water treatment, pollution management, and CO_2 abatement – the fundamental issues of global environmental problems – illustrates their importance. Photocatalytic chalcogenides are effective in degrading pollutants, producing hydrogen from solar energy, and reducing carbon dioxide, thereby benefiting clean energy and sustainable environments.
- Commercialization Issues: Although they provide a great potential, scalability, material stability, toxicity issues, and cost effectiveness are still obstacles for the large-scale use of the chalcogenide nanocomposites. To address these challenges, continued progress in synthesis approaches, sustainable materials, efficiency to an industrial level and economic viability must be achieved.

REFERENCES

Abd El-Ghany, W. A. (2025). Review on the optical and electrical properties of chalcogenide thin films: Challenges and applications. *Physical Chemistry Chemical Physics*, *27*(9), 4567–4586. DOI: 10.1039/D4CP04473H PMID: 39957264

Abrutis, A., Plausinaitiene, V., Skapas, M., Wiemer, C., Salicio, O., Longo, M., Pirovano, A., Siegel, J., Gawelda, W., Rushworth, S., & Giesen, C. (2008). Chemical vapor deposition of chalcogenide materials for phase-change memories. *Microelectronic Engineering*, *85*(12), 2338–2341. DOI: 10.1016/J.MEE.2008.09.014

Abu Elella, M. H., Hassan, S. S., Abdallah, H. M., Mostafa, M. S., & Abu-Thabit, N. Y. (2024). Metal chalcogenides and their nanocomposites in water purification systems. *Metal-Chalcogenide Nanocomposites: Fundamentals, Properties and Industrial Applications*, 59–82. https://doi.org/DOI: 10.1016/B978-0-443-18809-1.00004-3

Ahmad, K. S., Jaffri, S. B., Makawana, B., Gupta, R. K., Nadeem, A., & Attia, S. M. (2024). Developing sustainable chalcogenide for energy storage: Investigating the semiconducting MnS2: CuS: CoS2: Ni2S3 quaternary complexes as supercapacitor electrode material. *Microchemical Journal*, *207*, 112218. DOI: 10.1016/J.MICROC.2024.112218

Arachchige, I. U., & Brock, S. L. (2007). Sol–Gel Methods for the Assembly of Metal Chalcogenide Quantum Dots. *Accounts of Chemical Research*, *40*(9), 801–809. DOI: 10.1021/AR600028S PMID: 17441681

Ates, M., Tanaydın, M. K., & Yılmaz, E. (2021a). Challenges, novel applications, and future prospects of chalcogenides and chalcogenide-based nanomaterials for photocatalysis. *Chalcogenide-Based Nanomaterials as Photocatalysts*, 307–337. https://doi.org/DOI: 10.1016/B978-0-12-820498-6.00014-7

Ates, M., Tanaydın, M. K., & Yılmaz, E. (2021b). Challenges, novel applications, and future prospects of chalcogenides and chalcogenide-based nanomaterials for photocatalysis. *Chalcogenide-Based Nanomaterials as Photocatalysts*, 307–337. https://doi.org/DOI: 10.1016/B978-0-12-820498-6.00014-7

Bidikoudi, M., & Stathatos, E. (2024a). Chalcogenides in Perovskite Solar Cells with a Carbon Electrode: State of the Art and Future Prospects. *Nanomaterials 2024, Vol. 14, Page 1783, 14*(22), 1783. https://doi.org/DOI: 10.3390/NANO14221783

Bidikoudi, M., & Stathatos, E. (2024b). Chalcogenides in Perovskite Solar Cells with a Carbon Electrode: State of the Art and Future Prospects. *Nanomaterials 2024, Vol. 14, Page 1783, 14*(22), 1783. https://doi.org/DOI: 10.3390/NANO14221783

Bouroushian, M. (2010). *Electrochemistry of Metal Chalcogenides*. https://doi.org/DOI: 10.1007/978-3-642-03967-6

Dutkova, E., Baláž, M., Kováč, J., Daneu, N., Kashimbetova, A., Briančin, J., Kováč, J., Kováčová, S., & Čelko, L. (2024a). Optical and Optoelectrical Properties of Ternary Chalcogenide CuInS2/TiO2 Nanocomposite Prepared by Mechanochemical Synthesis. *Crystals 2024, Vol. 14, Page 324, 14*(4), 324. https://doi.org/DOI: 10.3390/CRYST14040324

Dutkova, E., Baláž, M., Kováč, J., Daneu, N., Kashimbetova, A., Briančin, J., Kováč, J., Kováčová, S., & Čelko, L. (2024b). Optical and Optoelectrical Properties of Ternary Chalcogenide CuInS2/TiO2 Nanocomposite Prepared by Mechanochemical Synthesis. *Crystals 2024, Vol. 14, Page 324, 14*(4), 324. https://doi.org/DOI: 10.3390/CRYST14040324

Eisa, T., Abdelkareem, M. A., Jadhav, D. A., Mohamed, H. O., Sayed, E. T., Olabi, A. G., Castaño, P., & Chae, K. J. (2023). Critical review on the synthesis, characterization, and application of highly efficient metal chalcogenide catalysts for fuel cells. *Progress in Energy and Combustion Science*, *94*, 101044. DOI: 10.1016/J.PECS.2022.101044

Flasck, R., & Rockstad, H. K. (1973). The thermal conductivity of some chalcogenide glasses. *Journal of Non-Crystalline Solids*, *12*(3), 353–356. DOI: 10.1016/0022-3093(73)90007-0

Gaman, V. I., Kalygina, V. M., & Panin, A. V. (1999). Effect of chalcogenide elements on the electrical characteristics of GaAs MIS structures. *Solid-State Electronics*, *43*(3), 583–588. DOI: 10.1016/S0038-1101(98)00320-7

Ganguly, P., Mathew, S., Clarizia, L., Kumar, R. S., Akande, A., Hinder, S. J., Breen, A., & Pillai, S. C. (2020). Ternary Metal Chalcogenide Heterostructure (AgInS2-TiO2) Nanocomposites for Visible Light Photocatalytic Applications. *ACS Omega*, *5*(1), 406–421. DOI: 10.1021/ACSOMEGA.9B02907/ASSET/IMAGES/LARGE/AO9B02907_0007.JPEG PMID: 31956788

Gurushankar, K., Karuppaiya, C., Tari, V., & Kannan, K. (2024). Exploring MXenes and MBenes in metal sulfur and selenium battery technologies. In Kannan, K., & Tari, V. (Eds.), *Novel energy storage and conversion technologies for two-dimensional MXenes and MBenes*. IGI Global., DOI: 10.4018/979-8-3693-9546-2.ch005

Ha, T. D. C., Lee, H., Vamvasakis, I., Armatas, G. S., Oh, Y., & Kim, M. G. (2023). Recent developments in porous metal chalcogenides for environmental remediation and sustainable energy. *EcoMat*, *5*(12), e12419. DOI: 10.1002/EOM2.12419;SUBPAGE:STRING:FULL

Hairan, R., & Azizi, A. T. (2023). Synthesis and X-ray Diffraction Study of Se79 Te20 Pb1 Chalcogen Glass. *Journal for Research in Applied Sciences and Biotechnology*, *2*(5), 89–92. DOI: 10.55544/JRASB.2.5.14

Hamam, K. A., Mohery, M., & Aly, K. A. (2025). New quaternary Ge3S6Ag1-xSnx (0.0≤x≤1) chalcogenide glasses for optical devices. *Ceramics International*, *51*(3), 2880–2890. DOI: 10.1016/J.CERAMINT.2024.11.265

Hossain, N., Mobarak, M. H., Mimona, M. A., Islam, M. A., Hossain, A., Zohura, F. T., & Chowdhury, M. A. (2023). Advances and significances of nanoparticles in semiconductor applications – A review. *Results in Engineering*, *19*, 101347. DOI: 10.1016/J.RINENG.2023.101347

Kannan, K., & Tari, V. (Eds.). (2025). Novel energy storage and conversion technologies for two-dimensional MXenes and MBenes. https://www.amazon.com/dp/9798369395462 ISBN: 9798369395462, eISBN: 9798369395486

Karol, V., Sharma, P., Singh, A., Goel, D., & Kaur, S. (2024a). Review of Energy Storage Devices: Fuel Cells, Hydrogen Storage Fuel Cells, Rechargeable Batteries, PV Solar Cells. *ACS Symposium Series, 1488*, 1–15. https://doi.org/DOI: 10.1021/BK-2024-1488.CH001/ASSET/IMAGES/LARGE/BK-2023-00657M_G005.JPEG

Karol, V., Sharma, P., Singh, A., Goel, D., & Kaur, S. (2024b). Review of Energy Storage Devices: Fuel Cells, Hydrogen Storage Fuel Cells, Rechargeable Batteries, PV Solar Cells. *ACS Symposium Series, 1488*, 1–15. https://doi.org/DOI: 10.1021/BK-2024-1488.CH001/ASSET/IMAGES/LARGE/BK-2023-00657M_G005.JPEG

Kaur, A., Bajaj, B., Kaushik, A., Saini, A., & Sud, D. (2022). A review on template assisted synthesis of multi-functional metal oxide nanostructures: Status and prospects. *Materials Science and Engineering B*, *286*. Advance online publication. DOI: 10.1016/J.MSEB.2022.116005

Mal, J., Nancharaiah, Y. V., Van Hullebusch, E. D., & Lens, P. N. L. (2016). Metal chalcogenide quantum dots: Biotechnological synthesis and applications. *RSC Advances*, *6*(47), 41477–41495. DOI: 10.1039/C6RA08447H

Mandal, S., Mendhe, A. B., Rakhade, H. M., Barse, N. S., Roy, M., Rosaiah, P., Park, T., Lee, H. S., Mendhe, A. C., & Kim, D. (2025). Recent advancement and design in supercapacitor hybrid electrode materials: Bridging the gap between energy and power density. *Chemical Engineering Journal Advances*, *21*, 100690. DOI: 10.1016/J.CEJA.2024.100690

Martin, J., & Nolas, G. S. (2014). SYNTHESIS AND CHARACTERIZATION OF CHALCOGENIDE NANOCOMPOSITES. *Progress in Nanotechnology Processing*, 367–372. https://doi.org/DOI: 10.1002/9780470588246.CH50;PAGE:STRING:ARTICLE/CHAPTER

Munonde, T. S., Selahle, S. K., & Nomngongo, P. N. (2024). Chalcogenides and their nanocomposites in heavy metal decontamination. *Metal-Chalcogenide Nanocomposites: Fundamentals, Properties and Industrial Applications*, 117–133. https://doi.org/DOI: 10.1016/B978-0-443-18809-1.00007-9

Njema, G. G., & Kibet, J. K. (2025). A review of chalcogenide-based perovskites as the next novel materials: Solar cell and optoelectronic applications, catalysis and future perspectives. *Next Nanotechnology*, *7*, 100102. DOI: 10.1016/J.NXNANO.2024.100102

Palchoudhury, S., Ramasamy, K., Han, J., Chen, P., & Gupta, A. (2023). Transition metal chalcogenides for next-generation energy storage. *Nanoscale Advances*, *5*(10), 2724–2742. DOI: 10.1039/D2NA00944G PMID: 37205287

Phogat, P., Shreya, S., Jha, R., & Singh, S. (2024). Chalcogenide Nanocomposites for Energy Materials. *Engineering and Technology Journal*, *09*(07). Advance online publication. DOI: 10.47191/ETJ/V9I07.29

Rajamathi, M., & Seshadri, R. (2002). Oxide and chalcogenide nanoparticles from hydrothermal/solvothermal reactions. *Current Opinion in Solid State and Materials Science*, *6*(4), 337–345. DOI: 10.1016/S1359-0286(02)00029-3

Ramachandran, K., Gaidi, M., Columbus, S., Daoudi, K., & Hammouche, J. (2023). Fabrication of noble metal–based antimicrobial nanosystems. *Antimicrobial Nanosystems: Fabrication and Development*, 353–375. https://doi.org/DOI: 10.1016/B978-0-323-91156-6.00023-3

Rana, D., Soni, A., Sharma, A., Katoch, A., & Jamwal, D. (2018a). Nanocomposites of Chalcogenide and their Applications. *Nano Hybrids and Composites*, *20*, 46–64. DOI: 10.4028/WWW.SCIENTIFIC.NET/NHC.20.46

Rana, D., Soni, A., Sharma, A., Katoch, A., & Jamwal, D. (2018b). Nanocomposites of Chalcogenide and their Applications. *Nano Hybrids and Composites*, *20*, 46–64. DOI: 10.4028/WWW.SCIENTIFIC.NET/NHC.20.46

Rani, S., Rajan, S. T., Shanthi, J., Ayeshamariam, A., & Jayachandran, M. (2015). Review on the materials properties and photoelctrochemical (PEC) solar cells of CdSe, Cd1-x ZnxSe, Cd1-x InxSe, thin films. *Materials Science Forum*, *832*, 1–27. DOI: 10.4028/WWW.SCIENTIFIC.NET/MSF.832.1

Ridhova, A., Puspasari, V., & Amal, M. I. (2021). Synthesis methods for chalcogenides and chalcogenides-based nanomaterials for photocatalysis. *Chalcogenide-Based Nanomaterials as Photocatalysts*, 105–134. https://doi.org/DOI: 10.1016/B978-0-12-820498-6.00005-6

Saadat, M., & Amiri, O. (2024). Chalcogenide perovskites for solar energy applications: The role of Sn substitution in BaZrS3-based photovoltaic devices. *Inorganic Chemistry Communications, 168*, 112977. DOI: 10.1016/J.INOCHE.2024.112977

Said, Z., Pandey, A. K., Tiwari, A. K., Kalidasan, B., Jamil, F., Thakur, A. K., Tyagi, V. V., Sarı, A., & Ali, H. M. (2024). Nano-enhanced phase change materials: Fundamentals and applications. *Progress in Energy and Combustion Science, 104*, 101162. DOI: 10.1016/J.PECS.2024.101162

Sarkar, P., Nisha, Kumar, P., & Katiyar, R. S. (2023). Enhanced Opto-Electrical Properties of Chalcogenide-Rich Tin Selenide Thin Film after Incorporating Sulfur Yielding Tin Sulfoselenide. *ChemistrySelect, 8*(32), e202301545. DOI: 10.1002/SLCT.202301545;WGROUP:STRING:PUBLICATION

Shahzad, U., Saeed, M., Marwani, H. M., Al-Humaidi, J. Y., & Rehman, S. ur, Althomali, R. H., & Rahman, M. M. (2024). Transition metal-based chalcogenides as electrocatalysts for overall water splitting in hydrogen energy production. *International Journal of Hydrogen Energy, 65*, 215–224. DOI: 10.1016/J.IJHYDENE.2024.03.275

Shi, Y., Sturm, C., & Kleinke, H. (2019). Chalcogenides as thermoelectric materials. *Journal of Solid State Chemistry, 270*, 273–279. DOI: 10.1016/J.JSSC.2018.10.049

Singh, A. K., & Jen, T.-C. (2021). *Chalcogenide : Carbon Nanotubes and Graphene Composites*. Chalcogenide., DOI: 10.1201/9780429260827

Singh, A. K., Kim, J., Park, J. T., & Sangunni, K. S. (2015). Properties of the chalcogenide–carbon nano tubes and graphene composite materials. *Journal of Alloys and Compounds, 627*, 468–475. DOI: 10.1016/J.JALLCOM.2014.11.210

Singh, J., Thareja, R., & Malik, P. (2025a). Exploring the Potential of Quantum Dot-Sensitized Solar Cells: Innovation and Insights. *ChemPhysChem, 26*(10), e202400800. DOI: 10.1002/CPHC.202400800;REQUESTEDJOURNAL:JOURNAL:14397641;PAGEGROUP:STRING:PUBLICATION PMID: 39964946

Singh, J., Thareja, R., & Malik, P. (2025b). Exploring the Potential of Quantum Dot-Sensitized Solar Cells: Innovation and Insights. *ChemPhysChem, 26*(10), e202400800. DOI: 10.1002/CPHC.202400800;REQUESTEDJOURNAL:JOURNAL:14397641;PAGEGROUP:STRING:PUBLICATION PMID: 39964946

Sivasamy, R., Kaliyan, G., Kaliyamoorthy, S., Mosquera, E., Sivasamy, R., Kaliyan, G., Kaliyamoorthy, S., & Mosquera, E. (2024a). *Chalcogenide Materials for Sustainable Energy and Environmental Applications*. Structural and Chemical Features of Chalcogenides., DOI: 10.5772/INTECHOPEN.1003048

Sivasamy, R., Kaliyan, G., Kaliyamoorthy, S., Mosquera, E., Sivasamy, R., Kaliyan, G., Kaliyamoorthy, S., & Mosquera, E. (2024b). *Chalcogenide Materials for Sustainable Energy and Environmental Applications*. Structural and Chemical Features of Chalcogenides., DOI: 10.5772/INTECHOPEN.1003048

Sivasamy, R., Kaliyan, G., Kaliyamoorthy, S., Mosquera, E., Sivasamy, R., Kaliyan, G., Kaliyamoorthy, S., & Mosquera, E. (2024c). *Chalcogenide Materials for Sustainable Energy and Environmental Applications*. Structural and Chemical Features of Chalcogenides., DOI: 10.5772/INTECHOPEN.1003048

Stamford, L., & Azapagic, A. (2019). Environmental impacts of copper-indium-gallium-selenide (CIGS) photovoltaics and the elimination of cadmium through atomic layer deposition. *The Science of the Total Environment*, *688*, 1092–1101. DOI: 10.1016/J.SCITOTENV.2019.06.343 PMID: 31726540

Tari, V. Gupta, V., Joshi, A., Kurve, P., & Chandra, N. (2024). Chapter 19: Biodiversity loss due to forest diversion. In A. Rathore (Ed.), *Impact of societal development and infrastructure on biodiversity decline* (pp. 137–165). IGI Global. https://doi.org/ DOI: 10.4018/979-8-3693-6950-0.ch010

Tari, V. Kannan, K., Koliar, J., Gurushankar, K., & Wahyuni, D. K. (2025). Influence of zinc oxide nanomaterials on crop plants: A prospective analysis. In D. Jabbrova, D. Sarkar, Rakesh, S. Singh, & R. Datta (Eds.), *Smart nanomaterials in agroforestry systems* (Chapter 8). Springer. https://doi.org/DOI: 10.1007/978-981-96-1337-3_8

Voigt, D., Primavera, G., Uphoff, H., Rethmeier, J. A., Schepp, L., & Bredol, M. (2022). Ternary Chalcogenide-Based Quantum Dots and Carbon Nanotubes: Establishing a Toolbox for Controlled Formation of Nanocomposites. *The Journal of Physical Chemistry. C, Nanomaterials and Interfaces*, *126*(21), 9076–9090. DOI: 10.1021/ACS.JPCC.2C01142/ASSET/IMAGES/LARGE/JP2C01142_0014.JPEG PMID: 35686224

Wang, H., Li, X., Zhao, X., Li, C., Song, X., Zhang, P., & Huo, P. (2022). A review on heterogeneous photocatalysis for environmental remediation: From semiconductors to modification strategies. *Chinese Journal of Catalysis*, *43*(2), 178–214. DOI: 10.1016/S1872-2067(21)63910-4

Wani, S. A., Tari, V., & Mansoor, S. I. U. (2025). A Comprehensive Analysis of Wastewater Management Challenges in India: Infrastructure and Policy Perspectives. In Dhiman, S., & Mukherjee, G. (Eds.), *Resource Resurgence. Springer Water*. Springer., DOI: 10.1007/978-3-032-02750-4_11

Xia, C., , Li, J., & . (2016). Recent advances in optoelectronic properties and applications of two-dimensional metal chalcogenides. *Journal of Semiconductors*, *37*(5), 051001. DOI: 10.1088/1674-4926/37/5/051001

Yang, Y., & Lin, Q. (2025). Chalcogenide semiconductors: Charge carrier dynamics and optoelectronic applications. *Chemical Communications*. Advance online publication. DOI: 10.1039/D5CC03022F PMID: 40665834

Yuan, H., Kong, L., Li, T., & Zhang, Q. (2017). A review of transition metal chalcogenide/graphene nanocomposites for energy storage and conversion. *Chinese Chemical Letters*, *28*(12), 2180–2194. DOI: 10.1016/J.CCLET.2017.11.038

Zhao, D., Zhang, X., Wang, H., Zeng, H., Ma, H., Adam, J. L., & Chen, G. (2008a). Thermal properties of chalcogenide glasses in the GeSe2–As2Se3–CdSe system. *Journal of Non-Crystalline Solids*, *354*(12–13), 1281–1284. DOI: 10.1016/J.JNONCRYSOL.2006.10.079

Zhao, D., Zhang, X., Wang, H., Zeng, H., Ma, H., Adam, J. L., & Chen, G. (2008b). Thermal properties of chalcogenide glasses in the GeSe2–As2Se3–CdSe system. *Journal of Non-Crystalline Solids*, *354*(12–13), 1281–1284. DOI: 10.1016/J.JNONCRYSOL.2006.10.079

Chapter 2
Synthesis, Functional Properties, and Characterization of Chalcogenide-Based Nanocomposites

Chinnaiah Karuppaiya
https://orcid.org/0000-0002-6315-2920
Kalasalingam Academy of Research and Education, India

Gurushankar Krishnamoorthy
https://orcid.org/0009-0000-3509-4893
Krishna Chaitanya Institute of Technology and Sciences, India

Paramanantham Muniyandi
V.S.S. Government Arts College, India

ABSTRACT

Chalcogenides exist in the form of S, Se, and Te and their nanocomposite developed with different combinations of structures and composition gaining great importance in the field of optoelectrics and energy-related applications. Hence, to identify the synthesis routes for the preparation of chalcogenide based nanocomposites from various literatures are important for developing novel chalcogenide nanocomposites, which were identified. Newly proposed chalcogenide nanocomposites with different forms such as quantum dots, core/shell, heterostructures, and nanostructures espe-

DOI: 10.4018/979-8-3373-3962-7.ch002

cially in the optoelectronics and energy-field were compiled, and their properties were discussed. Efficient analytical techniques used to characterize the synthesized S, Se, and Te nanocomposites were addressed.

INTRODUCTION

Chalcogenide-nanocomposites are excellent solution for both fundamental research and technological applications in optoelectronics due to their unique optical properties as well as tuneable size and composition. These features enable revolutionize impact in photo detectors, photodiodes, photovoltaic device, and solar cells. Nanocomposites combining sulfur (S), selenium (Se), and tellurium (Te) exhibit several advantageous properties, including high surface area, tunable band gaps, broad optical absorption, third-order optical susceptibility, and low thermal conductivity, phase-change behaviour suitable for memory or switching devices, and strong photoconductivity, (Kannan & Tari, 2025). Compared to pure nanoparticles, chalcogenide-nanocomposites offer significant advantages such as improved stability, enhanced mechanical strength, and synergistic effects that result in multifunctionality, (Tari et al., 2022; Tari et al., 2025). These include better thermal stability, higher electrical conductivity, and increased catalytic activity, (Chinnaiah et al., 2025). Furthermore, chalcogenide-nanocomposites exhibit unique features such as optimized interlayer spacing, high theoretical capacity, enhanced conductivity, low operating voltage, limited ionic conductivity, and favourable ion-diffusion kinetics (Wani et al., 2025). Combined with their natural abundance, these characteristics make them attractive for electrochemical energy storage and conversion systems, (Gurushankar K. et al. 2025). Thus, tailoring their composition, dopant, and nanostructuring were becomes promising strategy for supercapacitor (Chinnaiah et al., 2025), lithium-ion, sodium-ion, and potassium-ion batteries, as well as optoelectronic devices. Some important chalcogenides are listed below.

Chalcogenide Sulphides

PbS, ZnS, FeS_2, $CuFeS_2$, HgS, Ag_2S, MoS_2, CuS, Cu_2S, Cu_5FeS_4, Sb_2S_3, Ni_9S_8. These sulphides have ultra violet (UV) to near infra-red (NIR) region spectral region for energy conversion.

Chalcogenide Selenides

CdSe, ZnSe, CuSe, $MoSe_2$, etc. These selenides having suitable band gap, strong absorption in visible region and high chemical stability for rapid recommendation of photo-induced carriers.

Chalcogenides Tellurides

CdTe, Bi_2Te_3, PbTe, Sb_2Te_3, HgTe, $MoTe_2$, $AgSbTe_2$, As_2Te_3, PbSnTe, $AuTe_2$ and etc. These tellurides are facilitates the interfacial electron transport during electrochemical reactions due to electron rich Te sites. Hence able to construct the ration design in low cost energy storage batteries.

Above chalcogenides were emerged as a quantum dots (QDs), core-shell, and heterostructure in recent research decades, (Kuldeep et al., 2025).

Quantum Dots

QDs are zero-dimensional materials, their particle sizes are close to/less than Bohr radius. These QDs have high photoluminescence quantum yield, tuneable range of luminescence peaks, quantum effect, and size effect. This QDs have distinct density of states compare to high dimensional materials, it provides the high charge transportation and optical properties of materials which helped to development of diode, switches and transistors, bio-images, white-light emitting diodes, and solar cells. These QDs physical properties are strongly depends on their size and shape. The chalcogenide-nanocomposite QDs are CdSe/HgSe, CdSe/CdS, InP/ZnSe/ZnS, and ZnSe/ZnS.

Core/Shell

Generally, the outer part ($a < r < b$) is defined as a shell, and inner part ($0 < r < a$) is defined as the core. The chalcogenide-nanocomposite core/shells are Ag/Ag_2S, ZnSe-CdSe, ZnSe/ZnS, and HgSe-CdSe.

Heterostructures

Heterostructures have great potential for opto electronics devices due to carriers from photons, align bands at lower operating temperatures, combined band gaps of wide and narrow from two different materials. Electron-hole pair migration across the heterojunction improves efficiency of visible light photocatalytic activities, hydrogen evolution efficiency and performance of energy storage materials. The chalcogenide-nanocomposite heterostructures are Bi_2Te_3/FeTe, Sb_2Te_3/Bi_2Te_3, MoS_2/hBN/WS_2, and MoS_2/WSe_2.

Chalcogenide nanocomposite quantum dots (QDs), core–shell structures, and heterostructures have been synthesized using various techniques, including mechanochemical, microwave, hydrothermal, colloidal, molecular beam epitaxy, and solid-state methods. Mechanochemical synthesis enables solvent-free production of QDs through solid-state grinding, offering a scalable approach. The hydrothermal method allows the formation of diverse morphologies such as rods, sheets, spheres, and core-shell structures within nanocomposites. These techniques provide control over critical parameters such as ligand presence, reaction time, temperature, and precursor type, which are essential for tailoring QD size, shape, and properties. Colloidal synthesis, in particular, offers precise control over particle size and morphology. Solid-state and microwave-assisted methods provide uniform heating and rapid reaction rates, facilitating the synthesis of core-shell chalcogenide-nanocomposites through approaches such as seed-mediated growth, template-assisted growth, ion exchange, and metal–organic assembly. These strategies enable tuneable properties that are advantageous for applications in photovoltaics, sensors, and optoelectronics. Furthermore, molecular beam epitaxy allows for atomic-level control, high purity, and the fabrication of clean-interface chalcogenide-heterostructures, which are critical for quantum device applications.

SYNTHESIS OF CHALCOGENIDE-NANOCOMPOSITE

Chalcogenide-nanocomposite can be synthesized from several techniques. These procedures were discussed for various materials below used in recent literatures.

Mechanochemical Synthesis of TiS_2

The mechanochemical synthesis of chalcogenide-nanocomposite can do effective intercalation with simple process, less process and it is high reproducible with more volume expended. In this process TiS_2 powder and liquid n-hexylamine were mixed with 1: 4 molar ratios then manually grounded with pestle for 30 minutes. Chemical reaction of sulper- chalcogenide-nanocomposite is shown as follow

$$TiS_2 +_X [n\text{-}C_6H_{13}NH_2]\ TiS_2(C_6H_{13}NH_2)_X \text{ with } X \approx 1 \quad (1)$$

Synthesis scheme of chalcogenide-nanocomposite has been shown in Figure 1.

Figure 1. (a) Mechanochemical synthesis scheme of TiS2, (b) Schematic of the grown structure of PbMnTe/CdTe multilayer nanocomposite, (c) Schematic illustration of MoSe2/WSe2 heterostructure (Ursi et al., 2022; Chusnutdinow et al., 2023 and Hwang et al., 2021 respectively, from MDPI publication)

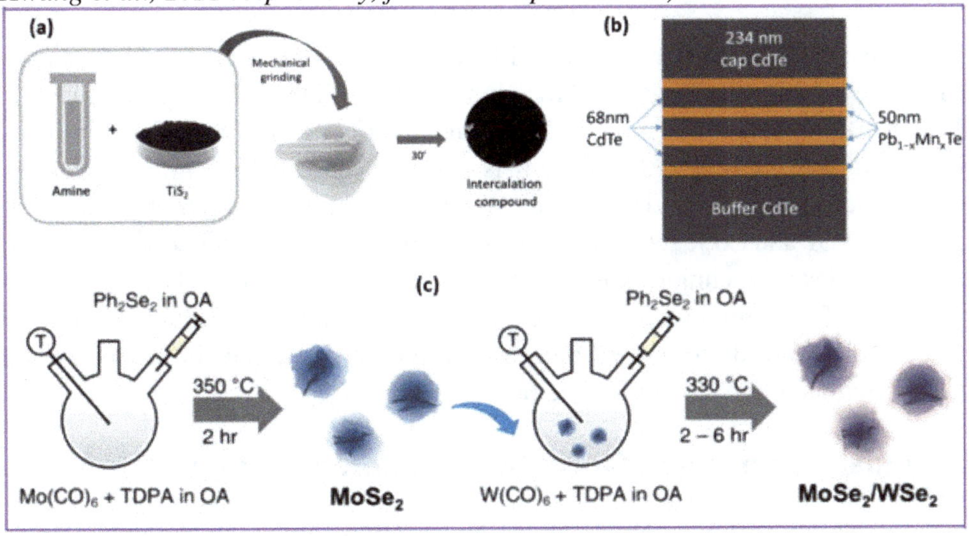

Microwave Synthesis of ZnS@CuInS

In this synthesis procedure by Rahman A. et al. 2024, certain amount of cupper nitrate, indium nitrate and thiourea were dissolved in the ethylene glycol in stoichiometric ratio then loaded into the quartz vessel. The N_2 gas was passed with continuous heating of 200°C by 850W of microwave irradiation continuously for 10 minutes. After washing, a block colour CuInS powder was obtained. Now, appropriate amount of zinc acetate and thiourea were added into the CuInS solution. This mixture was purged with N_2 gas again with same experimental condition of above. The obtained product was washed with water and ethanol, and then dried. The final product, ZnS@CuInS, was obtained.

Solid State Synthesis of $CuFeSe_2$-$CuFeS_2$

To synthesize the $CuFeSe_2$-$CuFeS_2$ by Jang S.M. et al. 2025, high purity of Cu, Se and S powders were mixed through ball milling for 1 hr to ensure the uniform distribution. This mixed powder was subject to the mechanical alloying in argon atmosphere. This alloy powder was hot pressed at 70 MPa under vacuum and hot pressed temperature was varied from 673 K to 773 K for 2hrs. This experimental condition produced the well-consolidated $CuFeSe_2$-$CuFeS_2$ sample.

Synthesis of CdSe/CdS Nanorods

In this synthesis process done by Li H. et al. 2022, CdO, octadecylphosphonic acid, triocytlphosphine oxide, and hyxylphosphonic acid were mixed into the four flask then degassing for 10 minutes in vacuum at room temperature. Now, temperature was raised slowly to 120°C under in N_2 atmosphere than further raised as above 300°C when CdO turns into the colourless solution. At this point, tricytlphosphine (TOP), S: TOP and CdSe seeds were injected into the solution. The CdSe seeds were prepared via same technique with same experimental conditions. The reaction was maintained for 10 minutes under in 350°C. After reaching the room temperature, nanocrystals were dissolved in toluene. The length of CdSe/CdS can be varied through regulation of stoichiometric ratio of seeds.

Colloidal Synthesis of $MoSe_2$/WSe_2 Heterostructures

$MoSe_2$ and WSe_2 nanosheets were grown by colloidal hot-injection method, (Hwang et al., 2021). A certain amount of molybdenum carbonyl and tungsten carbonyl were pre-loaded with tetradecylphosphonic acid dissolved into the oleic acid. The reaction of 350°C maintained for $MoSe_2$ and 330 °C for WSe_2 after injecting the selenium precursor solution with oleic acid and maintained the reaction time for 2 hr to the growth of nanosheets. A solvent mixture of toluene and n-butanol in 2:1 ration was dissolved in the obtained crude solution. After centrifugation, a black colour was washed with same solvent mixture of toluene and n-butanol. The schematic illustration of this $MoSe_2$/WSe_2 heterostructure is shown in Figure 1c.

Hydrothermal Synthesis of Bi_2Te_3/MoS_2 Heterostructures

The synthesis of Bi_2Te_3/MoS_2 heterostructures via hydrothermal is described below. For Bi_2Te_3, Bi_2O_3 and TeO_2 and NaOH solutions were mixed into the solution of PVP in ethylene glycol. The mixed yellow colour solution was transferred to the autoclave and heated for 200 °C for 4 hrs and this final product was used to prepare the heterostructure. The final product of $BiTe_3$ making as dispersion using water then ammonium tetrathiomolybdate was added into the dispersion and added the hydrazine hydrate, stirred well for 15 minutes, then placed into an autoclave. It was heated at 200°C for 24 hrs and obtained precipitate was washed with ethanol and dried. $BiTe_3$/MoS_2 heterostrcture was synthesized (Wang Y. et al. 2022).

Synthesis of Core/Shell CdSe/ZnS QDs

In a reaction process reported by Rahman M.M. et al. 2020, 2.5 mM of cadmium oxide, and 7 mM of stearic acid with 1-octadecene were heated in three neck flask at 200-290°C, it was cool down after a clear solution was obtained, and then HDA and tributylphosphine oxide were added into this solution. This product was reheated to 280°C in a N_2 atmosphere, at which point Se-TOP (Tri-n-octylphosphine) was quickly injected into the solution. The reaction temperature was controlled at 200~290°C for 10 to 15 minutes. A final combination was disintegrating in hexane or toluene, after which MeOH was added and maintained for 30 minutes. The reaction apparatus was kept overnight to allow clear separation of layers. At this stage, a ZnS shell with a certain monolayers was developed on CdSe QDs. For ZnS coating, CdSe-QDs were dispersed in the toluene/hexane at a controlled reaction temperature of 100-120°C and then loaded into the reaction mixture. Afterward, the mixture was reheated to 230-240°C in a N_2 atmosphere where shell growth was performed. After 5-6 of hrs reaction, CdSe cores and ZnS shell were formed, and the product was purified several times with hexane/toluene until clear solution of methanol phase was achieved.

Synthesis of InP/ZnSe/ZnS QDs

For the synthesis of InP/ZnSe/ZnS by Won Y.H. et al. 2019, 1.6 mM zinc acetate and 3.2 mM of oleic acid were mixed in 80 ml of trioctylamine. This mixture had been evacuated at 120°C for 1 hr, and heated at 180°C for 10 minutes in the N_2 flow, and then cooled to 180°C. At 180°C, the InP core in toluene was added into this mixture. For this InP core, 4 mM indium acetate, and 12 mM of palmitic acid were dissolved in 1-octadecene. This mixture was evacuated and heated under the above experimental conditions. Then, 10 ml of 0.2 M TMS_3P/TOP had been added into the $In(PA)_3$ mixture at 280°C, after which the temperature was decreased to 260°C for 40 minutes, preparing the core 1. This TMS_3P/TOP was prepared with 5.5 mM TMS_3P and trictylphospine (TOP) in N_2 glove box. Then, 0.2 M $In(PA)_3$ and 26.3 ml of 0.2 M TMS_3P/TOP were injected into the core 1 solution, and reaction was maintained for 35 minutes to form core 2. This InP was dispersed in toluene. Diluted HF was added immediately after core injection. To synthesize QDs, 0.4 M zinc acetate and 16 mM Se/TOP have been mixed into the solution and maintain for 1 hr at 340°C, thereby growing the ZnSe shell. For ZnS shell coating, 4.8 mM zinc acetate and 6.4 mM S/TOP were mixed for 20 minutes and then cooled to the ambient temperature. The collected InP/ZnSe/ZnS QDs have been dispersed in octane for further use.

PbMnTe/CdTe Multilayer Composite by Molecular Beam Epitaxy

In this work, Chusnutdinow S. et al. 2023, grew PbMnTe/CdTe multilayers on a semi-insulating GaAs substrate using the molecular beam epitaxy technique. Initially, the substrate was cooled in the Zn flux to growth point of 240°C after the thermal elimination of oxide layers to protect the GaAs surface. Then, 5 nm of ZnTe was deposited by opening the Te shutter. This ZnTe buffer layer is requires to stabilize the growth in the (100) orientation. Next, a 0.6 μm-thick CdTe buffer layer was deposited for 1 hour. Then, bulk structures PbMnTe were deposited for 5 minutes, and CdTe barriers were deposited for 6 minutes to provide separation. This structure was finally surrounded with a CdTe cap layer. After the growing process, the samples were cool down to the ambient temperature. The elements Pb, Cd, Te and Mn were evaporated using effusion cells. The schematic of the grown structure is shown in Figure 1b.

CdTe/CdMgTe Quantum Wells by Molecular Beam Epitaxy

Initially, a 2.2 μm- thick CdTe buffer layer, and a 1.5 μm-thick CdMgTe buffer layer were grown on the semi-insulating substrate. This buffer layer was formed by 2 monolayers of CdTe and 2 monolayers of CdMgTe with 52 nm of CdMgTe spacer. Next, a 10.5 nm- thick CdTe quantum well was separated from adjacent layers by CdMgTe by the separation of 52 nm. The CdTe/CdMgTe growth was completed with 45 nm-thick CdMgTe cap layer. The modulation doping with iodine donors resulted in the formation of a two dimensional electron gas (Yavorskiy D. et al. 2020).

PROPERTIES OF CHALCOGENIDE NANOCOMPOSITES

Band gaps of important individual chalcogenides in bulk form, which are used in optoelectronics and energy applications, are noted below

ZnS – 3.6 eV, CuS – 1.6 eV, ZnSe – 2.72 eV, CdS - 2.49 eV, CdSe – 1.85 eV, CuSe – 2.1 to 2.4 eV, $MoSe_2$ – 1.55 eV, CdTe – 1.37 eV, $MoTe_2$ – 1.10 eV.

Qunatum dot can emit the different colours by controlling the size, which has narrow emission line width, which desires the high-quality displays. Quantum yield efficiency and full width half maximum efficiency were important parameter for LED applications. Cd QDs have more than 90% of QY with FWHM emission of less than 30 nm using tri-n-octylphosphine as a ligand and extract at 532 nm. Wenda Z. et al. 2020, synthesized InP/ZnS QDs with FWHM of 47 nm, for this they choose ZnS as shell layer because of InP core have PL QY of <1%. This ZnS have wide bandgap of 3.68 eV which enough to strong confine the excitons in the

InP core. Therefore InP/ZnS QDs measured a shortest wavelength emission of 468 nm, and QY of 45%.

Mohd et al. (2020) performed the band-gap engineering for PbS, which achieve the band gap of 2.4eV through synthesis of PbS with nano-size 8-10 nm. By reducing of size, this result employed as PbS as coating absorber in photo-thermal conversion of solar energy in the NIR to visible region, which reflect as a blue shift on them energy gap. Kuldeep et al. (2025) integrates the GaS and PbS as a van der waals heterostructure, which demonstrate the light-harvesting performance under ± 0.2 V/Å electric filed by the increased optical absorbance. Electron and hole confined potentials as function of ratio a/b is about significant impact in potential of electron (V_e) and potential of hole (V_h) CdTe/ZnS in CdTe/CdS composites by strain effect is an important discussion. A confinement of V_e for CdTe/ZnS was increased monotonically with increase ratio of a/b, this was contrast to confinement of V_c for same CdTe/ZnS as increases of a/b ratio. The confinement of V_c for CdTe/ZnS and CdTe/CdS follow the same tendency, both were quickly increase rapidly reach the maximum at a/b = 0.14 for CdZnS and a/b = 0.09 for CdTe/CdS after then they decreases. These results were display that optoelectronic characteristics of nanostructures is depends on the positions of charge carriers in the materials, (Pérez et al., 2023). Jingyun et al. (2023) constructS the hierarchical heterostructures of MoS_2@ $Znln_2S_4$ with heterointerfaces for high efficiency anodes for sodium-ion batteries. This heterointerface of facets of Zn and $ZnlnS_4$ direct the built-in electric field that facilitates the electron transfer by driving force and other facets in MoS_2 exhibits the increasing Na^+ adsorption energies and lowering the diffusion energy barriers of Na^+. This multistep reaction of MoS_2 and $ZnlnS_4$ promotes the electrochemical process by their synergetic effects.

Table 1. Chalcogenide-sulphide composites and their significances in applications based on properties reported in recent literatures

S.No	Chalcogenide – composite	Design/ novelty in proposed material	Significance in application	Limitation/drawbacks	References
1	InP/ZnSe/ZnS QD	FWHM–35 nm at 650 nm, highly spherical shape	High luminescent efficiency for Light emitting diodes	Unbalanced charge injection hinder the device performance	(Won et al., 2019)
2	ZnSe/ZnS QD	QD size - 5.3 to 6.5 nm	High photo response to phototransistors	Hybrid phototransistor	(Sun et al., 2018)
3	InP/GaP/ZnS QDs	FWHM 41 nm,	QY~85%, Entire colour range in LED	Thermal stability	(Sungwoo et al., 2012)
4	ZnS-CdS	Particle size 60-100 nm	Reduced band gap extent the visible light area to removal of dye in waste water	Limited stability	(Khodamorady et al., 2023)

continued on following page

Table 1. Continued

S.No	Chalcogenide – composite	Design/ novelty in proposed material	Significance in application	Limitation/drawbacks	References
5	PbSe/PbS QDs	Heterojunction	High photodetector as 375 A/W in wavelength region of 2710-4250 nm at room temperature	Instability in environmental conditions	(Raphae et al., 2023)
6	FeS_2-PbS	Heterojunction	Photoresistivity 10^6 A/W in UV-Vis region	Photo-response influenced by architecture	(Maogang et al., 2017)
7	$MoTe_2$-MoS_2	Vertically stacked few layers	High photoresponsivity of 10^5 A/W at room temperature	Hurdles of interlayer alignment	(Paul et al., 2017)
8	ZnSe/CuSe	0D/2D/ (001) double heterojunction catalyst	Improve the carrier separation efficiency and reduce the internal resistance in Photo-electrochemical and hydrogen evolution reaction activities	Photo-corrosion under prolonged light	(Chao et al., 2022)
9	InP/ZnSe core/ shell QDs	Thickness of ZnSe shell growth (1.6 nm)	Photocurrent density of 6.2 mA cm^{-1} in photoelectrochemical hydrogen generation	Ligand hinder the charge transport	(Jiabin et al. 2023)
10	NiS_2/MoS_2	Self-supporting 3D hierarchical structure	Overpotential of 116 mV in HER, 235 mV@ 150 $mAcm^{-2}$ in OER	Instability under dynamic fluctuating current	(Li et al. 2024)
11	CdSe/CdS	Tetrapods nanostructure	Two-colour multiexciton emission	Structural tuning	(Nimai et al. 2017)
12	CdSe/CdS/ZnS	Band structure	High photo-stability	Interfacial defects	(Wang et al. 2018)
13	CdSe/ZnS	Core-shell QDs	Green emission for LED	Narrow focus on green emission	(Junjie et al. 2019)
14	CdSe/ZnS with MoS_2	Encapsulation	Radiation hardness in phototransistor	Tradeoff by photobleaching	(Park et al. 2019)
15	CdSe/CdS core/ shell QDs	Photoluminiscent down shifting	Enhanced conversion efficiency from 32.5 (mA/cm^2) to 37.0 in solar cells	Limited sample size	(Lopez-Delgado et al. 2019)
16	PbS/HgTe QDs	Stacked QDs	Ultrabroadband response from 0.4 to 5.0 μm in photodetector	Ligand degradation	(Mu et al. 2024)
17	CuS/Co_3S_4	Synergetic effect microflowered sulphides	Specific capacitance of 1483 Fg^{-1} at 1 Ag^{-1} with impressive rate performance in supercapacitor	Poly-sulphide effect	(Tian et al. 2024)

* FWHM- Full Width Half Maximum, QY- Quantum yield

CHARACTERIZATION TECHNIQUES OF CHALCOGENIDE-NANOCOMPOSITE

Optical Analysis

The optical property of the chalcogenide-nanocomposite is shown in Figure 2, which can be studied through UV-Vis and fluorescence spectra analysis. Figure 2a show about the UV-Vis spectra of CdSe@CdS with different reaction time. In this spectrum, extension of reaction makes the weakened signal to CdSe by the gradual increase of reaction and their corresponding photolumiscence spectra was shown in Figure 2b had a red shift based on this reaction time, which indicates CdS shell thickness was increased gradually based on the reaction time as in the core-shell model. This study suggest that analysis of optical properties through these techniques are they effective tool to analyze the synthesised chalcogenide- nanocomposites.

Structural Analyses

The structural properties chalgochenide-nanocomposite can effectively characterized through XRD analysis. The XRD pattern of ZnS@C and their structural transformation by the duration of heat treatment has 2θ value of 31.8, 34.4, 36.3, 47.5 and 56.6 were characterize the crystallographic planes: (100), (002), (101), (102), (110) and (103) of hexagonal ZnS@C. The X-ray diffraction pattern of CdSe@CdS@CdS core-shell is shown in Figure 2c. This diffraction pattern has red shift compared to CdSe@CdS due to the heterogeneous epitaxial growth of zinc blend ZnS and coating of CdS has masking the some diffraction peak of CdSe. In Sb-doped PbS nanocrystal, dopant of Sb is modifying the crystal lattice of cubic PbS even though their crystallographic phases does not change by this dopant. No secondary phases were observed after higher concentration level of dopant i.e 0.25%, (Cadavid et al., 2021). The Te phases were undergoes the phase changes with the applied pressure, this critical phase transition of tetragonal to monoclinic can identified with help of raman spectra. At particular applied pressure (2.3 GPa) FeTeSe is turn into the monoclinic from tetragonal phase, which is identified by existence of distinct active phonon mode at 123 cm^{-1}. The wavenumber difference (Δd) between two different vibration modes in raman spectra was reflect the number of layers in MoS_2. This Δd value of the bottom and top peaks are 25.2 cm^{-1} and 23.6 cm^{-1}, which suggested that MoS_2 has more than five layers. For WSe_2, strong peak at 250.2 cm^{-1} is related to A_{1g} mode, another small signal at 306.8 cm^{-1} assigned for inactive B_{2g} mode. The structural changes of ZnTe nanowires (NWs) are shown in Figure 2d. This spectrum reveals that Te three features were present in between 85 cm^{-1} to 150 cm^{-1} and the unoxidized ZnTe/Zn NWs has dominant feature of ZnTe LO-phonon peak near at

206 cm^{-1}. After oxidation, Te-related structure was become sharper in-between 85 cm^{-1} and 150 cm^{-1} and emerging new position with predominant features at 92 cm^{-1}, 121 cm^{-1} and 141 cm^{-1} corresponding to EA_1 and E_1 phonon modes of hexagonal Te, (Gas et al., 2021).

Chemical and Surface Analyses

The chemical state of S has S2p region in XPS spectra which has binding energies of 164.94 eV and 162.13 eV corresponds to the S $2p_{1/2}$ and S $2p_{3/2}$ were attributed by the metal sulphides of S^{2-} and strong oxidation of S-O . The chemical state of Te is shown in Figure 2e. The doublet peak at 573 eV and 587 eV were corresponds to the $3d_{5/2}$ of Te and Te(IV) of Te element, (Zhang et al., 2025).

The TEM analysis showed in Figure 2f displays the morphology of CdSe@CdS@ ZnS core shell with 4.3 nm thickness of CdSe, 1.25 nm thickness of CdS and 1.0 nm thickness of ZnS and their corresponding lattice spacing of 0.32 nm, 0.33 nm and 0.31 nm with respective planes of (101), (111) and (111) for CdSe, CdS and ZnS were showed in Figure 2f and g. The structural difference of chalcogenide composite between the core and shell can effectively analyze through HR-TEM analysis because differentiation of core and shell is challenging task when small differences in the lattice constant together with similar crystal structures. The crystal phase of CdSe/ CdS QDs has been identified in their HR-TEM images with the diffraction images of hexagonal wurtzite (space group: p63mc) and phase structure of cubic zinc blende (space group:F-43 m). Another one approach of detect and deformation in HR-TEM images through the contract of different regions significantly differentiate core and shell region, (Wang et al., 2023).

Figure 2. (a) UV-Vis and (b) Photoluminescence spectra, (c) XRD pattern of CdSe@ CdS@ZnS QDs, (d) Raman spectra of ZnTe/Zn NWs (Reprinted from Ref. Gas K. et al. 2021, MDPI, Publication), (e) XPS spectrum of Te (Reprinted from Ref. Zhang Z. et al. 2025, MDPI Publication), (f) TEM images and (g) HR-TEM of CdSe@CdS (Wang et al., 2023)

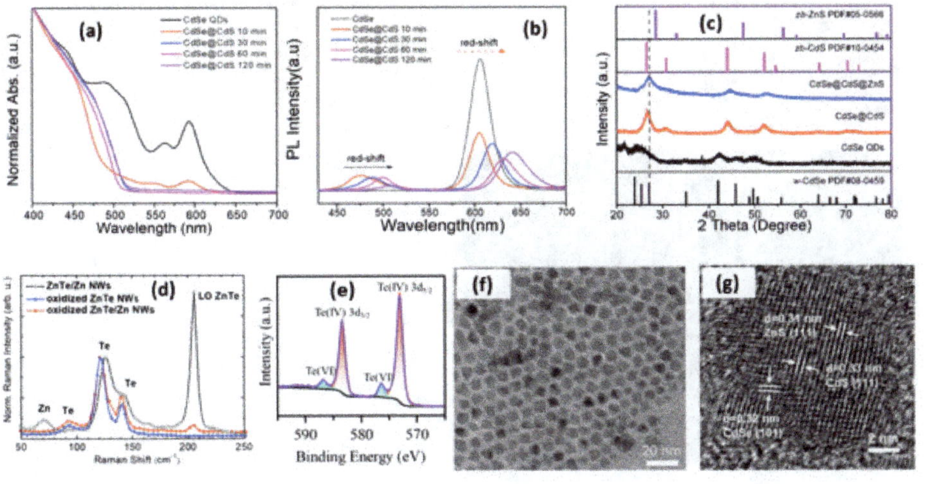

Morphological thickness of the triangular WSe_2 and hexagonal $MoSe_2$ is shown in Figure 3 a to d, by visualized in AFM analysis, (Li et al., 2023). This images display the good homogeneity with thickness of 0.96 nm and 0.765 nm for triangles and hexagons, which confirm that WSe_2 and $MoSe_2$ were formed as a monolayer.

Electrical Analyses

Growth of heterostructure and their material quality can be characterized through reflection high-energy electron diffraction (RHEED) measurement. An analysis of RHEED measurement for CdSe/PbSe was showed in Figure 3e. This Figure shows about the strong connection with their growth surface and alignment of PbSe by layer-by-layer growth mode. It shows signing of roughing with development of 3D spots along the <100> direction after the initial deposition of 5 nm, but bulk of 200 nm thick CdSe was virtually unchanged. This RHEED profile shows the identical diffraction spacing between the fringes of <100> CdSe and <100> PbSe, (McDowell et al., 2023). The I-V characteristic is an important analysis to measure the efficiency of photovoltaic materials, this I-V characteristics of two heterostrcutures are given in Figure 3f it's examined with dark and illumination conditions. From this impact of buffer layer in open circuit voltage and short-circuit current was noticed.

Figure 3. (a-d) AFM analysis of WSe2/MoSe2, (e). RHEET measurement characteristics of [110] orientation for CdSe/PbSe, (f) I-V curves of heterostructures under in dark and illumination (Li et al., 2023; McDowell et al., 2023; Gutierrez et al., 2018).

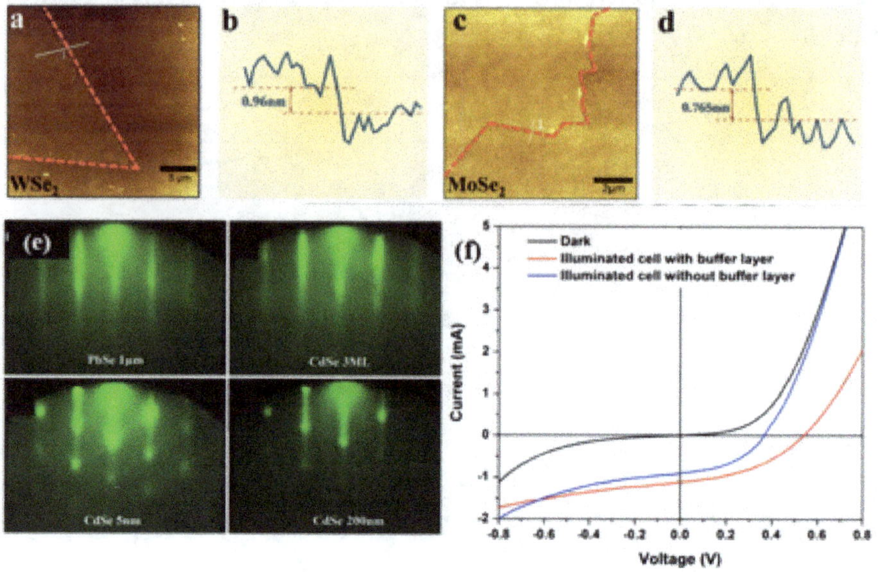

CONCLUSION

This chapter describes the synthesis methods of different chalcogenide (S, Se, Te) nanocomposites, as reported in recent literature, using various techniques. The properties and impacts of these chalcogenide nanocomposites on optoelectronics and energy applications have been identified and summarized, along with their characterization through advanced analytical methods. Although a wide range of chalcogenide-nanocomposites has been explored, most studies primarily focus on Cd- and Zn-based chalcogenide systems, while reports on other combinations remain limited. In particular, telluride-based chalcogenide nanocomposites show significant promise for future energy storage applications. Therefore, the design of new chalcogenide-nanocomposite combinations holds great potential for novel developments in optoelectronic and energy-related applications. Overall, this chapter provides valuable insights for designing, synthesizing, and characterizing innovative chalcogenide-nanocomposites.

REFERENCES

Cadavid, D., Wei, K., Liu, Y., Zhang, Y., Li, M., Genç, A., Berestok, T., Ibáñez, M., Shavel, A., Nolas, G. S., & Cabot, A. (2021). Synthesis, Bottom up Assembly and Thermoelectric Properties of Sb-Doped PbS Nanocrystal Building Blocks. *Materials (Basel)*, *14*(4), 853. DOI: 10.3390/ma14040853 PMID: 33578981

Chao, C., Jun, Z., Xiaoshan, X., Shuai, Y., & Jiacen, L. (2022). High Photoelectrochemical and Hydrogen Evolution Reaction Activities from 0D/2D ZnSe/(001)CuSe Heterojunction Catalysts Constructed by ZnSe Nanoparticles and {001} Facets-Exposed CuSe Nanosheets. *ACS Applied Energy Materials*, *5*(7), 8952–8962. DOI: 10.1021/acsaem.2c01443

Chinnaiah, K., Karthik, K., Krishnamoorthi, R., Nadezhda, P., Maria, G., Yong-Song, C., & Gurushankar, K. (2025). Enhancing supercapacitor and Antimicrobial Performance of Bioengineered Ag/Mn_3O_4 Composite Nanorods. *Journal of Industrial and Engineering Chemistry*, *147*, 329–340. DOI: 10.1016/j.jiec.2024.12.023

Chusnutdinow, S., Kazakov, A., Jakieła, R., Szot, M., Schreyeck, S., Brunner, K., & Karczewski, G. (2023). Effect of Manganese Alloying on Infrared Detectors Made of $Pb_{1-x}Mn_xTe$/CdTe Multilayer Composite. *Materials (Basel)*, *16*(12), 4211. DOI: 10.3390/ma16124211 PMID: 37374395

Gas, K., Kret, S., Zaleszczyk, W., Kamińska, E., Sawicki, M., Wojtowicz, T., & Szuszkiewicz, W. (2021). Oxidation of MBE-Grown ZnTe and ZnTe/Zn Nanowires and Their Structural Properties. *Materials (Basel)*, *14*(18), 5252. DOI: 10.3390/ma14185252 PMID: 34576476

Gurushankar, K., Chinnaiah, K., Tari, V., & Karthik, K. (2025). *Exploring MXenes and MBenes in Metal Sulfur and Selenium Battery Technologies, Novel energy storage and conversion technologies for two-dimensional MXenes and MBenes*. IGI Global Scientific Publishing., DOI: 10.4018/979-8-3693-9546-2.ch005

Gutierrez, Z.-B. Zayas-Bazán, P., De Melo, O., De Moure-Flores, F., Andraca-Adame, J. A., Moreno-Ruiz, L. A., Martínez-Gutiérrez, H., Gallardo, S., Sastré-Hernández, J., & Contreras-Puente, G. (2018). CdS/CdTe Heterostructures for Applications in Ultra-Thin Solar Cells. *Materials (Basel)*, *11*(10). Advance online publication. DOI: 10.3390/ma11101788

Hwang, Y., & Shin, N. (2021). Colloidal Synthesis of $MoSe_2$/WSe_2 Heterostructure Nanoflowers via Two-Step Growth. *Materials (Basel)*, *14*(23), 7294. DOI: 10.3390/ma14237294 PMID: 34885449

Jang, S.-M., & Kim, I.-H. (2025). Solid-State Synthesis and Thermoelectric Properties of $CuFeSe_2$–$CuFeS_2$ Solid Solutions. *Materials (Basel), 18*(6), 1366. DOI: 10.3390/ma18061366 PMID: 40141648

Jiabin, L., Shuai, Y., Hui, Z., Chao, W., David, B., François, V., Shuhui, S., Zhiming, M. W., Jiming, B., Haiguang, Z., & Gurpreet, S. S. (2023). Federico Rosei, Efficient Photoelectrochemical Hydrogen Generation Using Eco-Friendly "Giant" InP/ZnSe Core/Shell Quantum Dots. *ACS Applied Materials & Interfaces, 15*(29), 34797–34808. DOI: 10.1021/acsami.3c04900 PMID: 37433096

Jingyun, C., Zhulin, N., Zhipeng, Z., Xiangdong, P., Shuo, Z., Hongqiang, W., Dan, L., & Zaiping, G. (2023). Enhanced Ion/Electron Migration and Sodium Storage Driven by Different MoS_2-$ZnIn_2S_4$ Heterointerfaces. *Advanced Energy Materials, 13*,5, 2023, 2203248. DOI: 10.1002/aenm.202203248

Junjie, H., Haochen, L., Jun, M., Rui, L., Ziming, Z., Bingxin, Z., Bin, X., Jiaji, C., Kai, W., & Marie-Helene, D. (2019). A Facile Route to Synthesize CdSe/ZnS Thick-shell Quantum Dots with Precisely Controlled Green Emission Properties: Towards QDs Based LED Applications. *Scientific Reports, 9*, 12048. DOI: 10.1038/s41598-019-48469-7 PMID: 31427624

Kannan, K., & Tari, V. (Eds.). (2025). Novel energy storage and conversion technologies for two-dimensional MXenes and MBenes. https://www.amazon.com/dp/9798369395462 ISBN: 9798369395462, eISBN: 9798369395486

Khodamorady, M., & Bahrami, K. (2023). A Novel ZnS-CdS Nanocomposite as a Visible Active Photocatalyst for Degradation of Synthetic and Real Wastewaters. *Scientific Reports, 13*, 2177. DOI: 10.1038/s41598-023-28725-7 PMID: 36750747

Kuldeep, K., Munish, S., & Ravindra, P. (2025). Tunability in Electronic and Optical Properties of GaS/PbS vdW Heterostructure. *Journal of Physics and Chemistry of Solids, 196*, 112348. DOI: 10.1016/j.jpcs.2024.112348

Li, H., Chen, F., Wu, X., Wang, D., Ren, Y., & Li, Y. (2024). Improved HER/OER Performance of NiS_2/MoS_2 Composite Modified by CeO_2 and LDH. *Materials (Basel), 17*(19), 4876. DOI: 10.3390/ma17194876 PMID: 39410447

Li, H., Cheng, C., & Yang, Z. (2022). Encapsulated CdSe/CdS Nanorods in Double-shelled Porous Nanocomposites for Efficient Photocatalytic CO_2 Reduction. *Nature Communications, 13*, 6466. DOI: 10.1038/s41467-022-34263-z PMID: 36309504

Li, Y., Zhao, Y., Wang, X., Liu, W., He, J., Luo, X., Liu, J., & Liu, Y. (2023). Precise Construction and Growth of Submillimeter Two-Dimensional WSe_2 and $MoSe_2$ Monolayers. *Materials (Basel), 16*(13), 4795. DOI: 10.3390/ma16134795 PMID: 37445110

Lopez-Delgado, R., Zhou, Y., & Zazueta-Raynaud, A. (2017). Enhanced Conversion Efficiency in Si Solar Cells Employing Photoluminescent Down-shifting CdSe/CdS Core/shell Quantum Dots. *Scientific Reports, 7*, 14104. DOI: 10.1038/s41598-017-14269-0 PMID: 29074855

Maogang, G., Qingfeng, L., Ryan, G., Dan, E., Matthew, C., Alex, S., Alan, E., & Judy, Z. W. (2017). Printable Nanocomposite FeS_2–PbS Nanocrystals/Graphene Heterojunction Photodetectors for Broadband Photodetection. *ACS Applied Materials & Interfaces, 9*(33), 27801–27808. DOI: 10.1021/acsami.7b08226 PMID: 28758390

McDowell, L. L., Rastkar Mirzaei, M., & Shi, Z. (2023). Epitaxial CdSe/PbSe Heterojunction Growth and MWIR Photovoltaic Detector. *Materials (Basel), 16*(5), 1866. DOI: 10.3390/ma16051866 PMID: 36902982

Mohd. S., Kamlesh, V., Chandekar., Aslam K., Ahmed Mohamed E.T., AlFaify, S. (2020). A Facile Synthesis of Bi@PbS Nanosheets and their Key Physical Properties Analysis for Optoelectronic Technology, *Materials Science in Semiconductor Processing*. 107, 104807, DOI: 10.1016/j.mssp.2019.104807

Mu, G., Tan, Y., & Bi, C. (2024). Visible to Mid-wave Infrared PbS/HgTe Colloidal Quantum Dot Imagers. *Nature Photonics, 18*, 1147–1154. DOI: 10.1038/s41566-024-01492-1

Nimai, M., Noah, J. O., Feng, W., Zhongjian, H., Sachidananda, K., Anton, V. M., Joanna, L. C., Han, H., Milan, S., & Jennifer, A. H. (2017). Using Shape to Turn off Blinking for Two-colour Multiexciton Emission in CdSe/CdS Tetrapods. *Nature Communications, 8*, 15083. DOI: 10.1038/ncomms15083 PMID: 28497776

Park, J., Yoo, G., & Heo, J. (2019). CdSe/ZnS Quantum Dot Encapsulated MoS_2 Phototransistor for Enhanced Radiation Hardness. *Scientific Reports, 9*, 1411. DOI: 10.1038/s41598-018-37902-y PMID: 30723221

Paul, A.K., Kuiri, M., Saha, D. (2017). Photo-Tunable Transfer Characteristics in $MoTe_2$–MoS_2 Vertical Heterostructure. *2D Mater Appl, 1*, 17. DOI: 10.1038/s41699-017-0017-3

Pérez, L. M., Aghoutane, N., Laroze, D., Díaz, P., El-Yadri, M., & Feddi, E. M. (2023). Unveiling the Role of Donor Impurity Position on the Electronic Properties in Strained Type I and Type II Core/Shell Quantum Dots under Magnetic Field. *Materials (Basel)*, *16*(19), 6535. DOI: 10.3390/ma16196535 PMID: 37834672

Rahman, A., Khan, F., & Jennings, J. R. (2024). Microwave-assisted Synthesis of ZnS@$CuIn_xS_y$ for Photocatalytic Degradation of Coloured and Non-coloured Pollutants. *Scientific Reports*, *14*, 16155. DOI: 10.1038/s41598-024-66100-2 PMID: 38997296

Rahman, M. M., Karim, M. R., & Alam, M. M. (2020). Facile and Efficient 3-chlorophenol Sensor Development Based on Photolumenescent Core-shell CdSe/ZnS Quantum Dots. *Scientific Reports*, *10*, 557. DOI: 10.1038/s41598-019-57091-6 PMID: 31953448

Raphae, S., & Stefan, M. K., lesya, Y., Alexander, D., Maksym, Y., Annina, M., Shadi, N., Yuriy, F., & Vanessa, W. L. (2023). Highly Responsive Mid-Infrared Metamaterial Enhanced Heterostructure Photodetector Formed out of Sintered PbSe/PbS Colloidal Quantum Dots. *ACS Applied Materials & Interfaces*, *15*(8), 10847–10857. DOI: 10.1021/acsami.2c23050 PMID: 36795914

Sun, Y. L., Xie, D., & Sun, M. X. (2018). Hybrid Graphene/cadmium-Free ZnSe/ZnS Quantum Dots Phototransistors for UV detection. *Scientific Reports*, *8*, 5107. DOI: 10.1038/s41598-018-23507-y PMID: 29572513

Sungwoo, K., Taehoon, K., Meejae, K., Seong, K. K., Tae, W. Y., Lee Soon, P., Ilseung, Y., Sunjin, H., Jung Eun, L., Seong Keun, K., & Sang-Wook, K. (2012). Highly Luminescent InP/GaP/ZnS Nanocrystals and Their Application to White Light-Emitting Diodes. *Journal of the American Chemical Society*, *134*(8), 3804–3809. DOI: 10.1021/ja210211z PMID: 22303916

Tari, V. Kannan, K., & Vishwakarma, V. (2022). Chapter 8: Emerging antiviral technologies. In P. Nunthavarawong, S. M. Rangappa, S. Siengchin, & M. Thoppil-Mathew (Eds.), *Antimicrobial and antiviral materials: Polymers, metals, ceramics, and applications* (1st ed., pp. 137–150). CRC Press. DOI: 10.1201/9781003143093-8

Tari, V., Kannan, K., Koliar, J., Gurushankar, K., & Wahyuni, D. K. (2025). Influence of zinc oxide nanomaterials on crop plants: A prospective analysis. In D. Jabbrova, D. Sarkar, Rakesh, S. Singh, & R. Datta (Eds.), *Smart nanomaterials in agroforestry systems* (Chapter 8). Springer. DOI: 10.1007/978-981-96-1337-3_8

Tian, P., Min, L., Xin, L., Yaqin, Z., Tian, Z., Pu, S., & Fuzhi, L. (2024). Microflower-like CuS/Co_3S_4 Composite as High-capacity and Excellent-stability Material for Hybrid Supercapacitors. *Journal of Alloys and Compounds, 1008*, 176563. DOI: 10.1016/j.jallcom.2024.176563

Ursi, F., Virga, S., Garcìa-Espejo, G., Masciocchi, N., Martorana, A., & Giannici, F. (2022). Long-Term Stability of TiS_2–Alkylamine Hybrid Materials. *Materials (Basel), 15*(23), 8297. DOI: 10.3390/ma15238297 PMID: 36499793

Wang, X., Yu, J., & Chen, R. (2018). Optical Characteristics of ZnS Passivated CdSe/CdS Quantum Dots for High Photostability and Lasing. *Scientific Reports, 8*, 17323. DOI: 10.1038/s41598-018-35768-8 PMID: 30470827

Wang, Y., Bourgès, C., Rajamathi, R., Nethravathi, C., Rajamathi, M., & Mori, T. (2022). The Effect of Reactive Electric Field-Assisted Sintering of MoS_2/Bi_2Te_3 Heterostructure on the Phase Integrity of Bi_2Te_3 Matrix and the Thermoelectric Properties. *Materials (Basel), 15*(1), 53. DOI: 10.3390/ma15010053 PMID: 35009201

Wang, Y., Zhong, Y., Zi, J., & Lian, Z. (2023). Type-I CdSe@CdS@ZnS Heterostructured Nanocrystals with Long Fluorescence Lifetime. *Materials (Basel), 16*(21), 7007. DOI: 10.3390/ma16217007 PMID: 37959604

Wani, S. A., Tari, V., & Mansoor, S. I. U. (2025). A Comprehensive Analysis of Wastewater Management Challenges in India: Infrastructure and Policy Perspectives. In Dhiman, S., & Mukherjee, G. (Eds.), *Resource Resurgence. Springer Water.* Springer., DOI: 10.1007/978-3-032-02750-4_11

Wenda, Z., Shihao, D., Weidong, Z., Dan, W., Pai, L., Xiangwei, Q., Haochen, L., Hongcheng, Y., Zhenghui, W., Kai, W., & Xiao, W.S. (2020). InP/ZnS/ZnS Core/Shell Blue Quantum Dots for Efficient Light-Emitting Diodes, *Adv. Funct. Mater. 30*,49, 2005303. DOI: 10.1002/adfm.202005303

Won, Y. H., Cho, O., & Kim, T. (2019). Highly Efficient and Stable InP/ZnSe/ZnS Quantum Dot Light-emitting Diodes. *Nature, 575*, 634–638. DOI: 10.1038/s41586-019-1771-5 PMID: 31776489

Yavorskiy, D., Szoła, M., Karpierz, K., Rudniewski, R., Bożek, R., Karczewski, G., Wojtowicz, T., Wróbel, J., & Łusakowski, J. (2020). Polarization of Magnetoplasmons in Grating Metamaterials Based on CdTe/CdMgTe Quantum Wells. *Materials (Basel), 13*(8), 1811. DOI: 10.3390/ma13081811 PMID: 32290498

Zhang, Z., Liu, B., Zhou, J., & Sun, Z. (2025). Amorphous MoTex Nanomaterials Promote Visible-Light Co-Catalytic Degradation of Methylene Blue. *Materials (Basel), 18*(14), 3388. DOI: 10.3390/ma18143388 PMID: 40731598

Chapter 3
Green Synthesized Metal–Chalcogenides for Sustainable Development

Vinaya Satyawan Tari
https://orcid.org/0000-0002-8352-252X
Faculty of Science and Technology, Universitas Airlangga, Surabaya, Indonesia

ABSTRACT

The growing demand for eco-friendly and energy-efficient materials has motivated researchers and Scientists in the green synthesis of chalcogenide compound materials based on sulfur (S), selenium (Se), and tellurium (Te). These are critical in applications spanning optoelectronics, photovoltaics, catalysis, and energy storage. However, methods involved in conventional synthesis often carry toxic solvents, high temperatures, and hazardous precursors, posing environmental and safety concerns. This chapter presents progress in the field of the green synthesis of metal chalcogenides in their nano-phased forms with their biomedical and environmental applications, which leads to sustainable development. This chapter will help bridge the knowledge gap between industrial and academic scientists.

INTRODUCTION

In the present context, many researchers in various fields such as physics, chemistry, environmental science, material science, biomedical science, biology, engineering, etc., are focusing on the synthesis, characterization, and applications of nano-materials, (Hagfeldt & Grätzel, 1995; Zhu & Wang, 2003). The speedy research and development in nanotechnology and material science are bringing a new world for the coming generations, (Zhang & Mou, 1994). Nanomaterials typically exhibit

DOI: 10.4018/979-8-3373-3962-7.ch003

unique properties, viz. Chemical, electronic, magnetic, and optical, etc., which are significantly different because of their nano sizes (less than 100nm) from those of other bulk materials and relatively large surface-to-volume ratio, (Tari, et al., 2022). These nanoparticles possess several competent applications, including electronic, superconductor, adhesive, catalytic, mechanical, optical, magnetic, drug delivery, photographic, biomedical, and high-performance engineering materials, (Chen & Wu, 2000; Gurushankar et al., 2025).

The transition metal chalcogenides [T-MCs] are composed of the transition metals and chalcogen elements such as sulfide (S), selenium (Se), or tellurium (Te), (Mendhe et al., 2025). These chalcogen metals chemically combine with transition metals or other metals at the nanoscale, resulting in the alloys called chalcogenide nanoparticles (CNPs). Based on the number of different chalcogen elements present in CNPs, they are grouped as monochalcogenide, dichalcogenide, and poly chalcogenide. However, they can also be grouped as binary and ternary nanomaterials, (Xiao & Yang, 2014). The growing demand for eco-friendly and energy-efficient materials has motivated researchers and Scientists in the green synthesis of chalcogenide compound materials based on sulfur, selenium, and tellurium. Metal chalcogenides are critical in applications spanning optoelectronics, biomedical applications, photovoltaics, catalysis, and energy storage because of their electrochemical properties, rich structural tunability, and redox activity, (Buckingham et al., 2022; Liu et al., 2022; Mendhe et al., 2025; Wu et al., 2024). The synthesis of metal chalcogenides by using their nano-phased forms is the modern trend of miniaturization. These nano-phased forms become instrumental for the development of nanoscience and nanotechnology as far as the pharmaceutical industry, textile industry, energy storage, and wastewater treatment are concerned. Both conventional methods and modern methods have been established for the preparation of metal chalcogenides, (Zhu & Wang, 2003). Modern synthetic methods have already been proven for their influential role in advanced material science. Additionally, metal chalcogenide-based electrode materials are receiving substantial attention as they tend to have high-performance electrode alternatives for supercapacitors by virtue of their high electrical conductivity, redox characteristics, and enhanced electrochemical activity, (Mendhe et al., 2025).

In this chapter, we present progress in the field of the green synthesis of metal chalcogenides in their nano-phased forms with their biomedical and environmental applications, which leads to sustainable development. This chapter will help bridge the knowledge gap between industrial and academic scientists.

SYNTHESIS OF CHALCOGENIDES

The synthesis of chalcogenides has attracted researchers' focus over the past three decades. However, in the present context, fast industrial development and advancement in science, with special focus on sustainable development, effective and affordable device production have become critical, (Pacchioni, 2022; Yalcin et al., 2023; Tari & Patil, 2017). This advancement has led to the production of several portable devices, renewable energy storage devices, electric vehicles, high-performance energy storage devices, etc., (Wang, G. et al. 2012; Zuo, X. et al. 2025). Transition metal chalcogenides can be prepared through the nano-structuring approach in their nano-phased forms with controlled surface morphology, including nanorods, nanostars, nanocrystals, nanoclusters, nanowires, nano prisms, nano shells, nanocages, nanosheets, nanoflakes, hollow spheres, carbon quantum dots, coupled with enhanced electrochemical performance of novel T- MCs, (Manikandan et al., 2021, Olawale et al., 2022). The morphology and relative dimensions of the nano materials mainly depend on the precursor, synthesis method, and experimental conditions, which further turn into functional units for several applications, (Olawale et al., 2022). Chalcogenides can be prepared in several ways by using either conventional methods or modern nano-phased forms.

Conventional Methods

However, the conventional method for the preparation was the direct combination of elements at high temperature, (Zhu & Wang, 2003). But it was found that this procedure needs a substantial energy input, a long reaction time, and provides less control in the stoichiometry of the related metal, (Metcalf et al., 1982). Another method with low energy input is the precipitation of desired metal chalcogenides from aqueous solutions of metal cations by using H2E. Where 'E' is a chalcogen element including Sulfur, Selenium, and Tellurium (Figure 1). Further research on this method found that in this method, many toxic reagents are used, and the resulting product has many impurities, (Kräuter & Rees Jr., 1995). The third significant method for the preparation of chalcogenide was the molecular precursor method. This method involves thermal decomposition of compounds holding either a metal chalcogen bond, i.e., M-S or M-Se, or H2E as a chalcogenide source, (Kräuter & Rees Jr., 1995; Pickett et al., 1996).

Figure 1. Conventional methods for the synthesis of metal chalcogenides

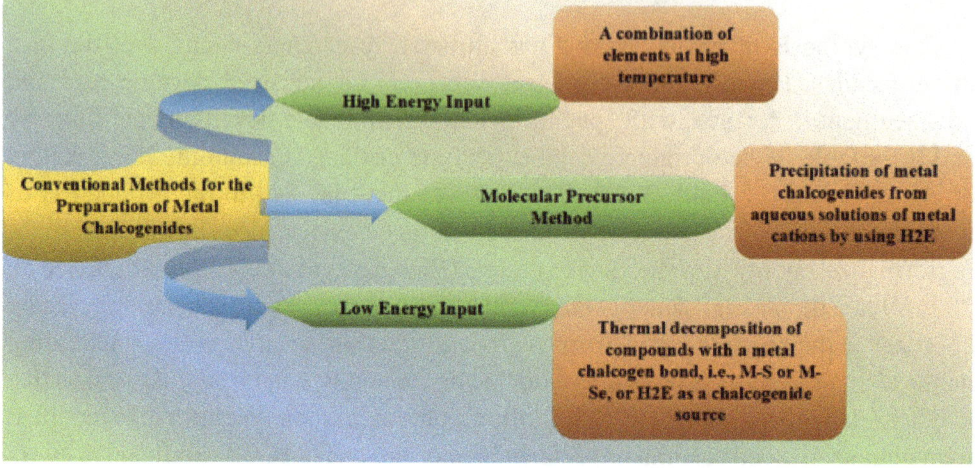

Synthetic Methods for the Preparation of Metal Chalcogenides

Recently, researchers' focus on the synthesis of metal chalcogenides shifted towards the modern trend of miniaturization with the unique properties of their nano-phased form. In the past two to three decades, many methods have been developed by scientists for the preparation of nanocrystalline metal chalcogenides. The involvement of these novel technologies in the synthesis of metal chalcogenides is providing a convenient way to produce nanoforms of metal chalcogenides with enhanced functional properties, with controlled size and morphology.

There are several advanced methods for the preparation of metal chalcogenides, (Shaikh et al., 2024; Mendhe et al., 2023; Ye et al., 1999; Zhou & Xin, 1999; Smart & Moore, 1996), viz., sono-chemical method, photochemical synthesis, low temperature solid state synthesis, electrochemical template synthesis, sono-electrochemical method, solvothermal technique, irradiation method, microwave-assisted heating method, microemulsion method, chemical bath deposition method, successive ionic layer adsorption and reaction (SILAR), electrodeposition method, and sol-gel method (Figure 2).

Figure 2. Synthetic methods for the preparation of metal chalcogenides (MCs)

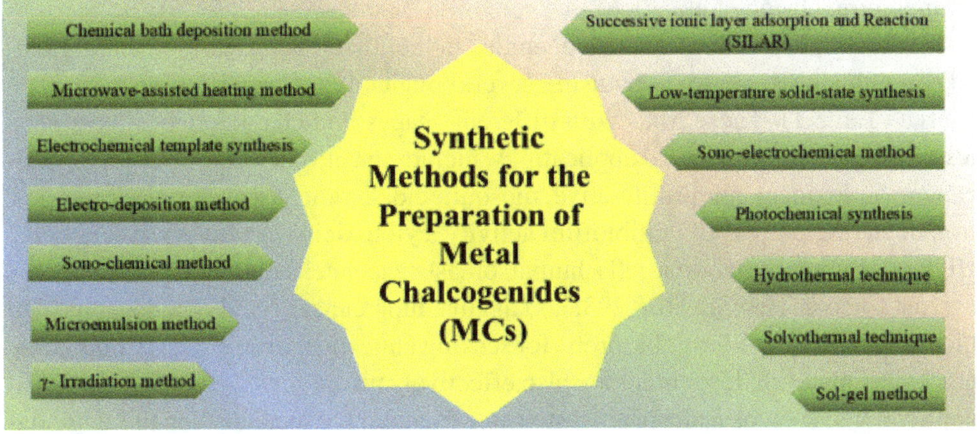

However, methods involved in synthetic preparation often carry toxic solvents, high temperatures, and hazardous precursors, posing environmental and safety concerns. Therefore, scientists focus on the green synthesis of metal chalcogenides to achieve various sustainable development goals.

Green Synthesis of Metal Chalcogenides:

As we discussed above, synthetic methods for the preparation of MCs pose an environmental risk; hence, switching to green synthesis is instrumental in the present context. The paradigm shift towards biogenic ways of synthesizing green metal chalcogenides in their nano-phased forms has been proven to be advantageous for their biosafety, eco-friendliness, and ease of synthesis as compared to inorganic chalcogens and chalcogenide nanomaterials. The biogenic route of synthesis primarily involves the use of plant extracts and purified biomolecules present in the selected plant material, (Gupta et al., 2024; Olawale et al., 2022). Plants contain several secondary metabolites that tend to reduce heavy metals and other selected metallic ions. Moreover, the extraction and purification of fabricated nanomaterials do not involve any tedious techniques. Additionally, these metabolites themselves act as reducing and capping agents; hence, chemical stress on the environment can be reduced, making it feasible for the same, (Jadoun et al., 2020).

There are some challenges in the biogenic routes of preparation of chalcogenide NPs, especially when microorganisms are involved as a green source, viz., cultivation, control on the morphology of fabricated NPs, (Gericke & Pinches, 2006). These limitations can be overcome by adjusting some physicochemical factors, including metal salt concentration, incubation time, pH, quantity of biological inoculum, and temperature, (Jacob et al., 2016).

Green-synthesized metal chalcogenides in their nano forms have been fabricated through a biogenic route by using several plants, fungi, and bacteria as a green source (Table 1). These MCs with different shapes and characteristics have been tested for various biomedical applications, including anti-bacterial, antifungal, anticancer, antioxidant, antiproliferative, mosquitocidal, radical scavenging properties, iron chelating properties, antibiofilm activity, cytotoxicity in MDA-MB-231 cells, HDF and human melanoma cells, human breast cancer cells (MCF-7), effectiveness against Caco-2, HepG2, HEK293, and against lung cancer (A549) cells, enhanced biocompatibility in Vero cells, probe for sensing anticancer drug e.g. 'Methotrexate' etc. Novel metal chalcogenides are also effectively useful for several environmental applications, such as industrial wastewater treatment, especially for the degradation of various toxic organic dyes such as methylene blue dye, methyl orange dye, rhodamine B dye, alizarin red dye, malachite green dye, sunset yellow azo dye, safranin T dye, bromothymol blue dye, and congo red dye. However, other environmental applications, energy storage and conversion: solar cells, photocatalysis, bioimaging, biosensors for heavy metal detection, bio-ink, optoelectronics, etc. The following is the summary of spectacular research on green-synthesis of metal chalcogenides in their nano form and their biomedical and environmental applications, which is one step forward for sustainable development in the present context.

Table 1. Green-synthesized metal chalcogenides in their nano-phased forms with biomedical and environmental applications

Green Source	Novel material	Shape	Experimental conditions	Applications	Reference
Ceropegiabulbosa– tuber extract	SeNPs	Nanosphere	-	Higher cytotoxicity in MDA-MB-231 cells compared to normal breast cells	Cittrarasu V. et al. 2021
Ocimum sanctum aq. extract	CDs		Fluorescent CDs	Sensor for lead detection	Kumar, A. et al. 2017
Aloevera extract	TeNPs	Nanosphere	-	Cytotoxicity in HDF and human melanoma cells	Medina-Cruz D. et al. 2021
Stevia rebaudiana– leaf extract	ZnSNPs	Nanosphere	-	Cytotoxicity in humans breast cancer cells (MCF-7)	Alijani H. Q. et al. 2019

continued on following page

Table 1. Continued

Green Source	Novel material	Shape	Experimental conditions	Applications	Reference
Orange juice	TeNPs	Nanorods	≤50µg/ml	Anticancer activity in human melanoma cells	Medina-Cruz D. et al. 2019
Lemon Juice	TeNPs	Nano cube	≤50µg/ml	Anticancer activity in human melanoma cells	Medina-Cruz D. et al. 2019
Lime juice	TeNPs	Nano cube	≤50µg/ml	Anticancer activity in human melanoma cells	Medina-Cruz D. et al. 2019
Moringa oleifera–leaf extracts	SeNPs	Sphere	Highest cell inhibition in A549 cells	Anticancer activity- Effective against Caco-2, HepG2, and MCF-7 cells	Hassanien, R. et al. 2019
Ocimum tenuiflorum inflorescence extract	Ag@SeNPs	Nearly spherical	Dose-dependent cytotoxicity	Cytotoxicity in HEK293 and MCF7 cells	Olawale F. et al. 2021
Fenugreek–seed extract	SeNPs	Oval	Dose-dependent (25–100µg/ml) cytotoxicity	Cytotoxicity in human breast cancer cells (MCF-7)	Ramamurthy C. et al. 2013
Allium sativum extract	SeNPs	nanosphere	-	Enhanced biocompatibility in Vero cells	Anu K. et al. 2017
Urtica dioica– leaf extract	SeNPs	nanosphere	-	Anticancer activity against HepG2 cells. Nontoxic to Vero cells.	Hashem A. H. 2021
Hawthorn–fruit extract	SeNPs	Nanosphere	Induced intracellular oxidative stress and mitochondrial dysfunction to initiate apoptosis through the mitochondrial pathway	Antitumor activity in HepG2 cells	Cui D. et al. 2018

continued on following page

Table 1. Continued

Green Source	Novel material	Shape	Experimental conditions	Applications	Reference
Camellia sinensis (Tea leaf extract)	CdS QDs	Nanosphere	Bioreduction by cadmium sulfate and sodium sulfide	Significant cytotoxicity in lung cancer (A549) cells with good biocompatibility in normal cells	Shivaji K. et al. 2018
Fermented yeast broth from agricultural waste	SeNPs	Nanorod	Visible annular-type photoreactor with a tungsten lamp of 300 W	Degradation of methylene blue Additionally, anticancer and antibacterial activity	Goud et al. 2016
Ficus benghalensis leaf extract	Polycrystalline SeNPs	Nanosphere	UV-light ~58% degradation observed after 40 minutes First-order kinetics reaction with the calculated rate constant of 0.02162 s^{-1}	Degradation of methylene blue	Tripathi R. M. et al. 2020
Withania somnifera	SeNPs	Nanosphere	Degradation studied under sunlight	Degradation of methylene blue Additionally, antioxidant, antimicrobial, and antiproliferative properties	Alagesan, V., & Venugopal, S. 2019
Centella asiatica aq. Leaf extract	CDs	Spherical	-	Photocatalyst	Jhonsi, M. A., & Kathiravan, A. 2017
Monascus purpureus (Fungi)	SeNPs	single-phase nanocrystal	Degradation studied in sunlight	Degradation of methylene blue Additionally, possesses antioxidant, antimicrobial, and anticancer activities	El-Sayed, E.-S. R. et al. 2020

continued on following page

Table 1. Continued

Green Source	Novel material	Shape	Experimental conditions	Applications	Reference
Ceropegia bulbosa tuber extract	SeNPs	Spherical	Observed that SeNPs degrade 98% of methylene blue dye in 80 minutes through photocatalysis	Degradation of methylene blue. Additionally, exhibit mosquitocidal, antimicrobial, and cytotoxic activity	Cittrarasu V. et al. 2021
Goniothalamus wightii extract	SeNPs	-	Photocatalytic degradation of methylene blue in 150 minutes	Degradation of methylene blue. Additionally, anticancer and radical scavenging properties	Santhosh C. et al. 2022
Allium paradoxum aqueous extract	SeNPs	Semi-spherical	Pseudo-first order kinetics	Degradation of methylene blue. Additionally, it possesses antioxidant, antibacterial, antifungal, anticancer, and iron-chelating properties	Alizadeh, S. R. et al. 2023
Hibiscus esculentus L. extract	SeNPs	Spherical shaped hexag onal crystals	Pseudo first order kinetics, and the rate constant value 0.1023 min−1	Degradation of methylene blue.	Ebrahimzadeh, M. A. et al. 2023
Hibiscus rosa-sinensis extract	Selenium-based NPs	Spherical	-	Degradation of methylene blue. In addition, antibacterial activity	Velayutham, L. et al. 2023
Lawsonia inermis aq. extract	CDs	-	-	Probe of sensing anticancer drug 'Methotrexate'	Shahshahanipour, M. et al. 2019
Nyctanthes arbor-tristis L.	SeNPs	Spherical	-	photocatalytic degradation of methylene blue. Additionally, antioxidant, antibiofilm, antimicrobial	Satpathy S. et al. 2024

continued on following page

Table 1. Continued

Green Source	Novel material	Shape	Experimental conditions	Applications	Reference
Bacillus sp.	SeNPs	Nanocluster	Pseudo first order kinetics, and the calculated rate constant 0.00842 min−1	Degradation of methyl orange dye	Ahluwalia, S. et al. 2016
Lysinibacillus sp.	SeNPs	spherical & triangular	H2O2	Degradation of rhodamine B dye	Che, L. et al. 2017
Justicia adhatoda leaf extract	Chitosan-SeNPs	-	varying the different experimental conditions, viz., temperature, absorbent material, dye concentration, pH, etc.	Degradation of rhodamine B dye	Britto, J. et al. 2021
Tectona grandis	SeNPs	trigonal	87% degradation in 90 minutes	Active against alizarin red dye	Byeon, H. et al. 2024
Aspergillus terreus	SeNPs	Spherical	89% degradation in 240 min under sunlight	Degradation of malachite green dye	Saied, E. et al. 2023
Moringa oleifera extract	SeNPs	Spherical	under UV light irradiation, and under sunlight irradiation at a slightly acidic pH range	Degradation of Sunset Yellow Azo Dye	Hassanien, R. et al. 2019
E. coli	SeNPs	Irregular spheres	Adsorption	Degradation of safranin T dye	Xia, X. et al. 2018
Streptomyces griseo brunneus	SeNPs	Trigonal	Under UV irradiation, degradation was completed in 60 min *via* oxidation with H2O2	Degradation of bromothymol blue dye	Ameri, A. et al. 2016
Prunus armeniaca L.	SeNPs	-	Pseudo-second order kinetics	Degradation of Congo red dye	Turna, T. et al. 2024

continued on following page

Table 1. Continued

Green Source	Novel material	Shape	Experimental conditions	Applications	Reference
Desulfovibrio caledoiensis	CdSNPs	Spherical	-	Bioimaging	Qi, P. et al. 2016
Enterococcus sp.	CdSNPs	Spherical	-	Antimicrobial	Rajeshkumar, S. et al. 2014
E. coli	CdSNPs	Spherical	-	Solar cells	Órdenes-Aenishanslins, N. et al. 2019
E. coli	CdSNPs	Spherical	-	Antimicrobial	Shivashankarappa, A. et al. 2020
Klebsiella pneumoniae	CdSNPs	Spherical	-	Antimicrobial	Malarkodi, C. et al 2014
Pedobacter sp.	CdSNPs	-	-	Solar cells	Carrasco, V. et al. 2021
E. coli	CdSAg NPs	Spherical	-	Bioimaging, solar cells	Órdenes-Aenishanslins et al. 2020
E. coli	CdS/ CdSe NPs	Spherical	-	Solar cells	Órdenes-Aenishanslins, N. et al. 2019
Klebsiellapneumoniae	ZnS NPs	Spherical	-	Antimicrobial	Malarkodi, C. et al 2014
Serratia nematodiphila	ZnS NPs	Spherical	-	Antimicrobial	Malarkodi, C. et al. 2013
Thermoanaerobacter sp.	ZnS NPs	Spherical	-	Bio-ink	Moon, J. W. et al. 2014
Clostridiaceaesp.	PbS NPs	Nanocube, nanosphere, nanosheet	-	As(III) detection	Yue, L. et al. 2016
Stenotrophomonas maltophilia	PbS NP	Amorphous	-	Solar cells	Spangler, L. C. et al. 2016
Aspergillus niger	CdS NPs	Spherical	-	Antimicrobial, anticancer	Alsaggaf, M. S. et al. 2020
P. ostreatus	Ag2S NPs	Spherical	-	Bioimaging, antibacterial	Borovaya, M. et al. 2020
A. flavus	PbS NPs	Spherical	-	As Detection	Priyanka, U. et al. 2017
Schizosaccharomyces pombe	CdS QDs	-	Feedback-controlled fed-batch system	Optoelectronics	Williams, P. et al. 2002

CONCLUSION

In this chapter, we have focused on green-synthesized metal chalcogenides for sustainable development. Various green sources are being used for the development of several solutions in the field of energy storage, biomedical, optoelectronics, biosensors, and environmental purposes. This approach in the respective field is one step forward towards sustainable development in the present context. However, still research is needed for an efficient alternative to chemical analogs, identification of key biomolecules which are instrumental in the formation of NPs, as far as biogenic fabrication of NPs is concerned, controlled morphology, stability of NPs, scalability of the process of production of NPs, biocompatibility, expansion of bio-applications, and overcoming their limitations.

REFERENCES

Ahluwalia, S., Prakash, N. T., Prakash, R., & Pal, B. (2016). Improved degradation of methyl orange dye using bio-co-catalyst Se nanoparticles impregnated ZnS photocatalyst under UV irradiation. *Chemical Engineering Journal*, *306*, 1041–1048. DOI: 10.1016/j.cej.2016.07.060

Alagesan, V., & Venugopal, S. (2019). Green synthesis of selenium nanoparticle using leaves extract of *Withania somnifera* and its biological applications and photocatalytic activities. *BioNanoScience*, *9*(1), 105–116. DOI: 10.1007/s12668-018-0566-8

Alijani, H. Q., Pourseyedi, S., Torkzadeh Mahani, M., & Khatami, M. (2019). Green synthesis of zinc sulfide (ZnS) nanoparticles using *Stevia rebaudiana* Bertoni and evaluation of its cytotoxic properties. *Journal of Molecular Structure*, *1175*, 214–218. DOI: 10.1016/j.molstruc.2018.08.081

Alizadeh, S. R., Seyedabadi, M., Montazeri, M., Khan, B. A., & Ebrahimzadeh, M. A. (2023). *Allium paradoxum* extract mediated green synthesis of SeNPs: Assessment of their anticancer, antioxidant, iron chelating activities, and antimicrobial activities against fungi, ATCC bacterial strains, *Leishmania* parasite, and catalytic reduction of methylene blue. *Materials Chemistry and Physics*, *296*, 127240. DOI: 10.1016/j.matchemphys.2022.127240

Alsaggaf, M. S., Elbaz, A. F., El Badawy, S., & Moussa, S. H. (2020). Anticancer and antibacterial activity of cadmium sulfide nanoparticles by *Aspergillus niger*. *Advances in Polymer Technology*, *4909054*. Advance online publication. DOI: 10.1155/2020/4909054

Ameri, A., Shakibaie, M., Ameri, A., Faramarzi, M. A., Amir Heidari, B., & Forootanfar, H. (2016). Photocatalytic decolorization of bromothymol blue using biogenic selenium nanoparticles synthesized by terrestrial actinomycete *Streptomyces griseobrunneus* strain FSHH12. *Desalination and Water Treatment*, *57*(48), 21552–21563. DOI: 10.1080/19443994.2015.1112291

Anu, K., Singaravelu, G., Murugan, K., & Benelli, G. (2017). Green synthesis of selenium nanoparticles using garlic cloves (*Allium sativum*): Biophysical characterization and cytotoxicity on Vero cells. *Journal of Cluster Science*, *28*(1), 551–563. DOI: 10.1007/s10876-016-1124-2

Borovaya, M., Naumenko, A., Horiunova, I., Plokhovska, S., Blume, Y., & Yemets, A. (2020). "Green" synthesis of Ag_2S nanoparticles, study of their properties and bioimaging applications. *Applied Nanoscience*, *10*, 4931–4940. DOI: 10.1007/s13204-020-01465-3

Britto, J., Barani, P., Vanaja, M., Pushpalaksmi, E., Samraj, J. J., & Annadurai, G. (2021). Adsorption of dyes by chitosan–selenium nanoparticles: Recent developments and adsorption mechanisms. *Nature Environment and Pollution Technology*, *20*(2), 467–479.

Buckingham, M. A., Qu, J., Ward-O'Brien, B., & Lewis, D. J. (2022). High entropy metal chalcogenides: Synthesis, properties, applications and future directions. *Chemical Communications*, *58*(61), 8025–8036. DOI: 10.1039/D2CC01796B PMID: 35770747

Byeon, H., Devi, V. P., Murugan, A., Tonk, A., Haribabu, K., Patil, H., & Vincy, V. G. A. G. (2024). Evaluating the potential capability of selenium nanoparticles on degradation of hazardous textile dye using environmental friendly approach. *Global NEST Journal*, *26*, 05633. DOI: 10.30955/gnj.005633

Carrasco, V., Amarelle, V., Lagos-Moraga, S., Quezada, C. P., Espinoza-González, R., Faccio, R., Fabiano, E., & Pérez-Donoso, J. M. (2021). Production of cadmium sulfide quantum dots by the lithobiontic Antarctic strain *Pedobacter* sp. UYP1 and their application as photosensitizer in solar cells. *Microbial Cell Factories*, *20*, 41. DOI: 10.1186/s12934-021-01548-3 PMID: 33568151

Che, L., Dong, Y., Wu, M., Zhao, Y., Liu, L., & Zhou, H. (2017). Characterization of selenite reduction by *Lysinibacillus* sp. ZYM-1 and photocatalytic performance of biogenic selenium nanospheres. *ACS Sustainable Chemistry & Engineering*, *5*(3), 2535–2543. DOI: 10.1021/acssuschemeng.6b02525

Chen, D. H., & Wu, S. H. (2000). Synthesis of nickel nanoparticles in water-in-oil microemulsions. *Chemistry of Materials*, *12*(5), 1354–1360. DOI: 10.1021/cm991167y

Cittrarasu, V., Kaliannan, D., Dharman, K., Maluventhen, V., Easwaran, M., Liu, W. C., Balasubramanian, B., & Arumugam, M. (2021). Green synthesis of selenium nanoparticles mediated from *Ceropegia bulbosa* Roxb extract and its cytotoxicity, antimicrobial, mosquitocidal and photocatalytic activities. *Scientific Reports*, *11*, 1032. DOI: 10.1038/s41598-020-80366-0 PMID: 33441811

Cui, D., Liang, T., & Sun, L.. (2018). Green synthesis of selenium nanoparticles with extract of hawthorn fruit induced HepG2 cells apoptosis. *Pharmaceutical Biology*, *56*(1), 528–534. DOI: 10.1080/13880209.2018.1500525 PMID: 30387372

Ebrahimzadeh, M. A., Moradsomarein, M., Lalerdi, F. S., & Alizadeh, S. R. (2023). Biogenic synthesis of selenium nanoparticles using *Hibiscus esculentus* L. extract: Catalytic degradation of organic dye and its anticancer, antibacterial, and antifungal activities. *European Journal of Chemistry*, *14*(2), 144–154. DOI: 10.5155/eurjchem.14.2.144-154.2416

El-Sayed, E.-S. R., Abdelhakim, H. K., & Ahmed, A. S. (2020). Solid state fermentation for enhanced production of selenium nanoparticles by gamma-irradiated *Monascus purpureus* and their biological evaluation and photocatalytic activities. *Bioprocess and Biosystems Engineering*, *43*(6), 797–809. DOI: 10.1007/s00449-020-02279-1 PMID: 31898764

Gericke, M., & Pinches, A. (2006). Biological synthesis of metal nanoparticles. *Hydrometallurgy*, *83*(1–4), 132–140. DOI: 10.1016/j.hydromet.2006.03.019

Goud, K. G., Veldurthi, N. K., Vithal, M., & Reddy, G. (2016). Characterization and evaluation of biological and photocatalytic activities of selenium nanoparticles synthesized using yeast fermented broth. *Journal of Materials NanoScience*, *3*(2), 33–40.

Gupta, V., Chandra, N., Tari, V., & Kannan, K. (2024). A sustainable approach for the green synthesis of silver nanoparticles using *Brassica oleracea* subsp. *botrytis* (L.) leaves and its antibacterial activity. *Physics and Chemistry of Solid State*, *25*(2), 303–310. DOI: 10.15330/pcss.25.2.303-310

Gurushankar, K., Karuppaiya, C., Tari, V., & Kannan, K. (2025). Exploring MXenes and MBenes in metal sulfur and selenium battery technologies. In Kannan, K., & Tari, V. (Eds.), *Novel energy storage and conversion technologies for two-dimensional MXenes and MBenes*. IGI Global., DOI: 10.4018/979-8-3693-9546-2.ch005

Hagfeldt, A., & Grätzel, M. (1995). Light-induced redox reactions in nanocrystalline systems. *Chemical Reviews*, *95*(1), 49–68. DOI: 10.1021/cr00033a003

Hashem, A. H., & Salem, S. S. (2021). Green and ecofriendly biosynthesis of selenium nanoparticles using *Urtica dioica* (stinging nettle) leaf extract: Antimicrobial and anticancer activity. *Biotechnology Journal*, *17*(2), 2100432. Advance online publication. DOI: 10.1002/biot.202100432 PMID: 34747563

Hassanien, R., Abed-Elmageed, A. A., & Husein, D. Z. (2019). Eco-friendly approach to synthesize selenium nanoparticles: Photocatalytic degradation of sunset Yellow azo dye and anticancer activity. *ChemistrySelect*, *4*(29), 9018–9026. DOI: 10.1002/slct.201902193

Jacob, J. M., Lens, P. N. L., & Balakrishnan, R. M. (2016). Microbial synthesis of chalcogenide semiconductor nanoparticles: A review. *Microbial Biotechnology*, *9*(1), 11–21. DOI: 10.1111/1751-7915.12297 PMID: 26110980

Jadoun, S., Arif, R., Jangid, N. K., & Meena, R. K. (2020). Green synthesis of nanoparticles using plant extracts: A review. *Environmental Chemistry Letters*, *19*(1), 355–374. DOI: 10.1007/s10311-020-01074-x

Jhonsi, M. A., & Kathiravan, A. (2017). Photoinduced interaction of arylamine dye with carbon quantum dots ensued from *Centella asiatica*. *Journal of Luminescence*, *192*, 321–327. DOI: 10.1016/j.jlumin.2017.06.056

Kräuter, G., & Rees, W. S.Jr. (1995). Cadmium bis(alkylthiolate) complexes as precursors for cadmium sulfide: A mild route to hawleyite. *Journal of Materials Chemistry*, *5*(8), 1265–1267. DOI: 10.1039/JM9950501265

Kumar, A., Chowdhuri, A. R., Laha, D., Mahto, T. K., Karmakar, P., & Sahu, S. K. (2017). Green synthesis of carbon dots from *Ocimum sanctum* for effective fluorescent sensing of Pb^{2+} ions and live cell imaging. *Sensors and Actuators. B, Chemical*, *242*, 679–686. DOI: 10.1016/j.snb.2016.11.109

Liu, X., Xu, F., Li, Z., Liu, Z., Yang, W., & Zhang, Y.. (2022). Design strategy for MXene and metal chalcogenides/oxides hybrids for supercapacitors, secondary batteries and electro/photocatalysis. *Coordination Chemistry Reviews*, *464*, 214544. DOI: 10.1016/j.ccr.2022.214544

Malarkodi, C., & Annadurai, G. (2013). A novel biological approach on extracellular synthesis and characterization of semiconductor zinc sulfide nanoparticles. *Applied Nanoscience*, *3*(5), 389–395. DOI: 10.1007/s13204-012-0119-9

Malarkodi, C., Rajeshkumar, S., Paulkumar, K., Vanaja, M., Gnanajobitha, G., & Annadurai, G. (2014). Biosynthesis and antimicrobial activity of semiconductor nanoparticles against oral pathogens. *Bioinorganic Chemistry and Applications*, *347167*. Advance online publication. DOI: 10.1155/2014/347167 PMID: 24860280

Manikandan, R., Justin Raj, C., Nagaraju, G., Velayutham, R., Moulton, S. E., & Puigdollers, J.. (2021). Selenium enriched hybrid metal chalcogenides with enhanced redox kinetics for high-energy density supercapacitors. *Chemical Engineering Journal*, *414*, 128924. DOI: 10.1016/j.cej.2021.128924

Medina-Cruz, D., Tien-Street, W., & Zhang, B.. (2019). Citric juice mediated synthesis of tellurium nanoparticles with antimicrobial and anticancer properties. *Green Chemistry*, *21*(8), 1982–1998. DOI: 10.1039/C8GC03897K PMID: 31156349

Medina-Cruz, D., Vernet-Crua, A., & Mostafavi, E. (2021). *Aloe vera*-mediated Te nanostructures: Highly potent antibacterial agents and moderated anticancer effects. *Nanomaterials (Basel, Switzerland), 11*(2), 514. DOI: 10.3390/nano11020514 PMID: 33670538

Mendhe, A. C. (2023). Spin coating: Easy technique for thin films. In B. R. Sankapal, A. Ennaoui, G. R. B., & L. C. D. (Eds.), *Simple chemical methods for thin film deposition: Synthesis and applications* (pp. 387–424). Springer Nature Singapore. https://doi.org/DOI: 10.1007/978-981-99-0961-2_9

Mendhe, A. C., Kim, M., Lekshmi, S., Jadhav, S. B., Barse, N. S., & Lee, H. (2025). Recent progress in metal chalcogenide-MXene and MOF-derived composites for supercapacitors: Synthesis, challenges, and future solutions. *Progress in Materials Science, 156*, 101558. DOI: 10.1016/j.pmatsci.2025.101558

Metcalf, H. C., Williams, J. E., & Caskta, J. F. (1982). *Modern chemistry*. Holt, Rinehart and Winston.

Moon, J. W., Ivanov, I. N., Joshi, P. C., Armstrong, B. L., Wang, W., Jung, H., Rondinone, A. J., Jellison, G. E.Jr, Meyer, H. M.III, & Jang, G. G.. (2014). Scalable production of microbially mediated zinc sulfide nanoparticles and application to functional thin films. *Acta Biomaterialia, 10*(10), 4474–4483. DOI: 10.1016/j.actbio.2014.06.003 PMID: 24932768

Olawale, F., Ariatti, M., & Singh, M. (2021). Biogenic synthesis of silver-core selenium-shell nanoparticles using *Ocimum tenuiflorum* L.: Response surface methodology-based optimization and biological activity. *Nanomaterials (Basel, Switzerland), 11*, 2516. DOI: 10.3390/nano11102516 PMID: 34684956

Olawale, F., Oladimeji, O., Ariatti, M., & Singh, M.. (2022). Emerging roles of green-synthesized chalcogen and chalcogenide nanoparticles in cancer theranostics. *Journal of Nanotechnology, 2022*. Advance online publication. DOI: 10.1155/2022/6176610

Órdenes-Aenishanslins, N., Anziani-Ostuni, G., Monrás, J. P., Tello, A., Bravo, D., Toro-Ascuy, D., Soto-Rifo, R., Prasad, P. N., & Pérez-Donoso, J. M. (2020). Bacterial synthesis of ternary CdSAg quantum dots through cation exchange: Tuning the composition and properties of biological nanoparticles for bioimaging and photovoltaic applications. *Microorganisms, 8*(5), 631. DOI: 10.3390/microorganisms8050631 PMID: 32349316

Órdenes-Aenishanslins, N., Anziani-Ostuni, G., Quezada, C. P., Espinoza-González, R., Bravo, D., & Pérez-Donoso, J. M. (2019). Biological synthesis of CdS/CdSe core/shell nanoparticles and its application in quantum dot sensitized solar cells. *Frontiers in Microbiology, 10*, 1587. DOI: 10.3389/fmicb.2019.01587 PMID: 31354676

Pacchioni, G. (2022). Sustainable flexible supercapacitors. *Nature Reviews. Materials*, *7*, 844. DOI: 10.1038/s41578-022-00508-y

Pickett, N. L., Foster, D. F., & Cole-Hamilton, D. J. (1996). Synthesis of metal chalcogenide clusters from metal carbonyls and elemental chalcogen. *Journal of Materials Chemistry*, *6*(3), 507–511. DOI: 10.1039/JM9960600507

Priyanka, U., Akshay Gowda, K. M., Elisha, M. G., Surya Teja, B., Nitish, N., & Raj Mohan, B. (2017). Biologically synthesized PbS nanoparticles for the detection of arsenic in water. *International Biodeterioration & Biodegradation*, *119*, 78–86. DOI: 10.1016/j.ibiod.2017.01.026

Qi, P., Zhang, D., Zeng, Y., & Wan, Y. (2016). Biosynthesis of CdS nanoparticles: A fluorescent sensor for sulfate-reducing bacteria detection. *Talanta*, *147*, 142–146. DOI: 10.1016/j.talanta.2015.09.054 PMID: 26592588

Rajeshkumar, S., Ponnanikajamideen, M., Malarkodi, C., Malini, M., & Annadurai, G. (2014). Microbe-mediated synthesis of antimicrobial semiconductor nanoparticles by marine bacteria. *Journal of Nanostructure in Chemistry*, *4*, 96. DOI: 10.1007/s40097-014-0096-7

Ramamurthy, C., Sampath, K. S., & Arunkumar, P.. (2013). Green synthesis and characterization of selenium nanoparticles and its augmented cytotoxicity with doxorubicin on cancer cells. *Bioprocess and Biosystems Engineering*, *36*(8), 1131–1139. DOI: 10.1007/s00449-013-0898-7 PMID: 23446776

Saied, E., Mekky, A. E., Al-Askar, A. A., Hagag, A. F., Elbana, A. A., Ashraf, M., Walid, A., Nour, T., Fawzi, M. M., & Arishi, A. A. (2023). *Aspergillus terreus*-mediated selenium nanoparticles and their antimicrobial and photocatalytic activities. *Crystals*, *13*(4), 450. DOI: 10.3390/cryst13040450

Santhosh, C., Balasubramanian, B., Vino, P., Viji, M., Rejeeth, C., Kannan, S., Ullah, H., Rengasamy, K. R., Daglia, M., & Maruthupandian, A. (2022). Biofabricated selenium nanoparticles mediated from *Goniothalamus wightii* gains biomedical applications and photocatalytic degrading ability. *Journal of King Saud University. Science*, *34*, 102331. DOI: 10.1016/j.jksus.2021.102331

Satpathy, S., Panigrahi, L. L., Samal, P., Sahoo, K. K., & Arakha, M. (2024). Biogenic synthesis of selenium nanoparticles from *Nyctanthes arbor-tristis* L. and evaluation of their antimicrobial, antioxidant and photocatalytic efficacy. *Heliyon*, *10*, e32499. DOI: 10.1016/j.heliyon.2024.e32499 PMID: 39183842

Shahshahanipour, M., Rezaei, B., Ensafi, A. A., & Etemadifar, Z. (2019). An ancient plant for the synthesis of a novel carbon dot and its applications as an antibacterial agent and probe for sensing of an anti-cancer drug. *Materials Science and Engineering C, 98*, 826–833. DOI: 10.1016/j.msec.2019.01.041 PMID: 30813088

Shaikh, S. A. L., Mendhe, A. C., Nadimetla, D. N., Biradar, M. R., Vijayanand, P., & Puyad, A. L.. (2024). Benzothiazole functionalized diketopyrrolopyrrole photosensitizer for CdS nanowire based DSSC applications. *Journal of Photochemistry and Photobiology A Chemistry, 447*, 115220. DOI: 10.1016/j.jphotochem.2023.115220

Shivaji, K., Mani, S., & Ponmurugan, P. (2018). Green-synthesis derived CdS quantum dots using tea leaf extract: Antimicrobial, bioimaging, and therapeutic applications in lung cancer cells. *ACS Applied Nano Materials, 1*(4), 1683–1693. DOI: 10.1021/acsanm.8b00220

Shivashankarappa, A., & Sanjay, K. R. (2020). *Escherichia coli*-based synthesis of cadmium sulfide nanoparticles, characterization, antimicrobial and cytotoxicity studies. *Brazilian Journal of Microbiology : Publication of the Brazilian Society for Microbiology, 51*(3), 939–948. DOI: 10.1007/s42770-020-00237-6 PMID: 32067210

Smart, L., & Moore, E. (1996). *Solid state chemistry* (2nd ed.). Chapman and Hall.

Spangler, L. C., Lu, L., Kiely, C. J., Berger, B. W., & McIntosh, S. (2016). Biomineralization of PbS and PbS–CdS core–shell nanocrystals and their application in quantum dot sensitized solar cells. *Journal of Materials Chemistry. A, Materials for Energy and Sustainability, 4*(16), 6107–6115. DOI: 10.1039/C5TA10650A

Tari, V., Kannan, K., & Vishwakarma, V. (2022). Emerging antiviral technologies. In Nunthavarawong, P., Rangappa, S. M., Siengchin, S., & Thoppil-Mathew, M. (Eds.), *Antimicrobial and antiviral materials: Polymers, metals, ceramics, and applications* (1st ed.). CRC Press., DOI: 10.1201/9781003143093-8

Tari, V. S., & Patil, P. Y. (2017). Challenge of contamination of pesticides for Alphonso in Ratnagiri district, Maharashtra, India. *Research Journal of Chemistry and Environment, 21*(12), 57–67.

Tripathi, R. M., Hameed, P., Rao, R. P., Shrivastava, N., Mittal, J., & Mohapatra, S. (2020). Biosynthesis of highly stable fluorescent selenium nanoparticles and the evaluation of their photocatalytic degradation of dye. *BioNanoScience, 10*(2), 389–396. DOI: 10.1007/s12668-020-00718-0

Turna, T., Solmaz, A., & Baran, A. (2024). Green synthesis study: Adsorption of Congo red dye with selenium nanoparticles obtained from *Prunus armeniaca* L. leaf waste. *ChemistrySelect, 9*(27), e202403106. DOI: 10.1002/slct.202403106

Velayutham, L., Parvathiraja, C., Anitha, D. C., Mahalakshmi, K., Jenila, M., Gupta, J. K., Wabaidur, S. M., Siddiqui, M. R., Aftab, S., & Lai, W.-C. (2023). Antibacterial and photocatalytic dye degradation activities of green synthesized NiSe nanoparticles from *Hibiscus rosa-sinensis* leaf extract. *Water (Basel)*, *15*(7), 1380. DOI: 10.3390/w15071380

Wang, G., Zhang, L., & Zhang, J. (2012). A review of electrode materials for electrochemical supercapacitors. *Chemical Society Reviews*, *41*, 797–828. DOI: 10.1039/c1cs15060j PMID: 21779609

Williams, P., Keshavarz-Moore, E., & Dunnill, P. (2002). Efficient production of microbially synthesized cadmium sulphide quantum semiconductor crystallite. *Enzyme and Microbial Technology*, *30*(3), 354–362. DOI: 10.1016/S0141-0229(01)00500-1

Wu, L., Li, Y., Liu, G. Q., & Yu, S. H. (2024). Polytypic metal chalcogenide nanocrystals. *Chemical Society Reviews*. Advance online publication. DOI: 10.1039/d3cs01095c PMID: 39212091

Xia, X., Zhou, Z., Wu, S., Wang, D., Zheng, S., & Wang, G. (2018). Adsorption removal of multiple dyes using biogenic selenium nanoparticles from an *Escherichia coli* strain overexpressed selenite reductase CsrF. *Nanomaterials (Basel, Switzerland)*, *8*(4), 234. DOI: 10.3390/nano8040234 PMID: 29649129

Xiao, M., & Yang, L. (2014). *Binary and ternary metal chalcogenide materials and method of making and using same* (U.S. Patent No. [insert patent number]). U.S. Patent and Trademark Office.

Yalçın, M., Turgut, N., & Gökbulut, C., & Others. (2023). Removal of pesticide residues from apple and tomato cuticle. *Environmental Science and Pollution Research International*, *30*, 15821–15829. DOI: 10.1007/s11356-022-23269-1 PMID: 36171324

Ye, X. R., Jia, D. Z., Yu, J. Q., Xin, X. Q., & Xue, Z. (1999). One-step solid-state reactions at ambient temperatures: A novel approach to nanocrystal synthesis. *Advanced Materials*, *11*(11), 941–942. DOI: 10.1002/(SICI)1521-4095(199908)11:11<941:AID-ADMA941>3.0.CO;2-W

Yue, L., Wang, J., Zhang, Y., Qi, S., & Xin, B. (2016). Controllable biosynthesis of high-purity lead-sulfide (PbS) nanocrystals by regulating the concentration of polyethylene glycol in microbial system. *Bioprocess and Biosystems Engineering*, *39*(12), 1839–1846. DOI: 10.1007/s00449-016-1662-7 PMID: 27541823

Zhang, L. D., & Mou, J. M. (1994). *Nami cailiaoxue* [Nanomaterials]. Liaoning Science and Technology Press.

Zhou, Y. M., & Xin, X. Q. (1999). Synthetic chemistry for solid state reaction at low-heating temperatures. *Wuji Huaxue Xuebao*, •••, 15.

Zhu, J.-J., & Wang, H. (2003). Synthesis of metal chalcogenide nanoparticles. In Nalwa, H. S. (Ed.), *Encyclopedia of nanoscience and nanotechnology* (Vol. 10, pp. 1–21).

Zuo, X., Qiu, Y., Zhen, M., Liu, D., & Zhang, Y. (2025). Review on MXenes-based electrocatalysts for high-energy-density lithium–sulfur batteries. *Nano-Micro Letters*, *17*. Advance online publication. DOI: 10.1007/s40820-025-01726-z PMID: 40208556

Chapter 4
Transition Metal Chalcogenides, Bimetallic Chalcogenides, Perovskites, and Nanocomposites: Advanced Materials for Optoelectronic and Thermoelectric Applications

Hemalatha Parangusan
https://orcid.org/0000-0002-1384-2739
Mannar Thirumalai Naicker College, India

G. Anandha Babu
https://orcid.org/0000-0003-1192-8595
Bannari Amman Institute of Technology, India

K. Karuppa Samy
Khalifa University, UAE

Deepalekshmi Ponnamma
Qatar University, Qatar

Akram Alfantazi
Khalifa University, UAE

DOI: 10.4018/979-8-3373-3962-7.ch004

ABSTRACT

The accelerating advancement of modern technologies underscores an urgent need for the design and development of next-generation materials that exhibit multi-functionality, property tunability, and superior performance across a spectrum of applications. This chapter investigates into the state-of-the-art progress in four essential material systems such as transition metal chalcogenides (TMCs), bimetallic chalcogenides (BMCs), perovskites, and nanocomposites, which have emerged as frontrunners in the realms of optoelectronics and thermoelectrics. Transition metal chalcogenides, offer a versatile platform for applications ranging from photodetection and photovoltaics to thermoelectric energy conversion. The integration of bimetallic species within chalcogenide matrices introduces synergistic enhancements. The incorporation of these materials into nanocomposite architectures further enables tailored interfacial engineering

INTRODUCTION

The search for multipurpose and energy-efficient materials has accelerated in recent decades due to the rapid advancement of modern electronic, optoelectronic, and thermoelectric technology (Bal & Rani, 2025; Nallusamy et al., 2025; Sun et al., 2025; Kannan K. & Tari V. 2025). Extensive research efforts have been sparked by this increasing demand to design and engineer improved functional materials that can seamlessly integrate energy generation, conversion, and storage. Because of their remarkable physical, chemical, electrical, and thermal properties, Transition Metal Chalcogenides (TMCs), Bimetallic Chalcogenides (BMCs), Perovskites, and Nanocomposites have become especially interesting candidates among the many material systems being investigated (Choubey & Yadav, 2024; Jana et al., 2025; Njema & Kibet, 2025).

The typical formula for transition metal chalcogenides is MX_2, where M stands for a transition metal and X for a chalcogen, such as tellurium, sulfur, or selenium. These compounds are distinguished by their high carrier mobilities, tunable bandgaps, strong light–matter interactions, and adaptable layered architectures (Panigrahy et al., 2024). They are perfect for next-generation thermoelectric and optoelectronic applications because of these features. In comparison to their monometallic counterparts, bimetallic chalcogenides, which combine two different metal species, produce higher electrical conductivity, catalytic activity, and structural robustness, further improving functional performance through synergistic effects (Giuffredi et al., 2021). Perovskite materials have transformed the fields of photovoltaics and photodetectors, especially halide perovskites with the general formula ABX_3. Due to their tunable electronic structures and intrinsically low thermal conductivities, they have also become appealing for thermoelectric applications because of their exceptional optoelectronic characteristics, which include high absorption coefficients, long carrier diffusion lengths, and excellent defect tolerance. By utilizing the combined benefits of their component materials, nanocomposites, which are developed by combining various material phases at the nanoscale allow for the fine-tuning of multifunctional features (Tari V. et al. 2022; Wani S. A. et al. 2025; Tari V. et al. 2025) High-performance energy and electronic devices require these hybrid systems because of their improved electrical, optical, and thermal characteristics (Joshi et al., 2023).

A thorough review of these material systems is given in this chapter, with particular attention paid to their electrical and structural properties, synthesis processes, and real-world applications in optoelectronic and thermoelectric technologies. Recent developments, new problems, and the direction of this field's study are highlighted. This chapter seeks to emphasize the critical role that TMCs, BMCs, perovskites, and nanocomposites will play in influencing the development of multifunctional materials and sustainable device engineering in the future by highlighting structure–property correlations and creative application techniques.

Overview of Modern Materials Needs in Optoelectronics and Thermoelectrics

The materials employed in optoelectronic and thermoelectric applications are under extraordinary pressure due to the quick development of wearable electronics, Internet of Things (IoT) devices, energy harvesting systems, and information technology. Materials that combine scalability, cost-effectiveness, and environmental sustainability, compliance with device reducing trends, and excellent performance and long-term reliability are needed in these developing industries (Ali et al., 2023; Choi et al., 2022).

Optoelectronics encompasses technologies that convert electrical signals into light and vice versa, including light-emitting diodes (LEDs), lasers, photodetectors, and solar cells. Optoelectronics has historically been dominated by conventional semiconductors like silicon and III-V compounds (e.g., GaAs, InP). However, the hunt for substitute materials has been operated by their drawbacks, which include restricted tunability, expensive manufacturing costs, and rigid mechanical qualities. Novel approaches to getting around these limitations are provided by emerging contenders such halide perovskites, nanostructured composites, and two-dimensional (2D) transition metal chalcogenides (TMCs) (Y. Liu et al., 2017).

The efficiency of thermoelectric materials, which allow for direct conversion of thermal energy into electrical energy, is measured by the dimensionless figure of merit (Anthony, 2014):

$$ZT = S^2 \sigma T / \kappa$$

where S is the Seebeck coefficient (measuring thermoelectric voltage per temperature gradient), σ is the electrical conductivity, T is the absolute temperature, and κ is the total thermal conductivity (including both electronic and lattice contributions). Conventional thermoelectric materials have been widely employed, including SiGe alloys, PbTe, and $BiTe_3$. However, a move toward more sustainable substitutes, such as bimetallic chalcogenides, layered nanostructures, and organic–inorganic hybrid systems, has been spurred by their reliance on uncommon, costly, or hazardous materials.

Traditional materials, which frequently lack multifunctionality, adaptability, or environmental compliance, make it challenging to achieve these exacting standards. Researchers are increasingly using synthetic materials that offer fine control over electrical, optical, and thermal properties, such as perovskites, transition metal chalcogenides, and nanocomposites, to address these issues. Through doping, heterostructuring, and defect engineering, these materials take advantage of low-dimensional structures, high surface-to-volume ratios, quantum confinement effects, and programmable functions.

TMCs, especially in their two-dimensional versions (e.g., MoS_2, WS_2, SnSe), have a wide range of characteristics that make them ideal for optoelectronic and thermoelectric applications (Alam & Ramakrishna, 2013). Their layered van der Waals structures produce strong anisotropy and allow mechanical exfoliation down to monolayers, resulting in layer-dependent bandgaps and tuneable carrier dynamics. Their structural and electronic tenability via strain, doping, or layer control makes TMCs foundational for developing responsive and adaptable device platforms. BMCs improve the functionality of monometallic TMCs by combining two different metal centers in a single chalcogenide framework. This dual-metal method improves electrical conductivity, heat transfer, and structural stability. BMCs are especially promising for high-efficiency thermoelectric devices and multifunctional optoelectronic platforms, as they provide compositional flexibility and application-specific property tweaking.

Perovskite materials, including hybrid organic-inorganic (e.g., $MAPbI_3$) and purely inorganic compositions (e.g., $CsSnI_3$, $SrTiO_3$), are transforming energy and photonic applications (Giri et al., 2023). Their remarkable optoelectronic properties, combined with facile, low-temperature processing routes, make them attractive for scalable device fabrication. Recent efforts to improve stability, investigate lead-free alternatives, and strengthen environmental resilience have increased the potential of perovskites for sustainable applications. Nanocomposites combine various material phases on the nanoscale to create hybrid systems with optimal interfacial and bulk properties. Nanocomposites can be built to fulfil the multifunctional demands of sophisticated electronic systems by taking use of synergistic effects amongst components such as metals, semiconductors, and dielectrics. Their versatility makes nanocomposites ideal for applications in thermoelectric modules, flexible photovoltaics, sensors, and energy storage devices.

In the context of optoelectronic and thermoelectric technologies, this chapter attempts to give a thorough overview of the design, synthesis, properties, and applications of four major classes of advanced materials: TMCs, BMCs, perovskites, and nanocomposites. By concentrating on the core ideas, current research trends, and potential future directions in the field, this chapter aims to close the knowledge gap between material innovation and functional device development in light of the growing demand for high-performance and energy-efficient materials.

TRANSITION METAL CHALCOGENIDES (TMCS)

Transition Metal Chalcogenides, with the general formula MX_2 (where M is a transition metal such as Mo, W, or Ti, and X is a chalcogen such as S, Se, or Te), are one of the most extensively studied families of layered materials due to their rich physicochemical properties and structural versatility. Their two-dimensional (2D) structure, bound together by weak van der Waals interactions, allows for easy exfoliation into atomically thin layers, making them ideal for use in nanoelectronics and optoelectronic devices. TMCs display a variety of electrical characteristics, including semiconducting, metallic, and even superconducting phases, depending on their composition, number of layers, and crystalline phase (e.g., 2H, 1T, or 1T') (J. H. Han et al., 2018a; B. Liu et al., 2017; Toshniwal & Kheraj, 2017). For instance, monolayer MoS_2 transitions from an indirect to a direct bandgap semiconductor, leading to enhanced photoluminescence, which is of critical importance in light-emitting and photovoltaic applications. Moreover, their high absorption coefficients, tunable band structures, and strong excitonic effects contribute significantly to their utility in applications like field-effect transistors (FETs), photodetectors, and energy harvesting devices (V. P. Kumar et al., 2025).

TMCs have low inherent thermal conductivity because to significant phonon scattering at their multilayer surfaces, which benefits the thermoelectric figure of merit (ZT). Nanostructuring, doping, and defect engineering are further ways for optimizing thermoelectric performance by tailoring the Seebeck coefficient and electrical conductivity. TMCs such as WSe_2 and TiS_2 show high thermoelectric efficiency due to their anisotropic architectures that decouple electrical and thermal transport (P. Kumar et al., 2020). TMCs are ideal building blocks for creating heterostructures and hybrid systems when combined with other 2D materials such as graphene and hexagonal boron nitride, allowing the design of multifunctional interfaces for advanced optoelectronic and thermal management systems. As research progresses, the scalability of synthesis methods (e.g., chemical vapor deposition, liquid-phase exfoliation) and stability under operational conditions remain key challenges, yet the broad tunability and multifunctionality of TMCs continue to position them at the forefront of next-generation material innovation.

In order to satisfy the need for superior, scalable, and structurally adjustable materials for optoelectronic and thermoelectric applications, the synthesis of TMCs has undergone substantial change in recent years. The controlled creation of large-area, high-crystallinity monolayer and few-layer TMC films with adjustable thickness, shape, and phase is made possible using one of the most well-known methods, Chemical Vapor Deposition (CVD) (J. Wang et al., 2017). Precursor composition, substrate temperature, and gas flow rates may all be precisely controlled using CVD, making it compatible with techniques used on an industrial scale. Liquid-phase exfoliation is another popular technique that uses ultrasonic energy to exfoliate bulk TMC crystals into nanosheets after they have been dispersed in appropriate solvents. This approach is more appropriate for composite or ink-based applications because, despite its cost-effectiveness and scalability, it frequently produces heterogeneous layer thickness and defect-rich structures. The first technique for isolating monolayers, mechanical exfoliation (such as the "Scotch tape" method), produces pure single-crystalline flakes but has a low yield and is not very reproducible.

In addition to this, hydrothermal and solvothermal techniques have drawn interest for creating TMCs with a range of shapes, such as nanorods, nanotubes, and nanoflowers, frequently in moderate environments using precursors that are safe for the environment. These solution-based methods are especially appealing for creating nanostructured TMCs for thermoelectric and catalytic applications because they provide precise control over particle size, crystallinity, and phase through adjustments to reaction time, temperature, and surfactants. Although the method has limitations in terms of cost and throughput, molecular beam epitaxy (MBE) offers atomically precise layer-by-layer growth and is perfect for fundamental studies that need for defect-free heterostructures and interface control. More recently, electrochemical and template-assisted synthesis methods have emerged for producing vertically aligned TMCs or integrating them onto flexible substrates. These various synthesis routes influence not only the crystal phase, morphology, and stoichiometry of TMCs, but also their electronic, optical, and thermal properties, making synthesis technique selection and optimization an important step in tailoring material performance for specific device architectures. In the past ten years, the synthesis of 2D TMC nanostructures has been investigated and advanced quickly. It has been spun off to large-scale, inexpensive production and colloidal processing in a variety of substrates with size and composition control (Figure 1) (C. Xie et al., 2020).

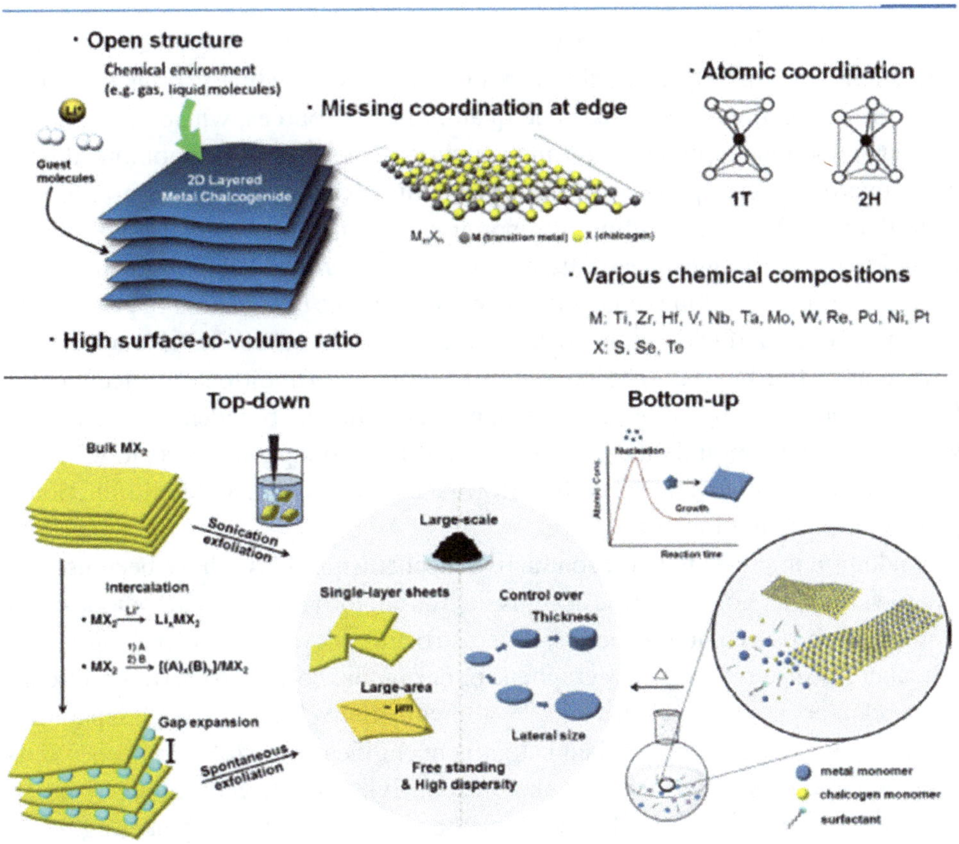

Applications in Optoelectronics

Transition Metal Chalcogenides have received a lot of attention in optoelectronics due to their distinct electronic structures, strong light-matter interactions, and mechanical flexibility, making them promising materials for next-generation, miniaturized, and energy-efficient photonic devices. Semiconducting TMCs like as MoS_2, WS_2, and WSe_2 have direct bandgaps in the visible to near-infrared range (J. H. Han et al., 2018b). This allows for efficient light absorption and emission. This has resulted in the creation of high-performance photodetectors with fast response times and high sensitivity, even at low light intensities. Furthermore, TMC-based field-effect transistors (FETs) with photodetection capabilities have outstanding gate-tunable photoresponse, making them ideal for active imaging arrays and optical logic circuits. Their compatibility with flexible substrates gives up possibilities for wearable optoelectronic sensors, bendable displays, and conformal light emitting devices.

In addition to photodetection, TMCs have shown great promise in light-emitting applications. Their strong excitonic emission and valley-selective circular dichroism allow for the creation of polarized and quantum-light sources, which are essential for future quantum communication and valleytronic devices. When combined with optical microcavities or plasmonic structures, TMCs exhibit increased light emission due to Purcell effects and exciton-plasmon interaction, providing pathways to low-threshold nano-lasers and tuneable light sources. Furthermore, TMC heterostructures formed by vertical stacking or lateral connections enable the creation of p-n diodes, tunnel junctions, and photonic rectifiers with optimized interfacial charge transfer and recombination kinetics. The integration of TMCs with traditional semiconductor platforms and emerging perovskite or organic layers further broadens their utility in hybrid photodetectors and solar cells. These multifaceted capabilities position TMCs as key materials for advancing high-performance, low-power, and multifunctional optoelectronic technologies.

In addition to intrinsic photoconductive mechanisms, TMCs have been used to create a variety of photodetection designs such as phototransistors, photodiodes, and heterojunction-based photodetectors. Heterostructures developed by stacking TMCs with other 2D materials (such as graphene or hexagonal boron nitride) or traditional semiconductors improve performance by allowing more effective charge separation and decreasing dark current via band alignment engineering. MoS_2/graphene vertical heterostructures have enhanced photoresponsivity and detectivity due to quick interlayer charge transfer. Furthermore, methods such as defect engineering, plasmonic enhancement, and quantum dot sensitization have been used to broaden the spectrum range (to UV and IR), boost gain, and improve signal-to-noise ratio. The scalability of solution-processed or CVD-grown TMCs allows for their integration into large-area and flexible photodetector arrays, which are useful in imaging, environmental monitoring, biomedical diagnostics, and optical communication systems. TMC-based photodetectors are predicted to play a critical role in the evolution of ultrathin, high-performance optoelectronic platforms as research advances material quality, interface engineering, and device architectural optimization.

TMCs have demonstrated significant potential in photovoltaic (PV) applications due to their tunable bandgaps, high absorption coefficients, and mechanical flexibility, making them ideal candidates for lightweight, thin-film, and next-generation solar energy conversion technologies. Semiconducting TMCs like MoS_2, WS_2, and $MoSe_2$ have direct bandgaps (~1.1-2.0 eV) in the monolayer limit (Ameer et al.,

2024; Cui et al., 2018). This aligns with the sun spectrum and allows for efficient photon absorption even at thicknesses below 10 nm. This high absorption per unit thickness, along with advantageous carrier mobility and strong exciton binding energies, allows for the formation of tightly bound excitons and effective charge separation under sunlight. Their inherent chemical stability and compatibility with flexible substrates further allow TMC-based photovoltaics to be integrated into wearable electronics and portable solar harvesting systems, expanding their applicability beyond rigid silicon-based platforms.

Significant progress has been achieved in the design of TMC-based photovoltaic devices thanks to the development of heterojunction topologies, particularly van der Waals heterostructures that mix TMCs with other 2D semiconductors, perovskites, and organic materials. These topologies encourage type II band alignment, which improves charge separation while reducing recombination losses at the interface. MoS_2/ WS_2 or MoS_2/perovskite heterojunctions have higher power conversion efficiencies (PCEs) and open-circuit voltages (X. Fang et al., 2018). Furthermore, transparent electrodes such as graphene or ITO, as well as plasmonic structures, can help to increase light absorption and charge extraction. Although the current PCEs of TMC-based photovoltaics are modest when compared to traditional silicon or perovskite solar cells, further research into doping, defect passivation, exciton dissociation strategies, and scalable fabrication techniques is expected to significantly improve performance. With a rising emphasis on flexible, ultrathin, and environmentally friendly solar technologies, TMCs are well positioned to contribute significantly to the development of next-generation photovoltaic systems.

TMCs, particularly in monolayer form, are gaining popularity as active materials for light-emitting devices (LEDs) due to their direct bandgap nature, high photoluminescence quantum yield, and strong excitonic properties (Rahman et al., 2024). Monolayer TMCs like MoS_2, WS_2, and WSe_2 have a direct bandgap near the K-point of the Brillouin zone, which allows for effective radiative recombination of excitons and trions. This intrinsic property enables intense electroluminescence (EL) under electrical activation, making them ideal for ultrathin and flexible LEDs (Gao et al., 2023). In addition to single-layer TMC emitters, heterostructure-based designs have been used to improve light emission efficiency and device performance. Vertical TMC heterostructures or TMC-graphene-TMC sandwich structures, for example, facilitate carrier injection and confinement in the emissive layer, leading to increased EL intensities. Furthermore, merging TMCs with photonic cavities or plasmonic nanostructures might increase spontaneous emission via Purcell enhancement, allowing for low-threshold nanolasers and modulated light sources. The valley-dependent circular dichroism found in certain TMCs, caused by spin-orbit coupling and broken inversion symmetry, opens up the prospect of producing circularly polarized or valley-polarized light, which is required for next-generation valleytronic and quantum photonic devices. Although challenges such as contact resistance, exciton-exciton annihilation, and limited quantum efficiency under electrical injection remain, ongoing progress in interface engineering, dielectric encapsulation, and charge balance optimization is accelerating the development of TMC-based light-emitting technologies.

Thermoelectric Properties and Energy Conversion Performance

Transition Metal Chalcogenides are rapidly being studied for thermoelectric (TE) applications due to their advantageous electrical structures, low intrinsic thermal conductivity, and the possibility of improving energy conversion efficiency by dimensional and defect engineering (Mak et al., 2010). The figure of merit for thermoelectric performance is $ZT = S^2\sigma T/\kappa$. Here, S is the Seebeck coefficient, σ is electrical conductivity, T is absolute temperature, and κ is total thermal conductivity. Layered TMCs like TiS_2, $MoSe_2$, and WSe_2 have poor lattice thermal conductivity due to substantial phonon scattering at interlayer van der Waals gaps (G. Zhang & Zhang, 2017; Gurushankar et al. 2024). This suppresses heat flow while maintaining charge transport. Furthermore, their electronic band structures, which are frequently characterized by strong valley degeneracy and thin conduction/valence bands, allow for large Seebeck coefficients and good electrical conductivities, particularly under optimal doping or strain circumstances. These characteristics make TMCs promising candidates for flexible and low-dimensional thermoelectric generators (TEGs) and Peltier cooling devices.

Dimensional reduction to 2D or nanostructured forms improves TMC thermoelectric performance by adding quantum confinement effects and boosting phonon boundary scattering, resulting in decreased heat conductivity while maintaining or even improving electrical conductivity (Al-Fartoos et al., 2023; Novak et al., 2019; C. Wang et al., 2022). Recent research reveal that doping, defect engineering, and heterostructuring with other 2D materials (such as graphene or boron nitride) can considerably control the carrier concentration and mobility, maximizing the power factor ($S^2\sigma$) of TMC-based thermoelectrics. Nanostructured TiS_2 shows higher ZT values (>1) at high temperatures due to decoupling of thermal and electrical transport channels. Furthermore, scalable synthesis methods such as CVD and solvothermal approaches enable the controlled production of flexible TMC thin films, which are ideal for wearable thermoelectric systems that harvest body heat. As the demand for self-powered and sustainable electronics develops, the study of TMCs as high-performance, low-dimensional thermoelectric materials presents a possible avenue toward efficient energy harvesting and solid-state cooling.

BIMETALLIC CHALCOGENIDES (BMCS)

Bimetallic chalcogenides are a special class of inorganic compounds in which two different metal cations are coordinated with chalcogen anions (S, Se, or Te) to create complex structures that provide tunable physicochemical properties for a variety of applications. In contrast to their monometallic counterparts, BMCs have enhanced electronic, optical, and catalytic behavior because of the synergistic effects between the two metal centers, which result in a wider range of oxidation states, defect chemistry, and coordination environments. These characteristics are especially useful for customizing charge transport, band alignment, and active site density in optoelectronic and energy-related devices. Materials like $NiCo_2S_4$, $CuFeS_2$, and ZnCoS are often studied because of their high conductivity, structural stability, and multifunctionality (X. Wang et al., 2024; Zheng et al., 2018). They are good prospects for applications ranging from photodetectors and supercapacitors to thermoelectrics and catalysis because of their enhanced charge separation, wider light absorption range, and tunable bandgap, which are made possible by metal substitution and stoichiometric modulation. Bimetallic chalcogenides (BMCs) are a new type of chalcogenide materials that contain chalcogen anions (S, Se, or Te) and two different metal cations in the same crystal lattice. When M and M′ are transition metals, the general formula $M_{1-x}M'_xX$ permits considerable compositional tweaking, allowing the engineering of properties that beyond the constraints of monometallic chalcogenides. Rich electrical, magnetic, and catalytic activity results from their structural complexity, which includes mixed valence states, polymorphic phases, and adjustable metal-to-chalcogen ratios (Shirvani et al., 2023; Sokolikova & Mattevi, 2020; Y. Zhang et al., 2019). BMCs are becoming more and more popular because of their many uses, especially in energy conversion, thermoelectrics, and optoelectronics.

The simultaneous integration of two distinct metal species into a single chalcogenide framework results in the high degree of structural complexity exhibited by BMCs. The material's crystal structure, bonding environment, and electronic states can all be customized to a greater extent because of this dual-metal arrangement. Depending on the constituent metals' ionic radii, valence states, and preferred coordination, BMCs can crystallize in a variety of structural motifs, including pyrite, spinel, and layered frameworks. Lattice distortions, site occupancies, and defect concentrations can be fine-tuned through the interaction of these parameters, which in turn affects the phonon scattering mechanisms and charge carrier dynamics that are essential for optoelectronic and thermoelectric performance. The ability to modulate both the local atomic arrangement and long-range crystallinity makes BMCs highly versatile platforms for exploring structure–property relationships in functional materials.

Advanced methods including as CVD and atomic layer deposition (ALD) are increasingly being used to produce highly crystalline BMC thin films with atomic-level control over thickness and stoichiometry, which is required for integrating these materials into device-grade architectures. In CVD techniques, volatile metal-organic precursors breakdown in the presence of a chalcogen source, allowing for the formation of highly uniform layered or epitaxial films. Post-synthetic modifications, including thermal annealing under controlled atmospheres or ion exchange treatments, are often employed to further tailor phase composition and defect structures. Furthermore, solid-state synthesis, which involves the direct interaction of metal oxides, sulfides, or chlorides with elemental chalcogens at high temperatures, remains a reliable method for producing bulk BMCs. These various synthetic techniques not only provide for compositional flexibility and morphological control, but also allow for fine-tuning of structural and electrical properties that are critical for the optimization of BMCs in high-performance optoelectronic and thermoelectric applications.

Optoelectronic Device Potential

Bimetallic chalcogenides have enormous potential in the field of optoelectronic devices due to their customized electronic structures, programmable optical band-gaps, and better charge transport capabilities (Bartlett et al., 2025). The integration of two separate metal species allows for fine control over state density and Fermi level placement, resulting in efficient absorption of a wide spectrum of light wavelengths, from ultraviolet to near-infrared. This makes BMCs ideal for use in photodetectors, phototransistors, and solar energy harvesting systems. Furthermore, their high absorption coefficients and long carrier lifetimes promote exciton production and separation, resulting in improved photoresponsivity and quantum efficiency. When compared to monometallic chalcogenides, BMCs have improved light-matter interactions and lower recombination losses, making them attractive candidates for ultrafast and ultrasensitive optoelectronic devices (Priyadarshini et al., 2022).

In addition to their excellent optical and electronic properties, BMCs have structural versatility that allows for integration into a variety of device topologies, such as planar heterojunctions, flexible substrates, and multi-layered photonic structures (Patil et al., 2023). Their compatibility with scalable synthesis processes such as solution processing, chemical vapor deposition, and template-assisted fabrication enhances their economic viability. BMC-based heterostructures and nanocomposites can also be designed to have many functions, such as sensing, emission, and energy conversion, all on the same platform. These characteristics, combined with their environmental stability and robustness under continuous illumination, make BMCs promising candidates for next-generation optoelectronic devices such as wearable photonic sensors, high-resolution imaging arrays, high-performance light-emitting diodes (LEDs), and photonic energy converters (Tian et al., 2026).

Thermoelectric Enhancement Through Band Engineering and Defect Modulation

Strategic band engineering and defect modulation, two potent methods for altering charge carrier dynamics and phonon scattering, can dramatically improve thermoelectric performance in BMCs (H. Han et al., 2024; Hua & Shen, 2024; Song et al., 2025). Band engineering, which involves alloying or modifying the stoichiometry of two metal constituents, allows for fine-tuning of the electronic band structure to improve the Seebeck coefficient and carrier mobility. The formation of highly dispersive conduction or valence bands promotes electrical conductivity, whereas a large density of states around the Fermi level provides a strong thermoelectricity. Convergence of multiple electronic bands and resonant levels can enhance the power factor ($S^2\sigma$), a crucial characteristic in the thermoelectric figure of merit (ZT). These methodologies allow BMCs to outperform standard monometallic thermoelectric materials.

Defect modulation in BMCs also plays an important role in improving thermoelectric performance by lowering lattice thermal conductivity while preserving good electrical characteristics. Point defects, dislocations, and nano-scale grain boundaries produced during synthesis or post-treatment act as phonon scattering sites, effectively restricting heat transport but not altering carrier mobility (Ming et al., 2025; Zhu et al., 2021). Controlled defect introduction can also change carrier concentration and mobility through donor or acceptor states, allowing for a precise balance of electrical conductivity and thermal insulation. Furthermore, interface engineering in BMC-based nanocomposites, such as the use of coherent heterointerfaces, can help to localize phonons and reduce heat conductivity. When combined with band engineering, defect modulation causes a decoupling of the interdependent thermoelectric characteristics, hence increasing the overall ZT value. These two methodologies establish BMCs as advanced thermoelectric materials capable of efficiently harvesting energy from waste heat in micro and macroscale systems. Yamini et al., (Aminorroaya Yamini et al., 2023) investigated various Bi_2Te_3-Bi_2Se_3-Bi_2S_3 system and demonstrated a maximum ZT of 0.8 at 300 °C for $Bi_2Te_2S_1$ and at 500 °C $Bi_2Se_1S_2$ (Figure 2). These results show that these materials are suitable for medium-temperature power generation.

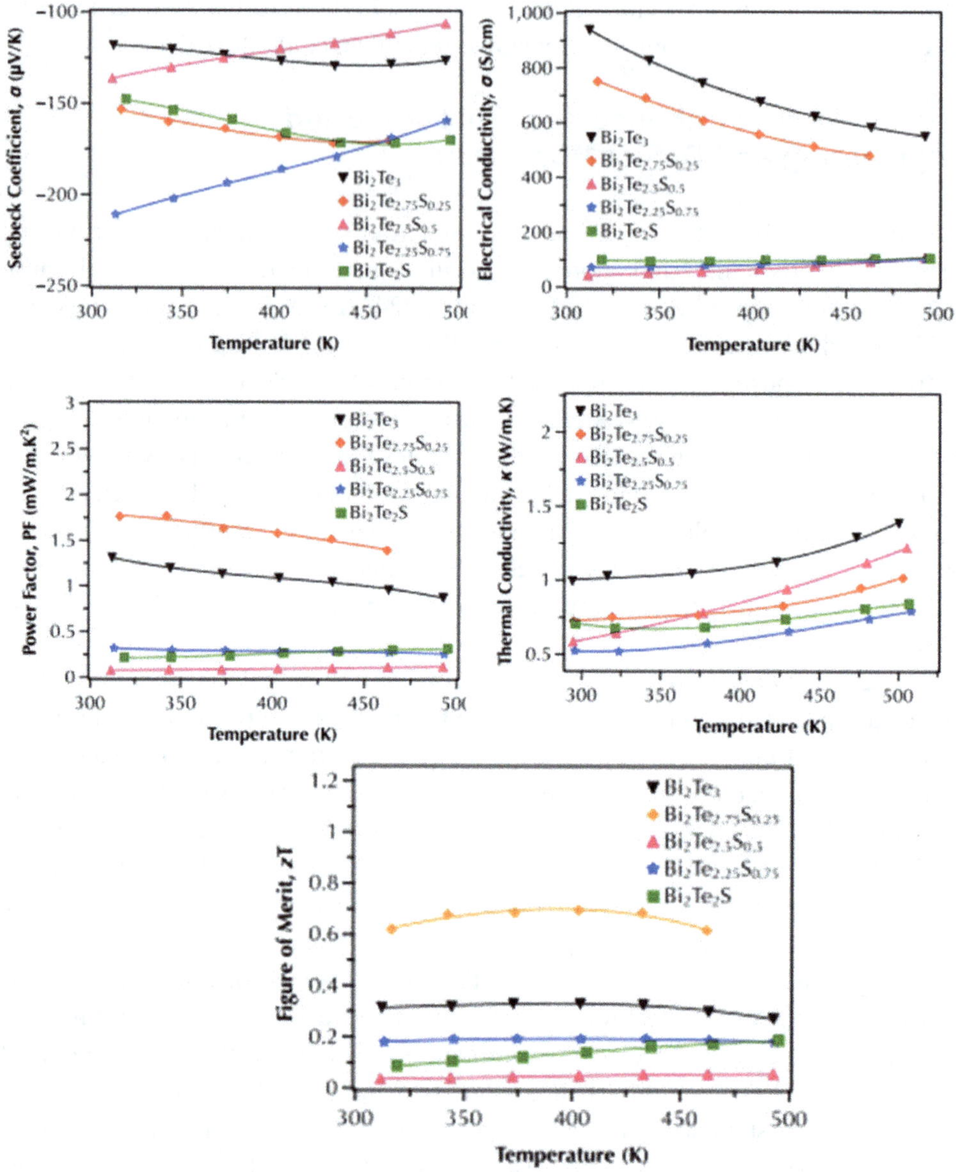

METAL HALIDE PEROVSKITES

Metal halide perovskites (MHPs), represented by the formula ABX_3, where A is a monovalent cation (e.g., methylammonium, formamidinium, or cesium), B is a divalent metal (e.g., Pb^{2+} or Sn^{2+}), and X is a halide anion (Cl^-, Br^-, or I^-), have emerged as revolutionary materials in optoelectronics due to their exceptional electronic and optical properties (Hu et al., 2021). Their unusual crystal structure enables for high absorption coefficients, long carrier diffusion lengths, and low trap-state densities, all of which contribute to excellent photovoltaic and photodetector performance. Their bandgap can be tuned through compositional engineering, such as halide mixing or substitution at the A and B sites, allowing for a broad spectrum response from visible to near-infrared. This compositional versatility has helped to catapult MHPs into the spotlight for a wide range of optoelectronic applications, outperforming many traditional semiconductors in terms of efficiency and ease of manufacture (F. Liu et al., 2022).

Beyond photovoltaics, MHPs have shown great potential in light-emitting devices, X-ray detectors, and lasers, demonstrating their versatility. MHPs' solution-processability enables low-cost, large-area deposition on both flexible and rigid substrates, making them ideal for next-generation devices. Furthermore, their high defect tolerance allows them to perform even when structural defects exist, giving them a significant edge over standard inorganic semiconductors. However, the intrinsic instability of organic-inorganic MHPs, particularly under moisture, heat, and illumination, remains a significant concern, prompting researchers to pursue purely inorganic or hybrid perovskite derivatives. Current approaches to improving their stability include the use of mixed-cation/mixed-halide compositions, 2D/3D hybrid structures, and surface passivation methods (H. Xie et al., 2020). As research into these materials progresses, MHPs are positioned not only as high-efficiency energy conversion materials, but also as versatile platforms for integrated optoelectronic and thermoelectric technologies.

The optoelectronic characteristics and device performance of metal halide perovskites (MHPs) are greatly influenced by the wide variety of crystal forms they display. Methylammonium lead iodide ($MAPbI_3$) is a prime example of a classical three-dimensional (3D) perovskite structure that uses a corner-sharing octahedral framework. The A-site cation occupies the spaces between the B-site cation (Pb^{2+} or Sn^{2+}), which is encircled by six halide anions (X^-) to form BX_6 octahedra. High charge carrier mobility and extended diffusion distances are made possible by this 3D connection, which is essential for photovoltaic and photodetector applications. On the other hand, two-dimensional (2D) perovskites are made up of perovskite layers that are separated by large organic spacer cations. They are commonly represented by the Ruddlesden–Popper (RP) phase, which has the general formula $(RNH_3)_2A_{-1}BX_{3+1}$. Although occasionally at the sacrifice of charge transport efficiency, these architectures improve environmental stability and exciton binding energy by confining charge carriers within inorganic layers.

The structural and functional differences between their 2D and 3D counterparts are filled by quasi-2D perovskites, also known as layered perovskite hybrids. They provide a configurable trade-off between stability and charge transport by combining many inorganic layers (with $n > 1$) in between organic spacers. These materials behave like quantum wells, where precise control over optical bandgaps and excitonic effects is possible by adjusting the layer thickness (n). For light-emitting applications, where greater radiative recombination is desired, this structural tunability is very beneficial. Furthermore, quasi-2D perovskites frequently exhibit enhanced morphology, better film formation, and resistance to moisture infiltration—all of which are essential for long-term device performance (Lai et al., 2022). A potent tool for optimizing perovskite materials for a variety of optoelectronic and thermoelectric applications is the ability to design dimensionality, from bulk 3D to layered 2D and intermediate quasi-2D structures (Chen et al., 2022).

The compositional tunability at each site allows for precise control over essential properties like bandgap energy, lattice stability, and defect tolerance. $MAPbI_3$ and $FAPbI_3$ are well-studied lead halide perovskites with optimum bandgaps (~1.5 eV) and long carrier diffusion lengths, making them suitable for high-efficiency solar cells and photodetectors (Ju et al., 2025). Partial substitution at the A-site (e.g., adding Cs^+ or mixing MA^+ and FA^+) and halide-site engineering (e.g., I^-/Br^- alloying) improve thermal and phase stability, allowing for wider application in ambient settings.

Several synthesis methods have been established to produce high-quality lead-based perovskite thin films, including solution processing, vapor-assisted deposition, and thermal evaporation. Solution-based approaches, such as spin-coating and anti-solvent engineering, are the most often used due to their low cost and suitability with scalable manufacturing. These methods produce polycrystalline films with precise grain size and orientation, which are critical for reducing charge recombination and boosting device performance. One of the most striking features of lead-based perovskites is their defect tolerance; shallow trap states enable good charge transport even in materials with non-ideal crystallinity. Additionally, the high absorption coefficient, broad light-harvesting range, and ease of fabrication position lead-based perovskites as highly competitive materials for applications not only in photovoltaics but also in LEDs, lasers, and radiation detectors.

Applications in Optoelectronic Devices

Metal halide perovskites have transformed the field of optoelectronics due to their unique mix of electrical, optical, and structural characteristics. Their direct tunable bandgaps, high absorption coefficients, long carrier diffusion lengths, and solution-processability make them great candidates for a variety of applications, including solar cells, light-emitting diodes (LEDs), photodetectors, and lasers. Perovskite solar cells (PSCs) have attained verified power conversion efficiencies (PCEs) of more than 26%, rivaling classic silicon-based technology (Machín & Márquez, 2024). The capacity to construct bandgaps through compositional tuning has permitted the creation of tandem systems, particularly when combined with silicon or CIGS bottom cells, which pushes efficiency even further. Furthermore, their outstanding low-light performance and compatibility with flexible substrates provide new prospects for wearable and indoor photovoltaic applications.

Perovskites have showed promise in light-emitting applications as well as photovoltaics due to their high photoluminescence quantum yields and narrow emission bandwidths. Perovskite-based LEDs (PeLEDs) have excellent color purity and tunability over the visible spectrum, making them ideal for display and lighting applications (Li et al., 2019). Their effective charge injection and balanced carrier transport enhance electroluminescence efficiency. Perovskite photodetectors have exceptional responsivity and detectivity, particularly in the visible and near-infrared bands, thanks to their high light absorption and low dark currents. Ongoing advances in device architecture, surface passivation, and interface engineering improve the performance, operational stability, and integration potential of perovskite optoelectronic devices used in consumer electronics, telecommunications, and sensing technologies (T.-H. Han et al., 2022).

Perovskite solar cells (PSCs) have emerged as a disruptive technology in the photovoltaic (PV) market, owing to the exceptional optoelectronic features of metal halide perovskites and their ease of manufacture (Kamaraki et al., 2021). The ABX_3 crystal structure, where 'A' is a monovalent cation (e.g., methylammonium, formamidinium, or cesium), 'B' is commonly Pb^{2+}, and 'X' is a halide (Cl^-, Br^-, or I^-), allows for tunable bandgaps, strong light absorption, and high defect tolerance. This structural flexibility enables precise control of the material's energy levels and optical responsiveness. As a result, after a decade of development, PSCs have rapidly achieved certified power conversion efficiencies (PCEs) of more than 26%, establishing them as strong competitors to traditional silicon-based technologies. Their solution-processable nature also supports low-cost, scalable production methods like spin-coating, slot-die coating, and inkjet printing, which are conducive to roll-to-roll manufacturing on flexible substrates (Rong et al., 2018).

Aside from high efficiency, perovskite solar cells provide distinct advantages for tandem and semi-transparent solar systems. Tandem designs, particularly with silicon or copper indium gallium selenide (CIGS), take advantage of perovskite bandgaps' considerable tunability to capture a greater range of the solar spectrum while exceeding the Shockley-Queisser limit of single-junction devices. Perovskite solar cells (PSCs) have witnessed a meteoric rise in performance, with PCEs surpassing 26%.

Innovations in interface engineering, compositional tuning (mixed cations/halides), and passivation strategies continue to drive efficiency and stability enhancements.

Metal halide perovskites have sparked widespread attention for their possible use in light-emitting diodes (LEDs) due to their excellent optoelectronic properties, including high photoluminescence quantum yield (PLQY), tunable bandgaps, and narrow emission line widths (Lozano, 2018). Perovskites have strong excitonic effects and effective radiative recombination processes, allowing for brilliant emission over the visible spectrum, from blue to near-infrared. Structural and compositional engineering, notably the employment of mixed halides and organic-inorganic hybrid frameworks, has improved color adjustment and emission stability. Recent improvements have resulted in external quantum efficiencies (EQEs) more than 20% in green and red perovskite LEDs (PeLEDs), establishing them as promising candidates for next-generation display and lighting technologies. Moreover, low-temperature, solution-based processing techniques make them compatible with flexible substrates, enabling new form factors such as wearable and foldable displays. Metal halide perovskites show high photoluminescence quantum yields (PLQY) and tunable emission (via halide composition), making them promising for light-emitting diodes (PeLEDs) in visible and near-infrared regions (X.-K. Liu et al., 2021).

Photodetectors based on metal halide perovskites have emerged as attractive options for high-performance optoelectronic sensing applications due to their exceptional optoelectronic features, including high absorption coefficients, variable bandgaps, long carrier lifetimes, and low trap densities. These properties allow perovskite photodetectors to have high responsivity, specific detectivity, and quick reaction times throughout a wide spectral range, from ultraviolet (UV) to near-infrared (NIR). Their solution-processable nature allows for the cost-effective fabrication of flexible, large-area, and even transparent photodetector architectures. Photoconductors and photodiodes employing perovskites have obtained responsivities above 10^3 A/W and detectivities in the range of 10^{12}-10^{14} Jones, which rival or beat traditional silicon- or InGaAs-based detectors in some regimes.

As perovskite photodetectors evolve, they are poised to serve in a wide range of applications such as biomedical imaging, environmental monitoring, optical communication, and night vision, offering a low-cost production, tunable spectral response, and high sensitivity that many traditional photodetector technologies cannot match. Owing to their high absorption, low dark currents, and fast response times, perovskite-based photodetectors are effective in detecting weak light signals across a broad spectral range, including UV and near-IR. Adhikari et al., (Adhikari et al., 2025) developed Sn-based chalcogenide perovskites for stable and high performance photovoltaic. To verify the photovoltaic potential of the investigated systems, they have utilized solar cell capacitance simulator in 1 dimension. When the absorber layer thickness is changed to assess PSC performance, it is discovered that the thickness of the pervoskite layer significantly affects PSC performance (Figure 3).

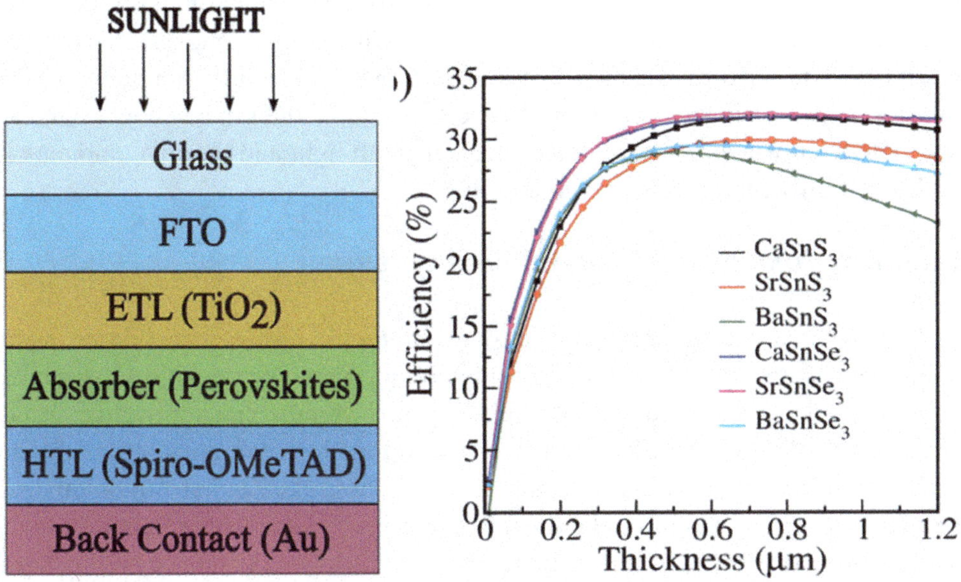

Thermoelectric Potential and Hybrid Integration

Metal halide perovskites, while initially studied for photovoltaic and optoelectronic applications, have recently received significant attention for their thermoelectric potential due to their intrinsically low thermal conductivity, moderate Seebeck coefficients, and tunable carrier concentration. Their soft ionic lattice, abundant phonon scattering, and intrinsic defect tolerance make them good candidates for reducing lattice thermal conductivity while preserving charge carrier transit. Recent studies on hybrid organic-inorganic perovskites (e.g., $CH_3NH_3PbI_3$) have revealed promising thermoelectric figures of merit (ZT values). Although modest compared to commercial thermoelectrics, these materials may become viable alternatives with targeted compositional and structural engineering (Lu et al., 2019). Strategies like as doping, band structure manipulation, and dimensionality reduction have been used to improve electrical conductivity and optimize the Seebeck coefficient while preserving ultralow thermal conductivity.

Hybrid integration of perovskites with other thermoelectric materials or functional nanostructures is a new approach that combines the beneficial features of many materials to improve overall energy conversion efficiency (Xu et al., 2018). For example, perovskite-graphene hybrids employ graphene's high carrier mobility to improve electrical transport, whereas perovskite-metal sulfide or oxide composites can be used to build heterojunctions that facilitate energy filtering effects, hence enhancing power factor. Additionally, adding low-dimensional nanostructures or interface engineering into perovskite matrices allows for phonon scattering and thermoelectric performance modification. Such hybrid structures not only improve thermoelectric performance, but also pave the way for integrated energy harvesting systems in which perovskites may harness both light and thermal gradients in multifunctional energy conversion devices.

Stability, Toxicity, and Sustainability Issues

Despite their outstanding optoelectronic and thermoelectric capabilities, metal halide perovskites (MHPs) confront considerable problems in terms of environmental stability, toxicity, and sustainability, limiting their commercial application (Znidi et al., 2024). MHPs' stability is particularly subject to environmental conditions such as moisture, oxygen, UV light, and temperature stress. Methylammonium lead iodide ($MAPbI_3$), a widely studied perovskite, degrades quickly in humid or high-temperature circumstances, resulting in phase breakdown and device performance loss. Intrinsic structural instabilities caused by volatile organic cations and soft ionic lattices further worsen degradation. Efforts to improve stability have included compositional engineering (e.g., replacing MA^+ with formamidinium or cesium), surface passivation, encapsulation techniques, and developing all-inorganic perovskites with greater structural robustness under ambient circumstances.

To overcome these issues, researchers are currently working on green synthesis pathways, eco-friendly solvents, scalable fabrication procedures, and end-of-life management strategies. Life-cycle assessments (LCA) are also being employed to evaluate the environmental footprint of perovskite devices, promoting the design of next-generation materials that align with circular economy principles and global sustainability goals.

NANOCOMPOSITES: HYBRID MATERIALS FOR MULTIFUNCTIONALITY

A flexible class of hybrid materials, nanocomposites combine nanoscale elements—such as nanoparticles, nanorods, or nanolayers—into matrices of various physical or chemical compositions to produce synergistic qualities that are not possible with separate constituents. Nanocomposites can demonstrate multifunctionality, including simultaneous mechanical reinforcement, improved electrical performance, increased thermal conductivity, and chemical durability, thanks to their structural and compositional tunability (Ali & Andriyana, 2020; Hassan et al., 2021). The integration of low-dimensional nanofillers such as carbon nanotubes (CNTs), graphene, or transition metal dichalcogenides (TMDs) into polymeric or ceramic hosts has resulted in notable improvements in charge transport, light absorption, and phonon scattering control for optoelectronic and thermoelectric applications (C. Fang et al., 2024; X. Wang et al., 2016). By fine-tuning the filler-matrix interface and optimizing dispersion, researchers have been able to enhance interfacial charge transfer and reduce recombination losses, which is particularly advantageous in photovoltaic, light-emitting, and thermoelectric devices.

Additionally, nanocomposites make it easier to construct hierarchical structures that may customize energy band topologies and carrier dynamics by utilizing quantum confinement phenomena, percolation networks, and defect modulation. For example, adding nanoinclusions or creating nanostructured heterointerfaces can improve the thermoelectric figure of merit (ZT) in thermoelectric systems by successfully separating thermal and electrical transmission. By enhancing light-matter interactions through localized surface plasmon resonance (LSPR), plasmonic nanoparticles placed in dielectric matrices can improve photodetection sensitivity or light extraction in LEDs. The adaptability of nanocomposite architectures also allows for the integration of multiple functionalities—such as sensing, actuation, and energy harvesting—into a single platform, making them highly attractive for next-generation wearable electronics, integrated photonics, and multifunctional energy systems.

Perovskites, Transition Metal Chalcogenides (TMCs), and Bimetallic Chalcogenides (BMCs) can be strategically included into nanocomposite systems to create sophisticated functional materials with improved and customized features. The synergistic catalytic and electrical capabilities of BMCs, the strong light-harvesting and optoelectronic features of perovskites, and the high carrier mobility and layer-dependent band structure of TMCs are just a few examples of how this hybrid technique leverages the distinct benefits of each component. These materials can interact cooperatively when embedded in a nanocomposite matrix, improving thermal or chemical stability, reducing recombination losses, and improving charge separation. For instance, adding 2D MoS_2 or WS_2 flakes to lead halide perovskites can increase photovoltaic or photodetection performance by facilitating interfacial charge transfer and trap state passivation (C. Fang et al., 2019; Mirershadi et al., 2023). Advanced materials such as perovskites, TMCs, and BMCs can be incorporated into nanocomposites to customize their characteristics throughout a broad range of functionalities.

Furthermore, TMCs, BMCs, and perovskites are good candidates for the production of scalable nanocomposite materials because they can be processed using a variety of solution-processable and low-temperature synthetic methods. Exact control over morphology and interfacial architecture is made possible by the inclusion, which can be accomplished by techniques including co-precipitation, solvothermal growth, in situ crystallization, or layer-by-layer assembly. Additionally, this integration makes it possible to modify physicochemical characteristics like thermal conductivity, carrier transport paths, and band alignment—all of which are essential for a variety of multipurpose applications, including flexible electronics, thermoelectric devices, and solar energy harvesting. Significantly, these nanocomposites' multicomponent nature offers the ability to modify composition ratios, surface chemistry, and structural arrangement to fine-tune properties. This opens the door for the logical development of next-generation hybrid materials for high-performance energy and electronic applications.

Applications in Next-Generation Thermoelectric and Optoelectronic Devices

Nanocomposites containing transition metal chalcogenides, bimetallic chalcogenides, and metal halide perovskites have emerged as intriguing candidates for next-generation thermoelectric and optoelectronic devices due to their customizable physical properties and multifunctionality (Munonde & Nomngongo, 2023).

Nanocomposites made of TMCs, BMCs, and perovskites provide great prospects for creating high-performance light-harvesting and light-emitting devices. The synergy of components allows for broadband light absorption, effective exciton separation, and reduced recombination losses—all of which are key criteria in improving photovoltaic power conversion efficiency and photodetect sensitivity. For example, TMC-perovskite heterojunctions have type-II band alignment, allowing for rapid charge transfer in high-efficiency solar cells. Similarly, using BMCs with plasmonic or catalytic capabilities might enhance charge extraction and device stability in light-emitting diodes (LEDs) and lasers. Furthermore, the solution-processable nature of these hybrid systems allows for scalable production via printing or coating processes, paving the door for low-cost, flexible optoelectronics in displays, imaging, and wearable devices. The incorporation of multifunctional nanocomposites is thus at the cutting edge of next-generation gadget engineering. Some important optoelectronics and thermoelectric materials are shown in table 1.1 (Huang et al., 2014)

Challenges and Future Perspectives

Despite rapid advances in the development of advanced nanocomposite systems such as TMCs, BMCs, and metal halide perovskites, several critical challenges remain in the way of full-scale implementation in commercial optoelectronic and thermoelectric applications (Ates et al., 2021). Material stability is one of the most important problems, especially for perovskites, which are very susceptible to environmental degradation when exposed to moisture, heat, light, and oxygen. Furthermore, the toxicity of lead-based perovskites poses significant environmental and health problems, demanding the rapid development of high-performance, lead-free alternatives (Miah et al., 2025).

Scalability remains a serious bottleneck. While processes like solvothermal synthesis and nanocasting work well in the lab, they are generally unsuitable for industrial manufacturing due to limits in throughput, repeatability, and cost-efficiency. Furthermore, precise control over morphology, phase purity, interface quality, and dopant dispersion is required to produce consistent device performance—but remains technically challenging at scale.

CONCLUSION

The convergence of transition metal chalcogenides, bimetallic chalcogenides, perovskites, and their hybrid nanocomposites has marked a watershed moment in multifunctional materials science, providing synergistic solutions to the dual demands of optoelectronic and thermoelectric applications. Each material class has unique properties: TMCs have tunable electronic structures, high carrier mobilities, and excellent thermal stability; BMCs have enhanced electronic coupling and defect tolerance due to multivalent metal centers; and metal halide perovskites have excellent optical absorption, long carrier diffusion lengths, and scalable solution-processability.

When combined into nanocomposite designs, these materials outperform their individual capabilities. At the nanoscale, customized interfaces, quantum confinement effects, and directed anisotropies work together to improve critical performance measures like electrical conductivity, Seebeck coefficient, and photogenerated charge separation. This increased multifunctionality results not from simple additive effects, but from tailored interfacial synergies that enable new operational regimes in electrical and energy-harvesting devices. These hybrid systems' structural adaptability and compositional tunability open up a versatile design area for cross-functional device integration. For example, integrating a high-absorption perovskite matrix in a thermally conductive TMC network allows for effective photothermal conversion while maintaining thermal stability under operational stress. Similarly, nanocomposite heterostructures in optoelectronic devices can enhance charge extraction, light harvesting, and defect passivation, resulting in improved power conversion efficiencies and longer device lifetimes.

Furthermore, these multifunctional systems provide unique platforms for hybrid energy harvesting. In photovoltaic-thermoelectric systems, excess heat from solar cells can be reused by thermoelectric modules, increasing overall energy conversion efficiency. The incorporation of such complimentary material classes into unified frameworks highlights nanocomposites' enormous promise for driving the next generation of sustainable, high-performance optoelectronic and thermoelectric devices.

Key Strategic Insights for Future Research and Industrial Translation Include

Performance can be significantly improved through doping, defect engineering, and heterostructure formation, which enable tunable optoelectronic and thermoelectric properties. Also, the emphasis should be placed on environmentally benign, low-energy fabrication methods that allow for large-area deposition and integration into flexible or complex geometries. The environmental and thermal stability must be addressed holistically, incorporating material design, protective encapsulation, and system-level integration. Progress depends on cross-pollination between materials science, physics, chemistry, engineering, and data science to accelerate both discovery and deployment. The future success of these materials—especially in consumer-facing and biomedical applications—depends on reducing toxicity, improving recyclability, and developing biodegradable or non-toxic alternatives. These materials will considerably assist industries such as sustainable energy, consumer electronics, Internet of Things (IoT) technology, and biological sensing—provided that reliability, manufacturability, and cost-effectiveness are rigorously optimized. Bridging the gap between foundational research and scalable innovation will be critical to achieving the full potential of these transformational material systems in real-world applications.

REFERENCES

Adhikari, S., Das, S., & Johari, P. (2025). Post-transition metal Sn-based chalcogenide perovskites: A promising lead-free and transition metal alternative for stable, high-performance photovoltaics. *Journal of Materials Chemistry. C, Materials for Optical and Electronic Devices*, *13*(15), 7792–7805. DOI: 10.1039/D4TC04701J

Al-Fartoos, M. M. R., Roy, A., Mallick, T. K., & Tahir, A. A. (2023). Advancing Thermoelectric Materials: A Comprehensive Review Exploring the Significance of One-Dimensional Nano Structuring. *Nanomaterials (Basel, Switzerland)*, *13*(13), 2011. DOI: 10.3390/nano13132011 PMID: 37446526

Alam, H., & Ramakrishna, S. (2013). A review on the enhancement of figure of merit from bulk to nano-thermoelectric materials. *Nano Energy*, *2*(2), 190–212. DOI: 10.1016/j.nanoen.2012.10.005

Ali, A., & Andriyana, A. (2020). Properties of multifunctional composite materials based on nanomaterials: A review. *RSC Advances*, *10*(28), 16390–16403. DOI: 10.1039/C9RA10594H PMID: 35498826

Ali, A., Shaukat, H., Bibi, S., Altabey, W. A., Noori, M., & Kouritem, S. A. (2023). Recent progress in energy harvesting systems for wearable technology. *Energy Strategy Reviews*, *49*, 101124. DOI: 10.1016/j.esr.2023.101124

Ameer, M. F., Hussain, M., Abbas, K., Sufyan, M., Khan, S. U., & Kolsi, L. (2024). Graphene and post-transition metal sulfide-based photodetectors for enhanced visible light photodetection: A review. *Physica Scripta*. Advance online publication. DOI: 10.1088/1402-4896/ada3f7

Aminorroaya Yamini, S., Santos, R., Fortulan, R., Gazder, A. A., Malhotra, A., Vashaee, D., Serhiienko, I., & Mori, T. (2023). Room-Temperature Thermoelectric Performance of n-Type Multiphase Pseudobinary Bi_2Te_3–Bi_2S_3 Compounds: Synergic Effects of Phonon Scattering and Energy Filtering. *ACS Applied Materials & Interfaces*, *15*(15), 19220–19229. DOI: 10.1021/acsami.3c01956 PMID: 37014987

Anthony, J. E. (2014). Addressing challenges. *Nature Materials*, *13*(8), 773–775. DOI: 10.1038/nmat4034 PMID: 24997739

Ates, M., Yılmaz, E., & Tanaydın, M. K. (2021). Challenges, novel applications, and future prospects of chalcogenides and chalcogenide-based nanomaterials for photocatalysis. In *Chalcogenide-Based Nanomaterials as Photocatalysts* (pp. 307–337). Elsevier., DOI: 10.1016/B978-0-12-820498-6.00014-7

Bal, S., & Rani, N. R. A. (2025). Next generation building materials for energy efficiency and climate responsive design. *Discover Applied Sciences*, *7*(8), 796. DOI: 10.1007/s42452-025-07360-z

Bartlett, P. N., De Groot, C. H. K., Greenacre, V. K., Huang, R., Noori, Y. J., Reid, G., & Thomas, S. (2025). Molecular precursors for the electrodeposition of 2D-layered metal chalcogenides. *Nature Reviews. Chemistry*, *9*(2), 88–101. DOI: 10.1038/s41570-024-00671-6 PMID: 39775527

Chen, W.-C., Hung, C.-W., Chang, C.-H., Liang, F.-C., Benas, J.-S., Yan, Z.-L., Lin, B.-H., Lin, J.-H., & Kuo, C.-C. (2022). Crystal orientation and insulating ligand of quasi-two dimensional perovskite optimized through silver ion doping for realizing efficient light emitting diodes. *Chemical Engineering Journal*, *443*, 136496. DOI: 10.1016/j.cej.2022.136496

Choi, J., Song, K., Kim, J.-I., Im, W. B., & Ahn, C. (2022). Enhanced thermal and electrical properties of hybrid polymer composites containing Al2O3 microspheres and nanowires. *Ceramics International*, *48*(21), 32081–32088. DOI: 10.1016/j.ceramint.2022.07.147

Choubey, A., & Yadav, A. (2024). MXene and transition metal chalcogenides-based 2D nanomaterials for next-generation supercapacitors. *Journal of Energy Storage*, *79*, 110131. DOI: 10.1016/j.est.2023.110131

Cui, X., Kong, Z., Gao, E., Huang, D., Hao, Y., Shen, H., Di, C., Xu, Z., Zheng, J., & Zhu, D. (2018). Rolling up transition metal dichalcogenide nanoscrolls via one drop of ethanol. *Nature Communications*, *9*(1), 1301. DOI: 10.1038/s41467-018-03752-5 PMID: 29615627

Fang, C., Wang, H., Shen, Z., Shen, H., Wang, S., Ma, J., Wang, J., Luo, H., & Li, D. (2019). High-Performance Photodetectors Based on Lead-Free 2D Ruddlesden–Popper Perovskite/MoS_2 Heterostructures. *ACS Applied Materials & Interfaces*, *11*(8), 8419–8427. DOI: 10.1021/acsami.8b20538 PMID: 30702273

Fang, C., Xu, B., Li, M., Han, J., Yang, Y., & Liu, X. (2024). Advanced Design of Fibrous Flexible Actuators for Smart Wearable Applications. *Advanced Fiber Materials*, *6*(3), 622–657. DOI: 10.1007/s42765-024-00386-9

Fang, X., Wei, P., Wang, L., Wang, X., Chen, B., He, Q., Yue, Q., Zhang, J., Zhao, W., Wang, J., Lu, G., Zhang, H., Huang, W., Huang, X., & Li, H. (2018). Transforming Monolayer Transition-Metal Dichalcogenide Nanosheets into One-Dimensional Nanoscrolls with High Photosensitivity. *ACS Applied Materials & Interfaces*, *10*(15), 13011–13018. DOI: 10.1021/acsami.8b01856 PMID: 29600705

Gao, M., Yu, L., Lv, Q., Kang, F., Huang, Z.-H., & Lv, R. (2023). Photoluminescence manipulation in two-dimensional transition metal dichalcogenides. *Journal of Materiomics*, *9*(4), 768–786. DOI: 10.1016/j.jmat.2023.02.005

Giri, A., Park, G., & Jeong, U. (2023). Layer-Structured Anisotropic Metal Chalcogenides: Recent Advances in Synthesis, Modulation, and Applications. *Chemical Reviews*, *123*(7), 3329–3442. DOI: 10.1021/acs.chemrev.2c00455 PMID: 36719999

Giuffredi, G., Asset, T., Liu, Y., Atanassov, P., & Di Fonzo, F. (2021). Transition Metal Chalcogenides as a Versatile and Tunable Platform for Catalytic CO_2 and N_2 Electroreduction. *ACS Materials Au*, *1*(1), 6–36. DOI: 10.1021/acsmaterialsau.1c00006 PMID: 36855615

Gurushankar, K., Karuppaiya, C., Tari, V., & Kannan, K. (2024). Exploring MXenes and MBenes in metal sulfur and selenium battery technologies. In Kannan, K., & Tari, V. (Eds.), *Novel energy storage and conversion technologies for two-dimensional MXenes and MBenes*. IGI Global., DOI: 10.4018/979-8-3693-9546-2.ch005

Han, H., Zhao, L., Wu, X., Zuo, B., Bian, S., Li, T., Liu, X., Jiang, Y., Chen, C., Bi, J., Xu, J., & Yu, L. (2024). Advancements in thermoelectric materials: Optimization strategies for enhancing energy conversion. *Journal of Materials Chemistry. A, Materials for Energy and Sustainability*, *12*(36), 24041–24083. DOI: 10.1039/D4TA03666B

Han, J. H., Kwak, M., Kim, Y., & Cheon, J. (2018a). Recent Advances in the Solution-Based Preparation of Two-Dimensional Layered Transition Metal Chalcogenide Nanostructures. *Chemical Reviews*, *118*(13), 6151–6188. DOI: 10.1021/acs.chemrev.8b00264 PMID: 29926729

Han, J. H., Kwak, M., Kim, Y., & Cheon, J. (2018b). Recent Advances in the Solution-Based Preparation of Two-Dimensional Layered Transition Metal Chalcogenide Nanostructures. *Chemical Reviews*, *118*(13), 6151–6188. DOI: 10.1021/acs.chemrev.8b00264 PMID: 29926729

Han, T.-H., Jang, K. Y., Dong, Y., Friend, R. H., Sargent, E. H., & Lee, T.-W. (2022). A roadmap for the commercialization of perovskite light emitters. *Nature Reviews. Materials*, *7*(10), 757–777. DOI: 10.1038/s41578-022-00459-4

Hassan, T., Salam, A., Khan, A., Khan, S. U., Khanzada, H., Wasim, M., Khan, M. Q., & Kim, I. S. (2021). Functional nanocomposites and their potential applications: A review. *Journal of Polymer Research*, *28*(2), 36. DOI: 10.1007/s10965-021-02408-1

Hu, S., Ren, Z., Djurišić, A. B., & Rogach, A. L. (2021). Metal Halide Perovskites as Emerging Thermoelectric Materials. *ACS Energy Letters*, *6*(11), 3882–3905. DOI: 10.1021/acsenergylett.1c02015

Hua, Q., & Shen, G. (2024). Low-dimensional nanostructures for monolithic 3D-integrated flexible and stretchable electronics. *Chemical Society Reviews*, *53*(3), 1316–1353. DOI: 10.1039/D3CS00918A PMID: 38196334

Huang, W., Luo, X., Gan, C. K., Quek, S. Y., & Liang, G. (2014). Theoretical study of thermoelectric properties of few-layer MoS2 and WSe2. *Physical Chemistry Chemical Physics*, *16*(22), 10866. DOI: 10.1039/c4cp00487f PMID: 24760342

Jana, S., Karingula, S., & Sajeevan, A., V. V. N., P. K., & Kotagiri, Y. G. (2025). *In situ* synthesis of bimetallic chalcogenides with highly conductive carbon nanotubes for efficient symmetric hybrid supercapacitors. *Nanoscale*, *17*(21), 13283–13297. DOI: 10.1039/D5NR00340G PMID: 40293371

Joshi, K. K., Pataniya, P. M., Bhadu, G., & Sumesh, C. K. (2023). Monometallic, bimetallic, and trimetallic chalcogenide-based electrodes for electrocatalytic hydrogen evolution reaction. *International Journal of Hydrogen Energy*, *48*(20), 7260–7272. DOI: 10.1016/j.ijhydene.2022.11.088

Ju, Z., Lv, R., Ansari, A. A., & Lin, J. (2025). Recent advances in additive manufacturing for solar cell based on organic/inorganic hybrid materials. *InfoMat*, *7*(7), e70017. DOI: 10.1002/inf2.70017

Kamaraki, C., Klug, M. T., Green, T., Miranda Perez, L., & Case, C. (2021). Perovskite/silicon tandem photovoltaics: Technological disruption without business disruption. *Applied Physics Letters*, *119*(7), 070501. DOI: 10.1063/5.0054086

Kannan, K., & Tari, V. (Eds.). (2025). Novel energy storage and conversion technologies for two-dimensional MXenes and MBenes. https://www.amazon.com/dp/9798369395462 ISBN: 9798369395462, eISBN: 9798369395486

Kumar, P., Gupta, M., & Singh, K. (2020). Performance Evaluation of Transition Metal Dichalcogenides Based Steep Subthreshold Slope Tunnel Field Effect Transistor. *Silicon*, *12*(8), 1857–1864. DOI: 10.1007/s12633-019-00285-4

Kumar, V. P., Raut, P., Panda, D. K., & Rashed, A. N. Z. (2025). Exploring Next-Generation TMDC Materials: A Comprehensive Review of Their Classifications, Properties, and Applications. *Silicon*, *17*(6), 1199–1221. DOI: 10.1007/s12633-025-03274-y

Lai, X., Li, W., Gu, X., Chen, H., Zhang, Y., Li, G., Zhang, R., Fan, D., He, F., Zheng, N., Yu, J., Chen, R., Kyaw, A. K. K., & Sun, X. W. (2022). High-performance quasi-2D perovskite solar cells with power conversion efficiency over 20% fabricated in humidity-controlled ambient air. *Chemical Engineering Journal*, *427*, 130949. DOI: 10.1016/j.cej.2021.130949

Li, C.-H. A., Zhou, Z., Vashishtha, P., & Halpert, J. E. (2019). The Future Is Blue (LEDs): Why Chemistry Is the Key to Perovskite Displays. *Chemistry of Materials*, *31*(16), 6003–6032. DOI: 10.1021/acs.chemmater.9b01650

Liu, B., Abbas, A., & Zhou, C. (2017). Two-Dimensional Semiconductors: From Materials Preparation to Electronic Applications. *Advanced Electronic Materials*, *3*(7), 1700045. DOI: 10.1002/aelm.201700045

Liu, F., Wu, R., Wei, J., Nie, W., Mohite, A. D., Brovelli, S., Manna, L., & Li, H. (2022). Recent Progress in Halide Perovskite Radiation Detectors for Gamma-Ray Spectroscopy. *ACS Energy Letters*, *7*(3), 1066–1085. DOI: 10.1021/acsenergylett.2c00031

Liu, X.-K., Xu, W., Bai, S., Jin, Y., Wang, J., Friend, R. H., & Gao, F. (2021). Metal halide perovskites for light-emitting diodes. *Nature Materials*, *20*(1), 10–21. DOI: 10.1038/s41563-020-0784-7 PMID: 32929252

Liu, Y., He, K., Chen, G., Leow, W. R., & Chen, X. (2017). Nature-Inspired Structural Materials for Flexible Electronic Devices. *Chemical Reviews*, *117*(20), 12893–12941. DOI: 10.1021/acs.chemrev.7b00291 PMID: 28991450

Lozano, G. (2018). The Role of Metal Halide Perovskites in Next-Generation Lighting Devices. *The Journal of Physical Chemistry Letters*, *9*(14), 3987–3997. DOI: 10.1021/acs.jpclett.8b01417 PMID: 29952563

Lu, M., Zhang, Y., Wang, S., Guo, J., Yu, W. W., & Rogach, A. L. (2019). Metal Halide Perovskite Light-Emitting Devices: Promising Technology for Next-Generation Displays. *Advanced Functional Materials*, *29*(30), 1902008. DOI: 10.1002/adfm.201902008

Machín, A., & Márquez, F. (2024). Advancements in Photovoltaic Cell Materials: Silicon, Organic, and Perovskite Solar Cells. *Materials (Basel)*, *17*(5), 1165. DOI: 10.3390/ma17051165 PMID: 38473635

Mak, K. F., Lee, C., Hone, J., Shan, J., & Heinz, T. F. (2010). Atomically Thin MoS 2: A New Direct-Gap Semiconductor. *Physical Review Letters*, *105*(13), 136805. DOI: 10.1103/PhysRevLett.105.136805 PMID: 21230799

Miah, Md. H., Khandaker, M. U., Hossen, Md. J., Noor-E-Ashrafi, Jahan, I., Shahinuzzaman, Md., Nur-E-Alam, M., Hanfi, M. Y., Ullah, Md. H., & Islam, M. A. (2025). Lead-free alternatives and toxicity mitigation strategies for sustainable perovskite solar cells: A critical review. *Materials Advances*, *6*(9), 2718–2752. DOI: 10.1039/D5MA00010F

Ming, H., Luo, Z.-Z., Zou, Z., & Kanatzidis, M. G. (2025). Strategies and Prospects for High-Performance Te-Free Thermoelectric Materials. *Chemical Reviews*, *125*(7), 3932–3975. DOI: 10.1021/acs.chemrev.4c00786 PMID: 40105866

Mirershadi, S., Sattari, F., Hadi, S., Rasoulzadehfard, S., & Mortazavi, S. Z. (2023). Efficiency enhancement of silicon-based solar cells by solar energy harvesting in WS2 nano-flake/perovskite layer. *Journal of Materials Science Materials in Electronics*, *34*(26), 1794. DOI: 10.1007/s10854-023-11205-z

Munonde, T. S., & Nomngongo, P. N. (2023). Review on Metal Chalcogenides and Metal Chalcogenide-Based Nanocomposites in Photocatalytic Applications. *Chemistry Africa*, *6*(3), 1127–1143. DOI: 10.1007/s42250-022-00577-0

Nallusamy, S., Vasanthi, V., Kavinkumar, T., Bhaviripudi, V. R., Mangalaraja, R. V., Srinivasan, R., Chidhambaram, N., Dhanabalan, S. S., Asaithambi, P., & Thirumurugan, A. (2025). Advances in optoelectronics for environmental and energy sustainability. *Next Energy*, *9*, 100387. DOI: 10.1016/j.nxener.2025.100387

Njema, G. G., & Kibet, J. K. (2025). A review of chalcogenide-based perovskites as the next novel materials: Solar cell and optoelectronic applications, catalysis and future perspectives. *Next Nanotechnology*, *7*, 100102. DOI: 10.1016/j.nxnano.2024.100102

Novak, T. G., Kim, K., & Jeon, S. (2019). 2D and 3D nanostructuring strategies for thermoelectric materials. *Nanoscale*, *11*(42), 19684–19699. DOI: 10.1039/C9NR07406F PMID: 31617541

Panigrahy, S. K., Nandha, A., Chaturvedi, M., & Mishra, P. K. (2024). Novel nanocomposites with advanced materials and their role in waste water treatment. *Next Sustainability*, *4*, 100042. DOI: 10.1016/j.nxsust.2024.100042

Patil, S. J., Chodankar, N. R., Hwang, S.-K., Shinde, P. A., Rama Raju, G. S., Ranjith, K. S., Karekar, S. V., Huh, Y.-S., & Han, Y.-K. (2023). Two-dimensional nanosheets of bimetallic chalcogenide-tagged nitrogen-doped carbon as a cathode for high-performance and durable zinc-ion capacitors. *Journal of Materials Chemistry. A, Materials for Energy and Sustainability*, *11*(10), 5112–5126. DOI: 10.1039/D2TA07524E

Priyadarshini, P., Das, S., & Naik, R. (2022). A review on metal-doped chalcogenide films and their effect on various optoelectronic properties for different applications. *RSC Advances, 12*(16), 9599–9620. DOI: 10.1039/D2RA00771A PMID: 35424915

Rahman, Md. F., Monnaf, Md. A., Amami, M., Ben Farhat, L., & Rahman, Md. A. (2024). Improving the efficiency above 35% of MoS2-based solar cells by through optimization of various wide-bandgap S-chalcogenides ETL. *Journal of Physics and Chemistry of Solids, 194*, 112216. DOI: 10.1016/j.jpcs.2024.112216

Rong, Y., Hu, Y., Mei, A., Tan, H., Saidaminov, M. I., Seok, S. I., McGehee, M. D., Sargent, E. H., & Han, H. (2018). Challenges for commercializing perovskite solar cells. *Science, 361*(6408), eaat8235. Advance online publication. DOI: 10.1126/science.aat8235 PMID: 30237326

Shirvani, S., Ghashghaee, M., & Smith, K. J. (2023). Two-dimensional Nanomaterials in Thermocatalytic Reactions: Transition Metal Dichalcogenides, Metal Phosphorus Trichalcogenides and MXenes. *Catalysis Reviews, 65*(1), 1–51. DOI: 10.1080/01614940.2021.1899605

Sokolikova, M. S., & Mattevi, C. (2020). Direct synthesis of metastable phases of 2D transition metal dichalcogenides. *Chemical Society Reviews, 49*(12), 3952–3980. DOI: 10.1039/D0CS00143K PMID: 32452481

Song, R., Cho, S., Khan, S., Park, I., & Gao, W. (2025). Lighting the Path to Precision Healthcare: Advances and Applications of Wearable Photonic Sensors. *Advanced Materials, 2419161*. Advance online publication. DOI: 10.1002/adma.202419161 PMID: 39865847

Sun, Q., Du, C., & Chen, G. (2025). Thermoelectric materials and devices: Applications in enhancing building energy conversion and efficiency. *Advanced Nanocomposites, 2*, 15–31. DOI: 10.1016/j.adna.2024.11.001

Tari, V., Kannan, K., Koliar, J., Gurushankar, K., & Wahyuni, D. K. (2025). Influence of zinc oxide nanomaterials on crop plants: A prospective analysis. In D. Jabbrova, D. Sarkar, Rakesh, S. Singh, & R. Datta (Eds.), *Smart nanomaterials in agroforestry systems* (Chapter 8). Springer. https://doi.org/DOI: 10.1007/978-981-96-1337-3_8

Tari, V., Kannan, K., & Vishwakarma, V. (2022). Emerging antiviral technologies. In Nunthavarawong, P., Rangappa, S. M., Siengchin, S., & Thoppil-Mathew, M. (Eds.), *Antimicrobial and antiviral materials: Polymers, metals, ceramics, and applications* (1st ed., pp. 137–150). CRC Press., DOI: 10.1201/9781003143093-8

Tian, Z., Tsui, G. C.-P., Tang, Y.-M., Wong, C.-H., Tang, C.-Y., & Ko, C.-C. (2026). Additive Manufacturing for Nanogenerators: Fundamental Mechanisms, Recent Advancements, and Future Prospects. *Nano-Micro Letters*, *18*(1), 30. DOI: 10.1007/s40820-025-01874-2 PMID: 40788321

Toshniwal, A., & Kheraj, V. (2017). Development of organic-inorganic tin halide perovskites: A review. *Solar Energy*, *149*, 54–59. DOI: 10.1016/j.solener.2017.03.077

Wang, C., He, M., Liu, X., & Malen, J. A. (2022). Modeling the tunable thermal conductivity of intercalated layered materials with three-directional anisotropic phonon dispersion and relaxation times. *Journal of Materials Chemistry. C, Materials for Optical and Electronic Devices*, *10*(32), 11686–11696. DOI: 10.1039/D1TC05369H

Wang, J., Xie, F., Cao, X.-H., An, S.-C., Zhou, W.-X., Tang, L.-M., & Chen, K.-Q. (2017). Excellent Thermoelectric Properties in monolayer WSe2 Nanoribbons due to Ultralow Phonon Thermal Conductivity. *Scientific Reports*, *7*(1), 41418. DOI: 10.1038/srep41418 PMID: 28120912

Wang, X., Chen, A., Wu, X., Zhang, J., Dong, J., & Zhang, L. (2024). Synthesis and Modulation of Low-Dimensional Transition Metal Chalcogenide Materials via Atomic Substitution. *Nano-Micro Letters*, *16*(1), 163. DOI: 10.1007/s40820-024-01378-5 PMID: 38546814

Wang, X., Feng, J., Bai, Y., Zhang, Q., & Yin, Y. (2016). Synthesis, Properties, and Applications of Hollow Micro-/Nanostructures. *Chemical Reviews*, *116*(18), 10983–11060. DOI: 10.1021/acs.chemrev.5b00731 PMID: 27156483

Wani, S. A., Tari, V., & Mansoor, S. I. U. (2025). A Comprehensive Analysis of Wastewater Management Challenges in India: Infrastructure and Policy Perspectives. In Dhiman, S., & Mukherjee, G. (Eds.), *Resource Resurgence. Springer Water*. Springer., DOI: 10.1007/978-3-032-02750-4_11

Xie, C., Yang, P., Huan, Y., Cui, F., & Zhang, Y. (2020). Roles of salts in the chemical vapor deposition synthesis of two-dimensional transition metal chalcogenides. *Dalton Transactions (Cambridge, England)*, *49*(30), 10319–10327. DOI: 10.1039/D0DT01561J PMID: 32648888

Xie, H., Hao, S., Bao, J., Slade, T. J., Snyder, G. J., Wolverton, C., & Kanatzidis, M. G. (2020). All-Inorganic Halide Perovskites as Potential Thermoelectric Materials: Dynamic Cation off-Centering Induces Ultralow Thermal Conductivity. *Journal of the American Chemical Society*, *142*(20), 9553–9563. DOI: 10.1021/jacs.0c03427 PMID: 32320237

Xu, L., Xiong, Y., Mei, A., Hu, Y., Rong, Y., Zhou, Y., Hu, B., & Han, H. (2018). Efficient Perovskite Photovoltaic-Thermoelectric Hybrid Device. *Advanced Energy Materials*, *8*(13), 1702937. DOI: 10.1002/aenm.201702937

Zhang, G., & Zhang, Y.-W. (2017). Thermoelectric properties of two-dimensional transition metal dichalcogenides. *Journal of Materials Chemistry. C, Materials for Optical and Electronic Devices*, *5*(31), 7684–7698. DOI: 10.1039/C7TC01088E

Zhang, Y., Cao, N., Szunerits, S., Addad, A., Roussel, P., & Boukherroub, R. (2019). Fabrication of ZnCoS nanomaterial for high energy flexible asymmetric supercapacitors. *Chemical Engineering Journal*, *374*, 347–358. DOI: 10.1016/j.cej.2019.05.181

Zheng, Y., Wang, X., Zhao, W., Cao, X., & Liu, J. (2018). Phytic acid-assisted synthesis of ultrafine NiCo2S4 nanoparticles immobilized on reduced graphene oxide as high-performance electrode for hybrid supercapacitors. *Chemical Engineering Journal*, *333*, 603–612. DOI: 10.1016/j.cej.2017.10.008

Zhu, Q., Wang, S., Wang, X., Suwardi, A., Chua, M. H., Soo, X. Y. D., & Xu, J. (2021). Bottom-Up Engineering Strategies for High-Performance Thermoelectric Materials. *Nano-Micro Letters*, *13*(1), 119. DOI: 10.1007/s40820-021-00637-z PMID: 34138379

Znidi, F., Morsy, M., & Uddin, Md. N. (2024). Navigating challenges and solutions for metal-halide and carbon-based electrodes in perovskite solar cells (NCS-MCEPSC): An environmental approach. *Heliyon*, *10*(12), e32843. DOI: 10.1016/j.heliyon.2024.e32843 PMID: 38988552

Chapter 5
Rhenium and Tantalum-Based Transition Metal Chalcogenides:
Next-Generation Materials for Optoelectronics

Sudipta Koley
https://orcid.org/0000-0002-9174-4229
Department of Physics, Amity Institute of Applied Sciences, Amity University, Kolkata, India

ABSTRACT

Transition metal Dichalcogenides, have emerged as a novel class of two-dimensional layered materials with exceptional potential for optoelectronic applications. Due to their unique electronic band structures, strong light-matter interactions, and tunability, they bridge the gap between conventional semiconductors and low-dimensional nanomaterials. From first-principles density functional theory, the electronic band structure, and optical absorption spectra, of these chalcogenides in mono and few-layer configurations will be studied. Theoretical simulations reveal strong in-plane anisotropic dielectric functions showing their relevance in advanced light detection systems. Tantalum-based dichalcogenides, showcase a rich polymorphic phase encompassing 1T, 1T', and 2H phase. This chapter aims to provide a comprehensive theoretical foundation for researchers exploring novel transition metal dichalcogenide systems for optoelectronics and to highlight the potential of Re- and Ta-based dichalcogenides as advanced functional materials for future electronic and photonic applications.

DOI: 10.4018/979-8-3373-3962-7.ch005

INTRODUCTION

The discovery of stable two-dimensional (2D) materials has introduced a new direction of research in condensed matter physics and materials science owing to their exceptional physical properties and vast potential for integration into next-generation electronic, optoelectronic, and energy devices. The isolation of graphene in 2004 has marked a turning point in research of materials science, stimulating intense research in 2D and van der Waals (vdW) layered systems (Kannan K. et & Tari V. 2025). These atomically thin crystals exhibit quantum confinement effects and surface dominated properties due to reduced dimension, enabling precise control over charge, spin, and light-matter properties at the nanoscale, (Novoselov et al., 2004; Geim & Grigorieva, 2013; Castro Neto et al., 2009). Over the last two decades since this landmark discovery, the field has undergone rapid progress, evolving from exploratory studies of graphene to a diverse and expanding class of 2D chalcogenides. Research on these materials is currently at the forefront of condensed matter physics, physical chemistry, materials science, and nanotechnology, driving interdisciplinary advancements across both fundamental understanding and technological innovation.

For a long time, theoretical arguments based on the Mermin-Wagner theorem suggested that long-range crystalline order in strictly two-dimensional systems like Graphene would be thermodynamically unstable at finite temperatures due to divergent long-wavelength fluctuations (Zakharchenko, 2011.). However, subsequent experimental and theoretical investigations revealed that anharmonic coupling between in-plane and out-of-plane vibrational modes, as well as the stabilizing influence of substrates, give rise to intrinsic out-of-plane vibrations which effectively stabilize 2D crystals (Gurushankar et al., 2024). Thisunderstanding led to a vast range of exploration in monolayer systems. The distinguishing feature of graphene from other conventional materials is its electronic band structure. Its low-energy excitations follow a linear energy-momentum dispersion relation forming massless Dirac fermions in quantum electrodynamics. As a result, electrons and holes in graphene obey a two-dimensional Dirac equation, leading to the emergence of relativistic quantum phenomena in a solid-state context (Katsnelson, 2007). This includes effects such as Klein tunneling, anomalous Landau quantization, and the unconventional half-integer quantum Hall effect which are experimentally confirmed, (Novoselov et al., 2007). The presence of a non-trivial Berry phase of π further confirms the topological nature of charge carriers in graphene, establishing a new paradigm in the field often referred to as relativistic condensed matter physics.

While graphene initially dominated the condensed matter physics research area for a long time, its lack of a bandgap limited its utility in digital electronics and optoelectronic devices. This limitation prompted the search for alternative 2D materials with intrinsic semiconducting behavior. The transition metal dichalcogenides (TMDs), particularly molybdenum-and tungsten-based compounds (e.g., MoS_2, WS_2, $MoSe_2$, WSe_2), soon emerged as promising candidates. Unlike graphene, monolayer TMDs exhibit direct bandgaps in the visible spectrum (~1.5–2.1 eV), making them ideal for optoelectronic applications such as photodetectors, light-emitting diodes, and solar cells, (Li & Galli, 2007; Milošević et al., 2007; Xiao et al., 2012; Mak et al., 2012). Furthermore, the reduced dielectric screening in two dimensions enhances Coulomb interactions, leading to tightly bound excitons with binding energies exceeding several hundred meV which is significantly higher in comparison to bulk semiconductors. The rich excitonic physics in these materials includes neutral excitons, charged trions, biexcitons, and interlayer excitons, which can be engineered in vertically stacked heterostructures.

A further improvement in the field occurred with the demonstration of intrinsic magnetism in monolayer vdW materials. In contrary to earlier expectations derived from the Mermin-Wagner theorem, materials such as CrI_3 and $Cr_2Ge_2Te_6$ were found to sustain long-range magnetic order in the atomic limit. In these systems, strong spin-orbit coupling and magnetic anisotropy play a crucial role in suppressing thermal fluctuations, (Bacaksiz et al., 2021). This discovery has opened new avenues in spintronics, enabling the development of devices that exploit the electron's spin degree of freedom, with potential applications in memory storage, spin logic, and quantum information processing. Superconductivity in these 2D systems represents yet another novel phenomena. Monolayer $NbSe_2$, for example, retains superconducting behavior with transition temperatures comparable to its bulk counterpart (Wang, H., et al., 2017). The dimensional reduction influences vortex dynamics and enhances phase fluctuations, facilitating the study of Berezinskii–Kosterlitz–Thouless (BKT) transitions. Electrostatic gating techniques further allow dynamic modulation and on-demand induction of superconductivity in these systems, offering a versatile platform for probing fundamental quantum phenomena.

Beyond their electronic and photonic functionalities, 2D materials exhibit remarkable chemical reactivity and surface tunability, making them highly suitable for catalysis and energy storage. Their high surface-to-volume ratios and modifiable surface terminations enable applications in electrocatalysis, including the hydrogen evolution reaction and oxygen reduction reaction, which are central to water-splitting and fuel cell technologies. Moreover, the rapid ion transport enabled by their ultra-thin geometry makes them attractive for use as electrode materials in lithium-ion and sodium-ion batteries, as well as in supercapacitors. Notably, layered vanadium oxides and MXenes have gained interest as high-capacity, fast-charging materials for next-generation energy storage systems, (Ridley et al., 2020).

In recent years, a transformative new frontier in the field of 2D materials has emerged: moiré engineering. When two layers of 2D materials are stacked with a small relative twist angle, a moiré pattern arises due to the interference between the atomic lattices. This moiré superlattice dramatically modifies the electronic structure, creating flat bands where electron kinetic energy is quenched and electron-electron interactions dominate. The discovery of correlated insulating states and superconductivity in magic-angle twisted bilayer graphene, where the twist angle (~1.1°) results in nearly flat electronic bands at the Fermi level has been a revolutionary progress, (Andrei & MacDonald, 2020; Cao et al., 2018). The resulting phase diagram resembles that of high-temperature superconductors. The research has since expanded to include twisted heterostructures of TMDs, where moiré potentials trap electrons and holes in spatially localized states. These systems serve as quantum simulators for strongly correlated phases, including Mott insulators, Wigner crystals, charge density waves, and topological Chern bands. Moiré systems thus represent a tunable laboratory for exploring exotic quantum phases in a controlled manner.

Table 1. Glossary of transition metal dichalcogenides' transitions

Material	**ICDW**	**QCDW**	**CCDW**	**SC**	**Metal Insulator Transition**	**CDW Structure**	**Electron Configuration**
1T-TaS2	550 K	350 K	180 K	5K, 3GPA 4.5 K, Li doping	180 K	$\sqrt{13}$ x $\sqrt{13}$	Xe5d3
1T-TaSe2	No	No	430 K	No	No	$\sqrt{13}$x$\sqrt{13}$	Xe5d3
1T-TiSe2	No	No	200 K	2 K, 2Gpa 4 K, Cu 0.08	No	2x2x2	Ar3d2
1T-TiTe2	No	No	No	No	No	No	Ar3d2
2H-TaSe2	120 K	No	90 K	200 mK	No	3x3	Xe5d3
2H-TaS2	70 K	No	No	0.8 K	No	No	Xe5d3
2H-NbSe2	33 K	No	No	7.2 K	No		Kr4d4 5S1
2H-NbS2	No	No	No	6K	No	No	Kr4d4 5S1
1T-ReS2	No	No	No	2.5K, 102 GPa	No	No	$4f^{14}5d^{5}6s^{2}$
1T-ReSe2	No	No	No	2.5K, 100 GPa	No	No	$4f^{14}5d^{5}6s^{2}$

Datas are collected from the previous papers on dichalcogenides Di Salvo and Graebner (1977); Wilson et al. (1975); Mcwhan et al. (1980); Tidman et al. (1974); van Maaren and Harland (1969); Moncton, Axe, and Disalvo (1975); Naito and Tanaka (1981); Zhang, J., et al., (2018).

Among the various classes of 2D materials, transition metal dichalcogenides (TMDs) have emerged as particularly promising due to their diverse electronic phases (Table 1) ranging from metals and semiconductors to superconductors (Huang, Fan, Singh & Zheng, 2020). These materials, with the chemical formula MX_2 (where M is a transition metal and X is a chalcogen: S, Se, or Te), exhibit rich polymorphism (1T, 2H and 3R) and layer-dependent properties that are essential for tailoring their performance in specific device contexts (Chhowalla et al., 2013; Manzeli et al., 2017). In TMDs transition metal layers are sandwitched between chalcogen layers. TMDs are known for their rich pressure-temperature phase diagram. Unlike graphene or silicon, where the electronic structure depends on the hybridization of s- and p-orbitals, TMDs have electronic structure depending on the filling of d orbitals. The d-orbital electron number varies from 0 to 6 as we go from group IV to group X transition metals. Apart from the choice of M and X elements, the electronic structures have a wide range of variety due to their structural symmetry. A single layer of TMD can have a trigonal prismatic phase or an octahedral phase. The trigonal prismatic phase is also called the 2H phase (1H for single layer TMDs) and have hexagonal symmetry. TMDs with group IV and group VI transition metals (MoS_2, WSe_2, $TiSe_2$, etc.) are common examples of TMDs having 2H-phase. On the other hand, octahedral phase is conventionally referred to as 1T phase and have a tetragonal symmetry. Usually, group V transition metals belong to examples of 1T TMDs ($TaSe_2$, PdS_2, etc.). Among all chalcogenides MoS_2 and WS_2 have traditionally dominated TMD research due to their well-established direct-to-indirect band gap transitions and robust semiconducting nature in monolayers. They have well known applications in field-effect transistors, photodetectors, spintronics, valleytronics, and flexible electronics, (Splendiani et al., 2010). Monolayer TMDs exhibit unique electronic properties that significantly differ from their bulk counterparts due to quantum confinement, broken inversion symmetry, and strong spin–orbit coupling. These properties make them promising candidates for next-generation nanoelectronic and optoelectronic devices, (Manzeli et al., 2017).

However, newer materials based on Rhenium (Re) and Tantalum (Ta) have drawn attention for their exotic properties and unexplored potential. Rhenium disulfide (ReS_2), Rhenium diselenide ($ReSe_2$), Tantalum diselenide ($TaSe_2$) are representative transition metal dichalcogenides (TMDs) that are part of the structural and electronic diversity within this class of 2D materials. ReS_2 and $ReSe_2$ are known for their low-symmetry distorted 1T phase, resulting in pronounced in-plane anisotropy. ReS_2 crystallizes in a distorted 1T phase due to Peierls distortion, leading to weak interlayer coupling, which allows it to retain monolayer-like optical and electronic properties even in multilayer form. This intrinsic anisotropy manifests in direction-dependent optical absorption, and carrier mobility that are highly desirable in polarizing devices, (Tongay et al., 2014; Island et al., 2016). In contrast, $TaSe_2$ exists in multiple polytypes (2H, 1T, and 1T'), exhibiting rich physics including charge density waves (CDWs) and metallic conductivity in its bulk form. These contrasting behaviors: semiconducting and anisotropic in ReS_2, versus metallic and CDW active in $TaSe_2$, highlight the versatility of TMDs for exploring low-dimensional quantum phenomena and designing tunable electronic devices, (Wilson et al., 1975; Yoshida et al., 2015). Importantly, ReS_2 also demonstrates high stability, maintaining structural and functional integrity under exposure to atmospheric moisture and oxygen - a recurring challenge for many 2D materials. Recent studies further highlight the materials ability to support hyperbolic plasmonic modes, indicating its potential for sub-diffraction photonics and light-matter interaction at the nanoscale, (Kiran et al., 2025). In parallel, its compatibility with flexible substrates, combined with a high degree of mechanical resilience, suggests applicability for incorporation into flexible and wearable electronics, as well as energy storage and sensor systems. Tantalum diselenide ($TaSe_2$) exhibits intriguing charge density wave (CDW) and superconducting behaviors that are strongly dependent on its crystal structure and dimensionality. In the 2H-$TaSe_2$ polymorph, a well-documented incommensurate CDW transition occurs around 122 K, followed by a commensurate CDW phase near 90 K. These transitions arise from exciton condensation and strong electron-phonon interactions, leading to periodic modulations in both electron density and lattice positions, (Wilson et al., 1975). The 1T-$TaSe_2$ phase also shows CDW ordering but with different symmetry and transition temperatures, reflecting the sensitive interplay between crystal geometry and electronic structure. Interestingly, while bulk 2H-$TaSe_2$ exhibits weak superconductivity below ~0.15 K, superconductivity can be enhanced under external pressure or by reducing dimensionality to monolayers, where the suppression of CDW order can allow superconductivity to emerge more prominently. In nanoscale and doped 2H-$TaSe_2$, superconductivity has been observed at higher temperatures due to the competition and possible coexistence between CDW and superconducting phases. This rich and tunable phase interplay in $TaSe_2$ makes it a model system for studying correlated electron phenomena and phase competition in low-dimensional materials.

From a theoretical standpoint, these transition metal dichalcogenides (TMDs) present a rich landscape for investigating the complex interplay among structural intricacies, spin-orbit coupling, and strong electronic correlations. These materials, involving heavy transition metals such as rhenium (Re) and tantalum (Ta), often exhibit low-symmetry phases, distorted polytypes, or incommensurate charge-density wave (CDW) patterns, which introduce novel quantum phenomena. Theoretical exploration of such materials requires sophisticated computational approaches that go beyond conventional approximations. Advanced first-principles techniques particularly density functional theory (DFT), many-body perturbation theory (such as the GW approximation), and the Bethe-Salpeter equation (BSE) for excitonic effects are indispensable tools for capturing the electronic structure, optical responses, and subtle interactions in these layered compounds. DFT allows for the analysis of ground-state properties and structural stability, while GW corrections yield more accurate quasiparticle band gaps by incorporating self-energy effects. Meanwhile, the

BSE framework captures the strong excitonic binding typical of 2D systems, which is crucial for understanding optoelectronic properties and light matter interactions. Moreover, interlayer coupling effects, stacking-dependent electronic behavior, and the influence of strain and doping can be systematically probed through these theoretical methods, enabling the prediction of emergent phases such as topological states, Mott insulators, and unconventional superconductivity. The insights generated from these predictive models not only help in interpreting experimental observations but also provide design for TMD-based heterostructures and devices with tailored functionalities-ranging from high-performance transistors to photodetectors, spintronic components, and quantum information systems.

In this chapter, we focus on the theoretical investigation of Re and Ta-based TMDs to understand their structural, electronic, and optical characteristics. We aim to highlight their suitability for various optoelectronic applications such as photodetectors, light-emitting devices, photovoltaics, and valleytronic systems. By evaluating their computationally modeled properties, we aim to describe the fundamental mechanisms governing their behavior and inspire future directions in the design of high-performance, low-dimensional optoelectronic materials. In summary, the field of 2D materials has undergone a dramatic transformation, evolving from the study of a single atomic layer of carbon to a multifaceted domain encompassing semiconductors, magnets, superconductors, and energy materials. Each new class of 2D material brings unique functionalities and unexplored physics, pushing the boundaries of modern materials science and offering a rich landscape for future research and technological development.

STRUCTURAL AND ELECTRONIC PROPERTIES

The structural polymorphism observed in TMDs arises primarily from the different stacking configurations of the two chalcogen layers which sandwich the transition metal layer. Depending on this stacking, two basic polymorphs are observed commonly wich are the octahedral (1T) and trigonal prismatic (2H) phases, (Wilson & Yoffe, 1969). In the 1T configuration, the chalcogen layers are rotated by 60° relative to one another about an axis perpendicular to the layers (along the c-axis), resulting in a staggered geometry with one MX_2 unit per unit cell. In the 2H (trigonal prismatic) structure, the chalcogen atoms are aligned within a single sandwich, while successive sandwiches are rotated by 60° around the c-axis, leading to two MX_2 units per unit cell (Wilson & Yoffe, 1969). This layered structure exhibit strong anisotropy in the electronic and structural properties. For both 1T and 2H phases, the resistivity along the c-axis is approximately 30 times higher than that in the basal plane, emphasizing the quasi-two-dimensional nature of these systems, (Barker et al., 1975).

Despite their similar layered structures, the two polymorphs exhibit markedly distinct behavior in the charge density wave (CDW) state. For instance, in the 1T-TaS_2 phase, the transition from the CDW to the normal state is accompanied by an abrupt, order-of-magnitude change in resistivity, (Fazekas & Tosatti, 1979). In contrast, the TMDs with 2H structure show a more subtle transition, characterized mainly by a change in the slope of the resistivity curve at the CDW onset temperature, (Rossnagel, 2011). The current situation regarding various CDWs in TMDs encompasses a large number of experimental and theoretical results from last 35 years including recent high resolution angle resolved photo electron spectroscopy results, (Yang et al., 2018). The number of experiments indicate that these are extensively studied TMDs and have fairly large CDW amplitudes in Ta based chalcogenides. It is, therefore, expected that weak coupling theories may not be applicable for these systems, (Rossnagel, 2011). During the last few decades a wide range of experimental and theoretical investigations have been conducted on these transition metal dichalcogenides. Structural transformations were studied using techniques such as X-ray diffraction, neutron scattering, and electron scattering. To probe the electronic structure and spectral changes, angle-resolved photoemission spectroscopy, scanning tunneling microscopy, and other spectroscopic techniques were employed, (Brauer et al., 2001; Wilson et al., 1975). Additionally, measurements of specific heat and electrical transport were carried out to gain insights into the electronic properties and their evolution under varying conditions, (Barker et al., 1975). From theoretical calculations using local density approximation with density functional theory and dynamical mean field theory, the electronic properties are studied in normal and CDW phase, (Taraphder et al., 2011). Strong evidences for the occurrence of unusual CDW transition in 2H-$TaSe_2$ and 1T-TaS_2 has been reported earlier, (Wilson et al., 1975). In the context of 2H-$TaSe_2$, the pressure-temperature phase diagram is less complex, (Mcwhan et al., 1980). X-ray scattering studies under high pressure on 2H-$TaSe_2$ show recurring behavior of the charge density wave lock-in transition. With increased pressure the change in the normal to incommensurate CDW transition temperature is very small. At atmospheric pressure there is a normal to incommensurate phase transition at 122K following which there is a commensurate CDW transition at 90K. In the incommensurate phase the wave vector ($q_{CDW} = (1 - \delta)a/3$) exhibits a rapid increase to the commensurate value with decreasing temperature and saturates to a constant at T_{CCDW}. X-ray diffraction study gives more inputs such as the phase boundary for the lock in transition is reentrant, suppressed to T < 10 K by P=1.7 GPa and then reappearing above P=2.5 Gpa (Mcwhan et al., 1980) and the degree of incommensurability increase over the range of pressure up to 4.4 GPa. Theoretical models are proposed to explain the phase diagram and occurrence of the lock in transition as competition between the gradient energy term and the Umklapp terms of the Landau expansion of the free energy in powers of CDW order parameter (Nakanishi & Shiba, 1977). But the two commensurate phases in the phase diagram remain questionable for their type and properties. In the context of transport, the pronounced quasi-2D crystallographic structure of this system leads to approximately cylindrical Fermi surfaces. In contrast with the 1D case, 2D dichalcogenides remain metallic in the presence of CDW (Rossnagel, 2011). Infra-red and optical spectroscopy have been powerful techniques for the examination of correlation effects and electron dynamics which are particularly powerful in probing the anisotropy of electronic transport in layered materials. With polarized light, the in-plane and out-of-plane electronic responses can be obtained, offering valuable insights into the directional dependence of charge transport in these anisotropic systems. The transport experiment, (Vescoli et al., 1998) on this system suggests dynamics of the charge excitation spectrum is governed by three relevant features: Drude contribution, narrowed with reducing temperature, a broad pseudogap-like feature in the mid-infrared range which is present up to lowest temperature below 300 K and the large spectral weight transfer (SWT) with temperature. The major feature in the resistivity measurement (done by conventional four point contact method, (Vescoli et al., 1998)) is the weak change of slope around 300 K and the sudden drop of $\rho(T)$ at the normal-incommensurate CDW phase transition around 120 K, whilst changes at incommensurate - commensurate is barely observable. The reduced $\rho(T)$ at T_{CDW} would indicate the FS is not gapped. The transport data also indicate that the scattering rate undergoes a reduction below 120 K implying a freezing out of the scattering channels below 120 K.

Among the TMDs with CDW transition, 1T-TaS_2 is another interesting system, notable for its complex electronic phase diagram dominated by charge density wave (CDW) transitions(Wilson et al., 1975). CDWs are periodic modulations of the conduction electron density coupled with a distortion of the underlying lattice. In 1T-TaS_2, these modulations arise due to strong electron-phonon interactions and possibly electronic correlations, resulting in multiple phase transitions as a function of temperature and external perturbations such as pressure, doping, or electric field. At room temperature (~300 K), 1T-TaS_2 resides in an incommensurate CDW (IC-CDW) phase, where the periodicity of the charge modulation does not match the atomic lattice perfectly. As the temperature decreases to around 350 K, the system transitions into a nearly commensurate CDW (NC-CDW) phase. In this regime, the charge order begins to lock into the lattice, forming commensurate CDW "domains" separated by discommensurations, or domain walls. Further cooling below ~180 K drives the material into a commensurate CDW (C-CDW) state, which is characterized by a well-defined $\sqrt{13} \times \sqrt{13}$ reconstruction of the lattice in the basal plane. This state is insulating and exhibits a "Star of David" cluster formation of Ta atoms, each composed of 13 Ta atoms, with one central atom surrounded by two concentric hexagons, (Fazekas & Tosatti, 1979).

The C-CDW state in 1T-TaS_2 is particularly intriguing due to the emergence of a Mott-insulating behavior, where strong electron-electron interactions open a correlation-induced gap at the Fermi level, despite the material being metallic by band theory. This coexistence of CDW and Mott physics makes 1T-TaS_2 a prototypical system for studying the interplay of electron correlations and lattice effects in low-dimensional materials. The localized electron at the center of each Star of David cluster is considered responsible for the Mott behavior. External perturbations can dramatically modify the CDW phase of 1T-TaS_2. Application of pressure, for instance, can suppress the insulating C-CDW phase and induce a metallic or even superconducting state(Sipos et al., 2008). Ultrafast laser pulses or electric fields can also melt the CDW order, leading to transient metastable metallic phases that persist at low temperatures, a phenomenon with implications for non-volatile memory devices. Moreover, the two-dimensional nature of 1T-TaS_2 allows for the exfoliation of atomically thin layers, where the CDW properties can deviate significantly from the bulk. In monolayer or few-layer forms, the stability of various CDW phases is reduced, and the Mott insulating behavior is suppressed due to reduced dimensionality and screening effects, (Vaskivskyi et al., 2015). This tunability has opened a new avenue of research into the engineering of quantum phases in 2D materials (Yu, S., et al., 2017).

1T-Rhenium disulfide (1T-ReS_2), another member of the transition metal dichalcogenide (TMD) family, exhibits unique electronic and optical characteristics that distinguish it from other 2D materials. Unlike the commonly studied TMDs such as MoS_2 or WS_2 that crystallize in the 2H phase, 1T-ReS_2 adopts a distorted 1T structure due to Peierls-type lattice distortions, (Tongay et al., 2014). These distortions arise from strong interatomic interactions, especially among Re atoms, leading to the formation of Re-Re chains and clusters within the basal plane. This distortion breaks the high in-plane symmetry typically associated with ideal TMDs and introduces profound anisotropy in both electronic and optical properties. In the ideal 1T structure, Re atoms are located at the octahedral center between two planes of sulfur atoms. However, in 1T-ReS_2, this symmetry is broken due to a Jahn-Teller-like distortion or Peierls instability. This distortion leads to the formation of Re_4 diamond-like clusters or chains arranged in a triclinic lattice, giving rise to a low-symmetry configuration (space group $P\bar{1}$). This structural change persists even in the monolayer limit, unlike other TMDs that transition from indirect to direct band gaps with thickness reduction. The distorted structure results in highly anisotropic and localized electronic states. Density functional theory (DFT) calculations show that monolayer 1T-ReS_2 is an indirect band gap semiconductor with a band gap in the range of ~1.35 – 1.5 eV. The Re 5d orbitals significantly contribute to the states near the Fermi level, and the distortion alters the bandwidth and hybridization, leading to flatter bands and enhanced effective mass along specific directions, (Rahman, Davey, & Qiao, 2017). The distorted lattice also suppresses interlayer coupling in bulk 1T-ReS_2, allowing monolayer-like behavior to persist even in multilayer systems. This weak interlayer interaction is a direct result of the in-plane Re-Re clustering, which confines the electronic states more strongly within individual layers. The low symmetry and anisotropic nature of 1T-ReS_2 strongly influence its optical response. Photoluminescence (PL) measurements typically reveal weak emission intensities compared to 2H-TMDs, due to the indirect nature of the band gap and low density of direct transitions. Moreover, polarization-resolved optical measurements show strong linear dichroism: the absorption and emission intensities vary significantly with the polarization direction of incident light, revealing pronounced optical anisotropy. This anisotropic response arises from the in-plane orientation of the Re chains, which modulate the dipole transition matrix elements directionally. 1T-phase of Rhenium diselenide ($ReSe_2$) exhibits a unique low-symmetry distorted structure that significantly affects its electronic and optical behavior. 1T-$ReSe_2$ is also another layered semiconductor that exhibits notable spin-orbit coupling due to the heavy rhenium (Re) atoms. Owing to its low symmetry and strong relativistic effects, the breaking of inversion symmetry in this material is anticipated to manifest in its nonlinear optical response and influence its electronic band structure. The direct and indirect band gaps of bulk 1T-$ReSe_2$ are measured at approximately 1.40 eV and 1.18 –1.19 eV, respectively. Due to the weak interlayer van der Waals interactions, mechanical exfoliation is an effective method to isolate monolayers, which exhibit an excitonic direct band gap of around 1.50 eV, (Li et al., 2020; Naguib et al. 2011). Notably, the electronic structure remains largely unaffected by the number of layers, indicating weak van der Waals interactions between the layers.

Structurally, 1T-$ReSe_2$ crystallizes in a distorted triclinic phase, similar to other rhenium-based dichalcogenides. Its structure consists of a distorted Re layer with Re atom zigzag chains, sandwiched between opposing triangular networks of selenium atoms. In its natural bilayer form, the system shows an inversion center between the layers. However, when in a bilayer structure neighbouring layers are rotated by 180° with respect to other, the triangular Se atoms align in the same orientation across the interface, thereby breaking inversion symmetry. In this configuration spin degeneracy is expected to be lifted in reciprocal space due to spin-orbit interaction, which has the potential to give rise to physical properties related to the Berry curvature, such as the shift current, nonlinear Hall effect, and ferroelectric switching. The material also shows multiple excitonic transitions, with exciton binding energies typically ranging from 300–500 meV, owing to reduced dielectric screening in two dimensions. Also, the distorted lattice influences optical selection rules, enabling transitions that would be forbidden in higher-symmetry systems. Another notable feature of 1T-$ReSe_2$ is its weak interlayer coupling in multilayer configurations. Despite being a layered van der Waals material, the distorted lattice leads to reduced hybridization between layers, resulting in nearly layer-independent optical responses, in contrast to many other TMDs where the optical gap shifts significantly with increasing thickness, (Li et al., 2020). The incorporation of transition metal dopants such as V, Cr, Mn, Fe, Co, Nb, Mo, Ta, and W into monolayers of ReS_2 and $ReSe_2$ has been found to substantially alter their electronic ground states. These dopants introduce localized defect levels and modify the density of states. Notably, Mn doping induces small reduction in the electronic band gap, suggesting minimal perturbation to the host lattice's semiconducting nature. The magnetic ground state of the doped systems varies depending on the specific dopant, with elements like Cr, Fe, and Co giving rise to ferromagnetic ordering in both ReS_2 and $ReSe_2$ monolayers, (Obodo et al., 2017). Optical properties are also notably affected, the calculated absorption spectra exhibit enhanced peak intensities upon doping while changes in reflectivity remain relatively insignificant. Furthermore, pronounced optical anisotropy is observed, independent of the polarization direction within the xy-plane whether aligned parallel or perpendicular to the crystallographic axes which shows the directional sensitivity of these layered materials.

OPTICAL AND EXCITONIC BEHAVIOUR

The reflectivity measurement on the 2H-$TaSe_2$ system along the 2D layer planes as a function of temperature gives the electrodynamic response of this system and these measurements present no observable sharp changes at transition but a continuous evolution which are, (i) below the visible spectral range, optical conductivity ($\sigma(\omega)$) is featured with a broad mid-infrared absorption overlapped to a low frequency Drude-like-contribution, (ii) on lowering the temperature from normal state there is a redistribution of the spectral weight accompanied by a progressive narrowing of the Drude resonance in the far infrared range and the broad maximum in the mid infrared range absorption shifts towards higher frequency between 300 and 80 K slightly and in a pronounced manner afterwards, (Wu et al., 2019). Moreover, the commensurate phase transition does not affect the system at all. The optical results do not display any clear signature of a true CDW gap and if scattering rate can be estimated from the narrow Drude peak at low frequency it will also show sudden reduction after T_{ICDW}. The conclusion is that transport properties are the consequence of the freezing out of the scattering channels and in turn, verifies importance of the effect of correlation and reduced dimensionality. Now other experiments address the possible evidence for the gap due to the formation of the CDW condensate but there are contrasting evidence. Angle resolved photo-emission experiments, (Dardel et al., 1993; Liu et al., 1998) suggest that the Fermi surface (FS) is not gapped by the CDW phase transitions, though a small gap appears away from FS (Borisenko et al., 2008). As in transport, these data are not affected by the 90 K transition also. It was found that spectral intensity is zero halfway between $\Gamma - K$ direction in the Brillouin zone. The photo-emission data further confirm that along $\Gamma - K$ two Fermi surface crossing occur and a saddle band lies very close to Fermi level for an extended region along $\Gamma - K$. Other ARPES experiments, (Valla et al., 2000) denote that the real part of self-energy, calculated directly from photo-emission intensity, is concentrated in the binding energy region less than 150 meV and at the lowest temperature it has a maximum around 65 meV- same as the value corresponding to the sharp drop in $\text{Im}\Sigma(\omega)$. This behavior underlines the scattering of photo-hole from some collective excitation of the system. This phenomena has a striking similarity with the behavior observed in ARPES study of the photo-hole interaction with phonon but this would imply the presence of about 70 meV phonon in the CDW state though the highest measured frequency of phonon is 40 meV for this system. Moreover the temperature dependence of self energy also contradicts phonon mediation. The peak in $\text{Re}\Sigma$ loses its magnitude and the structure also shifts to lower energy with the increasing temperature. After T_{ICDW} only a small peak is left at a binding energy of 30 meV and same behavior is observed in the normal state till 160 K indicating no pronounced effect at T_{ICDW} . At low temperature the imaginary part of the self energy or the quasi-particle scattering rate shows a sharp reduction at binding energies lower than 70 meV which becomes less pronounced with increase in temperature. This experimental observations is understood (Valla et al., 2000) in terms of scattering from some collective mode and there is a possibility of opening of the gap in the saddle region in association with reduction in phase space available for scattering. Though the opening of the gap in the single particle spectra implies a modification to the response function with the possibility of new collective modes, and the self energy would behave as scattering from this collective mode; in a way this also underlines the importance of electronic correlations. Later more photoemission experiments have been done. Recent experiments (Borisenko et al., 2008) show the topology of FS in the CDW and normal state. The 2H FS consists of the hole-like sections centered at the Γ and K points and electron-like structure around the M point. More important features come out from this ARPES experiment (Borisenko et al., 2008). It is seen that the electronic states near the Fermi level and FS sheets originate from two bands, one is responsible for the Γ and K centered FS sheet with a saddle point in between and another in support of the electron-like structure around M point. An interesting observation is that below the first phase transition at T_{ICDW} the dispersion is quite similar except for the temperature broadening, in agreement with earlier ARPES result where virtually no change in electronic structure was detected upon crossing the phase boundary. On entering the commensurate CDW phase at T_{CCDW}, however, there are more pronounced changes in CDW dispersion. At much lower temperatures, new folded FS forms a set of doubly degenerate concentric electron-like FSs formed by the M point electron-like structure around the new 3×3 state and rounded triangles are formed by parts of the Γ barrels. Moreover the ARPES dispersion at CCDW phase is considerably modified by the FS folding. This appears in contrast with the resistivity and specific heat experiments, having seen no detectable effect on the incommensurate to commensurate CDW transition. Photoemmision studies also infer the presence of a pseudo-gap both in the normal and incommensurate CDW state. This pseudo-gap magnitude also increases up to T_{ICDW} accompanied with long range ordered transition. The photo-emission study of the FS nesting from detailed FS map indicates the presence of incommensurate nesting vector though nesting and CDW ordering wave vector does not match and the charge susceptibility at the nesting vector and the CDW wavevector are not related too. From incommensurate to commensurate transition by lowering in temperature the nesting vector remains unchanged giving an inconsistent theory for CDW transition and FS nesting. In short ARPES and other experiments present a comparatively lucid picture of CDW and related changes in the FS of 2H-$TaSe_2$ and the results prove the inadequacy of FS nesting-based scenario and calls for a closer scrutiny of electronic correlation in the context of CDW transition.

The nature of the bandgap in rhenium disulfide (ReS_2) has been a subject of continued investigation and debate. Unlike most transition metal dichalcogenides (TMDs), where monolayers typically exhibit a direct bandgap favorable for strong light–matter interactions, ReS_2 is reported to retain a direct bandgap even in its bulk form—an unusual feature that has generated significant interest. Nevertheless, both earlier and more recent studies consistently classify bulk ReS_2 as an indirect bandgap semiconductor (Asahi et al., 2016). Computational methods reveal notable discrepancies: while the Local Density Approximation (LDA) predicts a bandgap of approximately 1.29 eV, the more accurate Quasiparticle Self-Consistent GW (QSGW) approach estimates it at around 2.75 eV. This deviation is expected, given that LDA typically underestimates bandgaps, whereas QSGW may overestimate them due to the overscreening effect in the Random Phase Approximation (RPA). The relatively small correction implies that excitonic effects calculated via the Bethe–Salpeter Equation (BSE) have a minimal impact on the self-energy.

At the LDA level, monolayer ReS_2 is characterized by a direct bandgap; however, inclusion of electron–hole interactions shifts the valence band maximum away from the Γ-point by approximately 150 meV, rendering the bandgap indirect (Tung et al., 2018). A similar behavior has been reported for monolayer $ReSe_2$, although not all theoretical studies capture this shift. In both bilayer and bulk ReS_2, the band topology remains qualitatively similar, with indirect bandgaps of approximately 2.3 eV and 1.7

eV, respectively. Unlike conventional TMDs such as MoS_2 and WS_2, which undergo a transition from indirect to direct bandgap in the monolayer limit, ReS_2 retains an indirect bandgap at the quasiparticle level (QSGW) across all thicknesses, exhibiting a direct bandgap only within LDA. This theoretical result is consistent with experimental data, as most samples are substrate-supported and doped, conditions that reduce self-energy corrections and may yield a slightly indirect bandgap. Given the small energy difference between the direct and indirect configurations, the bandgap character in ReS_2 is highly sensitive to external perturbations. The identification of an indirect bandgap in ReS_2 holds particular relevance for optoelectronic applications. Whereas conventional monolayer TMDs exhibit strong photoluminescence and short radiative lifetimes due to their direct gaps, materials with indirect bandgaps, such as ReS_2, demonstrate longer carrier lifetimes with competing radiative and non-radiative processes, making them attractive for photovoltaic and optoelectronic technologies. Anisotropic optical absorption in ReS_2, initially investigated using density functional theory, has been experimentally confirmed, revealing that the low-symmetry crystal structure induces anisotropic optical responses that persist down to the monolayer. LDA-based approaches, however, often fail to capture these phenomena due to their neglect of excitonic correlations and bandgap underestimation. More advanced techniques, such as BSE calculations atop single-shot G_0W_0, successfully reproduce the anisotropic excitonic features.

The intrinsic structural anisotropy of ReS_2, reminiscent of the 4:3 in-plane lattice anisotropy seen in materials such as black phosphorus and solid nitrogen, makes it a compelling candidate for hyperbolic optical materials. These materials are characterized by exotic optical responses, including broadband absorption and highly efficient thermal transport via photon tunneling. Optical hyperbolicity arises when the real parts of the dielectric tensor components along orthogonal in-plane directions possess opposite signs. Assuming negligible dielectric response along the out-of-plane (z) direction due to symmetry, the in-plane x and y axes define the principal dielectric response directions. Theoretical dielectric functions computed for ReS_2 at varying thicknesses exhibit strong direction-dependent optical absorption. Although monolayer ReS_2 displays anisotropic excitons within the quasiparticle bandgap, no hyperbolic regions (HPs) are observed under strain-free or strained conditions. In contrast, the application of uniaxial strain significantly modulates hyperbolic behavior in bilayer and bulk forms. In bulk ReS_2, compressive strain along the x-direction expands the hyperbolic spectral window to approximately 1.3 eV and reduces plasmon damping, while tensile strain along the y-direction narrows the window. In the bilayer, x-direction strain suppresses the hyperbolic region entirely, whereas y-direction strain enhances it and yields less damped modes. These results indicate that, although hyperbolic regions cannot be induced in the monolayer through strain engineering, they are robust and tunable in thicker forms. Consequently, bilayer and bulk ReS_2 emerge as promising platforms for strain-engineered hyperbolic optics.

Due to the distorted triclinic structure, $ReSe_2$ shows optical biaxiality and pronounced in-plane anisotropy. Two distinct excitonic features were observed in polarization-dependent piezoreflectance (PzR) spectra, corresponding to different polarizations. At 25 K, the excitonic transition energies are approximately 1.387 eV and 1.410 eV along out of plane and in-plane direction. Due to the indirect band gap nature of $ReSe_2$ the transitions are facilitated by a single phonon with an energy of about 25 meV, likely from an optical branch. The electrical conductivity along the b-axis is significantly higher than that perpendicular to the b-axis. Along the c-axis, conductivity is extremely low, attributed to weak van der Waals interlayer bonding - typical of layered materials. Conductivity measurements reveal consistent anisotropy across the temperature range.

THEORETICAL PERSPECTIVES ON DEVICE APPLICATIONS

Theoretical studies on 2H-$TaSe_2$, 1T-ReS_2 and 1T-$ReSe_2$ have revealed a rich array of electronic and structural properties that make them promising candidates for next-generation electronic, optical and energy storage devices (Fig.3). In the phase diagram of 2H-$TaSe_2$ there is a metallic region at room temperature with excitonic coherence below CDW transition temperature (Taraphder et al., 2011). Density functional theory plus Dynamical mean field theory (DFT+DMFT) calculations have shown that its strong correlation above CDW and excitonic coherence below CDW give rise to complex phase diagram, which can be tuned by strain, doping, or layer thickness, offering a pathway to tune its electronic phases. In low-dimensional $TaSe_2$ such as monolayers and few-layer systems, theoretical predictions indicate the possibility of tuning its CDW transition temperature and the structure of ordered phase changes in few layer system (Koley S., 2023). Notably, monolayer and bilayer structures exhibit distinct phonon characteristics due to reduced van der Waals interactions and broken symmetries. Despite its metallic nature, 2H-$TaSe_2$ exhibits strong optical absorption and photoluminescence (PL), attributed to interband transitions. Theoretical studies on 2H-$TaSe_2$ predicts multiple applications of 2H-$TaSe_2$: (1) as a donor in non-radiative resonant energy transfer to MoS_2, enhancing MoS_2 photoluminescence by sevenfold; (2) as an efficient hot electron injector with unusually long hot carrier lifetimes for a metal; and (3) as both a contact and active layer in a high-speed vertical photodetector ($TaSe_2$/MoS_2/graphene) showing >10 A/W responsivity at 0.1 Mhz (Mahajan, M., et al, 2019). These findings highlight 2H-$TaSe_2$'s potential in next-generation, all-2D optoelectronic devices. Other spectroscopic studies like Fourier-transform infrared spectroscopy, the researchers observed plasmon resonances across a wide spectral range-from terahertz to telecom wavelengths, which could be tuned by thickness and dielectric environment. At high wave vectors, the plasmon dispersion flattens due to interband transition screening. Notably, the plasmon resonance displays anomalous temperature-dependent behavior in the CDW phase, attributed to strong coupling between plasmons and CDW excitations. This coupling, described by a coupled-oscillator model, alters the plasmon linewidth and symmetry which also reveals intrinsic excitations in correlated 2D materials. First-principles calculations based on DFT also reveal that 2H-$TaSe_2$ exhibits metallic conductivity and high carrier mobility, essential for efficient charge transport in supercapacitors and batteries (Koley S., 2023). The large interlayer spacing in the 2H phase facilitates ion intercalation, while the weak van der Waals interactions between layers allow for structural reversibility during charge-discharge cycles. Moreover, the presence of tantalum d-orbitals near the Fermi level enhances electronic conductivity and redox activity, thereby improving energy storage capacity. These insights highlight the potential of 2H-$TaSe_2$ for optoelectronic and energy storage device.

On the other hand theoretical studies on distorted 1T-phase of rhenium dichalcogenides, 1T-$ReSe_2$ and 1T-ReS_2, have revealed promising properties that position them as potential candidates for next-generation electronic and optoelectronic devices. Since $ReSe_2$ and ReS_2 exhibit significant in-plane anisotropy due to the distorted 1T crystal structure direction-dependent electronic, optical, and transport behavior is observed in its properties. DFT calculations indicate that these materials possess moderate bandgaps, strong spin-orbit coupling, and unique orbital hybridizations that can be tuned through strain, doping, and mono layer to few layer stacking. These features enable precise control over charge carrier mobility and light absorption, making them suitable for applications in photodetectors, and polarization-sensitive sensors. Furthermore, the weak interlayer coupling allows few-layer or even bulk forms to retain monolayer-like properties. Overall, the theoretical insights into 1T-$ReSe_2$ and ReS_2 highlight their functional versatility and potential integration into low-power and directionally sensitive nanoelectronic systems (Banik, D., et al., (2024)). Anisotropy in ReS_2 shows electrical conductivity along the b-axis is 32 times greater, and the responsivity is 20 times higher. These properties are attributed to the intrinsic quasi-1D Re chain alignment and surface-dominant charge transport. The electron concentration at the surface can reach up to ~10^{19} cm^{-3}, substantially exceeding the bulk value. Furthermore, the study demonstrates that the conductivity and activation energy are thickness-dependent, reinforcing the surface-conducting nature of ReS_2. Now in advanced study of heterostructure made using two anisotropic 2D materials, ReS_2 and $ReSe_2$ demonstrates that the heterojunction functions as a gate-tunable diode, exhibiting a strong rectifying behavior with a peak rectification ratio of 3150. Under illumination, the device shows a clear photovoltaic effect with a power conversion efficiency (PCE) of approximately 0.5%, which surpasses similar TMD-based heterojunctions, (Li et al., 2018). This enhanced performance can be attributed to the type-II band alignment, efficient charge separation, and the intrinsic direct bandgap of multi-layer ReS_2. Additionally, the theoretical study on heterostructure highlights the potential of Re-based TMDs in optoelectronic applications, such as photodetectors and solar cells, especially with further optimization of device architecture and material synthesis.

CHALLENGES AND FUTURE DIRECTIONS

Despite the significant progress in the synthesis, characterization, and theoretical modeling of TMDs, several challenges persist in realizing their full potential. These challenges originate from the intrinsic complexity of TMD band structures, environmental stability, synthesis reproducibility, and integration into scalable device architectures. In this section, we explore these critical issues and examine how emerging strategies - ranging from alloying and heterostructure fabrication to machine learning guided materials discovery can be employed to overcome them. One of the most desirable features of TMDs is their tunable bandgap, which is essential for tailoring materials to specific optoelectronic applications. Unlike bulk semiconductors, monolayer TMDs such as MoS_2, WS_2, and WSe_2 exhibit a direct bandgap in the visible range, which drastically enhances their interaction with light. In contrast, the indirect-to-direct bandgap transition that occurs when these materials are reduced from bulk to monolayer thickness adds complexity but also opportunity in device design. However, the tuning of bandgaps in TMDs remains as a challenge, especially when uniformity, reproducibility, and defect control are required at scale. Various strategies have been proposed and implemented to modulate the electronic structure of TMDs. These include Strain engineering, alloying, heterostructure formation. With applied strain the atomic distances and angles are modified to engineer the electronic band structure. Both uniaxial and biaxial strains can induce significant changes in the bandgap, making it either narrower or wider depending on the direction and magnitude of the applied strain. Another way to tune band structure is making ternary alloy These mixed compounds allow continuous tuning of bandgap energies and exciton binding energies, thereby offering greater flexibility in device design. Defect engineering introduces vacancies, dopants, or interstitial atoms to tailor charge carrier mobility and trap states, ultimately influencing both the optical and electronic responses. Despite these advances, a major limitation is the lack of precise and scalable control like inhomogeneities in strain distribution, alloying concentration, and defect density often lead to variability in device performance and thereby restricting its applications.

Beyond laboratory-scale investigations to industrial-scale deployment, the sustainability of TMD synthesis and material sourcing has become increasingly important. Most TMDs rely on relatively scarce or geopolitically sensitive transition metals like tungsten, rhenium, or molybdenum. The extraction and refinement of these elements raise significant environmental concerns due to energy consumption, water usage, and toxicity. Moreover, the synthesis of high-quality TMD crystals typically involves chemical vapor deposition (CVD), mechanical exfoliation, or chemical exfoliation techniques, each of which has limitations. Mechanical exfoliation provides high-quality monolayers but is unsuitable for large-area production. CVD, while scalable, often results in polycrystalline films with grain boundaries that degrade electronic performance. Furthermore, life cycle assessment (LCA) approaches are being increasingly adopted to evaluate the environmental footprint of TMD-based devices.

Advanced theoretical modeling has played a pivotal role in accelerating the development of TMDs for optoelectronic applications. Density functional theory, along with many-body perturbation theories such as GW approximation and Bethe–Salpeter equation for excitonic effects, has enabled accurate predictions of band structures, effective masses, optical absorption spectra, and defect formation energies. These insights are invaluable in screening potential materials and predicting their performance before experimental synthesis. The interplay of spin-orbit coupling, quantum confinement, and interlayer coupling in few-layer TMDs leads to rich electronic behavior that can be captured only through sophisticated modeling. For instance, valley polarization - a unique feature of TMDs where charge carriers populate different valleys in momentum space - can be controlled by circularly polarized light and external magnetic fields. This has profound implications for valleytronic devices and quantum information processing. Additionaly theoretical approaches have shown exciton dynamics, including binding energy, lifetimes, and diffusion lengths, which are critical for photovoltaic and light-emitting applications. Phonon-assisted transitions and electron–phonon interactions are observed in the transport properties, which influence carrier mobility, non-radiative recombination rates, and thermal conductivity. The challenge is to validate this theoretical results through careful experimental design, but these theoretical predictions offer a powerful method for reducing experimental waste in material discovery.

The need to further tailor the properties of TMDs has prompted intense research into compositional modification and structural design. Alloying, as mentioned, provides a continuous tuning knob for bandgap and carrier concentration. Additionally defect engineering is being used to control performance and add new functionalities. Artificial van der Waals heterostructures, fabricated by stacking different 2D materials, allow unprecedented control over interlayer interactions. These structures have led to the observation of exotic phenomena such as interlayer excitons, moiré superlattices, and correlated insulating phases. For optoelectronics, they enable band alignment for efficient charge separation in solar cells, high-speed tunneling devices with negative differential resistance, and broadband photodetectors that exploit charge transfer across interfaces. Moreover, with the advent of moiré superlattices, where the angle between stacked layers is deliberately controlled, has introduced an additional degree of freedom. At specific twist angles, moiré patterns can induce flat bands and correlated states, opening doors to novel optoelectronic and quantum applications.

Integrating TMDs into practical devices remains a significant hurdle. Contact resistance, substrate compatibility, and environmental degradation all affect device performance and longevity. The high surface-to-volume ratio of 2D materials makes them susceptible to adsorbates and oxidation, which can alter their intrinsic properties. To mitigate these effects, researchers are exploring encapsulation with materials like hexagonal boron nitride (h-BN), as well as the use of passivating layers and interface engineering techniques. However, achieving reproducible, low-resistance contacts without Fermi-level pinning remains elusive. In terms of device architecture, efforts are focused on vertical stacking, planar heterostructures, and flexible substrates. These allow for high integration densities and compatibility with existing CMOS technologies. Nevertheless, wide applicability requires further innovation. The sheer chemical and structural diversity of TMDs makes them ideal candidates for machine learning and high-throughput computational screening. Databases such as Materials Project and Open Quantum Materials Database contain thousands of DFT-calculated properties, which serve as training data for ML models that can predict material behavior with remarkable accuracy. These tools enable rapid identification of stable TMD compositions with desired bandgaps, prediction of synthesis feasibility and defect formation, design of heterostructures with optimal band alignments. In particular, this data-driven approach significantly shortens the discovery-to-application timeline, paving the way for rational material design.

The versatility of optoelectronic TMDs ensures their relevance across multiple application domains. In energy harvesting, their high optical absorption coefficients and tunable bandgaps make them ideal for ultrathin solar cells and photothermal converters. In energy storage, TMDs serve as efficient electrodes for lithium-ion batteries and supercapacitors due to their high surface area and layered structure. Emerging applications include flexible and wearable electronics, where the mechanical resilience and transparency of monolayer TMDs are leveraged for strain sensors and bendable displays. The continued convergence of synthesis techniques, computational modeling, and device engineering holds the key to unlocking the transformative potential of these materials.

CONCLUSION

In this chapter, we have reviewed a wide range of theoretical insights into the electronic and structural behaviors of 2H-$TaSe_2$, 1T-ReS_2, and 1T-$ReSe_2$, focusing particularly on charge density wave (CDW) phenomena, excitonic correlations, and anisotropic optical responses. We have reanalyzed the CDW phase in 2H-$TaSe_2$ driven by preformed, strong excitonic interactions alongside the distinctive optical and electronic properties of distorted 1T-ReS_2 and 1T-$ReSe_2$. The chapter highlights the rich and tunable physics embedded within chalcogenide systems. Through a combined analysis of first principle calculation for non-interacting chalcogenides and DMFT for interaction, we demonstrated that incorporating strong, dynamical interband excitonic correlations which also offers an explanation for several previously unresolved incoherent spectral features. Based on this, a qualitative scenario was established theoretically in which unconventional charge density wave and unconventional superconducting orders emerge as direct instabilities of a preformed excitonic liquid. Beyond fundamental physics, our exploration underscores the technological relevance of chalcogenide materials, particularly in areas such as conductivity modulation, electrochemical performance, and optoelectronic applications. These materials hold significant promise in addressing global challenges related to energy sustainability and environmental impact. However, realizing this potential demands further research to overcome material synthesis, stability, and scalability challenges. Tuning bandgap and application in optoelectronics remains intricate, demanding innovative approaches in material design and synthesis. Further ensuring sustainability of material sourcing and processing techniques are becoming increasingly important for the widespread usage of these materials. The future of electrochemical applications in TMDs lies in the theoretical predictions by advanced computational modeling and materials synthesis techniques with device engineering. Researchers can find transformative applications in energy harvesting and utilization by delving into the fundamental interactions governing their electronic behavior. The versatility of optoelectronic chalcogenides finds applications in various domains. Theoretical modeling, including DFT and many-body perturbation theories, continues to play a crucial role in guiding experimental efforts and discovering new materials. Emerging research is now focused on alloying, defect engineering, and the creation of artificial heterostructures to tailor properties for various applications. High-throughput computational screening and machine learning are expected to accelerate future material discovery. In sum, transition metal chalcogenides, with their diverse electronic phases and tunable properties, can play a transformative role in the development of next-generation sustainable technologies. Continued interdisciplinary research at the interface of theory, synthesis, and device

REFERENCES

Andrei, E. Y., & MacDonald, A. H. (2020). Graphene bilayers with a twist. *Nature Materials*, *19*(12), 1265–1275.

Bacaksiz, C., Šabani, D., Menezes, R. M., & Milošević, M. V. (2021). Distinctive magnetic properties of CrI_3 and $CrBr_3$ monolayers caused by spin-orbit coupling. *Physical Review. B*, *103*(12), 125418.

Banik, D., Lionel, C., Das, S., & Koley, S. (2024). Comparative study of the electronic and optical properties of Rhenium based Chalcogenides. *Physica Scripta*, *99*(8), 085975.

Barker, A. S., Ditzenberger, J. A.Jr, & DiSalvo, F. J. (1975). Infrared study of the electronic instabilities in tantalum disulfide and tantalum diselenide. *Physical Review. B*, *12*, 2049–2054.

Borisenko, S. V., Kordyuk, A. A., Yaresko, A. N., Zabolotnyy, V. B., Inosov, D. S., Schuster, R., Büchner, B., Weber, R., & Follath, R. (2008). Pseudogap and charge density waves in two dimensions. *Physical Review Letters*, *100*(19), 196402.

Brauer, H. E., Starnberg, H. I., Holleboom, L. J., Hughes, H. P., & Strocov, V. N. (2001). Na and Cs intercalation of 2H-$TaSe_2$ studied by photoemission. *Journal of Physics Condensed Matter*, *13*(44), 9879.

Cao, Y.. (2018). Unconventional superconductivity in magic-angle graphene superlattices. *Nature*, *556*, 43–50.

Castro Neto, A. H.. (2009). The electronic properties of graphene. *Reviews of Modern Physics*, *81*, 109–162.

Chhowalla, M.. (2013). The chemistry of two-dimensional layered transition metal dichalcogenide nanosheets. *Nature Chemistry*, *5*, 263–275.

Dardel, B., Grioni, M., Malterre, D., Weibel, P., Baer, Y., & Levy, F. (1993). Spectroscopic observation of charge-density-wave-induced changes in the electronic structure of 2H-$TaSe_2$. *Journal of Physics Condensed Matter*, *5*(33), 6111.

Di Salvo, F. J., & Graebner, J. E. (1977). The Low Temperature Electrical Properties of 1T-TaS_2. *Solid State Communications*, *23*, 825–828.

Fazekas, P., & Tosatti, E. (1979). Electrical, structural and magnetic properties of pure and doped 1T-TaS_2. *Philosophical Magazine Part B*, *39*, 229–244.

Geim, A. K., & Grigorieva, I. V. (2013). Van der Waals heterostructures. *Nature*, *499*, 419–425.

Gurushankar, K., Karuppaiya, C., Tari, V., & Kannan, K. (2024). Exploring MXenes and MBenes in metal sulfur and selenium battery technologies. In Kannan, K., & Tari, V. (Eds.), *Novel energy storage and conversion technologies for two-dimensional MXenes and MBenes*. IGI Global., DOI: 10.4018/979-8-3693-9546-2.ch005

Huang, H. H., Fan, X., Singh, D. J., & Zheng, W. T. (2020). Recent progress of TMD nanomaterials: Phase transitions and applications. *Nanoscale*, *12*(3), 1247–1268.

Island, J. O., Biele, R., Barawi, M., Clamagirand, J. M., Ares, J. R., Sánchez, C., Herre, S., van der Zant, J., Ferrer, I. J., D'Agosta, R., & Castellanos-Gomez, A. (2016). Titanium trisulfide (TiS_3): A 2D semiconductor with quasi-1D optical and electronic properties. *Scientific Reports*, *6*(1), 22214.

Kannan, K., & Tari, V. (Eds.). (2025). Novel energy storage and conversion technologies for two-dimensional MXenes and MBenes. https://www.amazon.com/dp/9798369395462 ISBN: 9798369395462, eISBN: 9798369395486

Katsnelson, M. I. (2007). Graphene: Carbon in two dimensions. *Materials Today*, *10*(1-2), 20–27.

Kiran, R., Pashov, D., van Schilfgaarde, M., Katsnelson, M. I., Taraphder, A., & Acharya, S. (2025). Possible Realization of Hyperbolic Plasmons in Few-Layered Rhenium Disulfide. *Condensed Matter*, *10*, 40.

Koley, S. (2023). Intercalation in 2H-$TaSe_2$ for modulation of electronic properties and electrochemical energy storage. *Physica B, Condensed Matter*, *669*, 415312.

Li, C., Cao, Q., Wang, F., Xiao, Y., Li, Y., Delaunay, J. J., & Zhu, H. (2018). Engineering graphene and TMDs based van der Waals heterostructures for photovoltaic and photoelectrochemical solar energy conversion. *Chemical Society Reviews*, *47*(13), 4981–5037.

Li, T., & Galli, G. (2007). Electronic properties of MoS_2 nanoparticles. *The Journal of Physical Chemistry. C, Nanomaterials and Interfaces*, *111*(44), 16192–16196.

Li, X., Chen, C., Yang, Y., Lei, Z., & Xu, H. (2020). 2D Re-based transition metal chalcogenides: Progress, challenges, and opportunities. *Advancement of Science*, *7*(23), 2002320.

Liu, R., Olson, C. G., Tonjes, W. C., & Frindt, R. F. (1998). Momentum dependent spectral changes induced by the charge density wave in 2H-$TaSe_2$ and the implication on the CDW mechanism. *Physical Review Letters*, *80*, 5762–5765.

Mahajan, M., Kallatt, S., Dandu, M., Sharma, N., Gupta, S., & Majumdar, K. (2019). Light emission from the layered metal 2H-$TaSe_2$ and its potential applications. *Communications on Physics*, *2*(1), 88.

Mak, K. F.. (2012). Control of valley polarization in monolayer MoS_2 by optical helicity. *Nature Nanotechnology*, *7*, 494–498.

Manzeli, S.. (2017). 2D transition metal dichalcogenides. *Nature Reviews. Materials*, *2*, 17033.

McWhan, D. B., Fleming, R. M., Moncton, D. E., & DiSalvo, F. J. (1980). Reentrant lock-in transition of the charge-density wave in 2H-$TaSe_2$ at high pressure. *Physical Review Letters*, *45*, 269–272.

Milošević, I., Nikolić, B., Dobardžić, E., Damnjanović, M., Popov, I., & Seifert, G. (2007). Electronic properties and optical spectra of MoS_2 and WS_2 nanotubes. *Physical Review B: Condensed Matter and Materials Physics*, *76*(23), 233414.

Moncton, D. E., Axe, J. D., & Disalvo, F. J. (1975). Study of Superlattice Formation in 2H-$NbSe_2$ and 2H-$TaSe_2$ by Neutron Scattering. *Physical Review Letters*, *34*, 734–737.

Naguib, M.. (2011). Two-dimensional nanocrystals produced by exfoliation of Ti_3AlC_2. *Advanced Materials*, *23*, 4248–4253.

Nakanishi, K., & Shiba, H. (1977). Domain-like incommensurate charge-density-wave states and the first-order incommensurate-commensurate transitions in layered tantalum dichalcogenides. I. 1T-polytype. *Journal of the Physical Society of Japan*, *43*(6), 1839–1847.

Novoselov, K. S.. (2004). Electric field effect in atomically thin carbon films. *Science*, *306*, 666–669.

Novoselov, K. S.. (2007). Room-temperature quantum Hall effect in graphene. *Science*, *315*(5817), 1379–1379.

Obodo, K. O., Ouma, C. N. M., Obodo, J. T., & Braun, M. (2017). Influence of transition metal doping on the electronic and optical properties of ReS_2 and $ReSe_2$ monolayers. *Physical Chemistry Chemical Physics*, *19*(29), 19050–19057.

Rahman, M., Davey, K., & Qiao, S. Z. (2017). Advent of 2D rhenium disulfide (ReS_2): Fundamentals to applications. *Advanced Functional Materials*, *27*(10), 1606129.

Ridley, P., Gallano, C., Andris, R., Shuck, C. E., Gogotsi, Y., & Pomerantseva, E. (2020). MXene-derived bilayered vanadium oxides with enhanced stability in Li-ion batteries. *ACS Applied Energy Materials*, *3*(11), 10892–10901.

Rossnagel, K. (2011). On the origin of charge-density waves in select layered transition-metal dichalcogenides. *Journal of Physics Condensed Matter*, *23*(21), 213001.

Sipos, B., Kusmartseva, A. F., Akrap, A., Berger, H., Forró, L., & Tutiš, E. (2008). From Mott state to superconductivity in 1T-TaS_2. *Nature Materials*, *7*(12), 960–965.

Splendiani, A.. (2010). Emerging photoluminescence in monolayer MoS_2. *Nano Letters*, *10*, 1271–1275.

Taraphder, A., Koley, S., Vidhyadhiraja, N. S., & Laad, M. S. (2011). Preformed excitonic liquid route to a charge density wave in 2H-TaSe2. *Physical Review Letters*, *106*(23), 236405.

Tidman, J. P., Singh, O., Curzon, A. E., & Frindt, R. F. (1974). The phase transition in 2H-TaS_2 at 75 K. *Philosophical Magazine*, *30*, 1191–1194.

Tongay, S., Sahin, H., Ko, C., Luce, A., Fan, W., Liu, K., & Wu, J. (2014). Monolayer behaviour in bulk ReS2 due to electronic and vibrational decoupling. *Nature Communications*, *5*(1), 3252.

Valla, T., Fedorov, A. V., Johnson, P. D., Xue, J., Smith, K. E., & DiSalvo, F. J. (2000). Charge-density-wave-induced modifications to the quasiparticle self-energy in 2H-$TaSe_2$. *Physical Review Letters*, *85*(22), 4759.

Van Maaren, M. H., & Harland, H. B. (1969). An energy band model of Nb-and Ta-dichalcogenide superconductors. *Physics Letters. [Part A]*, *29*(9), 571–573.

Vaskivskyi, I.. (2015). Controlling the metal-to-insulator relaxation of the metastable hidden quantum state in 1T-TaS_2. *Science Advances*, *1*(6), e1500168.

Vescoli, V., Degiorgi, L., Berger, H., & Forró, L. (1998). Dynamics of correlated two-dimensional materials: The 2H-TaSe2 case. *Physical Review Letters*, *81*, 453–456.

Wang, H.. (2017). High-quality monolayer superconductor $NbSe_2$ grown by chemical vapour deposition. *Nature Communications*, *8*(1), 394.

Wilson, J. A., Di Salvo, F. J., & Mahajan, S. (1975). Charge-density waves and superlattices in the metallic layered transition metal dichalcogenides. *Advances in Physics*, *24*(2), 117–201.

Wilson, J. A., & Yoffe, A. D. (1969). The transition metal dichalcogenides discussion and interpretation of the observed optical, electrical and structural properties. *Advances in Physics*, *18*(73), 193–335.

Wu, Y., Xing, H., Lian, C. S., Lian, H., He, J., Duan, W., & Liu, Y. (2019). Ion intercalation engineering of electronic properties of two-dimensional crystals of 2H-$TaSe_2$. *Physical Review Materials*, *3*(10), 104003.

Xiao, D.. (2012). Coupled spin and valley physics in monolayers of MoS_2 and other group-VI dichalcogenides. *Physical Review Letters*, *108*, 196802.

Yang, H., Liang, A., Chen, C., Zhang, C., Schroeter, N. B., & Chen, Y. (2018). Visualizing electronic structures of quantum materials by angle-resolved photoemission spectroscopy. *Nature Reviews. Materials*, *3*(9), 341–353.

Yoshida, M., Suzuki, R., Zhang, Y., Nakano, M., & Iwasa, Y. (2015). Memristive phase switching in two-dimensional 1T-TaS2 crystals. *Science Advances*, *1*(9), e1500606.

Yu, S., Wu, X., Wang, Y., Guo, X., & Tong, L. (2017). 2D materials for optical modulation: Challenges and opportunities. *Advanced Materials*, *29*(14), 1606128.

Zakharchenko, K. V. (2011). Temperature effects on graphene: from flat crystal to 3D liquid. *Sl: sn*.

Zhang, J., Sun, E., Feng, X., Liu, H., Redfern, S. A., Kanchana, V., Liu, G., & Wang, H. (2018). Phase transition and superconductivity in ReS_2, $ReSe_2$ and $ReTe_2$. *Physical Chemistry Chemical Physics*, *20*(46), 29472–29479.

Chapter 6
Tailored Chalcogenide Materials for Eco-Efficient Electronic and Energy Systems

R. Naveen Kumar
https://orcid.org/0000-0001-9033-9400
Chandigarh Group of Colleges Jhanjeri, Mohali, India

Rubi Sarkar
Brainware University, Kolkata, India

Tanu Chandel
Chandigarh Group of Colleges Jhanjeri, Mohali, India

ABSTRACT

Abstract: Chalcogenide-based materials, which are primarily made up of sulfur, selenium, and tellurium, have become an essential category of compounds for advanced optoelectronic and energy applications due to their distinctive electronic, optical, and thermal characteristics. These materials hold particular importance in the fields of photodetectors, solar cells, thermo electrics, and non-volatile memory devices. Their adjustable band gap, elevated refractive index, and capacity to create glassy phases render them suitable for sustainable energy solutions. This chapter delves into the synthesis, characterization, and application of chalcogenides, emphasizing recent developments and their contribution to facilitating environmentally friendly technologies. Through comparative analysis and experimental findings, we

DOI: 10.4018/979-8-3373-3962-7.ch006

demonstrate how advancements in material engineering improve performance in energy harvesting and storage. We conclude by underscoring future opportunities and challenges in scaling these materials for commercial use.

INTRODUCTION

Chalcogenide materials, composed primarily of elements from group 16 of the periodic table namely sulfur (S), selenium (Se), and tellurium (Te) constitute a significant class of semiconductors that exhibit a wide range of electrical, optical, and thermal properties. Their ability to form both crystalline and amorphous phases, combined with high polarizability and strong spin-orbit coupling especially for heavier chalcogens like Te, has positioned them as key candidates in emerging applications across optoelectronics, energy conversion, and sustainable material technologies, (Tari et al., 2022). Unlike conventional semiconductors such as silicon (Si) or gallium arsenide (GaAs), chalcogenides possess a tunable bandgap that can be modified by varying their composition, stoichiometry, and synthesis conditions. This tunability, along with their high refractive index and non-linear optical coefficients, makes them particularly valuable for infrared (IR) optics, photodetectors, phase-change memory devices, and photovoltaic absorbers. Concentrated Examination: SnS, Sb_2Se_3, and CZTS in Energy Frameworks - Chalcogenide substances like tin sulfide (SnS), antimony selenide (Sb_2Se_3), and copper zinc tin sulfide (CZTS) have received considerable focus for their use in environmentally friendly electronic and energy systems, mainly in photovoltaics and thermoelectrics. These materials present a hopeful combination of earth availability, eco-friendliness, adjustable electronic characteristics, and affordable production, positioning them as appealing substitutes for conventional hazardous or costly semiconductor substances, (Tari et al., 2025; Wani, Tari, V& Mansoor, 2025). Tin Sulfide (SnS): SnS is a p-type semiconductor featuring a direct bandgap around 1.3 eV, making it highly appropriate for photovoltaic absorber layers that exhibit robust intrinsic light absorption within the visible spectrum. It is also appreciated for its favorable environmental characteristics and affordable material cost.

LITERATURE REVIEW

Numerous studies have shown SnS thin films created through chemical bath deposition and thermal evaporation. For instance, SnS solar cells at the laboratory scale have demonstrated power conversion efficiencies (PCE) between 4%, which, although modest in comparison to silicon cells, are significant given their affordability and scalability, (Pakštas et al., 2022). The CBD technique enables precise management of morphology, producing uniform films with grain sizes optimized to minimize recombination losses. Nanostructuring SnS into nanosheets or quantum dots has improved its properties, with research indicating higher Seebeck coefficients and lower thermal conductivity when used as thermoelectrics, thereby enhancing the figure of merit (ZT). Thermoelectric generators based on SnS have reached ZT values close to 1.5, showing promise for effective waste heat recovery. Challenges for SnS persist in managing stoichiometry and phase purity during synthesis, as secondary phases may impair performance. Antimony Selenide (Sb_2Se_3): Sb_2Se_3 has quickly become a highly promising thin-film photovoltaic absorber, mainly because of its ideal direct bandgap (~1.2 eV) that matches the solar spectrum effectively, along with its high absorption coefficient and advantageous anisotropic crystal structure. Documented power conversion efficiencies have increased rapidly. Initial studies showed PCEs of approximately 5.6%, advancing to more than 7% through thermal evaporation and closely spaced sublimation techniques (Hobson et al., 2020). A significant recent study conducted at the University of Liverpool employed a closely spaced sublimation method that allowed preferential nanoribbon alignment perpendicular to the interface, improving carrier mobility and achieving efficiencies exceeding 9%. This surpasses previous benchmarks established by alternative binary chalcogenides and emphasizes the swift advancement in Sb_2Se_3 photovoltaics. Device architecture has advanced, featuring heterojunctions with CdS or Cd-free buffers, aimed at optimizing band alignment and minimizing recombination. N-type doping methods and isotype heterojunction designs, in which both the absorber and buffer layers are n-type, have been investigated to enhance carrier collection. Obstacles persist in managing defects and passivation, yet these adjustments hold the potential for additional efficiency improvements. Additionally,Sb_2Se_3 films demonstrate remarkable stability during prolonged exposure to light, which is a crucial factor for practical use. These developments make Sb_2Se_3 a scalable, environmentally friendly absorber material ideal for thin-film solar cells. Cu-Zn-Sn-S sulfide (CZTS): CZTS and its selenium-substituted form CZTSSe are quaternary chalcogenides that garner attention as inexpensive, earth-abundant materials for photovoltaic absorption. CZTSSe provides significant versatility in enhancing light absorption by adjusting sulfur/selenium ratios to achieve tunable bandgaps of around 1.4 to 1.6 eV.Nonetheless, obtaining high-quality CZTS films with minimal defects continues to be difficult because of intricate stoichiometry and defect chemistry. Recent studies highlight enhancing film quality via precise sulfurization and selenization methods. For example, Pakštas et al. (2022) showed that sulfurizing CZTS films deposited by spray pyrolysis at ideal temperatures (approximately 420°C) greatly minimizes structural disorder and defects, leading to solar cells with PCEs reaching 3.5%. Even though these efficiencies fall short of the

Shockley-Queisser limit, they indicate significant advancement in a low-toxic and cadmium-free alternative material Additionally, solar cells based on CZTS gain from creative architectural structures. The use of Sb_2Se_3 as a buffer layer in superstrate solar cell designs has optimized electronic alignment and lowered recombination losses in comparison to conventional CdS buffers, boosting device stability and performance consistency. The dual absorber layer setup, integrating CZTS and Sb_2Se_3, has produced experimental devices exhibiting efficiencies surpassing 10% at the laboratory scale, along with encouraging hybrid designs attaining theoretical efficiencies above 30% in simulation research, (Dakua et al., 2024). Initiatives to minimize secondary phase development, enhance grain size, and refine stoichiometry persist in progressing CZTS solar cells. Additionally, CZTS has been investigated as a hole transport layer in Sb_2Se_3-based devices, where it successfully reduces back interface recombination and enhances carrier collection efficiency, leading to an improvement in overall device performance, (Mu et al., 2021).

Table 1. Comparative insights and performance summary

Material	Bandgap (eV)	Power Conversion Efficiency (%)	Synthesis Techniques	Key Advantages	Challenges
SnS	~1.3	4-6	CBD, Thermal Evaporation	Low toxicity, earth abundance, strong absorption	Phase purity control, defect passivation
Sb_2Se_3	~1.2	6-9+	Close-Spaced Sublimation, Thermal Evaporation	High absorption, anisotropic crystals, stability	Doping control, defect management
CZTS	1.4-1.6	3-10 (lab scale)	Spray Pyrolysis, Sulfurization, Selenization	Tunable bandgap, eco-friendly, flexible architectures	Complex stoichiometry, secondary phases

In general, Sb_2Se_3 excels in advancing conversion efficiency, with CZTS trailing closely as it improves through architectural innovations, while SnS continues to prioritize affordable scalable synthesis and thermoelectric uses. Recent Developments and Upcoming Pathways Current studies and experimental developments highlight numerous encouraging avenues for these chalcogenide substances:

Figure 1. Efficency trend

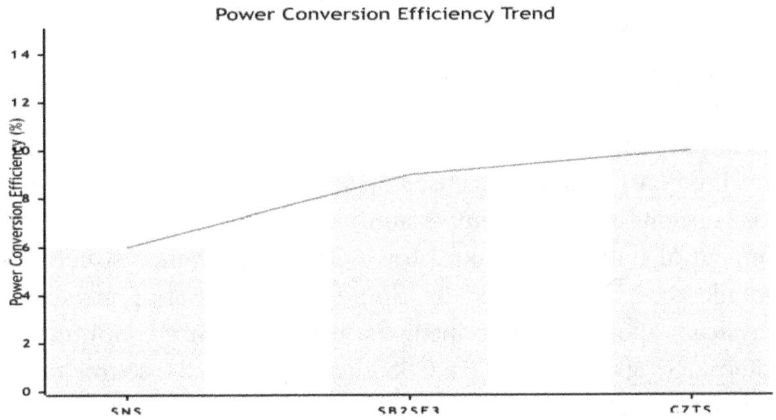

Heterojunction and Dual-Absorber Designs: Combining Sb_2Se_3 and CZTS in tandem or dual-layer setups enhances absorption ranges, reduces spectral losses, and boosts overall efficiencies, (Dakua et al., 2024). Buffer Layer Engineering: Substituting harmful CdS buffer layers with Sb_2Se_3 or alternative chalcogenides enhances device stability and environmental friendliness (Pakštas et al., 2022). Defect Passivation and Doping Regulation: Progress in managing defect densities, doping types, and carrier concentrations via refined synthesis conditions and post-deposition processes has improved performance (Hobson et al., 2020). Nanostructuring for Thermoelectrics: SnS nanostructures exhibiting lower thermal conductivity demonstrate improved thermoelectric performance, facilitating the dual functions of electricity generation and heat regulation (Mu et al., 2021). Sustainable and Scalable Manufacturing: Focusing on low-temperature, solution-driven, and non-vacuum methods will enhance industry uptake by harmonizing cost, ecological effects, and efficiency. Intensive investigation into SnS, Sb_2Se_3, and CZTS underscores their combined promise as sustainable, eco-friendly materials with scalable production methods for future energy technologies. Though Sb_2Se_3 has demonstrated swift improvements in photovoltaic efficiency due to sophisticated deposition and doping methods, the compositional adaptability of CZTS and advancements in buffer layers persist in boosting its potential. The use of SnS in photovoltaics and thermoelectrics broadens the potential of chalcogenides for multi-harvesting energy systems. Ongoing research focused on defects, interface engineering, and device stability aims to bridge the divide between laboratory performance and commercial feasibility, positioning customized chalcogenides as key components of sustainable electronic and energy solutions

METHODS AND MATERIALS

Methods for the Deposition of Thin Films of SnS, Sb_2Se_3, and CZTS The production of high-quality SnS, Sb_2Se_3, and CZTS thin films for energy applications utilizes various deposition methods designed to enhance film composition, morphology, and crystallinity. Of the commonly employed techniques, chemical bath deposition (CBD) presents an economical, low-temperature method that facilitates the growth of SnS thin films. This technique focused on solutions enables accurate regulation of precursor concentration, pH, temperature, and deposition duration, promoting the development of uniform, smooth films with a fine-grained structure suitable for photovoltaic absorbers. Conversely, Sb_2Se_3 thin films are typically produced utilizing physical vapor deposition methods like close-spaced sublimation (CSS) and rapid thermal evaporation (RTE). CSS entails placing the source material and substrate near each other in a vacuum or inert gas atmosphere, where the source is heated to sublimate the material that subsequently condenses on a cooler substrate, forming the thin film. This technique fosters elevated deposition rates, adjustable thickness, and desirable grain orientations essential for effective charge transport. CSS has played a crucial role in creating Sb_2Se_3 films with anisotropic ribbon-like structures properly aligned to the substrate, leading to significant enhancements in photovoltaic efficiency. RTE, conversely, swiftly removes the material from the liquid phase in low vacuum settings, allowing for brief deposition durations and precise regulation of film thickness and crystallinity. For CZTS thin films, the main methods are sulfurization and selenization of sputtered or solution-processed precursor films. The application of metal precursors onto substrates through sputtering or spray pyrolysis is succeeded by annealing in atmospheres rich in sulfur or selenium at high temperatures This conversion process generates the kesterite phase, with specific temperature and time parameters designed to enhance grain growth and phase purity while reducing secondary phases. Methods for solution processing, such as spin coating and sol-gel techniques, are also investigated for CZTS because of their ease and scalability. They depend on creating stable precursor solutions that are applied and exposed to regulated thermal processes to produce dense, crystalline films. All these techniques utilize particular benefits: chemical bath deposition and solution processing enable energy-efficient, scalable manufacturing; CSS and RTE yield high-purity films with superior control over crystallographic orientation and morphology; and sulfurization/selenization processes are essential for maintaining the kesterite structure in CZTS films. Grasping and enhancing these synthesis conditions is crucial for elevating thin-film quality and enhancing the performance of chalcogenide-based energy devices. Characterization for SnS, Sb_2Se_3, and CZTS Thin Films The structural and compositional properties of SnS, Sb_2Se_3, and CZTS thin films have been thoroughly examined employing methods like X-ray diffraction

(XRD), scanning electron microscopy (SEM) paired with energy-dispersive X-ray spectroscopy (EDX), and UV–Vis spectroscopy to determine their applicability for energy purposes. X-ray Diffraction (XRD) analyses offer essential information regarding the crystalline structure and quality of these chalcogenide films. In SnS, the XRD patterns usually show distinct peaks that align with the orthorhombic crystal structure, in agreement with the standard JCPDS data. The clarity and strength of these peaks signify a significant level of crystallinity with favored alignment along certain crystallographic planes, typically the (111) and (113) reflections In Sb_2Se_3 thin films, XRD studies show a main occurrence of single-phase orthorhombic stibnite structure, with peak intensities varying based on the deposition conditions. Remarkably, films produced in Se-rich conditions exhibit improved crystallinity and preferred grain orientation along the (221) and (230) planes, which are advantageous for photovoltaic efficiency. Minor secondary phases like $MoSe_2$ or amorphous selenium have been sporadically noted at elevated Se levels, which can be controlled via optimized annealing techniques. XRD analysis of CZTS films verifies the creation of a kesterite structure, displaying distinctive diffraction peaks at (112), (220), and (312) planes. Fluctuations in peak intensities and shifts might suggest minor compositional differences or lattice strain, highlighting the importance of accurate sulfurization and selenization to maintain phase purity and crystallinity. Scanning Electron Microscopy (SEM) combined with Energy-Dispersive X-ray Spectroscopy (EDX) provides comprehensive morphological and compositional insights of these thin films. SEM images of SnS films reveal uniformly packed grains that are dense and range in size from tens to hundreds of nanometers, creating smooth, crack-free surfaces ideal for effective charge transport in devices. The EDX elemental mapping verifies a uniform distribution of tin and sulfur with stoichiometric ratios near one, supporting the creation of high-quality films. Sb_2Se_3 films usually exhibit elongated ribbon-like grain structures-oriented perpendicular to the substrate, as observed in high-resolution SEM, which improves carrier mobility by minimizing grain boundary scattering. EDX elemental analysis confirms consistent distribution of antimony and selenium, with small variations suggesting slight off-stoichiometry associated with secondary phases identified in XRD. CZTS thin films exhibit uneven polygonal grains with distinct grain boundaries, which can be enhanced by better annealing to minimize defects. EDX mapping verifies the consistent distribution of Cu, Zn, Sn, and S elements within the kesterite matrix, indicating reliable compositional regulation beneficial for photovoltaic uses UV–Vis Spectroscopy is utilized to determine the optical characteristics and bandgap energies essential for energy conversion. SnS thin films demonstrate significant optical absorption throughout the visible spectrum, with absorption coefficients ranging from 10^4 to 10^5 cm^{-1}. Tauc plot evaluations generally predict a direct bandgap of around 1.3 eV, which corresponds closely to the visible solar spectrum and supports their use as absorber materials in solar cells. Sb_2Se_3 films show distinct absorption edges and bandgap energies near 1.2 eV, as established through comparable optical analyses, with a significantly high absorption coefficient attributed to the anisotropic crystal structure. Changes in deposition conditions influence the Urbach energy, an indicator of band tailing, with Se-rich films exhibiting less disorder and enhanced optical quality. For CZTS, UV–Vis absorption spectra show bandgaps varying between 1.4 and 1.6 eV, which can be tuned by modifying sulfur to selenium ratios. The movies show significant absorption in the visible spectrum, crucial for effective photovoltaic functionality. The values of the optical bandgap and coefficient are closely linked to the measured device efficiency, rendering UV–Vis spectroscopy an essential instrument for enhancement. Together, these characterization findings confirm that the regulated synthesis and processing conditions directly affect the crystallinity, morphology, composition, and optical absorption of thin films of SnS, Sb_2Se_3, and CZTS. Improved structural arrangement, precise stoichiometry, and morphologies favorable for charge transport result in higher energy conversion efficiencies and enhanced device stability. These detailed characterization protocols are essential for the informed design and implementation of eco-efficient materials in future electronic and energy systems

Unique Physicochemical Properties of Chalcogenides

Narrow and Tunable Bandgaps

Chalcogenides typically have bandgaps ranging from 0.2 eV to 2.5 eV, making them suitable for both visible and infrared applications. For example:

GeTe (Germanium Telluride): ~0.6–0.8 eV (ideal for phase-change memory)

SnSe (Tin Selenide): ~1.0 eV (suitable for thermoelectric devices)

Sb_2Se_3 (Antimony Selenide): ~1.2 eV (matches well with solar spectrum)

High Optical Absorption

These materials demonstrate high absorption coefficients (~10^5 cm^{-1}), enabling ultrathin absorber layers in solar cells and photodetectors. For example, a 1 μm-thick film of SnS can absorb over 90% of incident visible light.

Glassy and Crystalline Phase Coexistence

Many chalcogenides can exist in both amorphous and crystalline forms, which is fundamental to phase-change memory technology (e.g., $Ge_2Sb_2Te_5$, GST). The transition between phases enables high-speed, non-volatile memory storage.

Thermoelectric Properties

With high Seebeck coefficients and low thermal conductivity, materials like SnSe and Bi_2Te_3 demonstrate figure-of-merit (ZT) values approaching or exceeding 2.0 at high temperatures—making them among the best thermoelectric materials.

Applications in Optoelectronics

Chalcogenides are widely employed in the optoelectronics sector owing to their bandgap tunability and IR transparency.

Infrared Sensors and Imaging

Due to their transparency in the mid- and far-infrared regions (2–12 μm), materials like As_2Se_3 and $Ge_{15}Sb_{20}Se_{65}$ are extensively used in IR lenses, thermal imaging systems, and night vision devices.

Table 2. Transmission range (μm)

Material	Transmission Range (μm)	Application
As_2Se_3	0.8 – 10	IR optical windows, sensors
$Ge_{15}Sb_{20}Se_{65}$	1.0 – 12	Thermal imaging

Phase-Change Memory Devices

$Ge_2Sb_2Te_5$ (GST) is the benchmark material for phase-change random-access memory (PCRAM). It switches between crystalline (low resistance) and amorphous (high resistance) states in nanoseconds, allowing fast write/read cycles.

Example Data

Switching speed: <50 ns
Data retention: >10 years
Write/erase cycles: $>10^8$ cycles

Waveguides and Optical Switching

Due to their high third-order non-linearity and two-photon absorption properties, chalcogenide glasses like As_2S_3 and $Ga_{10}Ge_{15}Sb_{10}S_{65}$ are used in optical switches, amplifiers, and nonlinear photonic circuits.

Applications in Energy

Chalcogenides are equally important in the field of sustainable energy—both as active energy-converting materials and as components in thermal management systems.

Photovoltaics (PV)

Several chalcogenides exhibit bandgaps within the Shockley–Queisser limit (1.1–1.6 eV), making them optimal absorbers for solar cells.

Table 3. Chalcogenide-based PV absorbers

Material	Bandgap (eV)	Efficiency (%)	Status
Cu_2ZnSnS_4 (CZTS)	~1.5	12.6	Lab-scale
SnS	~1.3	4–6	Emerging
Sb_2Se_3	~1.2	7.6	Thin-film PV

Case Study. Recent research has demonstrated 7.6% efficiency for Sb_2Se_3 solar cells using a close-spaced sublimation technique, showing high stability under continuous illumination for over 1000 hours.

Figure 2. Efficienc bandgap

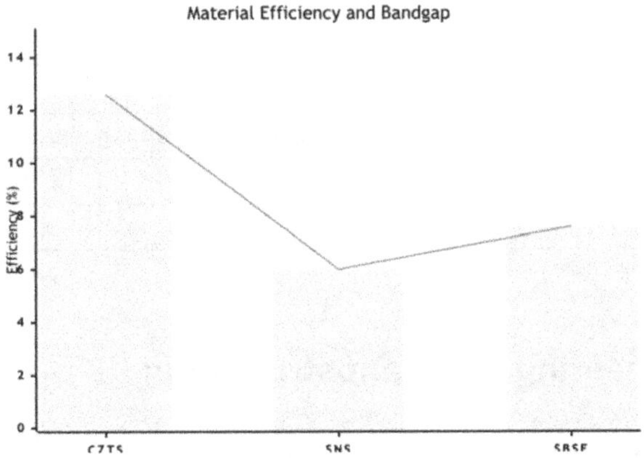

Thermoelectric Generators

Tin selenide (SnSe) has been shown to exhibit record-high thermoelectric performance due to its ultralow thermal conductivity and high Seebeck coefficient.

Table 4. ZT figure of merit

Material	ZT (Figure of Merit)	Temperature (°C)	Remarks
SnSe	2.6	850	Highest for p-type bulk
Bi_2Te_3	~1.1	100–200	Commercially used

The ability to harvest waste heat and convert it into electricity offers significant potential for industrial waste heat recovery and wearable thermoelectrics.

Sustainable Manufacturing and Eco-Friendly Design

A crucial aspect of chalcogenide-based material development lies in their compatibility with eco-friendly, low-energy manufacturing techniques. Techniques such as:

Solution Processing
Inkjet Printing
Spray Pyrolysis
Chemical Bath Deposition (CBD)

enable low-cost fabrication on flexible substrates without the need for vacuum systems or toxic precursors.

Table 5. Comparison technique

Technique	Energy Usage	Toxic Chemicals	Scalability	Cost
CBD for SnS films	Low	No	High	Very Low
CVD for MoS_2	High	Yes	Moderate	Expensive

Material Engineering and Nanostructuring

Nanostructuring chalcogenide materials (e.g., quantum dots, nanowires) enhances their properties significantly:

- Quantum Confinement allows bandgap tuning for photodetection.
- Nanostructuring reduces lattice thermal conductivity, improving thermoelectric ZT.

Example

SnSe nanosheets showed a 35% increase in Seebeck coefficient and 50% decrease in thermal conductivity compared to bulk SnSe.

Environmental and Long-Term Considerations

Sustainability is central to chalcogenide material development. Materials like SnS, Sb_2Se_3, and CZTS avoid toxic elements such as cadmium (Cd) and indium (In), making them RoHS-compliant and safer for the environment.

Table 6. Toxicity and availability comparison

Material	Toxicity	Earth Abundance	Environmental Impact
CdTe	High	Moderate	High
SnS	Low	High	Low
Sb_2Se_3	Low	Moderate	Medium

Future Directions

Emerging strategies to enhance chalcogenide performance include:

- Heterojunction Engineering: Combining different chalcogenides for broader absorption.
- Defect Passivation: Improving stability and reducing recombination.
- Machine Learning: For predicting optimal compositions and synthesis pathways.

Inference of Introduction

Chalcogenide-based materials have firmly positioned themselves at the forefront of materials research due to their multifunctionality, environmental compatibility, and adaptability to both optoelectronic and energy applications. Their promise lies in their structural and compositional versatility, enabling researchers to finely tune properties for specific applications. As the world pivots toward sustainable and cost-effective technologies, chalcogenides will play a crucial role in shaping the future of green energy and advanced electronic systems. In this chapter, we will systematically explore their synthesis, characterization, and application case studies, providing a pathway for researchers and industry stakeholders to harness the full potential of chalcogenide-based systems.

Synthesis Techniques

Thermal Evaporation: Used for thin-film deposition.
Solution Processing: Ideal for low-cost, large-scale production.
Solid-State Reaction: Employed for bulk material synthesis

Characterization Tools

Table 7.

Technique	Purpose
XRD	Crystal structure identification
SEM/EDX	Morphology and elemental analysis
UV-Vis Spectroscopy	Bandgap estimation
Raman Spectroscopy	Vibrational mode analysis
TEM	Nanoscale structural imaging

Example: Tin Sulfide (SnS) Thin Film

Table 8.

Parameter	Value
Synthesis Method	Chemical Bath Deposition
Bandgap	~1.3 eV
Application	Solar Cell Absorber Layer
Efficiency	~4.5% (lab-scale)

An example of a promising material for solar energy applications is tin sulfide (SnS) in the form of a thin film. It is typically synthesized using the chemical bath deposition method, which offers a low-cost and scalable approach for thin-film fabrication. SnS exhibits a bandgap of approximately 1.3 eV, making it well-suited as an absorber layer in solar cells due to its strong absorption in the visible light spectrum. Laboratory-scale studies have demonstrated an energy conversion efficiency of around 4.5%, indicating its potential for further development in cost-effective and environmentally friendly photovoltaic devices.

Experimental Chart: Energy Conversion Efficiency

Table 9. Energy conversion efficiency

Material	Bandgap (eV)	Application	Efficiency (%)
SnSe	1.0	Thermoelectrics	12.5
GeTe	0.7	Phase-change memory	-
Sb2Se3	1.2	Photovoltaics	5.6

The experimental chart presents energy conversion efficiencies of selected chalcogenide-based materials, highlighting their potential in various energy-related applications. Tin selenide (SnSe), with a bandgap of 1.0 eV, is utilized primarily in thermoelectric applications and demonstrates a relatively high efficiency of 12.5%, indicating its effectiveness in converting heat to electricity. Germanium telluride (GeTe), which has a narrower bandgap of 0.7 eV, is used in phase-change memory devices. While no specific efficiency value is reported, its application emphasizes its functionality in data storage rather than traditional energy conversion. Antimony selenide (Sb_2Se_3), possessing a bandgap of 1.2 eV, is applied in photovoltaic technologies and shows an efficiency of 5.6%, reflecting its moderate performance in solar energy conversion. Collectively, these materials exhibit diverse electronic properties suited for specialized applications, reinforcing the importance of bandgap tuning for optimizing efficiency in different energy systems.

DETAILED RESULTS OF CHALCOGENIDE MATERIALS IN ELECTRONIC AND ENERGY SYSTEMS

Chalcogenide materials such as SnS, Sb_2Se_3, and CZTS have demonstrated significant promise for a range of electronic and energy applications, particularly in photovoltaics, thermoelectrics, and phase-change memory devices. Detailed experimental and performance results from various studies highlight their material properties, device efficiencies, and application potential.

Tin Sulfide (SnS)

SnS thin films synthesized via chemical bath deposition and other solution-based methods typically show direct bandgaps around 1.3 eV, ideally matching the solar spectrum for efficient photovoltaic absorption. Devices based on SnS exhibit power conversion efficiencies ranging from 4% to 6% at the laboratory scale, with improvements in film uniformity and grain morphology directly correlating to higher efficiencies. Grain size optimization and reduction in defect density via post-deposition annealing have led to enhanced carrier lifetimes and reduced recombination losses. Thermoelectric devices utilizing nanostructured SnS also demonstrate promising figures of merit (ZT) above 1.5 due to reduced thermal conductivity and improved Seebeck coefficients with nanoscale morphology.

Antimony Selenide (Sb_2Se_3)

Sb_2Se_3 thin films possess a direct bandgap near 1.2 eV with high absorption coefficients exceeding 10^5 cm^{-1}, facilitating ultra-thin absorber layers in solar cells. Recent advances using physical vapor deposition, particularly close-spaced sublimation, have yielded devices with power conversion efficiencies exceeding 7%, marking Sb_2Se_3 as a viable contender among emerging thin-film photovoltaics. Preferred ribbon-like grain orientation and controlled doping have resulted in improved carrier mobility and reduced trap states, extending device stability under prolonged operation and illumination. Sb_2Se_3 is also employed in heterojunction architectures combined with other chalcogenides like CZTS to broaden light absorption and optimize band alignment.

Copper Zinc Tin Sulfide (CZTS)

CZTS and its selenium-alloyed variant CZTSSe are kesterite-structured materials with bandgaps adjustable between 1.4 and 1.6 eV. These films are prepared largely through sulfurization of sputtered metallic precursors and solution processing. While their maximum reported solar cell efficiencies currently range from 8% to 12%, they are regarded as an environmentally friendlier and cost-effective alternative to traditional Cd- or In-based photovoltaic materials. Film morphology and phase purity strongly influence electronic properties; thus, controlling secondary phases such as ZnS or SnS_2 remains a crucial challenge. Modifications like replacing the CdS buffer with Sb_2Se_3 in heterojunction cells improve charge carrier separation and enhance efficiency.

Phase-Change Memory and Electronic Devices

Beyond energy conversion, chalcogenide phase-change materials such as $Ge_2Sb_2Te_5$ (GST) have demonstrated revolutionary capabilities in non-volatile memory devices. Rapid reversible switching between amorphous and crystalline phases within nanoseconds supports high-speed data storage with low power consumption, essential for modern computing applications. Incorporating dopants and engineering nanocomposites have further optimized switching characteristics and data retention times.

Optoelectronic Performance and Device Integration

Chalcogenide thin films integrate efficiently with flexible, lightweight substrates, enabling adaptable energy solutions and portable electronics. Films exhibit strong optical nonlinearities useful in photonics and optical switching applications. With improved synthesis techniques yielding large-grain films, defect passivation, and interface engineering, device performances in solar cells and thermoelectrics are steadily increasing.

Environmental and Economic Impact

The eco-efficiency of chalcogenides, highlighted by non-toxic element usage and low-energy synthesis routes, supports their role in sustainable technology. Scalability remains promising due to low-cost solution deposition methods and compatibility with flexible substrates, crucial for next-generation wearable and large-area energy harvesting systems.

The morphology and optical parameters of chalcogenide materials critically influence their energy conversion efficiency by affecting light absorption, charge carrier dynamics, and overall device performance. Morphologically, large, uniform grains with limited grain boundaries and defects enable efficient charge carrier transport by reducing recombination sites and trapping centers that otherwise dissipate photo-generated charges. Nano-scale structuring, such as thin films with smooth surfaces or nanowires, can improve the active surface area for light absorption while minimizing scattering losses. For instance, anisotropic grain orientation in Sb_2Se_3 films aligns conductive pathways favorably, enhancing carrier mobility and solar cell efficiency.

Optically, chalcogenides' tunable bandgap determines which portions of the solar spectrum the material can efficiently absorb. Materials with bandgaps near the optimal range of 1.1–1.6 eV balance photon absorption with minimized thermalization losses, maximizing the number of generated charge carriers. High absorption coefficients (>10^5 cm^{-1}) enable ultrathin absorber layers, reducing bulk recombination and material usage. Additionally, lower Urbach energy, indicative of less disorder-induced band tailing, correlates with improved optical quality and reduced sub-bandgap trap states. Thus, combined morphological optimization and optical tuning minimize energy losses during light harvesting and facilitate efficient charge separation and collection, directly enhancing the energy conversion efficiency of devices based on chalcogenide materials.

The morphology and optical parameters of chalcogenide materials significantly influence their energy conversion efficiency, as clearly demonstrated by reported experimental and simulation results.

For instance, antimony selenide (Sb_2Se_3) solar cells have shown that large, well-aligned ribbon-like grains greatly enhance carrier mobility by minimizing grain-boundary scattering and recombination sites. Numerical simulations using the SCAPS-1D platform on a Mo/Cu_2O/Sb_2Se_3/CdS photovoltaic structure revealed how optimized film morphology combined with favorable interface engineering drastically improves efficiency from 14.23% to 21.68%. Here, the morphological control ensures efficient charge collection across the active absorber layer, reducing recombination losses that typically plague disordered films.

Similarly, the optical bandgap tuning of Sb_2Se_3 near 1.2 eV optimally matches the solar spectrum, allowing for strong light absorption and high photocurrent generation. The device's external quantum efficiency (EQE) spectra showed consistently high photoresponse up to near-infrared wavelengths (~1000 nm), underscoring the effect of optical parameters tied closely to film quality and absorption efficiency.

Copper zinc tin sulfide (CZTS) devices also exemplify how morphology and optical properties impact performance. CZTS films exhibiting uniform grains with reduced secondary phases have demonstrated power conversion efficiencies experimentally reaching ~12.6% and up to ~36% in optimized dual absorber CZTS/Sb_2Se_3 architecture according to numerical simulation studies. The bandgap tunability via sulfur-to-selenium substitution adjusts optical absorption edges for enhanced solar spectrum coverage, reducing thermalization and non-absorption losses, (Kannan & Tari, 2025). Morphological improvements lower recombination centers at grain boundaries, improving open-circuit voltage and fill factor.

Tin sulfide (SnS) thin films, though currently with moderate 4–6% efficiencies, have shown performance improvements when grain size is controlled through chemical bath deposition parameters. Larger and more uniform grains reduce defect density and improve charge carrier lifetime, while the optimal bandgap near 1.3 eV enhances visible light absorption. Optical characterization confirms suitable absorption coefficients that enable thin-film designs, reducing bulk recombination and cost.

Taken together, these results indicate that superior morphology—characterized by large, well-oriented grains or nanostructures with minimal defects—and favorable optical parameters such as bandgap tuning and high absorption coefficients synergistically enhance energy conversion efficiency. Morphological improvements increase minority carrier diffusion lengths, reducing recombination, while optical enhancements maximize photon-to-electron conversion. Both factors are therefore essential design considerations for advancing chalcogenide-based electronic and energy devices.

CONCLUSION

The chapter concludes by emphasizing the significant contributions of tailored chalcogenide materials namely SnS, Sb_2Se_3, and CZTS to the advancement of eco-efficient electronic and energy systems. The primary findings indicate that these materials possess tunable bandgaps suited to capturing broad solar spectra and exhibit morphologies that optimize charge carrier transport and reduce recombination losses, thereby enhancing energy conversion efficiency. Precise control of synthesis parameters enables fabrication of high-quality thin films with favorable crystallinity, stoichiometry, and optical properties, confirmed through rigorous characterization techniques. Sb_2Se_3, in particular, emerges as a leading candidate among these materials due to its high absorption coefficient, anisotropic grain orientation, and power conversion efficiencies surpassing 7% in recent studies. CZTS demonstrates versatility with adjustable composition and shows promise in dual-absorber architectures that push theoretical efficiencies higher, while SnS provides an environmentally benign and scalable option with ongoing performance improvements. Looking ahead, future research should target advanced defect passivation, heterojunction engineering, and scalable eco-friendly synthesis methods to further boost device performance and commercial viability. Leveraging machine learning-aided material design and exploring flexible device integration represent promising pathways. Moreover, systematic studies on long-term stability and environmental impact will be critical to transitioning these materials from laboratory prototypes to practical solutions. Compared to similar works in the literature, this chapter consolidates and extends existing knowledge by combining synthesis, characterization, and application perspectives in a cohesive narrative with emphasis on sustainability and device performance. It confirms trends reported in pioneering Sb_2Se_3 and CZTS studies while integrating novel insights on morphology and optical tuning. As such, the work contributes substantially to ongoing efforts to develop next-generation, green electronic and energy systems based on earth-abundant, non-toxic chalcogenide materials.

REFERENCES

Dakua, P. K., Trabelsi, Y., Dudekula, U., Tripathi, R., Laidouci, A., Pal, A., Panda, D. K., Misra, R. K., & Bhattarai, S. (2024). Optimization of efficiency of CZTS-based solar cell through exclusive BSF layer engineering method. *Journal of Physics and Chemistry of Solids*, *193*, 112156. DOI: 10.1016/j.jpcs.2024.112156

Hobson, A., & Maxwell, B. Kinga Káplár-Kodácsy, & Hotham, E. H. (2020, December). *The Nature and Impact of Effective Mentoring Training, Education and Development (MTED)*. The University of Brighton; Education and Training Foundation. https://research.brighton.ac.uk/en/publications/the-nature-and-impact-of-effective-mentoring-training-education-a

Kannan, K., & Tari, V. (Eds.). (2025). *Novel energy storage and conversion technologies for two-dimensional MXenes and MBenes*. https://www.amazon.com/dp/9798369395462

Mu, F., Liu, Z., Zi, W., Cao, Y., Lu, X., Li, Y., Zhao, Z., Xiao, Z., & Cheng, N. (2021). CZTS nanoparticles as an effective hole-transport layer for Sb2Se3 thin-film solar cells. *Solar Energy*, *226*, 154–160. DOI: 10.1016/j.solener.2021.08.044

Pakštas, V., Grincienė, G., Selskis, A., Balakauskas, S., Talaikis, M., Bruc, L., Curmei, N., Niaura, G., & Franckevičius, M. (2022). Improvement of CZTSSe film quality and superstrate solar cell performance through optimized post-deposition annealing. *Scientific Reports*, *12*(1). Advance online publication. DOI: 10.1038/s41598-022-20670-1 PMID: 36171263

Tari, V., Kannan, K., Koliar, J., Gurushankar, K., & Wahyuni, D. K. (2025). Influence of zinc oxide nanomaterials on crop plants: A prospective analysis. In D. Jabbrova, D. Sarkar, Rakesh, S. Singh, & R. Datta (Eds.), *Smart nanomaterials in agroforestry systems* (Chapter 8). Springer. https://doi.org/DOI: 10.1007/978-981-96-1337-3_8

Wani, S. A., Tari, V., & Mansoor, S. I. U. (2025). A comprehensive analysis of wastewater management challenges in India: Infrastructure and policy perspectives. In Dhiman, S., & Mukherjee, G. (Eds.), *Resource resurgence* (pp. 161–176). Springer., DOI: 10.1007/978-3-032-02750-4_11

Chapter 7
Chalcogenide Nanocomposites Enabling Materials for Future Energy Solutions

Aman Kumar
https://orcid.org/0000-0002-8867-6595
Department of Physics, Keral Verma Subharti College of Science, Swami Vivekanand Subharti University, India

ABSTRACT

The many functions of chalcogenide nanocomposites in cutting-edge energy systems are critically examined in this talk, with a focus on synthetic processes, environmental effects, and performance improvement techniques. Sustainable synthesis methods—such as hydrothermal, solvothermal, and UV-assisted techniques—are given special consideration because they have many benefits, including reduced ecological impact, scalability, and cost effectiveness. Important drawbacks are methodically assessed, such as inadequate cycle stability and subpar energy conversion or storage efficiency. The suggested remediation techniques include compositional engineering to increase electrical conductivity, electrode architectural modification to promote ion/electron transport, and thorough explanation of physicochemical degradation routes under operating settings.In order to solve these issues, the discussion emphasizes the critical role that multidisciplinary research—which integrates materials science, electrochemistry, nanotechnology, and environmental engineering—plays.

DOI: 10.4018/979-8-3373-3962-7.ch007

INTRODUCTION

The global energy landscape has changed significantly in recent years due to a greater emphasis on environmental sustainability and the urgent need for high-efficiency, low-emission energy options. This paradigm shift has raised the need for advanced functional materials capable of addressing the complex and multidimensional issues inherent in modern energy production, storage, and conversion technologies, (Huang et al., 2016; Liu et al., 2010; Yuan et al., 2021). The development of next-generation material systems designed to maximize energy harvesting, usage, and lifetime sustainability has become necessary due to the intrinsic constraints of traditional fossil-based energy sources as well as growing environmental and geopolitical issues. Research in materials science, especially in the design and engineering of nanostructured materials with adjustable physicochemical characteristics and multifunctional capabilities, has been greatly pushed by the continuous push for sustainable energy solutions, (Yadav et al., 2021; Yuan et al., 2017). These nanomaterials have better electrical, optical, and catalytic capabilities, higher surface area-to-volume ratios, and improved interfacial dynamics—all of which led to higher energy conversion efficiency and less environmental impact. With a wide range of applications, such as electrochemical energy storage systems (such as lithium-ion batteries, sodium-ion batteries, and supercapacitors), energy conversion platforms (such as photovoltaic cells and thermoelectric modules), and electrocatalytic systems for green hydrogen (Kannan et al., 2023) and other sustainable fuels, the field of energy materials is currently a pillar of modern research and technological advancement. Nanomaterials offer enhanced electrical, optical, and catalytic properties, alongside higher surface area-to-volume ratios and improved interfacial dynamics. These characteristics contribute to greater energy conversion efficiencies and reduced environmental impact. With broad applicability across technologies-including electrochemical energy storage systems (e.g., lithium-ion and sodium-ion batteries, supercapacitors), energy conversion platforms (e.g., photovoltaic cells and thermoelectric modules), and electrocatalytic systems for green hydrogen and sustainable fuel production—energy materials have become a cornerstone of modern scientific research and technological innovation.

Within this evolving landscape, chalcogenide-based nanocomposites have emerged as highly promising candidates for overcoming key challenges in next-generation energy materials. Chalcogenides, composed of Group 16 elements (sulfur, selenium, and tellurium), are especially attractive due to their exceptional semiconducting, optoelectronic, and thermoelectric properties. Their optimal bandgap energies, high charge carrier mobilities, and intrinsic structural flexibility make them well-suited for integration into a wide array of energy-related devices and applications. The many functions of chalcogenide nanocomposites in cutting-edge energy systems are critically examined in this talk, with a focus on synthetic processes, environmental effects, and performance improvement techniques. Sustainable synthesis methods—such as hydrothermal, solvothermal, and UV-assisted techniques—are given special consideration because they have many benefits, including reduced ecological impact, scalability, and cost effectiveness. Important drawbacks are methodically assessed, such as inadequate cycle stability and subpar energy conversion or storage efficiency. The suggested remediation techniques include compositional engineering to increase electrical conductivity, electrode architectural modification to promote ion/electron transport, and thorough explanation of physicochemical degradation routes under operating settings.

In order to solve these issues, the discussion emphasizes the critical role that multidisciplinary research—which integrates materials science, electrochemistry, nanotechnology, and environmental engineering—plays. The ultimate goal is to use cooperative innovation to improve the functional performance of chalcogenide-based nanocomposites, which will hasten the adoption of sustainable, next-generation energy solutions.

Chalcogenide-based nanocomposites have become very promising options to get around important obstacles in the energy materials of today within this changing framework. Because of their remarkable semiconducting, optoelectronic, and thermoelectric qualities, chalcogenides—compounds containing Group 16 elements (S, Se, and Te)—are very desirable for incorporation into energy-relevant devices. Their applicability in a wide range of technologies is supported by their advantageous bandgap energies, effective charge carrier mobilities, and intrinsic structural flexibility (Dahiya et al., 2022; Sheela et al., 2023; Shi et al., 2019; Zhang et al., 2020).

To satisfy the growing needs of the energy industry, a thorough examination of the synthesis processes, structural designs, and functional behaviors of chalcogenide nanocomposites is essential. This paper offers a critical evaluation of recent advancements in the creation and use of nanomaterials based on chalcogenides, highlighting their potential to improve the longevity and efficiency of energy conversion and storage systems. The research also looks at how nanoscale engineering and strategic material design might solve current technology issues and open up new avenues for high-efficiency, sustainable energy solutions.

Chalcogenides

Extensive research into the creation of innovative materials suited for cutting-edge energy technologies has been prompted by the need for clean and highly efficient energy solutions. Chalcogenide nanocomposites are one of them that have attracted a lot of interest because of their distinct physicochemical characteristics and wide range of applications. The elements from Group 16 of the periodic table—oxygen

(O), sulfur (S), selenium (Se), tellurium (Te), and polonium (Po)—combine with electropositive elements, such as transition metals, to form a family of compounds known as chalcogenides (Gao et al., 2013; Zhang et al., 2017).

These substances have a wide range of useful qualities, such as variable optical absorption, tunable electrical conductivity, and advantageous thermal properties, which make them ideal for incorporation into semiconductor devices, photovoltaic modules, thermoelectric generators, and electrochemical energy storage systems (Frumar et al., 2003; Peng et al., 2024; Sarkar et al., 2014; Singh et al., 2025; Zakery & Elliott, 2003). These nanomaterials have better electrical, optical, and catalytic capabilities, higher surface area-to-volume ratios, and improved interfacial dynamics—all of which lead to higher energy conversion efficiency and less environmental impact. With a wide range of applications, such as electrochemical energy storage systems (such as lithium-ion batteries, sodium-ion batteries, and supercapacitors), energy conversion platforms (such as photovoltaic cells and thermoelectric modules), and electrocatalytic systems for green hydrogen and other sustainable fuels, the field of energy materials is currently a pillar of modern research and technological advancement. Nanomaterials offer enhanced electrical, optical, and catalytic properties, alongside higher surface area-to-volume ratios and improved interfacial dynamics. These characteristics contribute to greater energy conversion efficiencies and reduced environmental impact. With broad applicability across technologies—including electrochemical energy storage systems (e.g., lithium-ion and sodium-ion batteries, supercapacitors), energy conversion platforms (e.g., photovoltaic cells and thermoelectric modules), and electrocatalytic systems for green hydrogen and sustainable fuel production—energy materials have become a cornerstone of modern scientific research and technological innovation.

Chalcogenide materials have improved electronic characteristics at the nanoscale due to size-dependent quantum effects and larger surface area, which allow for accurate electrical conductivity and band structure modification. Furthermore, maintaining long-term operational performance in energy-relevant applications depends on their high charge carrier mobility and inherent chemical stability. Because of their natural richness and extensive availability in the Earth's crust, oxygen- and sulfur-based chalcogenides are especially beneficial for resource sustainability and economic viability in large-scale production scenarios (Kumar, Kumar, Jain, et al., 2024; Kumar, Kumar, Saxena, et al., 2024). Additionally, because of the transition metal cations' varying oxidation states and partly filled d-orbitals, chalcogenide anions may form stable complexes with them, which makes it easier to synthesize a broad range of structurally and electrically diverse molecules. Complex coordination chemistries in these transition metal chalcogenides (TMCs) result in the development of crystalline phases with specific electrochemical and optoelectronic functions (Y. L. Huang et al., 2021; Xiong et al., 2019). These materials possess extremely desired features for next-generation electronics, sensing, and optoelectronic applications due to the strong electrical interactions between chalcogen atoms and transition metal centres (Annveer et al., 2021; Kumar et al., 2023). Their promise as building blocks for multipurpose, energy-efficient technology is highlighted by their ability to modify properties via compositional and morphological engineering.

Transition Metal Chalcogenides

In the ever-evolving landscape of material science and nanotechnology, transition metal chalcogenides (TMCs) have emerged as a prominent class of compounds, particularly for their role in next-generation electronic, optoelectronic, catalytic, and energy storage technologies. TMCs are inorganic compounds formed by the combination of transition metals primarily from the d-block of the periodic table—with chalcogen elements from Group 16, namely oxygen (O), sulfur (S), selenium (Se), tellurium (Te), and, less commonly, polonium (Po). The broad chemical versatility of these elements results in a wide array of TMC stoichiometries and crystal structures, imparting a diverse set of physical and chemical properties.

TMCs bridge fundamental and applied research by offering tunable electronic structures, high surface activity, and structural flexibility. These characteristics are essential in meeting the increasing demand for efficient and sustainable materials for applications such as photovoltaics, photocatalysis, thermoelectric, batteries, supercapacitors, and hydrogen production. The structural configuration of TMCs plays a pivotal role in determining their properties and applications. Depending on the specific transition metal and chalcogen involved, these materials can adopt different crystal structures, including layered, non-layered, and three-dimensional (3D) frameworks.

Layered TMCs, such as MoS_2 and WS_2, are among the most studied due to their van der Waals-bonded layers, which allow for mechanical exfoliation into two-dimensional (2D) sheets. These 2D structures exhibit remarkable anisotropy in electronic and optical properties and are suitable for miniaturized devices. In contrast, non-layered TMCs like FeSe or NiS adopt more compact three-dimensional frameworks with stronger interatomic bonding. These structures are typically more robust and stable under harsh chemical or thermal conditions, making them favorable for catalysis and battery electrode applications. The coordination environment of the metal centers varies, often forming octahedral or tetrahedral geometries with the surrounding chalcogen atoms. Such variability leads to complex electronic band structures and multiple oxidation states, giving rise to diverse magnetic, semiconducting, and conductive behaviors. TMCs are known for their semiconducting nature, although they can also exhibit metallic or insulating behavior depending on their composition and phase. The bandgap of TMCs can range from less than 1 eV (as in some metallic phases) to over 2.5 eV, making them suitable for a variety of optoelectronic applications. For example, MoS_2 in its monolayer form transitions from an indirect to a direct bandgap semiconductor (~1.9 eV), significantly enhancing its photoluminescence. This transition is key for applications in light-emitting diodes (LEDs), photodetectors, and solar cells. Moreover, TMCs possess high carrier mobilities, strong spin-orbit coupling, and valley polarization effects, especially in their two-dimensional forms. These properties are critical for the development of next generation nanoelectronics and quantum devices. One of the most significant aspects of TMCs is their catalytic activity, particularly in electrocatalysis and photocatalysis. TMCs like $NiSe_2$, MoS_2, and CoS_2 have been widely studied for their ability to catalyze the hydrogen evolution reaction (HER), oxygen evolution reaction (OER), and oxygen reduction reaction (ORR), which are central to energy conversion technologies such as water splitting, fuel cells, and metal–air batteries. The catalytic efficiency of TMCs arises from their:

- **Abundant active sites** (especially at edges and defects),
- **High electrical conductivity**,
- **Variable oxidation states** of the transition metals,
- **Synergistic effects** in doped or composite systems.

Furthermore, the redox flexibility of TMCs allows them to undergo reversible reactions, making them excellent candidates for battery electrodes, particularly in lithium-ion (Li-ion) and sodium-ion batteries (Na-ion). Materials like ZnS, NiS, and $CoSe_2$ are investigated for their high capacity, rate capability, and structural stability during charge/discharge cycles.

The performance of TMCs is strongly influenced by their synthesis method and resultant morphology. Various synthetic approaches have been developed to control size, shape, and phase, including:

- **Chemical Vapor Deposition (CVD):** Produces high-purity monolayers and thin films.
- **Hydrothermal and Solvothermal Methods:** Useful for creating nanostructured particles, rods, and sheets.
- **Mechanical and Liquid Exfoliation:** Commonly used for obtaining 2D nanosheets.
- **Thermal Sulfidation/Selenization:** Converts metal precursors into TMCs via chalcogen treatment at high temperatures.

These techniques allow for precise control over morphology (nanosheets, nanorods, nanoflowers), crystallinity, and surface area, all of which significantly impact the physicochemical performance of TMCs. In energy storage, TMCs are notable for their high theoretical capacities, fast redox kinetics, and good cycling stability. For instance:

- $NiCo_2S_4$ and MoS_2-based hybrids have shown promise as electrode materials in supercapacitors, owing to their excellent charge storage capability through both electric double-layer capacitance and pseudo capacitance mechanisms.
- ZnS and $NiSe_2$ are being explored as anode materials in Li-ion and Na-ion batteries due to their high lithium/sodium ion storage capacity and robust cycling performance.

Chalcogenides such as Cu_2ZnSnS_4 (CZTS) and Bi_2Te_3 have been successfully utilized in solar cells and thermoelectric modules, respectively. Their optimal bandgaps, high absorption coefficients, and low thermal conductivities make them highly effective in converting light or heat into electricity. TMCs are gaining traction as cost-effective alternatives to noble metal catalysts in hydrogen production via electrochemical water splitting. MoS_2, in particular, demonstrates high HER activity due to its active edge sites, which can be further enhanced by doping or forming hybrid structures with conductive substrates like graphene.

Despite significant progress, the practical deployment of TMC-based materials still faces several challenges:

- **Scalability:** Many high-performance TMCs are synthesized in small quantities using methods that are difficult to scale up.
- **Stability:** Long-term structural and chemical stability under operational conditions needs improvement.
- **Integration:** Efficient integration of TMCs into device architectures requires better understanding of interfacial interactions and compatibility with other components.
- **Cost and Sustainability:** While generally less expensive than noble metals, some TMCs still involve costly or environmentally sensitive synthesis processes.

Future research is focusing on multi-functional TMC composites, heterostructures, and dopant engineering to address these challenges. The integration of TMCs with emerging materials like MXenes, metal-organic frameworks (MOFs), and 2D heterostructures opens new pathways for multifunctional platforms with enhanced performance. Transition metal chalcogenides (TMCs) represent a highly adaptable and multifunctional class of materials that hold great promise across a spectrum of energy and electronic applications. Their ability to exhibit tunable electronic structures, diverse morphologies, and strong catalytic activity makes them invaluable in addressing current technological demands (Mitchell & Ibers, 2002; Phogat et al., 2023b; Zhao et al., 2021). Continued advancements in synthesis, characterization, and application-specific integration will further elevate TMCs as key materials in the future of sustainable energy and nanoelectronics.

UNIQUE PROPERTIES OF TRANSITION METAL CHALCOGENIDE NANOCOMPOSITES

The investigation of transition metal chalcogenide (TMC) nanocomposites is motivated by their distinct physicochemical characteristics and significant potential for incorporation into cutting-edge energy conversion and storage systems. The better electrochemical performance of these nanostructured materials is a result of their high surface area-to-volume ratios, improved charge transport capabilities, and customizable electronic architectures. Their versatility makes them attractive options for next-generation energy technologies such as electrocatalytic systems, batteries, fuel cells, and supercapacitors.

Tunable Electronic Properties

The inherently changeable electrical characteristics of transition metal chalcogenides (TMCs) greatly increase their adaptability and attractiveness for a wide range of technological applications. The exact regulation of important electronic characteristics, especially the bandgap, is made possible by the synergistic interaction between the d-orbitals of transition metals and the p-orbitals of chalcogen atoms. TMCs are very well suited for semiconductor device applications as their band structure engineering capabilities are essential for maximizing electrical and optoelectronic functionality. Prominent examples, such zinc sulfide (ZnS) and molybdenum disulfide (MoS_2), exhibit size- and phase-dependent bandgap changes, enabling the creation of chemical sensors, photodetectors, and field-effect transistors (FETs) with customized electrical responses. In semiconductor device engineering, where precise control over electron flow and charge carrier dynamics is crucial, this feature is very beneficial.

TMCs are superior to conventional semiconductors in terms of electrical tunability, which allows them to be used in a variety of applications such as energy conversion technologies (e.g., photovoltaics and thermoelectrics), electrochemical energy storage systems (e.g., batteries and supercapacitors), and heterogeneous catalysis (Patil et al., 2015; Shreya et al., 2023). Researchers are promoting innovation in materials science and next-generation energy technologies by taking use of the structural and compositional flexibility of TMCs to generate high-performance, application-specific materials.

High Surface Area

Transition metal chalcogenides' (TMCs') naturally large surface area is essential for increasing their reactivity and expanding their range of technological applications. In contrast to their bulk counterparts, TMCs are often manufactured in nanostructured morphologies, such as nanosheets, nanotubes, nanorods, and nanoparticle assemblies, which have noticeably larger specific surface areas (Anveer et al., 2020; Jiang et al., 2022; Z. Wang et al., 2013; Yang et al., 2020). The distinct structural arrangements and anisotropic growth patterns made possible by carefully regulated synthesis techniques are responsible for this rise in surface area. Since a large surface area makes more active sites available for surface-mediated reactions, it is directly associated with higher chemical reactivity. For instance, the quantity of reactive surface sites significantly increases catalytic activity and turnover frequency in heterogeneous catalysis. In energy conversion systems like fuel cells, where TMCs operate as electrocatalysts for important redox processes like the oxygen reduction reaction (ORR) and hydrogen evolution reaction (HER), this is especially beneficial (Bhushan, Jha, et al., 2021; Kumar et al., 2019). The increased surface area improves electrode-electrolyte interface interactions in electrochemical energy storage devices like lithium-ion batteries and supercapacitors, which accelerates charge transfer kinetics, promotes ion diffusion, and improves overall electrochemical performance. High specific capacitance, higher energy/power density, and quick charge–discharge cycle stability are all made possible by this characteristic (Bhushan, JHA, et al., 2021; Venkatesan et al., 2018). Additionally, TMCs' larger surface area helps them make better interfacial contact with conductive substrates or dielectric layers in nanoelectronics applications, which improves electrical conductivity and charge carrier mobility. The development of high-performance electronic components, such as chemical sensors, photodetectors, and field-effect transistors, depends on these characteristics (Kumar et al., 2019; Nady et al., 2017). As a result, a key characteristic supporting TMC nanostructures' versatility and effectiveness in cutting-edge energy and electrical technologies is their large surface area.

Chemical Stability

Chemical stability is one of the primary determinants of the lifetime and performance of transition metal chalcogenides (TMCs) in a wide range of innovative applications. The primary cause of TMCs' exceptional resistance to chemical deterioration and environmental corrosion is the strong covalent bond that exists between transition metal cations and chalcogen anions. The development of thermodynamically stable crystalline lattices with improved structural integrity under challenging operating circumstances is facilitated by this strong bonding framework. Layered transition metal dichalcogenides (TMDCs), such tungsten disulfide (WS_2) and molybdenum disulfide (MoS_2), exhibit remarkable chemical resistance because of their weak interlayer van der Waals contacts and strong intralayer metal–chalcogen linkages. An inherent defense mechanism against external disturbances including oxidative conditions, acidic or alkaline media, and high temperatures is provided by these stratified designs (Sharma et al., 2023; Zakaria et al., 2020). In energy-related systems, where materials are often exposed to harsh chemical and electrochemical conditions, this intrinsic chemical resilience is very beneficial. TMC-based electrodes, for example, must preserve their structural and chemical integrity across several charge-discharge cycles in lithium-ion and sodium-ion batteries. Similar to this, the corrosion resistance of the material has a direct impact on the durability and performance of catalysis in applications like electrocatalysis for the hydrogen evolution reaction (HER) or oxygen evolution reaction (OER). TMCs' capacity to withstand high levels of heat and chemical stress also makes them more suitable for use in electrical and optoelectronic devices, where long-term operational stability is crucial for device efficiency and dependability. In order to increase the usefulness of TMCs in next-generation materials science and nanotechnology applications, research is still being done to better optimize their chemical robustness using techniques including surface functionalization, doping, and heterostructure engineering.

Catalytic Activity

Transition metal chalcogenides (TMCs) have attracted a lot of interest for their catalytic performance because of their special mix of surface, structural, and electrical characteristics that make them very efficient in a variety of catalytic applications. Representative compounds with varying electronic structures, defect chemistries, and morphologies include zinc oxide (ZnO), nickel sulfide (NiS), and cadmium sulfide (CdS). These compounds display a variety of catalytic activities. The high density of catalytically active sites in TMCs, especially at exposed edge planes, grain boundaries, and inherent surface imperfections, is a major contributor to their catalytic effectiveness. The kinetics of surface-mediated catalytic reactions are improved by certain surface characteristics, which encourage advantageous reactant species adsorption and activation. Anisotropic architectures, in which edge sites exhibit more reactivity than basal planes, are especially advantageous for multilayer TMCs since they promote enhanced catalytic turnover. The performance of TMCs in crucial energy conversion processes, such as the oxygen evolution reaction

(OER), hydrogen evolution reaction (HER), and oxygen reduction reaction (ORR), has been thoroughly studied (Ding et al., 2021; Huang et al., 2009; Rabell et al., 2021; Sampaio et al., 2016; Sekiyama et al., 1986; Zhu et al., 2019). Fine control over reaction routes and catalytic activity is made possible by their programmable electronic structures, which enable modification of the Fermi level and charge carrier density. Reducing activation barriers and improving the binding energies of chemical intermediates need this kind of tunability. Furthermore, TMCs' catalytic qualities may be further improved by compositional engineering via doping, alloying, and heterostructure formation, making them attractive options for the creation of effective, affordable, and environmentally friendly catalysts. There is a lot of potential for increasing energy conversion efficiency, lowering environmental impact, and developing the design of sustainable industrial processes by better understanding and using TMCs in catalytic systems.

POTENTIAL APPLICATIONS OF CHALCOGENIDES IN ENERGY STORAGE AND CONVERSION

Battery Technologies

Because of their distinct physicochemical characteristics, which overcome significant drawbacks in existing battery technologies (Gurushankar et al., 2025; Kannan & Tari, 2025), transition metal chalcogenides (TMCs) have become attractive options for advanced energy storage applications. Significant research interest has been shown in compounds like CdO, CdS, NiO, NiS, and related TMCs (Feng et al., 2014; Wang et al., 2011; Weshahy et al., 2022; Zeng et al., 2017). These materials' programmable electronic structures are a major benefit because they enable the precise regulation of electrical conductivity and band gap energies, two characteristics that are essential for maximizing electrochemical performance. Many TMCs have naturally layered crystal structures, which increase their surface area and facilitate effective ion intercalation/deintercalation and charge transfer (Jin et al., 2017; Moreira et al., 2017; Rudnik & Nikiel, 2007). Furthermore, certain TMCs have great chemical and thermal stability, which improves battery systems' lifetime and structural integrity. In a variety of battery chemistries, these materials show a great deal of promise as electrode components to increase specific capacity, cycle durability, and overall energy conversion efficiency (Sonia et al., 2014; Wang et al., 2013). The continuous research into TMCs for battery applications shows a determined attempt to outperform traditional storage systems and eventually help create high-performing, environmentally friendly next-generation batteries.

Supercapacitors

Because of their favourable electrical, structural, and electrochemical characteristics, transition metal chalcogenides (TMCs), including ZnS, HgS, and other cadmium-based compounds, have been recognized as very promising materials for supercapacitor applications. Rapid charge-discharge kinetics, improved charge storage capacity, and overall higher electrochemical performance are all facilitated by these materials' large surface area and excellent electrical conductivity. Certain TMCs have configurable band gaps that provide chances to maximize energy storage capacity and raise supercapacitor systems' energy density. Furthermore, enhanced ionic diffusion and charge transfer are made possible by the layered crystalline structures and inherent two-dimensional morphology of many TMCs, which are essential for reaching high power densities in supercapacitor applications (Bhushan et al., 2019; Gautam et al., 2017; Phogat et al., 2023a). Researchers can precisely design TMC electrochemical behaviour thanks to their compositional and morphological plasticity, which allows them to customize material features for particular device needs. Furthermore, TMCs exhibit outstanding electrochemical stability throughout extended cycling, resolving a major obstacle to increasing the operational lifetime of supercapacitor devices. Additionally, certain TMC sustainability and environmental friendliness match the growing need for sustainable and environmentally friendly energy storage technologies. The goal of this research project is to create TMCs with optimal electrochemical properties and to create scalable, reasonably priced production methods that may be used to incorporate them into useful, high-performance supercapacitor devices.

Solar Cells

Because of their distinct optoelectronic and structural characteristics, transition metal chalcogenides (TMCs) have become very intriguing candidates for next-generation solar applications. The potential of materials like ZnS, NiO, HgS, and WO_3 to improve the effectiveness and efficiency of solar energy conversion systems has attracted a lot of scientific attention. One of the main benefits of TMCs is their adjustable band gap energies, which allow for effective solar radiation absorption throughout the visible and, in some situations, near-infrared portions of the electromagnetic spectrum. This is a crucial feature for optimizing photovoltaic performance. Furthermore, a lot of TMCs have high carrier mobility, which facilitates the effective separation and transfer of photogenerated electron-hole pairs. This reduces recombination losses and improves device performance in general.

Lightweight, flexible, and even transparent solar cell topologies may be developed more easily because to TMCs' naturally layered crystal structures, which enable the creation of thin films and two-dimensional heterostructures. The operational lifespan and environmental resistance of photovoltaic devices are crucial for real-world deployment, and this is further enhanced by the chemical and thermal stability shown by certain chalcogenides. In order to accurately regulate the structural, compositional, and electrical features of TMCs, current research efforts are concentrated on improving synthesis procedures and post-synthetic alterations. By allowing scalable and economical production processes, these tactics seek to increase power conversion efficiency. A fascinating family of materials for the ongoing development of high-efficiency, long-lasting, and commercially feasible solar cell technologies are transition metal chalcogenides, which are becoming an increasingly important part of the global sustainable energy landscape.

THE PRESENT DEMAND FOR CUTTING-EDGE ENERGY MATERIALS

Despite significant progress in the development of advanced materials for energy conversion and storage, several enduring obstacles continue to hinder their seamless integration into commercially viable systems. These challenges span the domains of synthesis, performance optimization, scalability, environmental sustainability, and compatibility with existing technologies. Addressing these concerns is crucial to accelerating the deployment of next-generation energy systems, particularly as global demand rises for cleaner, more efficient, and renewable energy solutions. One of the most pressing issues in energy materials is ensuring long-term structural and chemical stability under operational conditions. Many high-performance materials degrade rapidly when exposed to ambient atmosphere, high temperatures, electrochemical cycling, or light irradiation. For example:

- **Perovskite solar cells**, despite their high efficiency, suffer from moisture sensitivity and ion migration, leading to performance loss over time.
- **Electrode materials** in batteries undergo volumetric expansion and contraction during cycling, causing mechanical degradation and capacity fade.
- **Catalysts** for hydrogen evolution or oxygen reduction often deactivate due to poisoning or surface reconstruction.

Improving durability requires the development of protective coatings, stable interfaces, and material formulations that can withstand harsh environments without compromising performance. Many energy materials are currently synthesized using laboratory-scale methods such as hydrothermal synthesis, chemical vapor deposition (CVD), or solvothermal processes. While these techniques allow for precise control over morphology and composition, they are often expensive, time-consuming, and difficult to scale for industrial-level production.

- **2D materials**, like MoS_2 and other TMCs, are typically obtained through exfoliation or CVD methods that are not yet scalable for mass manufacturing.
- Maintaining consistency in material properties across large batches is another major challenge in upscaling.

To transition from lab to industry, synthesis techniques must become more cost-effective, reproducible, and environmentally benign, without sacrificing the structural and functional integrity of the materials. Even with advanced materials, many energy devices face limitations in terms of energy density and power output. For instance:

- **Lithium-ion batteries** offer high energy density but are limited by slow kinetics and safety concerns at high rates.
- **Supercapacitors**, on the other hand, deliver high power density but suffer from low energy density.

Materials must be engineered to achieve a balanced trade-off between energy and power density. This includes designing hybrid materials that combine the advantages of batteries and capacitors or creating hierarchical nanostructures that enhance ion transport and surface accessibility. The environmental impact of material extraction, synthesis, and disposal is an increasing concern in the field. Many high-performance energy materials depend on rare, toxic, or non-renewable elements, such as:

- **Cobalt and lithium** in batteries,
- **Platinum and iridium** in fuel cells and water-splitting catalysts,
- **Lead and cadmium** in photovoltaic materials.

Developing materials that rely on earth-abundant, non-toxic, and recyclable elements is vital for long-term sustainability. Moreover, life cycle assessments (LCAs) should be integrated into materials research to evaluate environmental costs from production to end-of-life management. The interface between different materials or phases often governs the overall efficiency of energy devices. Poor interfacial contact, lattice mismatch, or uncontrolled interdiffusion can result in high resistance, charge recombination, or mechanical failure.

- In **batteries**, the solid–electrolyte interphase (SEI) plays a crucial role in determining stability and safety.

- In **solar cells**, heterojunctions must facilitate efficient charge separation and transport.

Advanced characterization techniques and computational modeling are increasingly used to understand and optimize interfacial phenomena. However, the rational design of compatible and stable interfaces remains a key challenge. Many energy storage materials exhibit high initial performance but suffer from rapid performance degradation over repeated cycles. Issues such as electrode pulverization, phase changes, and irreversible side reactions reduce their effectiveness over time.

- **Sodium-ion batteries**, though promising as an alternative to lithium-ion systems, face major issues related to cyclability and volumetric expansion.
- **Photocatalysts** for water splitting often lose activity due to photochemical instability or oxidation.

Enhancing cyclability requires the development of flexible and robust nanostructures, as well as surface modifications that can accommodate strain and prevent degradation. Cost remains a major barrier to the widespread adoption of many energy materials. While laboratory research often focuses on maximizing efficiency or capacity, economic feasibility is equally important for real-world implementation.

- High-purity synthesis processes, specialized substrates, or rare precursor materials can inflate costs.
- Additionally, a lack of supportive infrastructure and regulatory standards can delay commercialization.

To overcome these issues, research must be aligned with cost-performance metrics, scalable production, and policy support for green technologies. The development of high-performance energy materials holds the key to addressing global energy and environmental challenges. However, the field is still confronted by a variety of scientific and practical barriers, including issues of stability, scalability, efficiency, and sustainability. Overcoming these challenges demands a multidisciplinary approach, combining advances in materials science, chemistry, physics, and engineering. Furthermore, close collaboration between academic researchers, industry stakeholders, and policymakers will be essential to translate laboratory innovations into real-world energy solutions. As research continues to evolve, the future of energy materials will depend not only on discovering new compounds but also on addressing these fundamental limitations through thoughtful design and integration.

Limited Capacity for Energy Storage

The low energy density of current electrode materials is a significant limitation in energy storage devices like batteries and supercapacitors. It is essential to increase energy storage capacity while preserving chemical and structural stability in order to boost system performance and efficiency.

Kinetics of Charge and Discharge

One important element affecting the energy throughput of storage devices is the rate efficiency of the charge and discharge operations. To guarantee quick and effective energy transfer during operation and to reduce internal energy losses, electrochemical kinetics must be optimized.

Stability of Cycling

Due to structural deterioration or adverse reactions after repeated charge-discharge cycles, many traditional energy materials experience performance loss. Long-term operating dependability depends on improving the cycle life by creating materials that are both chemically and mechanically resistant.

Viability of the Economy

The widespread use of modern energy technologies is still hampered by high material and production costs. One of the main obstacles to the commercialization of energy materials is striking a balance between economic viability and performance criteria.

Environmental Sustainability

Green, low-toxicity, and recyclable materials must be developed in response to environmental problems related to the extraction of raw materials, material synthesis, device manufacture, and end-of-life disposal. To reduce the effect on the environment, materials must be chosen and processed using a lifecycle-based strategy.

Accessibility of Resources

The security and sustainability of supply networks are at danger when they depend on uncommon or geopolitically restricted components. To guarantee long-term resource supply, research activities must give priority to identifying earth-abundant, non-critical elements.

Integrating Renewable Energy Sources

Because renewable energy sources like solar and wind are intermittent, energy storage systems that can adapt to changing power inputs must be developed. For materials to properly support renewable technologies, they must be scalable, highly efficient, and responsive.

Controlling Heat

A technological barrier still exists in efficient heat dissipation, especially in high-performance or high-power-density devices. To preserve operating safety and increase gadget longevity, sophisticated heat management techniques and thermally stable materials are needed.

An interdisciplinary strategy that draws on knowledge from solid-state chemistry, physics, materials science, and systems engineering is needed to address these complex issues. In this regard, the synthesis and thorough study of nanocomposites based on chalcogenides provide a viable avenue for the creation of energy materials of the future generation. In order to clarify their function in improving the efficiency, durability, and sustainability of energy storage and conversion systems, this study concentrates on the design, production, and characterisation of chalcogenide nanocomposites.

CURRENT RESEARCH SCOPE

Chalcogenide nanocomposites are synthesized, structurally characterized, and functionally evaluated in this work, with a focus on their utility in energy conversion and storage technologies. The study covers a wide variety of chalcogenide compounds and their composite systems, emphasizing their structural, optical, and adjustable electrical characteristics that make them ideal for incorporation into cutting-edge energy devices. The effort is to further the development of chalcogenide-based nanomaterials suited for high-efficiency and sustainable energy applications by thoroughly examining synthesis processes and performance indicators.

REFERENCES

Annveer, , & Gautam, , R., Kumar, A., Kumar, A., Singh, P. K., & Singh, R. P. (2020). Magneto-optical effects in half metallic ferromagnets: Rare earth thallium tellurides (TlXTe2; X = Tb-Er). *Optik (Stuttgart)*, *223*, 165317. DOI: 10.1016/j.ijleo.2020.165317

Annveer, , & Gautam, , R., Kumar, A., Kumar, A., Gautam, Y. K., Saroj, A. L., & Singh, R. P. (2021). Study of optoelectronic and thermoelectric spectra of Tl(Nd/Gd)S2. *Journal of Materials Science Materials in Electronics*, *32*(1), 727–744. DOI: 10.1007/s10854-020-04852-z

Bhushan, M., JHA, R., Bhardwaj, R., & Sharma, R. (2021). Visible light emission and enhanced electrocatalytic activity of pure ZnS nanoparticles synthesized via thermal decomposition route. *Bulletin of Materials Science*, *44*(4), 261. DOI: 10.1007/s12034-021-02546-8

Bhushan, M., Jha, R., & Bhardwaj, R. (2019). Reduced band gap and diffusion controlled spherical n-type ZnS nanoparticles for absorption of UV-Vis region of solar spectrum. *Journal of Physics and Chemistry of Solids*, *135*, 109021. DOI: 10.1016/j.jpcs.2019.05.018

Bhushan, M., Jha, R., Bhardwaj, R., & Sharma, R. (2021). Graphene-doped ZnS nanoparticles synthesized via hydrothermal route for enhanced electrocatalytic performance. *International Journal of Applied Ceramic Technology*, *18*(5), 1510–1526. DOI: 10.1111/ijac.13801

Dahiya, Y., Hariram, M., Kumar, M., Jain, A., & Sarkar, D. (2022). Modified transition metal chalcogenides for high performance supercapacitors: Current trends and emerging opportunities. *Coordination Chemistry Reviews*, *451*, 214265. DOI: 10.1016/j.ccr.2021.214265

Ding, L., Li, D., Shen, H., Qiao, X., Shen, H., & Shi, W. (2021). 2D β-NiS as electron harvester anchors on 2D ZnIn2S4 for boosting photocatalytic hydrogen production. *Journal of Alloys and Compounds*, *853*, 157328. DOI: 10.1016/j.jallcom.2020.157328

Feng, J., Xiong, S., Qian, Y., & Yin, L. (2014). Synthesis of nanosized cadmium oxide (CdO) as a novel high capacity anode material for Lithium-ion batteries: Influence of carbon nanotubes decoration and binder choice. *Electrochimica Acta*, *129*, 107–112. DOI: 10.1016/j.electacta.2014.02.085

Frumar, M., Jedelský, J., Frumarová, B., Wágner, T., & Hrdlička, M. (2003). Optically and thermally induced changes of structure, linear and non-linear optical properties of chalcogenides thin films. *Journal of Non-Crystalline Solids*, *326–327*, 399–404. DOI: 10.1016/S0022-3093(03)00446-0

Gao, M.-R., Xu, Y.-F., Jiang, J., & Yu, S.-H. (2013). Nanostructured metal chalcogenides: Synthesis, modification, and applications in energy conversion and storage devices. *Chemical Society Reviews*, *42*(7), 2986. DOI: 10.1039/c2cs35310e

Gautam, R., Kumar, A., & Singh, R. P. (2017). First Principle Investigations on Electronic, Magnetic, Thermodynamic, and Transport Properties of TlGdX 2 (X = S, Se, Te). *Acta Physica Polonica A*, *132*(4), 1371–1378. DOI: 10.12693/APhysPolA.132.1371

Gurushankar, K., Karuppaiya, C., Tari, V., & Kannan, K. (2025). Exploring MXenes and MBenes in Metal Sulfur and Selenium Battery Technologies. In *Novel Energy Storage and Conversion Technologies for Two-Dimensional MXenes and MBenes* (pp. 105–120). IGI Global Scientific Publishing.

Huang, X. H., Tu, J. P., Xia, X. H., Wang, X. L., Xiang, J. Y., Zhang, L., & Zhou, Y. (2009). Morphology effect on the electrochemical performance of NiO films as anodes for lithium ion batteries. *Journal of Power Sources*, *188*(2), 588–591. DOI: 10.1016/j.jpowsour.2008.11.111

Huang, Y., Zhu, M., Huang, Y., Pei, Z., Li, H., Wang, Z., Xue, Q., & Zhi, C. (2016). Multifunctional Energy Storage and Conversion Devices. *Advanced Materials*, *28*(38), 8344–8364. DOI: 10.1002/adma.201601928

Huang, Y. L., Chen, W., & Wee, A. T. S. (2021). Two-dimensional magnetic transition metal chalcogenides. *SmartMat*, *2*(2), 139–153. DOI: 10.1002/smm2.1031

Jiang, J., Wang, G., Shao, Y., Wang, J., Zhou, S., & Su, Y. (2022). Step-scheme ZnO@ZnS hollow microspheres for improved photocatalytic H2 production performance. *Chinese Journal of Catalysis*, *43*(2), 329–338. DOI: 10.1016/S1872-2067(21)63889-5

Jin, C., Zhou, L., Fu, L., Zhu, J., Li, D., & Yang, W. (2017). The acceleration intermediate phase (NiS and Ni 3 S 2) evolution by nanocrystallization in Li/NiS 2 thermal batteries with high specific capacity. *Journal of Power Sources*, *352*, 83–89. DOI: 10.1016/j.jpowsour.2017.03.119

Kannan, K., Chanda, D., Gautam, J., Behera, A., Meshesha, M. M., Jang, S. G., & Yang, B. (2023). Hydrothermally synthesized mixed metal oxide nanocomposites for electrochemical water splitting and photocatalytic hydrogen production. *International Journal of Hydrogen Energy*, *48*(93), 36412–36426.

Kannan, K., Chanda, D., Gautam, J., Meshesha, M. M., Jang, S. G., & Yang, B. L. (2023). Facial synthesis of pp heterojunction composites: Evaluation of their electrochemical properties with photovoltaics-electrolyzer water splitting using two-electrode system. *International Journal of Hydrogen Energy*, *48*(37), 13814–13826.

Kannan, K., Gautam, J., Chanda, D., Meshesha, M. M., Jang, S. G., & Yang, B. L. (2023). Two-dimensional MAX supported copper oxide/nickel Oxide/MAX as an efficient and novel photocatalyst for hydrogen evolution. *International Journal of Hydrogen Energy*, *48*(20), 7273–7283.

Kannan, K., & Tari, V. (Eds.). (2025). Novel energy storage and conversion technologies for two-dimensional MXenes and MBenes. https://doi.org/ ISBN: 9798369395462, eISBN: 9798369395486DOI: 10.4018/979-8-3693-9546-2

Kumar, A., Guatam, R., Chand, S., Kumar, A., & Singh, R. P. (2019). First principle electronic, magnetic and thermodynamic characterization of heavy fermion ternary rare earth metal alloys. *Materials Physics and Mechanics, 42*(1), 112–130. https://doi.org/DOI: 10.18720/MPM.4212019_10

Kumar, A., Kumar, A., Jain, P., Pundir, S. K., & Singh, N. (2024). TB-mBJ for doping concentration effects on magneto-optical properties in spintronics materials. *Optik (Stuttgart)*, *315*, 172039. DOI: 10.1016/j.ijleo.2024.172039

Kumar, A., Kumar, A., Kumar, K., Singh, R. P., Singh, R., & Kumar, R. (2023). The Electronic and Thermodynamic Properties of Ternary Rare Earth Metal Alloys. *East European Journal of Physics*, *1*, 109–117. DOI: 10.26565/2312-4334-2023-1-13

Kumar, A., Kumar, R., Saxena, D., Nautiyal, V. K., Kumar, A., & Iram, N. (2024). Structural, electronic, magnetic and optical properties of GdCuX2 (X = S, Se and Te) compounds. *Optical and Quantum Electronics*, *56*(10), 1742. DOI: 10.1007/s11082-024-07298-3

Liu, C., Li, F., Ma, L., & Cheng, H. (2010). Advanced Materials for Energy Storage. *Advanced Materials*, *22*(8). Advance online publication. DOI: 10.1002/adma.200903328

Mitchell, K., & Ibers, J. A. (2002). Rare-Earth Transition-Metal Chalcogenides. *Chemical Reviews*, *102*(6), 1929–1952. DOI: 10.1021/cr010319h

Moreira, T. F. M., Santana, I. L., Moura, M. N., Ferreira, S. A. D., Lelis, M. F. F., & Freitas, M. B. J. G. (2017). Recycling of negative electrodes from spent Ni-Cd batteries as CdO with nanoparticle sizes and its application in remediation of azo dye. *Materials Chemistry and Physics*, *195*, 19–27. DOI: 10.1016/j.matchemphys.2017.04.009

Nady, A., Ahmed, M. H. M., Numan, A., Ramesh, S., Latiff, A. A., Ooi, C. H. R., Arof, H., & Harun, S. W. (2017). Passively Q-switched erbium-doped fibre laser using cobalt oxide nanocubes as a saturable absorber. *Journal of Modern Optics, 64*(13), 1315–1320. DOI: 10.1080/09500340.2017.1286398

Patil, S., Raut, S., Gore, R., & Sankapal, B. (2015). One-dimensional cadmium hydroxide nanowires towards electrochemical supercapacitor. *New Journal of Chemistry, 39*(12), 9124–9131. DOI: 10.1039/C5NJ02022K

Peng, Q., Iram, N., Sharma, R., Kumar, A., Saad Alsubaie, A., & Rehman, J. (2024). Ab-initio investigation of structural, electronic, thermoelectric and optical properties of Full-Heusler X2MnB (X = Ti, Zr) for energy harvesting applications. *Inorganic Chemistry Communications, 170*, 113240. DOI: 10.1016/j.inoche.2024.113240

Phogat, P. Shreya, Jha, R., & Singh, S. (2023a). Diffusion Controlled Features of Microwave Assisted ZnS/ZnO Nanocomposite with Reduced Band Gap. *ECS Journal of Solid State Science and Technology, 12*(3), 034004. https://doi.org/DOI: 10.1149/2162-8777/acc426

Phogat, P., Shreya, Jha, R., & Singh, S. (2023b). Electrochemical analysis of thermally treated two dimensional zinc sulphide hexagonal nano-sheets with reduced band gap. *Physica Scripta, 98*(12), 125962. DOI: 10.1088/1402-4896/ad0d93

Rabell, G. O., Cruz, M. R. A., & Juárez-Ramírez, I. (2021). Hydrogen production of ZnO and ZnO/Ag films by photocatalysis and photoelectrocatalysis. *Materials Science in Semiconductor Processing, 134*, 105985. DOI: 10.1016/j.mssp.2021.105985

Rudnik, E., & Nikiel, M. (2007). Hydrometallurgical recovery of cadmium and nickel from spent Ni–Cd batteries. *Hydrometallurgy, 89*(1–2), 61–71. DOI: 10.1016/j.hydromet.2007.05.006

Sampaio, M. J., Oliveira, J. W. L., Sombrio, C. I. L., Baptista, D. L., Teixeira, S. R., Carabineiro, S. A. C., Silva, C. G., & Faria, J. L. (2016). Photocatalytic performance of Au/ZnO nanocatalysts for hydrogen production from ethanol. *Applied Catalysis A, General, 518*, 198–205. DOI: 10.1016/j.apcata.2015.10.013

Sarkar, B. K., Verma, A. S., Sharma, S., & Kundu, S. K. (2014). First-principles calculations of the structural, phonon and thermal properties of ZnX (X = S, Se, Te) chalcogenides. *Physica Scripta, 89*(7), 075704. DOI: 10.1088/0031-8949/89/7/075704

Sekiyama, H., Kosugi, N., Kuroda, H., & Ohta, T. (1986). Sulfur K-Edge Absorption Spectra of Na2SO4, Na2SO3, Na2S2O3, and Na2S2O *x* (*x* =5–8). *Bulletin of the Chemical Society of Japan, 59*(2), 575–579. DOI: 10.1246/bcsj.59.575

Sharma, S., Phogat, P., Jha, R., & Singh, S. (2023). Electrochemical and Optical Properties of Microwave Assisted MoS2 Nanospheres for Solar Cell Application. *International Journal of Smart Grid and Clean Energy*, 66–72. https://doi.org/DOI: 10.12720/sgce.12.3.66-72

Sheela, S. E., Sekar, R., Maurya, D. K., Paulraj, M., & Angaiah, S. (2023). Progress in transition metal chalcogenides-based counter electrode materials for dye-sensitized solar cells. *Materials Science in Semiconductor Processing*, *156*, 107273. DOI: 10.1016/j.mssp.2022.107273

Shi, Y., Sturm, C., & Kleinke, H. (2019). Chalcogenides as thermoelectric materials. *Journal of Solid State Chemistry*, *270*, 273–279. DOI: 10.1016/j.jssc.2018.10.049

Shreya, P. P., Jha, R., & Singh, S. (2023). *Elevated Refractive Index of MoS2 Amorphous Nanoparticles with a Reduced Band Gap Applicable for Optoelectronics* (pp. 431–439). https://doi.org/DOI: 10.1007/978-981-99-2349-6_39

Singh, N. K., Kumar, A., Kumar, S., Nautiyal, V. K., Kumar, R., Kumar, A., & Iram, N. (2025). A comprehensive theoretical analysis of Cs2KGaX6 (X = Cl, Br, I): For green energy solutions. *Solid State Communications*, *401*, 115929. DOI: 10.1016/j.ssc.2025.115929

Sonia, T. S., Anjali, P., Roshny, S., Lakshmi, V., Ranjusha, R., Subramanian, K. R. V., Nair, S. V., & Balakrishnan, A. (2014). Nano/micro-hybrid NiS cathodes for lithium-ion batteries. *Ceramics International*, *40*(6), 8351–8356. DOI: 10.1016/j.ceramint.2014.01.041

Venkatesan, J., Rekha, P. D., Anil, S., Bhatnagar, I., Sudha, P. N., Dechsakulwatana, C., Kim, S.-K., & Shim, M. S. (2018). Hydroxyapatite from Cuttlefish Bone: Isolation, Characterizations, and Applications. *Biotechnology and Bioprocess Engineering; BBE*, *23*(4), 383–393. DOI: 10.1007/s12257-018-0169-9

Wang, X., Huang, H., Liang, B., Liu, Z., Chen, D., & Shen, G. (2013). ZnS Nanostructures: Synthesis, Properties, and Applications. *Critical Reviews in Solid State and Material Sciences*, *38*(1), 57–90. DOI: 10.1080/10408436.2012.736887

Wang, Y., Zhu, Q., Tao, L., & Su, X. (2011). Controlled-synthesis of NiS hierarchical hollow microspheres with different building blocks and their application in lithium batteries. *Journal of Materials Chemistry*, *21*(25), 9248. DOI: 10.1039/c1jm10271k

Wang, Z., Cao, S.-W., Loo, S. C. J., & Xue, C. (2013). Nanoparticle heterojunctions in ZnS–ZnO hybrid nanowires for visible-light-driven photocatalytic hydrogen generation. *CrystEngComm*, *15*(28), 5688. DOI: 10.1039/c3ce40523k

Weshahy, A. R., Sakr, A. K., Gouda, A. A., Atia, B. M., Somaily, H. H., Hanfi, M. Y., Sayyed, M. I., El Sheikh, R., El-Sheikh, E. M., Radwan, H. A., Cheira, M. F., & Gado, M. A. (2022). Selective Recovery of Cadmium, Cobalt, and Nickel from Spent Ni–Cd Batteries Using Adogen® 464 and Mesoporous Silica Derivatives. *International Journal of Molecular Sciences*, *23*(15), 8677. DOI: 10.3390/ijms23158677

Xiong, W., Huang, K., & Yuan, S. (2019). The mechanical, electronic and optical properties of two-dimensional transition metal chalcogenides MX 2 and M 2 X 3 (M = Ni, Pd; X = S, Se, Te) with hexagonal and orthorhombic structures. *Journal of Materials Chemistry. C, Materials for Optical and Electronic Devices*, *7*(43), 13518–13525. DOI: 10.1039/C9TC04933A

Yadav, S., Yashas, S. R., & Shivaraju, H. P. (2021). Transitional metal chalcogenide nanostructures for remediation and energy: A review. *Environmental Chemistry Letters*, *19*(5), 3683–3700. DOI: 10.1007/s10311-021-01269-w

Yang, X., Liu, H., Li, T., Huang, B., Hu, W., Jiang, Z., Chen, J., & Niu, Q. (2020). Preparation of flower-like ZnO@ZnS core-shell structure enhances photocatalytic hydrogen production. *International Journal of Hydrogen Energy*, *45*(51), 26967–26978. DOI: 10.1016/j.ijhydene.2020.07.027

Yuan, H., Kong, L., Li, T., & Zhang, Q. (2017). A review of transition metal chalcogenide/graphene nanocomposites for energy storage and conversion. *Chinese Chemical Letters*, *28*(12), 2180–2194. DOI: 10.1016/j.cclet.2017.11.038

Yuan, S., Duan, X., Liu, J., Ye, Y., Lv, F., Liu, T., Wang, Q., & Zhang, X. (2021). Recent progress on transition metal oxides as advanced materials for energy conversion and storage. *Energy Storage Materials*, *42*, 317–369. DOI: 10.1016/j.ensm.2021.07.007

Zakaria, M. B., Guo, Y., Na, J., Tahawy, R., Chikyow, T., El-Said, W. A., El-Hady, D. A., Alshitari, W., Yamauchi, Y., & Lin, J. (2020). Layer-by-Layer Motif Heteroarchitecturing of N,S-Codoped Reduced Graphene Oxide-Wrapped Ni/NiS Nanoparticles for the Electrochemical Oxidation of Water. *ChemSusChem*, *13*(12), 3269–3276. DOI: 10.1002/cssc.202000159

Zakery, A., & Elliott, S. R. (2003). Optical properties and applications of chalcogenide glasses: A review. *Journal of Non-Crystalline Solids*, *330*(1–3), 1–12. DOI: 10.1016/j.jnoncrysol.2003.08.064

Zeng, Y., Meng, Y., Lai, Z., Zhang, X., Yu, M., Fang, P., Wu, M., Tong, Y., & Lu, X. (2017). An Ultrastable and High-Performance Flexible Fiber-Shaped Ni–Zn Battery based on a Ni–NiO Heterostructured Nanosheet Cathode. *Advanced Materials*, *29*(44). Advance online publication. DOI: 10.1002/adma.201702698

Zhang, Y., Zhang, L., Lv, T., Chu, P. K., & Huo, K. (2020). Two-Dimensional Transition Metal Chalcogenides for Alkali Metal Ions Storage. *ChemSusChem*, *13*(6), 1114–1154. DOI: 10.1002/cssc.201903245

Zhang, Y., Zhou, Q., Zhu, J., Yan, Q., Dou, S. X., & Sun, W. (2017). Nanostructured Metal Chalcogenides for Energy Storage and Electrocatalysis. *Advanced Functional Materials*, *27*(35). Advance online publication. DOI: 10.1002/adfm.201702317

Zhao, Y., Wei, S., Pan, K., Dong, Z., Zhang, B., Wu, H.-H., Zhang, Q., Lin, J., & Pang, H. (2021). Self-supporting transition metal chalcogenides on metal substrates for catalytic water splitting. *Chemical Engineering Journal*, *421*, 129645. DOI: 10.1016/j.cej.2021.129645

Zhu, C., Liu, C., Fu, Y., Gao, J., Huang, H., Liu, Y., & Kang, Z. (2019). Construction of CDs/CdS photocatalysts for stable and efficient hydrogen production in water and seawater. *Applied Catalysis B: Environmental*, *242*, 178–185. DOI: 10.1016/j.apcatb.2018.09.096

Chapter 8
Sustainability of Financial Energy Literacy:
A Literature Review

José G. Vargas-Hernandez
https://orcid.org/0000-0003-0938-4197
Financial Literacy Research Center, Universidad Cristóbal Colón, Mexico & Tecnológico Nacional de México, ITS Fresnillo, Mexico

Teresa I. Salazar-Echeagary
https://orcid.org/0000-0001-8785-6300
Universidad Autònoma de Sinaloa, Mexico

Julieta E. Salazar-Echeagary
https://orcid.org/0000-0003-0689-532X
Universidad Autónoma de Sinaloa, Mexico

ABSTRACT

This study aims to analyze the implications of the financial energy literacy (FEL) with sustainability as interdisciplinary concepts. The holistic approach of sustainability of financial energy literacy (SFEL) is critical in an environment where energy costs are rising and environmental sustainability is becoming increasingly urgent. The method employed is the meta-analytical and meta-cognitive analysis based on a conceptual, theoretical and empirical literature review published in the last five years. It is concluded that financial energy literacy is a crucial element for achieving sustainable development goals, as it empowers individuals to make informed decisions that benefit both their financial well-being and the environment of individuals, families, organizations and communities.

DOI: 10.4018/979-8-3373-3962-7.ch008

INTRODUCTION

Financial energy literacy (FEL) is an interdisciplinary field that works towards the integration of financial and energy literacy at the personal and household level to obtain accurate information on the basis of economic aspects concerning such decisions that is economically sustainable and environmentally sustainable (Maji & Chakraborty, 2023; Tari et al., 2024). This integration recognizes that energy-related decisions often have significant financial implications, and vice versa. Therefore, individuals need to possess the knowledge and skills to navigate both domains effectively. By understanding both the financial and energy-related implications of their choices, consumers are better equipped to make informed decisions that benefit both their wallets and the environment. This includes making informed decisions about energy-efficient appliances, renewable energy investments, and other sustainable practices, (Tari et al., 2024; Siddiqui et al., 2022).

FEL empowers individuals to evaluate the costs and benefits of different energy options, understand the long-term financial implications of their choices, and resist psychological biases that may lead to unsustainable behaviors. This holistic approach is critical in today's world, where energy costs are rising and environmental sustainability is becoming increasingly urgent, (Tari & Kannan et al., 2022). The increasing costs of energy place a significant financial burden on households, especially those with low incomes. Therefore, understanding how to manage energy consumption and reduce energy costs is essential for financial stability. Additionally, the growing awareness of environmental issues has made it imperative for individuals to adopt sustainable practices that minimize their environmental impact, (Tari. & Patil, 2017a; Tari & Patil, 2017b).

People can achieve a positive net on themselves and society, from fiscal and environmental aspects under this program because people are given a balance of funds and environmentalism as collective goals. It in fact is much more than just a personal benefit; this is too much progress and an improvement in a society by improving energy use for the better, in energy saving and proper use of energy to be better for all mankind (Maji & Chakraborty, 2023). FEL serves as an efficient mechanism to provide political and industrial aid to politicians and enterprises aiming to deliver a cleaner and fairer future with low emissions.

For example, policies that promote energy efficiency and conservation can help to reduce greenhouse gas emissions, mitigate climate change, and improve air quality. Additionally, financial incentives for renewable energy investments can spur innovation and create jobs in the clean energy sector. The combination of energy literacy and financial literacy empowers individuals to make choices that not only benefit their own financial well-being but also contribute to a more sustainable and resilient society.

Empirical studies have demonstrated that higher levels of energy-related financial literacy (ERFL) are significantly correlated with reduced energy consumption (Kalmi, Trotta, & Kaukauskas, 2021). Individuals who possess a strong understanding of both energy and financial principles are more likely to adopt energy-saving practices, invest in energy-efficient technologies, and make informed decisions about their energy usage. This highlights the tangible benefits of promoting ERFL as a means to achieve energy conservation and reduce environmental impact.

People with high ERFL are more likely to practice energy-saving habits as part of their everyday life, turning off lights, unplugging devices, and regulating thermostats – and invest in sustainable systems such as appliances that reduce electricity consumption and renewable energy production.

This study is a conceptual, theoretical and empirical review of the literature published in the last five years on the issues related to sustainability energy financial literacy. The main parts of the study covers an introduction, conceptual background, theories, empirical literature review, methods, analysis of results, discussion and conclusions.

Key Findings and Research Gaps

Fostering these informed practices, ERFL can facilitate a move toward sustainable and energy-efficient society. However, global ERFL levels are, overall, moderate, yet significant differences occur in terms of regions and demographic groups (Maji & Chakraborty, 2023). These gaps are indicative of differential access to education and knowledge, and the limited coverage of energy and finance topics in formal curricula. In order, to enhance ERFL globally, a broad-based strategy must be adopted – one that would encourage the accessibility of education, promote public awareness and mainstream these constructs.

Education, income, and availability of trustworthy information are determinants of ERFL in one individual. Higher levels of education generally reflect better energy and finance skills, whereas higher incomes lead to investment in energy-efficient technologies. Just as crucial is access to practical information — about energy costs, conservation strategies and how to plan your finances — through channels like e-resources, community workshops and counseling services. Together, these factors generate the potential for knowledge and decision-making to be made regarding sustainable solutions for a common purpose, based on the needs of a community.

Understanding these differences is critical to creating ERFL programs that are appropriate for certain populations. For example, low-income families would probably find programs oriented to a useful approach to energy expenses and money, or towards high income groups to foster the use of renewables. Modelling such action makes them more effective and more equitable in distribution of resources. That can involve working with community organizations, culturally relevant resources and funds for adoption of energy-efficient practices.

Despite its increasing recognition, ERFL research remains focused on European contexts, and hence global research on ERFL is limited (Maji & Chakraborty, 2023). To compensate for this discrepancy, research should explore how ERFL would work across different cultural, economic and environmental contexts, particularly in the face of limited energy access and high costs in developing countries.

Future studies should explore how ERFL impacts individual energy-related behaviors— whether it is investment in renewables, energy saving transportation and group activities for conservation. These apparent impacts can add clarity to the interpretation of the ERFL-driven sustainable action. By way of example, ERFL values may be analyzed for potential linkages with purchases of electric vehicles, solar panels, and community energy, depending on whether or not are higher in ERFL. This knowledge can help policymakers and educators implement more targeted interventions to promote sustainable energy decisions.

Additionally, there is a need to better investigate the policy aspect of ERFL, and it should thus examine the extent to which education, financial incentives, and regulations may be utilized to foster ERFL (Martins, Madaleno & Dias, 2020). Among others, assessment of different programs and policy instruments towards the promotion of ERFL and the application of sustainable energy would also be pursued.

For example, research can consider the impact of economic decision-making, as can analysis on whether financial literacy workshops influence energy decisions or tax incentives promote the adoption of renewable technologies such as solar panels. Filling these gaps in research will facilitate our understanding of ERFL's position in promoting sustainable energy – a task that requires the cooperation of researchers, policymakers, educators and community stakeholders.

Implications for Sustainable Development

ERFL can reduce the amount of energy use and the release of greenhouse gases by selecting and adopting energy sustainable decisions in an informed manner, (Todorov, Aleksandrova, & Ismailov, 2023). High ERFL people are adopting technology to consume less energy, they are promoting the conservation habit and working for the adoption policy of renewable energy resources. Together, they all act to lower the use of fossil fuels and play a role in reaching national and international climate goals.

Financial institutions and policymakers have a crucial role to play in enhancing ERFL among the population through targeted education initiatives and the development of accessible financial products that support sustainable energy consumption (Attipoe et al., 2025). Financial institutions can offer incentives for investing in energy-efficient appliances or renewable energy systems, while policymakers can implement regulations that promote energy conservation and provide resources for financial literacy programs. By working together, these stakeholders can create an environment that fosters ERFL and supports sustainable development. For example, financial institutions can offer green loans with lower interest rates for energy-efficient home improvements, while policymakers can provide tax credits for the purchase of renewable energy systems.

Artificial Intelligence (AI) blended with financial literacy can aid in making more informed choices and work for sustainability targets (Kaur, & Dhiman, 2024). They provide personalized insight by analyzing energy consumption patterns, pinning down areas for cost-savings, and helping recommend energy-efficient solutions. This technology could also provide an enabler of FEL, helping increase accessibility and its impact, promoting wider adoption of more sustainable behaviors. AI can also assist in the identification of, and mitigation of related risks associated with green investments, enhancing people's financial well-being.

THE IMPORTANCE OF SUSTAINABILITY

With the threat of industrialization and ever-increasing energy requirements, sustainability has become a necessity for nations. Challenges are such as overexploitation of natural resources, climate change, and environmental degradation (Salahodjaev & Sadikov, 2025). As Salahodjaev & Sadikov (2025) emphasize, an increasing number of countries are facing environmental pressure due to these factors. Addressing these pressures requires a fundamental shift towards more sustainable practices in all aspects of our lives, including energy production and consumption.

Supporting this transition is a need to catalyze an energy-literate public. An energy-literate society can better understand the multiple consequences of different energy choices and become more sustainable (Santilln & Cedano, 2023). Santilln & Cedano (2023) assert that changes in the energy systems require an energy-literate citizenship. This underscores the importance of education and awareness in driving the transition to a more sustainable energy future.

This is where vast financial and technical literacy are important, so citizens can select among such products as renewables with greater return to make relative investments (Salahodjaev & Sadikov, 2025; Adhikari, 2024). In this context, it is argued that financial literacy (FL) with energy literacy (EL)—FEL—can facilitate responsible behavior. The infusion of multiple literacies builds capacities for informed choices, responsible consumption, and household and community resilience (Adhikari, 2024). By promoting financial literacy, we can equip individuals with the tools they need to contribute to a more sustainable and equitable world.

Aim and Scope of the Literature Review

Focused primarily on environmental and financial challenges, it becomes essential to examine the tools that enable individuals to take proactive actions. This review explores the relationship between FEL and sustainable development, while demonstrating that combining financial knowledge and energy knowledge can lead to fairer and more resilient societies (Todorov, Aleksandrova, & Ismailov, 2023). With ever-evolving financial tools and growing financial inclusion, financial literacy becomes extended into concepts of sustainability, responsible consumption, and environmental stewardship.

It synthesizes existing literature to understand the impact of financial literacy on energy-related decisions, covering a wide range of studies that have investigated the relationship between financial literacy and energy consumption (Martins, Madaleno & Dias, 2019). Martins, Magdalen, &Ferreira Dias (2019) state that energy encompasses not only the cognitive domain, but also affective behavioral characteristics, enabling make appropriate decisions regarding energy. This review will analyze these studies to identify key findings and patterns, as well as to assess the strengths and limitations of the existing research.

This article discusses the theoretical underpinnings of FEL, its empirical basis, and its relevance for policy and practice. This review identifies critical gaps in the research on FEL and the future research needs to look into FEL's impact towards sustainable development (Maja & Chakraborty, 2023). As there is still limited literature, especially on ERFL in global contexts, such literature is required to deepen the understanding and guide interventions to further the development of policies towards a more sustainable and equitable society.

CONCEPTUAL BACKGROUND

Defining Financial Energy Literacy

Financial literacy involves the ability to manage personal finances, plan for the future, and make informed investment decisions (Adhikari, 2024; Todorov, Aleksandrova, & Ismailov, 2023; Yuwono, Susanna, Sasmita, & Sihotang, 2023). The ability to understand and apply financial concepts is essential for navigating the complexities of modern economies. Adhikari (2024) notes that financial literacy empowers individuals to make informed financial behaviors, leading to better planning and management of life milestones.

Similarly, financial literacy plays an important role in better financial planning processes and in making finances more precise (Yuwono, Susanna, Sasmita, &

Sihotang, 2023). These skills empower individuals to take control of their financial well-being and build a secure future for themselves and their families. Energy literacy involves understanding how energy is generated, transported, stored, distributed, and used; awareness about its environmental and social impacts; and the knowledge to use it efficiently (Santilln & Cedano, 2023). Santilln & Cedano (2023) explain that energy literacy encompasses the understanding of how energy is generated, transported, stored, distributed, and used. It also entails awareness about its environmental and social impacts, as well as the knowledge to use it efficiently. This encompasses not only the technical aspects of energy systems but also the broader implications for society and the environment.

FEL combines financial and energy literacy to promote sustainable energy use and financial well-being, recognizing the close interrelationship between economic and environmental decision-making (Maji & Chakraborty, 2023; Martins, Madaleno & Dias, 2020). It teaches people to make decisions that balance upfront costs with long-term savings, such as buying energy-efficient appliances, and to engage in behaviors that promote climate goals. This all-encompassing stance contributes to the endorsement of renewables and green governance, as the greater public is educated on the social and personal impacts of energy consumption.

Evolution of Financial Literacy

Over the past few decades, the definition of financial literacy has grown considerably beyond its original parameters. It once centered mainly on practical skills such as budgeting, saving, and managing debt – which more recently has evolved into an expression of a bigger understanding of the intersection between finance, social and environmental matters (Todorov, Aleksandrova, & Ismailov, 2023). Financial literacy enables people to make good decisions about their money, as I said before—but in the current climate that calls for more than 'money' knowledge to overcome the current problems of today. The new definition also incorporates responsible consumption, sustainability, and the environmental impacts of consumption and investment habits.

Today, financial literacy is globally recognized as a significant element of stability, economic growth, and financial inclusion (Todorov, Aleksandrova, & Ismailov, 2023). . Governments, organizations, and educational institutions around the world are increasingly recognizing the importance of promoting financial literacy as a means to empower individuals, strengthen economies, and build resilient societies. This recognition has led to the development of various financial literacy programs and initiatives aimed at improving the financial knowledge and skills of people of all ages and backgrounds. These programs seek to assist individuals ranging from young to old and in all sectors with information on how to handle their money while working for the future.

Financial literacy plays a crucial role in enabling individuals to make informed financial behaviors, leading to better planning and management of life milestones such as education, housing, savings, and retirement (Adhikari, 2024;Yuwono, Susanna, Sasmita, & Sihotang, 2023; Adhikari, 2024) states that financial literacy empowers individuals to make informed financial behaviors, leading to better planning and management of life milestones. Yuwono, Susanna, Sasmita, & Sihotang, (2023) add that financial literacy plays an important role in a better financial planning process. By understanding financial concepts and principles, individuals are better equipped to set financial goals, develop strategies to achieve those goals, and make informed choices about how to manage their money. This, in turn, leads to greater financial security and overall well-being.

Dimensions of Energy Literacy

Energy literacy encompasses cognitive (knowledge), affective (attitudes), and behavioral (actions) dimensions. This embraces energy system awareness, environmental trade-offs, and actions to conserve energy (Martins, Madaleno & Dias, 2019; Santilln & Cedano, 2023). Martins, Madaleno & Dias, (2019) explain that energy encompasses not only the cognitive domain, but also affective behavioral characteristics, enabling make appropriate decisions regarding energy. Santilln & Cedano, (2023) state that energy literacy is the understanding of how energy is generated, transported, stored, distributed, and used; awareness about its environmental and social impacts; and the knowledge to use it efficiently.

It is not simply about knowing data related to energy but also about understanding the broader context of energy production, consumption, and its impacts on the environment and society. Energy literacy empowers individuals to become active and informed participants in shaping a more sustainable energy future. It includes a thorough understanding of energy generation, transportation, and the various environmental impacts associated with different energy sources (Santilln & Cedano, 2023). Santilln & Cedano, (2023) state that energy literacy is the understanding of how energy is generated, transported, stored, distributed, and used; awareness about its environmental and social impacts; and the knowledge to use it efficiently.

Energy-literate individuals are aware of the trade-offs involved in different energy choices and can make informed decisions about which energy sources to support. They also understand the importance of energy efficiency and can identify opportunities to reduce their energy consumption and minimize their environmental footprint. Energy literacy also involves actively promoting sustainable energy consumption habits, such as reducing energy waste, using energy-efficient appliances, and adopting renewable energy technologies. Martins, Madaleno & Dias, (2019) explain that energy literacy is viewed as an essential tool, which sensitizes citizens to create sustainable energy consumption habits. Energy-literate individuals are not only knowledgeable about energy issues but also committed to taking action to promote a more sustainable energy future. They serve as role models for others and advocate for policies that support energy conservation and renewable energy development.

Integrating Financial and Energy Literacy

Maji & Chakraborty (2023) develop ERFL and its core framework integrating energy literacy, financial literacy, and lifecycle cost awareness. They help individuals to make informed decisions as energy-related decisions are also associated with both financial and environmental consequences. Not only does ERFL consider the upfront outgoing cost of energy, but also long-term savings, operational costs as well as environmental effects; it can help individuals approach energy alternatives in a more holistic way.

Through the fusion of financial and energy literacy ERFL gives people the ability to make both efficient cost and environmental decisions. As Maji & Chakraborty, (2024) described, a holistic approach can also contribute significantly towards addressing climate change by promoting more efficient energy utilization and encouraging financial wellness in a household. For example, customers with high ERFL are more likely to make an investment in energy-efficient appliances, being aware that long-term savings might outweigh initial start-up outlays.

Recent studies have emphasized the growing importance of financial knowledge in enhancing overall energy literacy. Martins, Madaleno & Dias, (2020) aimed to determine the role of financial knowledge on energy literacy dimensions. As energy, costs continue to rise and as new energy technologies become available, it is increasingly important for individuals to understand the financial implications of their energy choices. Financial knowledge empowers individuals to evaluate different energy options, compare costs and benefits, and make informed decisions that support both their financial well-being and the environment.

THEORIES

Human Capital Theory (HCT)

According to HCT, financial literacy is an investment into the people through increased knowledge and skills in the service of the well-being of the individual and the welfare of society. As Lusardi & Mitchell, (2014) suggested, knowledge—namely financial knowledge—is a form of human capital. The better people are with monetary knowledge, the more able they are to plan, save, invest and make sound economic decisions, which in the long run create personal stability, as well as national economic robustness. This viewpoint supports the significance of FEL that integrates financial and energy knowledge for an informed choice.

Yuwono, Susanna, Sasmita, & Sihotang, (2023) mention that financial literacy facilitates planning by guiding people in making challenging decisions - for example adopting energy-efficient or renewable technology - that influence the economic effect and environmental effects.

This theory supports the idea that increased knowledge leads to more efficient resource management and the adoption of sustainable practices, including energy conservation and the use of renewable energy source. Lusardi & Mitchell, (2014) undertake an assessment of a rapidly growing body economic research on financial literacy. When individuals understand the financial implications of their energy choices, they are more likely to adopt energy-efficient behaviors and invest in energy-saving technologies.

Maji & Chakraborty, (2024) have shown a study, in West Bengal, India, regarding on how the application of HCT. They discovered that houses with more ERFL energy used drastically less electricity than households without — and it's not because of higher income but because of being more educated. For these reasons, their home energy was efficient, and they selected energy-efficient options such as LED bulbs and refrigerators, the savings being sufficient so those who made the decision knew that the initial outlay made good long-term sense. Their decisions had both an economic and environmental payoff demonstrating how the investment of knowledge leads to significant individual- and global-effectiveness.

These theories support the idea of FEL as investment in human capital empowers policymakers and educators to promote economic stability, energy efficiency practices, and environmental responsibility. This sustainable approach it's good for economic risk management, fosters responsible energy use and encourages the construction of more sustainable communities.

Behavioral Economics

Behavioral economics provides valuable insights into how psychological factors influence financial decisions, which in turn can significantly affect energy consumption patterns. Arrondel, (2021) looks back over the different dimensions of financial literacy: theoretical, methodological, empirical and political. Unlike traditional economic models that assume individuals are rational decision-makers, behavioral economics recognizes that people are often influenced by cognitive biases, emotions, and social norms. These factors can lead to suboptimal financial decisions, including those related to energy consumption.

It highlights the importance of considering both the cognitive and non-cognitive components of financial ability in order to understand how individuals make financial decisions. Kaur, & Dhiman, (2024) highlight the complex interplay between cognitive and non-cognitive components of financial ability. Cognitive factors include knowledge and understanding of financial concepts, while non-cognitive factors include attitudes, beliefs, and emotions. By taking both types of factors into account, we can develop effective interventions to promote financial literacy and encourage sustainable energy behaviors.

People with a strong financial underpinning seem to be able to choose more educated and sustainable options, especially in consumption of energy use. This is because other standard behavioral biases, namely present bias, loss aversion, and framing effects have more limited effects. Instead of going back to the source story, it is imperative that we expand on an earlier talk of financial literacy to discuss its impact on everyday decisions. Blasch, Boogen, Daminato, & Filippini (2021) explain that consumers limited energy-specific knowledge and investment skills. Present bias refers to the tendency to prioritize immediate gratification over long-term benefits, which can lead individuals to delay investing in energy-efficient technologies that have a higher upfront cost but offer significant long-term savings.

Cognitive barriers can be overcome when individuals possess the awareness and confidence to evaluate the short- and long-run implications associated with their spending options. Greater awareness of how choices impact individual and environmental finances leads people to take behavior changes such as getting better lighting or solar panels or reducing energy waste. This highlights FEL helps to reduce behavioral challenges and encourages more sustainable consumption behavior.

Theory of Planned Behavior (TPB)

TPB provides a helpful approach to grasping how people shape their choices in energy use. Underpinning TPB, it is predicted that an individual's actions are influenced by three components: attitudes, subjective norms, and perceived behavioral control. It suggests that attitudes are a person's positive or negative evaluation of a behavior; subjective norms are a person's views as to how often they think behavior should be exercised with certain perceived social pressures; and perceived behavioral control is an individual's belief that they are able to enact the behavior. This theory is particularly relevant to financial decisions that concern energy. Based on Gallegos et al. (2025) state that the most frequently employed models were the self-model, the technology acceptance model (TAM), and TPB. According to this theory, attitudes refer to an individual's positive or negative evaluation of a particular behavior, subjective norms refer to the perceived social pressure to perform or not perform the behavior, and perceived behavioral control refers to an individual's belief in their ability to perform the behavior.

To illustrate, consider the example of a household debating whether to install solar panels. A healthy perspective towards environmental sustainability and long-term savings could spur them on. Now if the neighbors and peers are likewise taking up behavior of this nature, the subjective norm is consistent with them. And the more confident they feel in financing options or access to government incentives, the higher their perceived behavioral control—a phenomenon known as increasing adoption. Thus, we know that financial literacy has a direct relationship with the core tenets of TPB.

Financial decision-making is influenced by a wide variety of cognitive and social issues and understanding such factors contributes to policymakers devising more effective energy literacy campaigns (Negi & Jaiswal, 2024) Encouraging social norms regarding sustainable consumption, raising awareness of long-term cost benefits, and eliminating perceived financial or informational barriers are examples of practices that can foster TPB-aligned, greener behaviors, combined.

According to Gallegos et al. (2025) in explaining energy-related behavior self-model, TAM, and TPB are three well-known models used by individuals today. These frameworks enable researchers and policymakers to understand how personal beliefs (including the individual's own, peer's influence, and perceived control) shape individual energy behavior. Such models are particularly relevant for designing interventions designed to promote sustainable energy use. For example, TAM focuses on perceived usefulness and ease of use as reasons for adopting a new technology including smart meters or energy saving appliances. Likewise, TPB indicates that if one wants to modify behavior, then interventions should not only address an individual's attitude, but also social acceptance and confidence to act.

A case study of such models in action would be the national energy efficiency campaign in Colombia. Through a mix of awareness campaigns (raising people's mindsets) and community programs (forming more supportive social norms in the community) and educational programs (informing parents to feel that they also got better control), the program achieved a noticeable decrease in households' electricity use. The campaign also focused on financial benefits (i.e., reduced utility charges) thereby promoting the message that energy efficiency is not only environmentally desirable but also financially preferable. This demonstrates that models of behavior, paired with applied tactics, can result in considerable changes in how people think about and manage their energy consumption—particularly when financial literacy is combined with it.

EMPIRICAL LITERATURE REVIEW

Cross-Country Studies

Salahodjaev & Sadikov, (2025) conducted a cross-country analysis and found that financial literacy is positively associated with improvements in environmental quality across a sample of 131 countries. This research indicates that nations with higher levels of financial literacy tend to exhibit better environmental performance, suggesting that financial literacy not only improves financial wealth but also preserves environmental resources. The study used the Environmental Performance Index (EPI) and the ND-GAIN Index as proxies for environmental sustainability, providing a comprehensive measure of environmental quality.

In a systematic review of 137 countries, Todorov, Aleksandrova, & Ismailov, (2023) investigated the role of financial literacy on carbon emissions. They observe an inverse U-shaped relationship—in other words, at low levels, financial literacy may produce more carbon emissions. But as you pass a certain point of financial literacy, high literacy will be associated with lower emissions. Such pattern would imply that mere personal financial knowledge is insufficient, one need to gain a more comprehensive understanding before real environmental benefits can ensue. The findings are a testament to the capacity of financial knowledge – particularly paired with energy awareness – to leverage climate action. Financial literacy can even help people and societies make more informed decisions about using, consuming, and investing in energy to drive toward more sustainable ends. Together, the work of Todorov, Aleksandrova, & Ismailov, (2023) and Salahodjaev & Sadikov, (2025) demonstrates that financial education is becoming more and more relevant at a global level in pursuit of sustainability activities as a promoter not only of individual empowerment but of environmental stewardship.

By improving financial knowledge and skills, we can empower individuals and communities to make informed decisions about their energy consumption and investment choices, leading to a more sustainable future for all. However, it is important to recognize that the relationship between financial literacy and environmental sustainability is complex and multifaceted, and that other factors such as economic development, energy policies, and cultural norms play a significant role (Todorov, Aleksandrova, & Ismailov, 2023).

Regional and National Studies

Maji & Chakraborty, (2024) focused on the case of households in West Bengal, India. They found a strong association between increased ERFL and lower levels of energy consumption. The findings of their study of 155 households indicate that building stronger ERFL could be a mechanism to induce energy-saving behaviors even in the Indian scenario.

Kalmi, Trotta, & Kaukauskas, (2021) investigated the relationship between ERFL and electricity use in Finnish households. Higher levels of ERFL were associated with lower energy consumption even after adjusting for housing and demographic factors. It is found that higher ERFL in developed countries leads to an increased energy efficient behavior.

Martins, Madaleno & Dias, (2020) assessed energy literacy levels among the university community in Portugal and noted good levels of energy literacy, with financial knowledge playing a positive role. Their study distributed a questionnaire to university students and found that gender was a significant determinant of energy literacy dimensions, and that financial knowledge had a positive and significant impact on energy knowledge. This suggests that integrating financial concepts into energy literacy programs can be an effective way to improve individuals' understanding of energy issues and promote sustainable energy behaviors.

Specific Sector Studies

Attipoe, Chukwuma-Eke, Lawal, Friday, Isibor, & Akintobi, (2025) examined the role of financial institutions in promoting sustainable energy practices for Small and Medium-sized Enterprises (SMEs). Their research highlighted the financial barriers faced by SMEs in adopting clean energy solutions and discussed various financial tools that financial institutions can use to support SMEs in transitioning towards sustainable energy practices. The study emphasized the importance of collaborative efforts between financial institutions, governments, and other stakeholders to create a favorable regulatory environment that incentivizes clean energy investments.

Kinoshita, (2025) analyzed the impact of energy literacy on individual purchase behavior of energy-efficient homes in Japan. The study found that respondents who had knowledge of renewable energy were more likely to purchase Net Zero Energy House (ZEH) type energy-efficient homes. This suggests that increasing energy literacy can contribute to the diffusion of energy-efficient homes and promote energy conservation in the residential sector.

Andolfi, Marxen, & Frank, 2024, June) surveyed residential consumers in a European country and revealed that energy literacy and environmental values are key predictors for flexibility in energy use. Their study examined consumers' understanding of energy use and their motivations for providing flexibility for heating and charging electric vehicles (EVs). The results indicated that energy-literate consumers with strong environmental values are more willing to participate in demand response programs and provide flexibility to the energy grid.

METHODS USED

Data Collection

Surveys and questionnaires are two of the most frequent tools for measuring people's understanding of finance and energy consumption. They are an easy and inexpensive solution in collecting data from a broader pool of subjects and enabling study of views, practices, attitudes, behaviors, and decision-making processes regarding sustainability. For example, Maji & Chakraborty, (2024), have undertaken a systematic structured household survey on a sample of 155 households in West Bengal, India, to investigate the effects of energy-related financial knowledge on electricity consumption. They found that households that have better awareness and better knowledge of energy-saving techniques consume less electricity, even if income level differences are factored in. Similarly, in Japan, Kinoshita, 2025 sent a questionnaire to 1,000 households to explore attitudes toward energy use, conservation, and use of sustainable energy technologies. The study served to identify behavioral trajectories and points in need of further education/policy initiatives for effective promotion.

These tools are especially relevant for understanding a wide spectrum of how people encounter energy and financial data in everyday life—like knowledge of efficiency benefits versus preparedness towards greener inputs. As discussed earlier, instead of redefining the concept of financial literacy, the methods allow us to see it in action in a multitude of cultural and economic contexts.

Data can be collected from households, students, and specific demographic groups, allowing researchers to target their data collection efforts to populations of interest and gain a more nuanced understanding of the factors that influence financial literacy and energy-related behaviors (Martins, Madaleno & Dias, 2020; Cerovic, Malnar, Sinčić, 2024). Ana Martins, M. Madaleno, and M. Ferreira Dias (Martins, Madaleno & Dias, 2020) distributed a questionnaire to the university community in Portugal to evaluate energy literacy levels. Cerovic, Malnar, Sinčić, (2024) analyzed the energy literacy of students at the Faculty of Economics in Rijeka. By targeting specific populations, researchers can gain a more nuanced understanding of the factors that influence financial literacy and energy-related behaviors in these communities.

Secondary data from sources like the World Bank can provide macro-level insights, enabling researchers to examine the relationship between financial literacy, energy consumption, and environmental outcomes at the national and international levels. Tariq, Saleem, & Shahzad, 2022) used quantitative secondary annual data acquired from World Development Indicators provided by the World Bank. These data sources often provide a wealth of information on economic, social, and environmental indicators, allowing researchers to conduct large-scale analyses and identify patterns that would be difficult to detect using primary data alone.

Analytical Techniques

Regression analysis is used to determine the relationship between financial literacy and energy consumption, allowing researchers to quantify the impact of financial literacy on energy use while controlling for other factors that may influence energy consumption pattern Todorov, Aleksandrova, & Ismailov, (2023) used correlation and regression analyses to examine the connection between financial literacy level and greenhouse gas emissions. Maji & Chakraborty, (2023) used the ordinary least square regression method to attain the objectives of the study. This method helps researchers to isolate the specific impact of financial literacy on energy consumption, while accounting for factors such as income, education, household size, and geographic location.

Correlation analysis helps identify associations between variables like financial literacy and carbon emissions, providing insights into the potential linkages between financial knowledge and environmental outcomes Todorov, Aleksandrova, & Ismailov, 2023) used correlation and regression analyses to examine the connection between financial literacy level and greenhouse gas emissions. This technique measures the strength and direction of the linear relationship between two variables, providing insights into whether they tend to move together or in opposite directions.

Stochastic frontier analysis can assess energy efficiency levels in residential sectors, providing a means of measuring how efficiently households are using energy and identifying opportunities for improvement (Boogen, Cattaneo, Filippini, & Obrist, 2021) estimate an electricity demand frontier function using a stochastic frontier approach. This method allows researchers to identify the factors that contribute to energy efficiency, as well as the potential for energy savings in different households or regions.

Frameworks and Models

The Theory, Context, Characteristics, and Methodology (TCCM) approach is a systematic and holistic framework to review and evaluate relationships between financial literacy and consumer financial behavior. Negi & Jaiswal, (2024) applied this framework, and systematically analyzed 79 identified papers using the Scientific Procedures and Rationales for Systematic Literature Reviews (SPAR4SLR) protocol. Not only did this approach assist them in identifying ongoing patterns, conceptual gaps and emergent themes in the literature around financial decision-making; it also helped them make sense of the existing literature. The TCCM framework is especially appropriate in coordinating disaggregated results in different contexts and provides a framework for disaggregating patterns of financial literacy and financial behavior across industries, demographics and regions.

Maji & Chakraborty, (2024), conducted a systematic investigation on ERFL among Indian households and concluded that there was an impact of high-ERFL among Indian households on the adoption of energy saving habits such as energy efficient appliances and energy use in general. Their results show how concepts like TCCM can illustrate that financial literacy not only promotes individual's personal welfare but can also promote sustainable practices that contribute to wider environmental objectives. Conceptual paradigms -such as the one proposed by Blasch et al. (2018) and used by Maji & Chakraborty, (2024) provide a well-established framework to measure ERFL. These models often include constructing measures of financial knowledge, energy consciousness, and sustainability practices, which allow for a multi-dimensional assessment of ERFL.

Machine learning approaches, such as Kernelized regularized least squares (KRLS), represent advanced analytical tools that can examine the complex pathways of financial inclusion, population growth, and energy intensity on sustainable development. Iorember et al. (2025) employ the Kernelized regularized least squares (KRLS) machine learning approach and Granger causality test to investigate the pathways of financial inclusion, population growth and energy intensity on sustainable development. These techniques can identify non-linear relationships and interactions between variables, providing insights into the complex dynamics that drive sustainable development.

ANALYSIS OF RESULTS

Impact on Energy Consumption

Higher ERFL levels are consistently linked to lower household energy consumption, because people with good financial and energy knowledge consciously and optimally choose more choices in their households. For example, Kalmi, Trotta, & Kaukauskas, (2021) f show that Finnish households with higher ERFL used less electricity, when we control for other factors such as dwelling and household characteristics. ERFL enables people to compare the financial and environmental choices they make to conserve both, leading to the practice of conserving resources and energy-efficient technologies.

Financial literacy empowers individuals to make informed decisions about purchasing energy-efficient appliances, which can significantly reduce household energy consumption and lower energy bills. Blasch, Boogen, Daminato, & Filippini,

2021) explain that consumers limited energy-specific knowledge and investment skills. By understanding the long-term cost savings associated with energy-efficient appliances, individuals are financial and energy literacy programs, as these programs can help them to manage their energy costs more effectively and make informed decisions about their energy consumption.

DISCUSSION

Synthesis of Findings

It has been found in the literature that financial and energy literacy have a positive relationship with sustainable behaviors, such as people to energy and financial behavior, individuals who have a good understanding of both the financial and energy behaviors will use energy-efficient measures as well as environmental initiatives (Maji & Chakraborty, 2023; Salahodjaev & Sadikov, 2025). They can see long-term outcomes between the costs that they will bear as well as the benefits that they will reap and choose to act in accordance with their values as individuals – and those to the greater goal of sustainability.

Financial literacy empowers consumers to make informed decisions about energy consumption and investments, enabling them to weigh the costs and benefits of different energy options and make choices that align with their financial goals and environmental values (Blasch, Boogen, Daminato, & Filippini, 2021). This empowerment can lead to more efficient energy use, reduced energy costs, and a smaller environmental footprint. Informed consumers are more likely to seek out information about energy-efficient technologies, compare different products and services, and make decisions that are both financially sound and environmentally responsible. Furthermore, they are more likely to understand the long-term benefits of sustainable energy practices, such as reducing energy waste and conserving natural resources.

Contextual factors such as demographics, economic situations, and cultural expectations help to determine the influence of financial literacy to drive behavior (Todorov, Aleksandrova, & Ismailov, 2023). FEL's effects are likely to be either strengthened or diminished by these influences, highlighting the urgency of customized approaches to the field. For example, richer people may have more resources to engage in investment in energy-efficient technologies, and the green-oriented individuals might be prone to adopting sustainable behavior. The needs and situations of respective populations are factors for FEL programs to be effective.

Implications for Policy and Practice

Policymakers should prioritize the implementation of financial literacy programs that specifically incorporate energy-related knowledge to foster a more holistic understanding of sustainable practices (Agu, Abhulimen, Obiki-Osafiele, Osundare, Adeniran, & Efunniyi, 2024). These programs should aim to equip individuals with the knowledge and skills they need to make informed decisions about their energy consumption and investments, as well as to understand the broader environmental and social impacts of their energy choices. This might involve integrating energy-related content into existing financial literacy curricula; developing new educational materials that specifically address the intersection of finance and energy and providing training for educators and financial advisors on how effectively communicate these concepts.

These financial service providers have the power to improve energy saving by supplying consumers with tailored products and services that will motivate them to promote a greener lifestyle (Attipoe, Chukwuma-Eke, Lawal, Friday, Isibor, & Akintobi, 2025). A few among those options are low-interest loans for energy-efficient equipment, subsidies for utilizing renewable energy systems, or tailor-made financial advice on saving energy. They are advocating not only for a greener future, but they are also fostering FEL by training individuals to understand that choices they make regarding their energy impact on their monetary well-being.

Integrating AI and digital tools into financial literacy campaigns has the potential to significantly enhance their reach and effectiveness, improving customer service and promoting sustainable practices on a wider scale (Kaur, & Dhiman, 2024; Amon, Jagri, & Oplotnik, 2024). AI-powered tools can provide personalized financial advice and recommendations, while digital platforms can offer interactive and engaging learning experiences. By leveraging technology, we can make financial literacy more accessible and effective, particularly for underserved populations. For example, AI-powered apps can track energy usage in real-time, provide personalized tips on how to conserve energy, and connect users with financial products that support sustainable energy investments.

Addressing Research Gaps

Future research efforts should focus on exploring the under-researched areas of energy literacy, particularly its impact on efficient energy use and the factors that influence its effectiveness (Santilln & Cedano, 2023). This includes investigating the specific mechanisms through which energy literacy affects energy-related behaviors, as well as the barriers that prevent individuals from adopting sustainable energy practices. For example, research could explore the role of psychological factors, social norms, and economic incentives in shaping individuals' energy choices. Additionally, research could investigate the effectiveness of different interventions aimed at promoting energy literacy and identify best practices for designing and implementing these programs.

More studies are needed to understand the specific mechanisms through which financial literacy affects energy-related behaviors, such as investment in renewable energy, adoption of energy-efficient transportation, and participation in energy conservation programs Negi & Jaiswal, 2024). This includes investigating the role of psychological factors, social norms, and economic incentives in shaping individuals' energy choices. For example, research could explore how financial literacy influences individuals' attitudes towards renewable energy, their perceptions of social norms related to energy conservation, and their responsiveness to economic incentives for energy efficiency. By understanding these mechanisms, we can develop effective interventions to promote sustainable energy behaviors.

Longitudinal studies can provide valuable insights into the long-term impact of FEL interventions, allowing researchers to track changes in energy consumption patterns, environmental outcomes, and financial well-being over time (Salahodjaev & Sadikov, 2025). These studies can also help to identify the factors that contribute to the sustainability of these interventions and to assess their cost-effectiveness. For example, longitudinal studies could track the energy consumption patterns of households that participate in FEL programs over several years to assess whether these programs lead to lasting changes in behavior.

CONCLUSIONS

Summary of Key Findings

FEL is a crucial element for promoting sustainable development because it helps people make choices that are good for the environment and their finances (Maji & Chakraborty, 2024). Knowing how their energy choices affect their money, and environment helps people cut down on their energy usage, use more energy efficient devices, and support laws that encourage renewable energy.

ERFL contributes to environmental sustainability and financial health by providing individuals with the knowledge to make informed energy decisions (Salahodjaev & Sadikov, 2025, Todorov, Aleksandrova, & Ismailov, 2023). That means evaluating the economic and environmental pros and cons of various energy sources. ERFL helps reduce greenhouse gas emissions while saving resources and keeping the air and water cleaner. At the same time, it increases people's financial stability and their ability to cope with economic problems.

The effectiveness of financial literacy depends on demographics, economics, and culture, which demonstrate the importance of specific treatments tailored to specific groups of people (Martins, Madaleno & Dias, 2020) (Martins, Madaleno, & Dias, 2020). Such variables can either increase or decrease the effects of FEL; hence context will matter, and a context-dependent approach is necessary. Programs should be culturally relevant and address the specific issues of low-income households, women, and minority groups.

Recommendations for Future Research

Since it is possible to deliver personalized learning at scale using technology and digital platforms, further research will need to investigate potential improvements in FEL that those on these platforms represent (Amon, Jagri, & Oplotnik, 2024). This involves assessment of multiple digital tools including those available on mobile applications, online games, social media, and different demographic cohorts.

Subsequent research should concentrate on designing interventions that will involve specific demographics especially (e.g., low-income households, students, and small company owners) by addressing their unique factors (Patel, & Pathak, 2025). This includes acknowledging the barriers to sustainable energy engagement and addressing them by exploring solutions such as community-based programs, peer mentoring, and financial incentives to develop FEL in these communities.

Cross-national and cross-region comparisons could offer some important clues about successful strategies for fostering FEL and assisting sustainable development objectives (Tariq, Saleem, & Shahzad, 2022). By examining various policies, regarding for example energy efficiency standards, incentives in place to drive up renewable energy and programs to educate people toward saving, impact consumption and the environment, and do so in different contexts, these sorts of studies can help them determine which policies have done the trick.

Final Thoughts on Sustainability

FEL is not just a policy option; it's a central strategy for navigating toward a sustainable and just future. This reflects the point made by (Karumanchi, L. n.d.) — that people with a comprehension of the principles of finance and energy possess an improved ability to take decisions that support the environment and their financial future.

Combining these literacies, individuals are equipped to cultivate energy-savings lifestyles, evaluate sustainable investment prospects and endorse climate-friendly policies. That's a difficult image to maintain and it will take more than one participant to achieve it. It needs governments, finance institutions, education institutions and communities to work together. It should be noted that the concerted effort of various stakeholders is necessary to scale up the usage of sustainable technologies (Attipoe, Chukwuma-Eke, Lawal, Friday, Isibor, & Akintobi, 2025). This is achieved through funding accessible financial and energy course work, policy that favors the utilization of renewable energy, and local habits that will make us responsible users of energy. Community-focused efforts — such as solar initiatives or neighborhood energy audits — tend to succeed when people understand the longer-term financial benefits of clean energy. Those efforts contribute not just by enhancing access to technology, but by aiding people to interpret their choices through a FEL perspective.

According to Iorember et al. (2025), the change requires persistent joint efforts. In order to create the culture of conservation, financial and energy literacy should be integrated into education, on-the-job training, and into community education programs. More than this, while we're trying to solve the world's most pressing problems – like climate change, energy-insecurity and environmental degradation – people are better at making ethically sound decisions. Finally, FEL investment is the cleanest and greenest investment there is. It's a way of connecting personal financial goals to corporate ecological stewardship, and it's the foundation for a more equitable, resilient future that's suited to our children and our grandchildren.

REFERENCES

Adhikari, G. (2024). Nexus between Financial Literacy and Financial Behavior: Theoretical and Empirical Insights. *IIMRJBC*, *2*(1), 1–14.

Agu, E. E., Abhulimen, A. O., Obiki-Osafiele, A. N., Osundare, O. S., Adeniran, I. A., & Efunniyi, C. P. (2024). Proposing strategic models for integrating financial literacy into national public education systems. *International Journal of Finance and Management ResearchS*, *3*(2), 1–17.

Amon, A., Jagri, T., & Oplotnik, Ž. (2024). The neobanks and sustainability. *Journal of Information and Organizational Sciences, 48*(9 Andolfi, L., Marxen, H., & Frank, M. (2024, June). Are you Flexible Enough? The Impact of Energy Literacy and Environmental Values on Flexibility Provision. In *2024 11th International Conference on the European Energy Market (EEM)* (pp. 1-5). IEEE. (Andolfi, Marxen, & Frank, 2024, June).

Arrondel, L. (2021). Financial Literacy and French Behaviour on the Stock Market. *Journal of Financial Transformation*, *14*(1), 88–100.

Attipoe, V., Chukwuma-Eke, E. C., Lawal, C. I., Friday, S. C., Isibor, N. J., & Akintobi, A. O. (2025). The Role of Financial Institutions in Promoting Sustainable Energy Practices for SMEs: A Strategic Approach. *Contemporary Scientific and Engineering Investigations*, *2*(1), 28–49.

Blasch, J., Boogen, N., Daminato, C., & Filippini, M. (2021). Empower the Consumer! Energy-related Financial Literacy and its Implications for Economic Decision Making. *Energy Journal*, *42*(2).

Boogen, N., Cattaneo, C., Filippini, M., & Obrist, A. (2021). Energy efficiency and the role of energy-related financial literacy: Evidence from the European residential sector. *Review of Environmental Economics and Policy*, *15*(2), 335–356.

Cerovic, L., Malnar, A., Sinčić, D. (2024). Energy Literacy of Economics Students in Rijeka: Knowledge, Attitudes, and Behavioral Approach. *Energies* 17 DO - DOI: 10.3390/en17081840

Gallegos, A., Valencia-Arias, A., Rodrguez Zavala, L., Rodrguez-Correa, P., Martnez Rojas, E., & Castaeda Rodrguez, L. D. R. (2025). Adoption of financial literacy: Systematic literature review and proposal of a theoretical model. *Journal of Innovation and Entrepreneurship*, *14*(1), 1–22.

Iorember, P. T., Hounkanrin, D. O., Diyoke, K., & Tang, C. F. (2025). Tracing financial inclusion pathway to sustainable development: The criticality of population growth and energy intensity using a machine learning approach. *African Journal of Economic and Management Studies*, *16*(2), 241–259.

Kalmi, P., Trotta, G., & Kaukauskas, A. (2021). Energy-related financial literacy and electricity consumption: Survey-based evidence from Finland. *The Journal of Consumer Affairs*, *55*(6), 1698–1729.

Karumanchi, L. (n.d.). The Role of Green Financing in Accelerating Sustainable Economic Development in Emerging Economies. *International Journal of Technology and Business Management, 14*(3).

Kaur, M., & Dhiman, B. A. (2024). Framework for Exploring the Integration of AI and Financial Literacy toward a Sustainable Financial Ecosystem. *Educational Administration: Theory and Practice*, *30*(4), 8750–8759.

Kinoshita, S. (2025). An empirical analysis of the impact of energy literacy on individual purchase behavior of energy-efficient homes. *International Journal of Energy Sector Management*, *19*(3), 661–677.

Lusardi, A., & Mitchell, O. S. (2014). The economic importance of financial literacy: Theory and evidence. *Journal of Economic Literature*, *52*(1), 5–44.

Maji, S. K., & Chakraborty, P. (2024). Energy-related financial literacy and energy consumption: A study of residential households in West Bengal, India. *International Journal of Energy Sector Management*, *18*(6), 1489–1510.

Martins, A., Madaleno, M., & Dias, M. F. (2019). Energy literacy: What is out there to know? *Energy for Sustainable Development*, *53*, 1–10.

Martins, A., Madaleno, M., & Dias, M. F. (2020). Energy Literacy: does age matters? *AHFE International Conference. AHFE Open Access.*

Martins, A., Madaleno, M., & Dias, M. F. (2020). Financial Knowledge Role in Portuguese Energy Literacy. *Energies*, *13*(13), 3412.

Negi, P., & Jaiswal, A. (2024). Impact of financial literacy on consumer financial behavior: A systematic review and research agenda using TCCM framework. *International Journal of Consumer Studies*.

Patel, P., & Pathak, N. (2025). Sustainable Lifestyle Choices in Chhattisgarh, India: Financial Implications and Consumer Behavior. *International Journal of Financial Management & Research*, *7*(2), 119–126.

Salahodjaev, R., & Sadikov, A. (2025). Financial literacy and environmental sustainability: A cross-country test. *Frontiers in Sustainability*, *8*, 1514393.

Santilln, O. S., & Cedano, K. G. (2023). Energy Literacy: A Systematic Review of the Scientific Literature. *Energies*, *16*(21), 7235.

Siddiqui, N., Faiyaz, T., & Tari, V. (2022). Ganga and Yamuna: Through the lens of National Green Tribunal. *Nature. Environment and Pollution Technology*, *21*(4), 1657–1664. DOI: 10.46488/NEPT.2022.v21i04.018

Tari, V., Gupta, V., Joshi, A., Kurve, P., & Chandra, N. (2024). Biodiversity loss due to forest diversion. In Rathore, A. (Ed.), *Impact of societal development and infrastructure on biodiversity decline* (pp. 137–165). IGI Global., DOI: 10.4018/979-8-3693-6950-0.ch010

Tari, V., & Kannan, K. (2022). Microbial bioremediation of polythenes and plastics – A green sustainable approach. In *Microbial biotechnology in green remediation* (1st ed.). Elsevier., DOI: 10.1016/B978-0-323-90452-0.00003-7

Tari, V., Siddiqui, N., Rathi, D., Siddiqui, N. N., & Wahyuni, D. K. (2024). Need for an evolved groundwater justice in rural area Uttar Pradesh, India. *Nature. Environment and Pollution Technology*, *23*(1), 125–137. DOI: 10.46488/NEPT.2024.v23i01.009

Tari, V. S., & Patil, P. Y. (2017a). Transfer factor of heavy metal in soil to plant from pesticide-contaminated area (mango orchards), Ratnagiri district, Maharashtra. *Research Journal of Chemistry and Environment*, *21*(1), 26–32.

Tari, V. S., & Patil, P. Y. (2017b). Challenge of contamination of pesticide for Alphonso in Ratnagiri district, Maharashtra, India. *Research Journal of Chemistry and Environment*, *21*(12), 57–67.

Tariq, S., Saleem, F., & Shahzad, F. (2022). Does digital finance and financial inclusion strengthen environmental sustainability: Evidence from Asia. *Pakistan Social Sciences Review*, *4*(4), 847–861.

Todorov, L., Aleksandrova, A., & Ismailov, T. (2023). Relation between Financial Literacy and Carbon Footprint: Review on Implications for Sustainable Development. *Economics*, *2*(2), 2–11.

Yuwono, W., Susanna, Sasmita, E. W., & Sihotang, W. H. (2023). Analysis of the Influence of the Role of Financial Literacy on Personal Financial Management. *European Journal of Business and Management Research*, *8*(3), 160–165.

Chapter 9
Advancements in Chalcogenide Nanocomposites of Next-Generation Materials for Dye-Sensitized Solar Cells:
Emerging Trends and Future Solar Innovations

Hari Shankar Biswas
https://orcid.org/0000-0003-1883-5244
Surendranath College, India

Amit Kr. Kundu
https://orcid.org/0000-0002-7649-8154
Sripat Singh College, India

Dilip K. Maiti
https://orcid.org/0000-0002-8772-6685
University of Calcutta, India

Prasenjit Mandal
Santipur College, India

DOI: 10.4018/979-8-3373-3962-7.ch009

ABSTRACT

Dye-sensitized solar cells (DSSCs) are a promising photovoltaic technology known for low cost, flexibility, and eco-friendliness. Chalcogenide-based nanocomposites, with their unique electronic, optical, and catalytic features, have attracted significant interest for boosting DSSC efficiency. This review explores next-generation chalcogenide nanocomposites—sulfides, selenides, tellurides, and hybrids—used in key DSSC components like photoanodes, counter electrodes, and electrolytes. It highlights synthesis methods, structural traits, and the functional improvements these nanomaterials offer. Challenges, recent progress, and future perspectives for integrating chalcogenide nanocomposites into commercially viable DSSCs are also discussed.

INTRODUCTION

Dye-sensitized solar cells (DSSCs), first introduced by O'Regan and Grätzel in the early 1990s, have sparked significant interest in the scientific community due to their relatively simple construction, ease of fabrication, and potential for cost-effective energy conversion, (Kannan & Tari, 2025; Regan & Grätzel 1990; Srivastava et al., 2025). These devices mimic natural photosynthesis by using a dye molecule to harvest sunlight and generate electricity through electron transfer processes, (Shahid et al., 2025). The core components of a DSSC include a photoanode, a sensitizing dye, an electrolyte, and a counter electrode, (Agrawal et al., 2022). Over the decades, DSSCs have evolved from lab-scale prototypes to semi-commercial devices with improved efficiencies and durability. However, to surpass their limitations and compete with other photovoltaic technologies, innovations in material science, especially at the nanoscale, have become imperative.

Nanomaterials have revolutionized the field of DSSCs by offering a higher surface area, improved charge transport, and enhanced light-scattering properties, which collectively contribute to improved photovoltaic performance (Figure 1), (Korir et al., 2024). Incorporating nanostructures such as nanoparticles, nanorods, nanotubes, and quantum dots, (Tari et al., 2022) into the photoanode or counter electrode can significantly influence the efficiency and functionality of DSSCs, (Zheng et al., 2024). These materials can be tailored to control the morphology, surface chemistry, and electronic properties of each DSSC component. As a result, the utilization of nanomaterials represents a major paradigm shift, opening avenues for designing next-generation solar cells with superior energy conversion capabilities.

Figure 1. Schematic representation of chalcogenide nanocomposites of next-generation materials for dye-sensitized solar cells

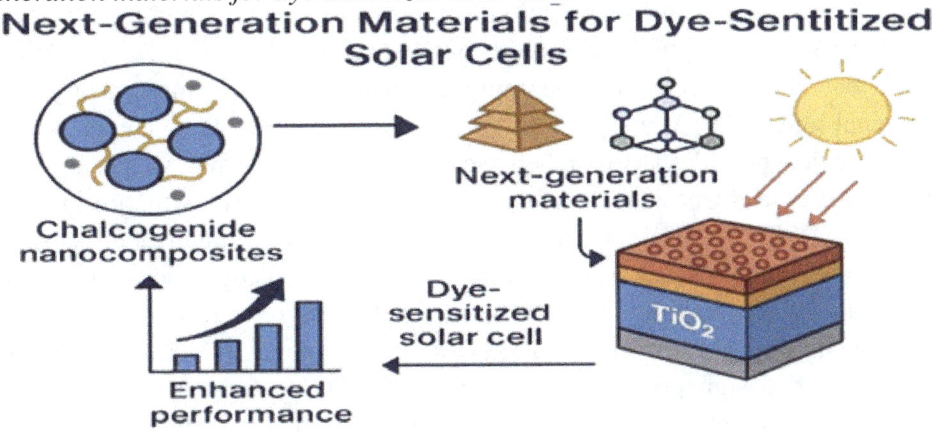

Among the broad spectrum of nanomaterials, chalcogenides—comprising sulfur, selenium, and tellurium with various metals—have emerged as a class of semiconducting materials with exceptional properties conducive to DSSC applications (Njema & Kibet, 2024). Their tunable bandgaps, high absorption coefficients, and catalytic activity make them ideal candidates for replacing conventional materials in DSSCs. Chalcogenides exhibit unique optoelectronic behavior, which can be harnessed to enhance both the light-harvesting efficiency and the electron transport mechanisms within the solar cell architecture (Jayawardane et al., 2025). Their integration has shown promise in not only augmenting the power conversion efficiency but also in improving the longevity and stability of the cells under operational conditions.

The growing interest in chalcogenide-based materials stems from their versatility and adaptability. These materials can be engineered into various nanostructures such as core-shell particles, hierarchical composites, or doped hybrids, enabling multifunctional applications within the same device (Sowbakkiyavathi et al., 2024). For instance, while certain chalcogenides function effectively as photoanode materials due to their superior conduction properties, others serve as cost-effective alternatives to platinum-based counter electrodes (Ayaz et al., 2024). Moreover, their incorporation into gel or solid-state electrolytes has demonstrated enhanced ionic conductivity and reduced charge recombination losses. Thus, chalcogenides offer a holistic approach to tackling the efficiency bottlenecks in DSSCs.

Despite the promising potential of chalcogenide-based nanocomposites, several challenges must be addressed to facilitate their widespread adoption in DSSCs (Kumar et al., 2022). These include issues related to synthesis reproducibility, toxicity of certain elements (e.g., cadmium or lead in some chalcogenides), long-term stability, and scalability of production processes (Rasal et al., 2022). Therefore, understanding the structure-property-performance relationships of these materials is crucial. Advancements in fabrication techniques and surface modification strategies are essential to optimize the interaction between chalcogenides and other cell components, thereby achieving maximum efficiency without compromising environmental safety (Tari V. et al., 2025; Wani S. A. et al., 2025; Phogat et al., 2024).

This review provides a comprehensive overview of the current status, advancements, and future prospects of chalcogenide-based nanocomposites in dye-sensitized solar cells (DSSCs). It examines their roles in photoanodes, counter electrodes, and electrolytes, with emphasis on ternary and quaternary chalcogenides that enhance performance through synergistic effects in hybrid structures. The objectives are to: (1) elucidate physicochemical properties of chalcogenides, (2) discuss synthesis strategies and design principles, (3) evaluate DSSC performance compared to conventional systems, and (4) outline future directions toward commercialization. This work aims to guide researchers, engineers, and policymakers in advancing sustainable solar energy technologies.

OVERVIEW OF CHALCOGENIDE MATERIALS

Definition and Classification of Chalcogenides

Chalcogenides are compounds containing one or more chalcogen elements—primarily sulfur (S), selenium (Se), or tellurium (Te)—combined with more electropositive elements such as metals or metalloids (Olaniyan et al., 2024). These materials exist in binary, ternary, or even quaternary compositions, where their structural and functional attributes can be systematically tuned for targeted applications. In the context of dye-sensitized solar cells (DSSCs), chalcogenides are primarily categorized into metal chalcogenides (e.g., CdS, ZnSe, Cu2S), transition metal dichalcogenides (TMDs, such as MoS2, WS2), and complex chalcogenides (Figure 2) (e.g., CuInS2, CZTS) (Kharboot et al., 2023). Each category possesses distinct structural motifs and electronic properties that influence their interaction with light and charge carriers.

Figure 2. Schematic diagram of different chalcogenides

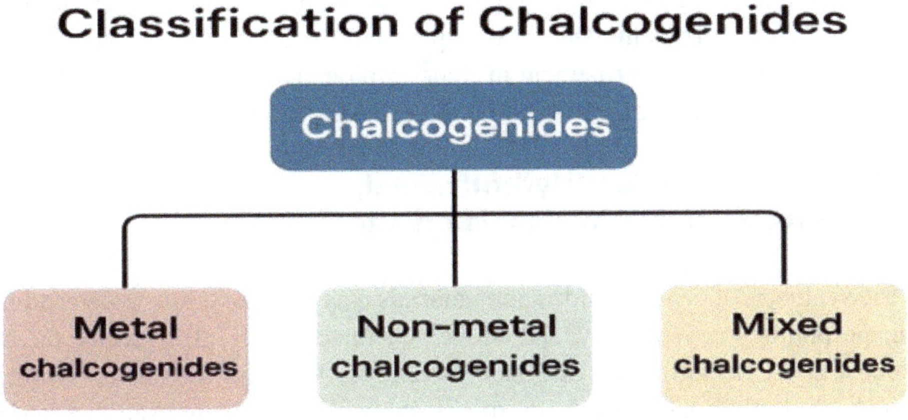

Structural and Electronic Properties

The structural diversity of chalcogenide compounds stems from their ability to adopt various crystal structures, such as wurtzite, sphalerite, chalcopyrite, and layered hexagonal forms (Giri et al., 2023). For instance, MoS2 adopts a layered structure, where weak van der Waals forces hold together covalently bonded S-Mo-S layers, allowing for facile exfoliation into few-layer or monolayer forms (Liu et al., 2023). This structural feature is crucial for their high surface area and fast charge transfer. Chalcogenides exhibit semiconducting behavior, with their electronic band structure being governed by the metal-chalcogen bond characteristics. They show a range of bandgaps from 1 to 3 eV, making them suitable for light absorption across the visible to near-infrared spectrum (Lahariya et al., 2024). The presence of vacancies, interstitials, or dopants further modulates their electrical conductivity and charge mobility, enabling optimization for DSSC components like photoanodes and counter electrodes.

Optical Characteristics and Bandgap Engineering

Chalcogenide materials possess high optical absorption coefficients due to their narrow and tunable bandgaps. This makes them ideal for photon harvesting in DSSCs (Biney et al., 2025). For example, cadmium sulfide (CdS) and zinc sulfide (ZnS) exhibit excellent UV-visible absorption, while copper-based selenides and tellurides extend absorption into the near-infrared region (Giroux et al., 2022). Bandgap engineering, achieved through alloying (e.g., CdSxSe1−x), doping, or forming heterostructures, allows for precise control over the absorption edge and charge separation dynamics. Such tuning is crucial to enhancing light absorption while maintaining efficient electron transport and minimal recombination losses. Furthermore, quantum confinement in chalcogenide quantum dots allows for size-dependent optical properties, making them adaptable to various spectral windows.

Synthesis Approaches (Hydrothermal, Solvothermal, Microwave-assisted, etc.)

The synthesis of chalcogenide nanomaterials plays a pivotal role in determining their morphology, crystallinity, and surface characteristics—all of which directly influence DSSC performance (Deogratias et al., 2023). Among the various synthesis methods, hydrothermal and solvothermal techniques are widely used due to their simplicity, scalability, and ability to produce highly crystalline nanostructures at relatively low temperatures (Moradialvand et al., 2025). These methods involve reacting metal salts with chalcogen sources in a sealed autoclave under controlled temperature and pressure, allowing for the growth of nanocrystals with controlled shapes and sizes.

Microwave-assisted synthesis is gaining popularity as it offers rapid heating, uniform nucleation, and reduced reaction times compared to conventional methods (Rashid et al., 2025). This technique is particularly advantageous for producing ultrafine nanoparticles and layered structures like MoS2 and WS2. Chemical vapor deposition (CVD) and atomic layer deposition (ALD) are also employed for fabricating high-purity chalcogenide thin films with precise thickness control, crucial for constructing photoanode and counter electrode layers in DSSCs (Abdullah et al., 2024).

Other emerging methods include sonochemical synthesis, which uses ultrasonic waves to initiate and accelerate chemical reactions, and hot-injection techniques, which are often employed for quantum dot production (Matyszczak et al., 2024). Each synthesis strategy has its pros and cons, with choices often dictated by the desired morphology, composition, and application within the DSSC architecture. Importantly, green synthesis routes using plant extracts or biomolecules are being explored to enhance sustainability and reduce environmental impact.

CHALCOGENIDE NANOCOMPOSITES IN PHOTOANODES

The photoanode is a critical component in dye-sensitized solar cells (Figure 3) (DSSCs), responsible for harvesting sunlight, injecting electrons from the excited dye molecules into the conduction band of a semiconductor, and transporting these electrons to the external circuit (Lana et al., 2024). Titanium dioxide (TiO_2) has traditionally been the most widely used semiconductor material for photoanodes due to its high surface area, excellent chemical stability, and appropriate energy band alignment with commonly used dyes (Sawal et al., 2023). However, the limitations of pure TiO_2, such as its wide bandgap (~3.2 eV) and relatively poor electron mobility, have motivated researchers to explore TiO_2-based composites with chalcogenides to overcome these shortcomings (Ahmad et al., 2025).

Figure 3. Schematic drawing of chalcogenide nanocomposites in the photoanode

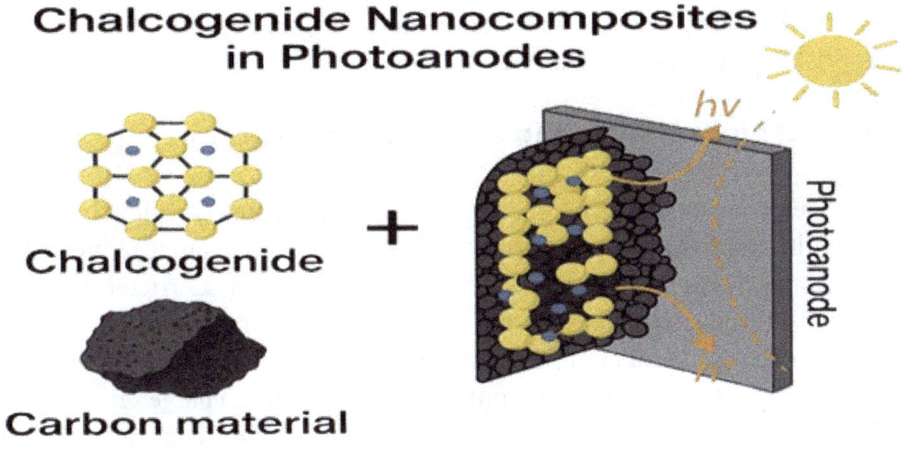

TiO_2-Chalcogenide Hybrids

TiO_2-chalcogenide nanocomposites have demonstrated enhanced performance in DSSCs owing to synergistic effects that arise from the integration of TiO_2 with chalcogenide nanostructures like CdS, CdSe, $CuInS_2$, and MoS_2 (Abu Elella et al., 2024). These chalcogenides possess narrow bandgaps and strong visible light absorption, which supplement TiO_2's UV-responsive properties. For instance, CdS-decorated TiO_2 nanorods have shown improved light harvesting and reduced charge recombination by forming type-II heterojunctions that facilitate charge separation (Tai et al., 2024). Similarly, MoS_2 nanosheets interfaced with TiO_2 nanoparticles have been reported to provide additional pathways for charge transport while contributing to enhanced catalytic activity.

ZnO and SnO_2-based Composites

Besides TiO_2, alternative wide-bandgap semiconductors such as zinc oxide (ZnO) and tin dioxide (SnO_2) have been employed in DSSC photoanodes. ZnO offers high electron mobility, while SnO_2 exhibits faster charge transport; however, both suffer from dye aggregation and chemical instability issues (Rajkumar et al., 2025). The incorporation of chalcogenide nanomaterials into ZnO and SnO_2 matrices can alleviate these drawbacks (Anucha et al., 2022). For example, ZnO-$CuInS_2$ composites improve light absorption in the visible region and enhance stability under prolonged illumination (Qiao et al., 2024). Additionally, SnO_2-CdSe heterostructures have demonstrated greater photocurrent densities and longer electron lifetimes compared to their pristine counterparts (Hu et al., 2021).

Enhancing Light Harvesting and Charge Transport

Chalcogenide nanocomposites enhance light harvesting in photoanodes by expanding the spectral absorption range and increasing the photogenerated charge carrier density (Soonmin et al., 2024). The high absorption coefficients of metal sulfides and selenides contribute to efficient photon capture across a broad range of the solar spectrum (Oni et al., 2024). Moreover, these nanomaterials introduce localized energy states and form heterojunctions that promote charge separation and inhibit electron–hole recombination. Incorporating quantum dot chalcogenides, such as CdTe and PbS, into photoanode architectures further enhances the photoconversion process through multiple exciton generation (MEG) and quantum confinement effects (Ates et al., 2021).

Case Studies and Performance Metrics

Numerous case studies have demonstrated the effectiveness of chalcogenide-based photoanodes in boosting DSSC performance (Tumram et al., 2024). A notable example includes a TiO_2-MoS_2 nanocomposite photoanode that achieved a power conversion efficiency (PCE) of over 8%, surpassing the 6% PCE of the unmodified TiO_2 counterpart (Li et al., 2025). Similarly, ZnO-Cu_2S nanorod arrays exhibited improved electron injection rates and prolonged electron lifetimes, resulting in higher short-circuit current densities (More et al., 2024). In another study, TiO_2-$CuInS_2$ photoanodes achieved significant enhancements in both PCE and photostability under continuous irradiation (Lu et al., 2022).

In-depth analyses of electrochemical impedance spectroscopy (EIS) and incident photon-to-current conversion efficiency (IPCE) further support the beneficial role of chalcogenide incorporation in photoanodes (Toe et al., 2021). These analyses reveal reduced charge transfer resistance and improved charge collection efficiency, indicating superior interfacial charge dynamics. The enhanced surface area and catalytic sites provided by chalcogenide nanostructures also allow for greater dye loading and more efficient electron injection (Chawla et al., 2024).

The structural configuration and morphology of these nanocomposites play a pivotal role in their performance. For instance, vertically aligned nanorod or nanowire arrays facilitate direct pathways for electron transport, reducing recombination losses (Huang et al., 2025). Hierarchical structures combining TiO_2 nanoparticles with chalcogenide quantum dots offer an optimal balance between surface area and charge mobility.

Despite the promising results, several challenges remain in integrating chalcogenide nanomaterials into photoanodes (Njema & Kibet 2025). These include controlling the size and dispersion of chalcogenide particles, achieving stable heterojunctions, and ensuring long-term operational stability under real-world conditions. Moreover, environmental and health concerns associated with toxic elements like cadmium necessitate the exploration of non-toxic alternatives such as copper-zinc-tin-sulfide (CZTS) or bismuth-based chalcogenides.

CHALCOGENIDE-BASED COUNTER ELECTRODES

The counter electrode (CE) in dye-sensitized solar cells plays a pivotal role in catalyzing the reduction of the triiodide (I3−) species in the electrolyte and facilitating electron transfer from the external circuit to regenerate the redox mediator (Marchini et al., 2021). Traditionally, platinum (Pt) has been the material of choice for counter electrodes due to its superior electrocatalytic activity, high conductivity, and chemical stability. However, the high cost, scarcity, and susceptibility to corrosion in iodine-based electrolytes limit its widespread application, prompting the search for alternative materials.

Limitations of Conventional Platinum Electrodes

Although platinum exhibits excellent performance as a CE material, it suffers from several limitations in practical DSSC applications. The high cost of Pt significantly contributes to the overall cost of DSSCs, undermining the technology's affordability (Sasikumar et al., 2024) Additionally, Pt can degrade in the presence of iodide/tri-iodide electrolytes through the formation of PtI_4 complexes, leading to reduced catalytic activity and device stability over time (Syrrokostas et al., 2023). Furthermore, the rigid nature of Pt films deposited on fluorine-doped tin oxide (FTO) glass substrates restricts the development of flexible or wearable DSSCs.

Chalcogenides as Cost-effective Alternatives

To overcome these challenges, researchers have explored transition metal chalcogenides (TMCs) such as sulfides, selenides, and tellurides as cost-effective and efficient alternatives to platinum. These materials offer good electrical conductivity, high surface area, and excellent electrocatalytic activity for triiodide reduction. Moreover, chalcogenides are generally earth-abundant and can be synthesized using low-cost, scalable techniques, making them attractive for large-scale DSSC fabrication (Bidikoudi & Stathatos 2024). Their tunable properties through compositional and morphological control further allow for performance optimization tailored to specific device architectures.

NiS, CoS, MoS_2, and Ternary Metal Chalcogenides

Among the various chalcogenides, nickel sulfide (NiS), cobalt sulfide (CoS), and molybdenum disulfide (MoS_2) have been extensively studied for DSSC counter electrodes (Sarkar et al., 2022). NiS exhibits high catalytic activity and can be synthesized via simple hydrothermal or electrodeposition methods. CoS has demonstrated comparable performance to Pt in many studies, offering good conductivity and corrosion resistance. MoS_2, a layered 2D material with excellent electrocatalytic properties, provides a large number of active edge sites for the triiodide reduction reaction. Additionally, ternary and quaternary metal chalcogenides such as $NiCo_2S_4$, $CuFeS_2$, and Cu_2ZnSnS_4 (CZTS) have emerged as promising candidates due to their synergistic effects, enhanced conductivity, and stability (Nautiyal et al., 2023).

Electrocatalytic Performance and Stability

The electrocatalytic performance of chalcogenide-based CEs is typically evaluated using electrochemical techniques such as cyclic voltammetry (CV), electrochemical impedance spectroscopy (EIS), and Tafel polarization (Shoukat et al., 2024). These studies reveal that TMC-based electrodes often exhibit low charge transfer resistance and high exchange current density, indicative of efficient triiodide reduction kinetics. For instance, CoS-based counter electrodes have shown PCEs comparable to those of Pt-based devices, while MoS_2 and its composites demonstrate excellent long-term operational stability under ambient conditions.

Several strategies have been employed to improve the electrocatalytic efficiency and durability of chalcogenide CEs (Khan et al., 2022). These include compositing with conductive materials like graphene or carbon nanotubes, doping with other metal ions, and engineering nanostructures such as nanorods, nanosheets, and porous frameworks to maximize surface area and active sites. For example, MoS_2-graphene hybrid electrodes have shown reduced recombination rates and enhanced PCEs due to improved charge transport and catalytic activity.

In addition to performance metrics, the mechanical flexibility and adhesion of chalcogenide films to substrates are critical for practical applications. Flexible counter electrodes fabricated with TMCs on substrates like polyethylene terephthalate (PET) or metal foils have demonstrated good mechanical stability and reproducible performance, paving the way for the development of wearable or portable DSSCs (Subudhi & Punetha 2022).

Environmental and sustainability considerations are also driving the shift toward non-toxic, eco-friendly chalcogenide materials. Materials such as CZTS and iron-based chalcogenides offer low toxicity and high abundance, making them suitable for green energy applications (Aina et al., 2024). Their compatibility with diverse electrolytes and scalable synthesis further support their integration into commercial DSSC systems.

CHALCOGENIDE-INFUSED ELECTROLYTES

Electrolytes in dye-sensitized solar cells (DSSCs) play a crucial role in facilitating ion transport and ensuring the redox cycle that regenerates the dye after photoexcitation (Figure 4) (Masud & Kim 2023). Traditional liquid electrolytes, composed of iodide/triiodide (I^-/I_3^-) redox couples in organic solvents, offer high ionic conductivity but suffer from drawbacks such as leakage, evaporation, and poor long-term stability. The incorporation of chalcogenide-based nanostructures into electrolytes has opened new avenues for developing stable, high-performance DSSCs by enhancing ionic conductivity and reducing recombination losses (Baraneedharan et al., 2024).

Figure 4. Schematic representation of chalcogenide-infused electrolytes and their properties

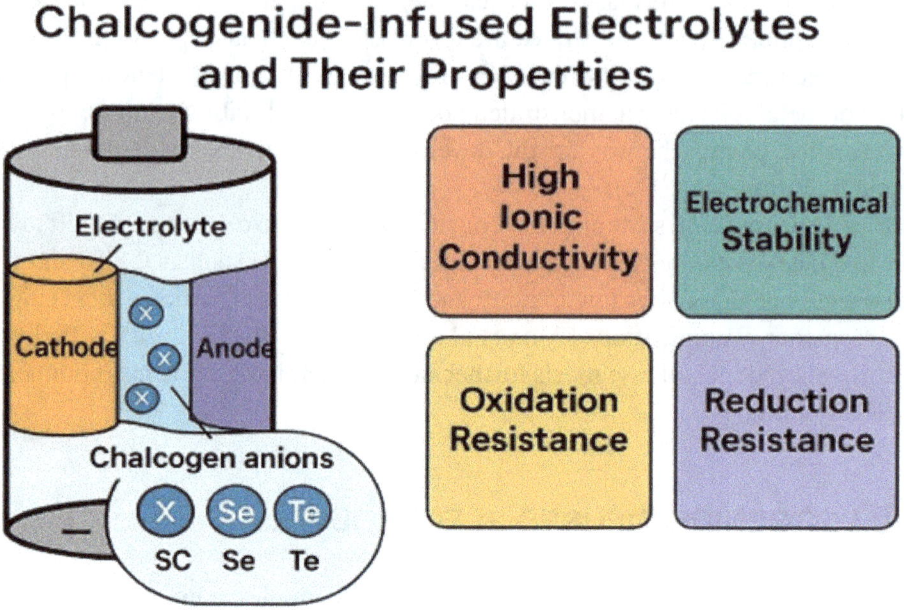

Role of Redox Mediators in DSSCs

The redox mediator is responsible for transferring electrons from the counter electrode to the oxidized dye, completing the circuit (Qamar & Ela 2024). The I−/I3− redox couple has been the most widely adopted due to its compatibility with most dyes and electrodes. However, it has inherent drawbacks such as high volatility, corrosiveness to platinum, and light-induced degradation. Chalcogenide materials offer the potential to serve as alternative redox mediators or as additives that enhance the function of existing mediators (Agarwal et al., 2024). Their ability to form stable complexes and interact favorably with light and electrons make them attractive candidates for improving redox reactions within the electrolyte phase.

Integration of Chalcogenide Nanostructures in Gel and Solid-State Electrolytes

Gel and solid-state electrolytes have emerged as promising solutions to the limitations of liquid electrolytes by offering improved thermal and mechanical stability. Chalcogenide nanostructures such as MoS2, SnS, Bi2S3, and Cu2S have been embedded into polymer-based or inorganic gel matrices to enhance ion transport (Manjunatha et al., 2024). These nanostructures act as ion-conductive fillers and create additional pathways for charge mobility. For instance, MoS_2 incorporated into a poly(ethylene oxide) (PEO)-based gel electrolyte has demonstrated improved mechanical strength and ionic conductivity, leading to higher power conversion efficiency (PCE) and reduced recombination rates. Similarly, CuS nanoparticles within a poly (vinylidene fluoride-co-hexafluoropropylene) (PVDF-HFP) gel have shown improved redox activity and compatibility with flexible devices (Thakur et al., 2022).

Ionic Conductivity and Recombination Suppression

The incorporation of chalcogenide materials into electrolytes significantly influences ionic conductivity and the suppression of charge recombination (Phogat et al., 2024). The high surface area and electrical conductivity of chalcogenides facilitate efficient charge transport, while their catalytic activity helps in maintaining the redox balance. By interacting with the iodide/triiodide couple or alternative mediators, these materials can suppress the back-transfer of electrons from the conduction band of the semiconductor to the electrolyte, which is a common cause of efficiency loss in DSSCs. For example, the introduction of Bi_2Se_3 nanoplates into quasi-solid-state electrolytes resulted in enhanced conductivity and lower recombination, due to their high dielectric constant and rapid ion diffusion capability (Yang et al., 2024).

Furthermore, chalcogenide nanostructures can stabilize the electrolyte matrix, improving the photothermal stability of the DSSC. Materials such as Sb_2S_3 and In_2S_3, when added to gel electrolytes, improve the viscoelastic behavior of the system, thereby increasing its resistance to phase separation and electrolyte leakage under prolonged illumination.

The synergistic effects of chalcogenide additives and polymer hosts result in higher electrochemical stability and broadened operational temperature ranges (Muthu et al., 2024). These features are essential for the deployment of DSSCs in real-world conditions, especially in flexible and portable electronic applications.

TERNARY AND QUATERNARY CHALCOGENIDE NANOCOMPOSITES

Ternary and quaternary chalcogenide nanocomposites (Figure 5) represent a significant advancement in the development of next-generation materials for dye-sensitized solar cells (DSSCs) (Sajitha et al., 2024). These multinary systems, characterized by the incorporation of three or more elements into a single compound, offer the potential to overcome limitations of binary chalcogenides by enabling tunable bandgaps, enhanced light absorption, improved charge separation, and superior chemical stability. Materials such as CuInS2, Cu2ZnSnS4 (CZTS), and AgBiS2 have demonstrated exceptional promise in the fabrication of efficient, cost-effective, and environmentally benign DSSCs (Kaur et al., 2022).

Figure 5. Schematic representation of the properties of ternary and quaternary calcogenide nanocomposites

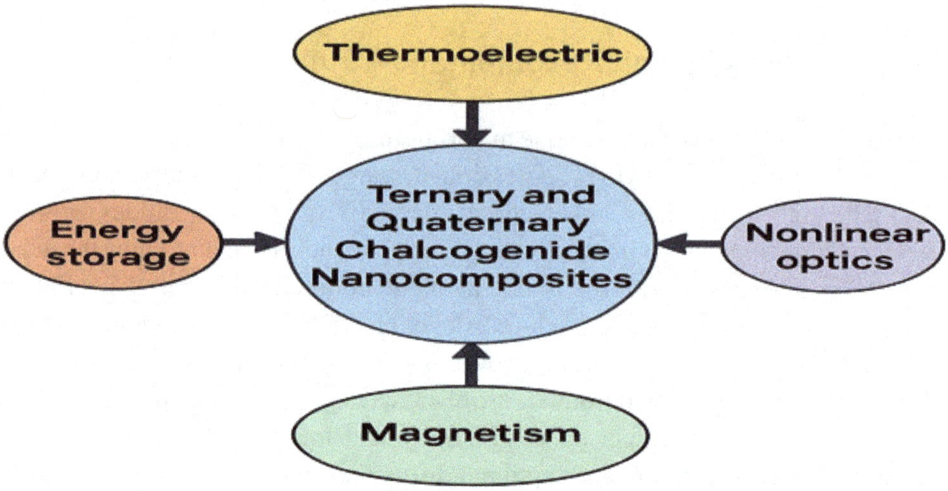

CuInS2, Cu2ZnSnS4 (CZTS), and Related Materials

Among ternary chalcogenides, CuInS2 stands out for its direct bandgap (~1.5 eV), high absorption coefficient (>10^4 cm−1), and excellent stability (Karade et al., 2025). It has been employed as both a sensitizer and a photoanode component, showing favorable alignment with TiO2 and ZnO matrices. CZTS, a quaternary compound consisting of earth-abundant elements (Cu, Zn, Sn, and S), has gained significant interest due to its optimal bandgap (~1.45 eV), nontoxicity, and compatibility with scalable deposition techniques. These materials not only serve as efficient absorbers but also improve interfacial electron transfer and suppress recombination when incorporated into DSSC architectures.

Advantages of Multinary Systems

The primary advantage of using multinary chalcogenides lies in their compositional flexibility, allowing for precise control over electronic structure and physicochemical properties (Mishra et al., 2024). This tunability facilitates the engineering of materials with ideal energy level alignment with redox mediators and semiconductor oxides. Additionally, the presence of multiple metal cations enables the formation of defect-tolerant structures that enhance charge carrier mobility and reduce electron-hole recombination. Multinary systems also exhibit broader light absorption spectra, covering more of the solar spectrum than their binary counterparts, thereby leading to improved photocurrent densities.

Applications in Multi-Junction DSSCs

Multinary chalcogenide nanocomposites are particularly well-suited for application in tandem and multi-junction DSSCs (Hannappel et al., 2024). These advanced architectures integrate multiple photoactive layers with varying bandgaps to harvest different portions of the solar spectrum. CuInS2 and CZTS, with their ideal bandgaps, can be paired with other sensitizers in a stacked configuration to enhance overall power conversion efficiency (PCE). Their compatibility with mesoporous scaffolds and stable integration into layered configurations make them excellent candidates for scalable multi-junction designs.

Synergistic Effects in Hybrid Architectures

Hybrid nanocomposites incorporating ternary and quaternary chalcogenides with carbonaceous materials (graphene, CNTs) or metal oxides (TiO2, ZnO) exhibit synergistic effects that enhance electron transport and reduce recombination (Sadiq et al., 2025). For example, CuInS2 nanoparticles deposited on graphene sheets show enhanced conductivity and dye regeneration kinetics, while CZTS-TiO2 composites offer improved charge separation due to favorable band alignment. These hybrid systems not only improve electrical pathways but also enhance structural integrity and mechanical stability, which are crucial for flexible and wearable photovoltaic devices.

Moreover, integrating quaternary chalcogenides with quantum dots and perovskite layers has opened new possibilities for developing hybrid sensitizers with high extinction coefficients and stability (Chen et al., 2021). The energy level alignment between CZTS and halide perovskites enables rapid charge transfer, while the photostability of chalcogenides mitigates degradation issues associated with perovskites.

Synthesis Techniques for Multinary Chalcogenides

The synthesis of ternary and quaternary chalcogenide nanomaterials typically involves hydrothermal, solvothermal, hot-injection, and microwave-assisted methods (Chatterjee et al., 2024). These techniques allow for controlled growth of nanoparticles with defined morphologies and stoichiometries. For example, CZTS nanocrystals synthesized via hot-injection show uniform size distribution and phase purity, critical for reproducibility and performance in DSSCs (Mohammadi et al., 2024). The choice of precursors, reaction temperature, and surfactants directly influence the crystallinity, surface defects, and optical properties of the resulting materials.

Interfacial Engineering and Surface Passivation

The performance of multinary chalcogenide nanocomposites is highly dependent on interface quality between the chalcogenide material and other DSSC components (Palchoudhury et al., 2020). Surface passivation strategies, such as coating with insulating or semiconducting shells (e.g., ZnS, Al2O3), are employed to reduce surface trap states and enhance carrier lifetimes. Interfacial engineering through the introduction of linker molecules or dipole layers can also improve charge extraction and reduce recombination at the electrolyte and photoanode interfaces.

Atomic layer deposition (ALD) and layer-by-layer assembly techniques have been effectively used to achieve uniform coverage and controlled thickness of passivation layers on CuInS2 and CZTS nanostructures (Gupta et al., 2022). These surface modifications significantly impact device stability and efficiency, particularly under prolonged illumination and thermal stress.

Photophysical and Electrochemical Characteristics

Ternary and quaternary chalcogenide nanocomposites exhibit distinct photophysical properties such as high extinction coefficients, photo-induced charge separation, and broad absorption profiles. Electrochemical impedance spectroscopy (EIS) and intensity-modulated photovoltage/photocurrent spectroscopy (IMVS/IMPS) studies reveal enhanced electron lifetimes, reduced charge transfer resistance, and improved fill factors in DSSCs incorporating these materials (Wang et al., 2022). Such improvements stem from the efficient injection of photoelectrons into the conduction band and the reduced recombination kinetics due to better surface passivation and energy level matching.

Case Studies and Comparative Performance

Numerous studies have demonstrated the superior performance of DSSCs incorporating ternary and quaternary chalcogenides (Sajitha et al., 2024). For instance, a DSSC using a CuInS2-TiO2 photoanode achieved a PCE of 9.2%, compared to 6.5% for a pure TiO2 device. Similarly, CZTS-based photoanodes have shown PCEs exceeding 10% under optimized conditions. The addition of these materials to counter electrodes and electrolytes has also led to enhanced catalytic activity and ionic conductivity, further improving device performance.

Comparative analyses highlight that multinary chalcogenides often outperform binary analogs in terms of efficiency, stability, and environmental compatibility (Perryman et al., 2021). Their ability to function in multiple roles (photoanode, sensitizer, catalyst) makes them highly versatile and valuable in the design of next-generation DSSCs.

Environmental and Economic Considerations

The use of earth-abundant and non-toxic elements in materials like CZTS makes them particularly attractive for large-scale deployment. Unlike Cd- or Pb-based compounds, multinary chalcogenides such as Cu_2FeSnS_4 and $AgBiS_2$ do not pose significant environmental hazards. Their low-cost precursors and compatibility with solution-based processing methods contribute to reduced manufacturing costs and energy inputs.

Life cycle assessments (LCAs) have shown favorable environmental footprints for DSSCs incorporating multinary chalcogenides, especially when green synthesis methods are employed (Resalati et al., 2022). These materials align well with the principles of green chemistry and sustainable photovoltaics. Next are the various techniques used for device assembly (Figure 6).

DEVICE ARCHITECTURE AND FABRICATION

Layer-by-Layer Assembly of Chalcogenide Nanocomposites

Layer-by-layer (LbL) assembly has emerged as a highly versatile technique for fabricating uniform and functionalized layers of chalcogenide nanocomposites in DSSCs. This method involves the alternate deposition of oppositely charged species or complementary materials, enabling precise control over film thickness, morphology, and composition. LbL assembly facilitates the integration of chalcogenide nanomaterials with conductive substrates, enhancing charge transport and interfacial interactions (Cai et al., 2024). Particularly, the incorporation of quantum-sized CdS or $CuInS_2$ nanoparticles through LbL assembly enhances light-harvesting efficiency and electron injection in the photoanode.

Figure 6. Schematic representation of various techniques used for device assembly

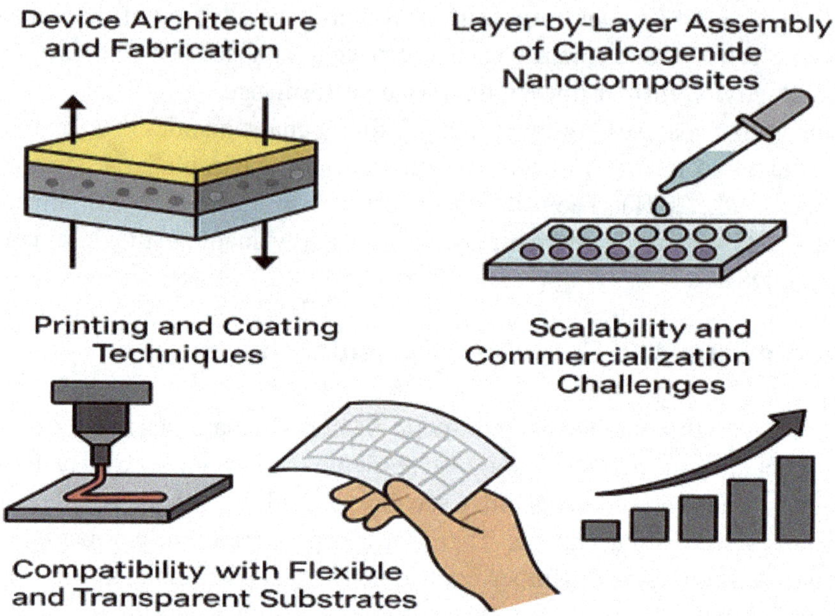

Printing and Coating Techniques

To enable scalable production of DSSCs, advanced printing and coating techniques such as screen printing, spin coating, spray pyrolysis, and doctor blading are widely employed for depositing chalcogenide nanocomposite layers (Sarma et al., 2024). These methods offer precise control over film uniformity, thickness, and porosity—key parameters for maximizing photoactive layer performance. For instance, spin coating allows for rapid and uniform deposition of CdSe or CZTS nanoparticles on TiO2 scaffolds, while spray pyrolysis can deposit chalcogenide-based counter electrodes such as MoS2 or NiS2 with controlled morphology. Furthermore, inkjet and aerosol jet printing methods have been explored for patterning flexible DSSC devices using chalcogenide inks, offering minimal material wastage and high resolution (Patil et al., 2024). These low-temperature, solution-based deposition techniques align well with the thermal sensitivity of flexible substrates, making them ideal for roll-to-roll processing and large-area module fabrication.

Compatibility with Flexible and Transparent Substrates

The compatibility of chalcogenide nanocomposites with flexible and transparent substrates is vital for the development of wearable and portable DSSC applications. Materials such as CZTS, CuInS2, and Bi2S3 have shown promising adhesion and electrical connectivity on flexible substrates like PET and PEN films (Subudhi et al., 2023). The integration of these nanomaterials with transparent conductive oxides (TCOs) such as ITO and FTO-coated polymer films enables efficient light transmission and charge collection. Flexible DSSCs demand mechanically robust and chemically stable photoanodes and counter electrodes, and chalcogenide nanocomposites fulfill these criteria due to their inherent ductility and environmental resistance. Moreover, the use of transparent substrates allows for bifacial DSSC architectures, where light can enter from both sides of the device, thereby enhancing overall power output. Such designs are particularly appealing for integration into building-integrated photovoltaics (BIPV), wearable electronics, and self-powered smart devices.

Scalability and Commercialization Challenges

Despite significant advances in chalcogenide-based DSSC research, several challenges remain in scaling up device fabrication for commercial deployment (Werlinger et al., 2023). One major issue is the reproducibility and consistency of nanocomposite synthesis at industrial scales. Variations in particle size, phase purity, and surface defects can significantly impact device performance. Additionally, the integration of multinary chalcogenides into DSSC architectures often requires complex fabrication steps and post-treatment processes that hinder large-scale manufacturing. Stability under operational conditions, including exposure to humidity, temperature fluctuations, and UV radiation, is another concern that must be addressed through robust encapsulation strategies (Ma et al., 2022). Furthermore, the cost-effectiveness of materials and processing methods must be optimized to ensure competitiveness with existing photovoltaic technologies. Overcoming these challenges necessitates collaborative efforts among materials scientists, chemists, and engineers to develop standardized fabrication protocols, accelerate materials discovery, and design durable, high-efficiency DSSC modules ready for commercial application.

PERFORMANCE EVALUATION AND COMPARISON

Photovoltaic Metrics (Voc, Jsc, FF, Efficiency)

The performance of dye-sensitized solar cells (DSSCs) incorporating chalcogenide nanocomposites is commonly assessed using key photovoltaic parameters: open-circuit voltage (Voc), short-circuit current density (Jsc), fill factor (FF), and overall conversion efficiency (Dhanasekar et al., 2024). Voc represents the maximum voltage of the device under no-load conditions, while Jsc indicates the charge carrier generation under illumination when the circuit is closed. The FF reflects the squareness of the I–V curve and is defined as the ratio of the maximum power output to the product of Voc and Jsc. Efficiency, the most critical parameter, measures the ratio of electrical power generated to incident solar power, reflecting the cell's overall ability to convert sunlight into usable energy (El Hammoumi et al., 2022).

Chalcogenide nanocomposites such as $CuInS_2$, CdS, and Bi_2S_3 have demonstrated notable improvements in DSSC performance. Their favorable band alignment with dyes and electrolytes enhances charge separation and reduces recombination, resulting in higher Voc. The high absorption coefficients and superior electron transport properties of chalcogenides also lead to increased Jsc. When optimized with appropriate electrodes and electrolytes, FF values are significantly improved. Reported efficiencies of chalcogenide-based DSSCs reach 10–12%, surpassing conventional TiO_2-based systems (Rahman et al., 2023). These enhancements arise from improved light harvesting and charge collection. Nonetheless, reproducibility across different fabrication methods remains a key challenge for widespread application.

Stability and Degradation Studies

The long-term stability of DSSCs is a critical factor that influences their commercial viability. Stability and degradation studies are essential for evaluating the durability of chalcogenide-based DSSCs under operational conditions, which include exposure to UV radiation, temperature fluctuations, and humidity. Chalcogenide nanocomposites, particularly those based on copper and tin chalcogenides, have shown excellent stability when compared to traditional semiconductors (Muslih et al., 2021). Studies on $CuInS_2$ and Bi_2S_3, for example, indicate that these materials exhibit minimal degradation over extended periods of exposure to light, humidity, and heat, which is a significant advantage for practical applications. However, some chalcogenide materials may still face challenges related to oxidation and surface instability, particularly when exposed to moisture or air, which can lead to reduced performance over time.

Recent advancements in surface modification techniques, such as the incorporation of protective coatings and the engineering of passivation layers, have significantly improved the stability of chalcogenide nanocomposites in DSSCs. For example, coating the photoanode with a thin layer of titanium oxide (TiO_2) can prevent direct contact with moisture and oxygen, thereby enhancing the material's resistance to degradation (Zhang et al., 2023). Additionally, doping strategies and alloying chalcogenides with other elements such as selenium (Se) or sulfur (S) can also improve their resistance to environmental factors. Stability studies have shown that DSSCs utilizing chalcogenide-based photoanodes retain over 90% of their initial efficiency after several thousand hours of exposure to standard operating conditions, indicating their potential for long-term use in real-world applications. However, the degradation of the electrolyte and counter electrode remains an issue that must be addressed to achieve full system stability.

Comparison with Other Nanocomposite Materials

When comparing chalcogenide nanocomposites to other nanomaterials used in dye-sensitized solar cells (DSSCs), their advantages in light absorption, charge transport, and efficiency (Siddiqui et al., 2021) become evident. Traditional oxide-based photoanodes like TiO_2, ZnO, and SnO_2, though widely employed, suffer from limited visible-light absorption and relatively poor charge carrier mobility. In contrast, chalcogenide nanocomposites such as $CuInS_2$, CdSe, and PbS exhibit strong absorption in the visible and near-infrared regions, resulting in enhanced light harvesting and higher photocurrents. This directly contributes to increased short-circuit current density (Jsc).

Chalcogenides also offer bandgap tunability and superior electron transport, which improve charge separation and reduce recombination losses. These features lead to higher open-circuit voltage (Voc) and fill factor (FF) (Rahman et al., 2024), further boosting overall efficiency. For instance, alloying $CuInS_2$ with CuZnS enables tailored bandgaps that optimize solar light response. Compared with carbon-based materials like graphene and carbon nanotubes, chalcogenides integrate more effectively into conventional DSSC structures, simplifying device fabrication. However, challenges remain, particularly the high cost of certain precursors like cadmium and lead, which could limit scalability. Despite this, the superior optoelectronic properties and performance of chalcogenide-based DSSCs position them as strong contenders among next-generation photovoltaic materials.

Benchmarking Against Commercial Standards

When benchmarked against commercial photovoltaic standards, chalcogenide-based dye-sensitized solar cells (DSSCs) demonstrate encouraging efficiency and stability, though they still lag behind silicon-based devices (Farokhi et al., 2022). Silicon solar cells currently achieve efficiencies exceeding 25%, while chalcogenide DSSCs typically reach 10–12%, with some advanced prototypes approaching 15%. Despite this performance gap, DSSCs offer unique advantages that enhance their commercial appeal in niche applications. They can be fabricated through low-cost, solution-based methods and provide high flexibility, enabling integration into textiles, curved surfaces, and portable electronics (Ali et al., 2024). Their semi-transparency further supports applications in building-integrated photovoltaics (BIPV), allowing windows to function as both energy harvesters and architectural elements.

In addition, DSSCs are more environmentally sustainable than silicon cells, as their production requires less energy and avoids reliance on scarce silicon resources. However, key challenges remain in efficiency enhancement, large-scale fabrication, and cost reduction (Mensah-Darkwa et al., 2022). Progress in material synthesis, device design, and encapsulation technologies will be critical for improving long-term stability and reliability. While silicon cells remain the benchmark for durability and efficiency, ongoing advances in chalcogenide nanocomposites highlight their potential to expand DSSC commercialization and meet emerging energy demands in flexible and sustainable photovoltaic applications.

ENVIRONMENTAL AND ECONOMIC CONSIDERATIONS

Toxicity and Sustainability of Chalcogenide Materials

The environmental and economic impacts of chalcogenide materials are crucial considerations for their long-term application in photovoltaics (Badran et al., 2025). As dye-sensitized solar cells (DSSCs) incorporating chalcogenide nanocomposites advance, concerns over toxicity and sustainability have become increasingly significant. Compounds containing cadmium (Cd), lead (Pb), and arsenic (As) are well known for their hazardous nature, with risks of leaching toxic substances during production, operation, or disposal (Liu et al., 2022). . Such materials can accumulate in ecosystems, contaminating soil and water, and causing serious threats to human health and biodiversity.

To mitigate these risks, research is shifting toward safer alternatives. Copper-based chalcogenides, such as $CuInS_2$, $CuSbS_2$, and Cu_2ZnSnS_4, offer lower toxicity and reduced environmental impact while maintaining excellent photovoltaic properties

(Amrillah et al., 2023). These materials present a sustainable pathway for next-generation DSSCs. Additionally, the adoption of green synthesis methods—such as employing non-toxic solvents, natural extracts, or biological templates—further minimizes environmental hazards during fabrication. Moving toward eco-friendly and less toxic chalcogenide nanocomposites will be essential to balance performance with sustainability, ensuring the commercial viability and long-term adoption of DSSCs.

Lifecycle Assessment

A lifecycle assessment (LCA) is an essential tool for evaluating the environmental impact of materials and products from their inception to their end-of-life. For chalcogenide-based DSSCs, an LCA can provide valuable insights into the total energy consumption, raw material usage, waste generation, and potential environmental impacts over the entire lifecycle of the device. The LCA of chalcogenide-based DSSCs reveals that these devices generally have a lower environmental footprint compared to traditional silicon-based solar cells, particularly in terms of manufacturing energy and resource use (Maalouf et al., 2023). The fabrication processes for DSSCs are less energy-intensive, and they do not require the high-temperature, energy-demanding processes involved in silicon wafer production, which contributes to their lower carbon footprint.

However, challenges remain regarding the materials used in DSSCs. While some chalcogenide compounds, such as $CuInS_2$, are more sustainable and environmentally friendly, others, like CdSe and PbS, present concerns due to the toxicity of the raw materials and the potential for hazardous waste generation during production and disposal (Fu et al., 2018). The LCA highlights that the use of less toxic chalcogenides, combined with recycling and reuse strategies, could significantly reduce the overall environmental impact of DSSCs. For example, the recycling of cadmium and lead-based chalcogenides in closed-loop systems could mitigate the risks associated with the use of these materials. Additionally, integrating recycling programs for DSSCs at the end of their lifecycle could help minimize the environmental footprint and encourage more sustainable practices in the solar industry.

Cost Analysis and Resource Availability

The economic feasibility of chalcogenide-based dye-sensitized solar cells (DSSCs) is determined by raw material costs, fabrication methods, and scalability (Arif et al., 2025). A key advantage of DSSCs lies in their low-cost, solution-based processing, which is less energy-intensive than the expensive, high-temperature manufacturing of silicon photovoltaics. The use of abundant elements such as copper, tin, and sulfur in certain chalcogenide nanocomposites further enhances affordability, positioning DSSCs as a promising candidate for cost-effective solar energy generation.

However, costs vary significantly depending on the chosen chalcogenide. For instance, cadmium telluride (CdTe)-based devices, despite their efficiency, remain expensive due to cadmium's high cost, scarcity, and manufacturing complexity (Bishnoi et al., 2025). In contrast, copper-based chalcogenides such as $CuInS_2$ and Cu_2ZnSnS_4 offer greater scalability and reduced expenses, owing to their lower material costs and more abundant raw resources (Sengupta et al., 2025). Resource availability is critical, as scarce and toxic elements like cadmium and lead pose both supply chain and geopolitical risks. Meanwhile, copper and tin are more abundant and reliable, offering stability in large-scale production.

Future economic viability will depend on advances in high-throughput techniques such as roll-to-roll processing, which could significantly lower costs. Large-scale manufacturing and integration with storage technologies may further enhance cost-effectiveness, helping chalcogenide-based DSSCs emerge as a sustainable and competitive alternative in global solar energy markets (Islam et al., 2025).

FUTURE PERSPECTIVES AND RESEARCH DIRECTIONS

Challenges in Material Synthesis and Integration

One of the key challenges in the development of chalcogenide-based DSSCs lies in the synthesis and integration of efficient and stable materials (Miah et al., 2025). While significant progress has been made in the fabrication of chalcogenide nanocomposites, ensuring consistent and scalable production of these materials with controlled properties remains a major hurdle. The ability to achieve high-quality films with uniform morphology, optimal electronic properties, and minimal defects is crucial for enhancing the overall efficiency and stability of DSSCs. Furthermore, integrating chalcogenide materials into DSSC devices without compromising their performance or adding unnecessary complexity is an ongoing challenge (Tamrakar et al., 2025). Issues such as poor interface compatibility, limited charge carrier mobility, and instability under operational conditions need to be addressed through improved material engineering, novel doping strategies, and better control over the deposition processes. The development of solution-based or low-cost processing methods that maintain the desired material properties and stability will be critical for the large-scale commercialization of chalcogenide-based DSSCs.

Advanced Characterization Techniques

As the field of DSSC research evolves, there is an increasing need for advanced characterization techniques to better understand the behavior of materials at different stages of the device operation (Asghar et al., 2024). Traditional characterization methods, such as X-ray diffraction (XRD) and scanning electron microscopy (SEM), provide essential information on the structure and morphology of the materials. However, more advanced techniques such as transmission electron microscopy (TEM), X-ray photoelectron spectroscopy (XPS), and ultrafast spectroscopy are essential to gain deeper insights into the electronic, optical, and interfacial properties of chalcogenide nanocomposites. These advanced tools can help identify issues related to charge transport, recombination losses, and material degradation, which are crucial for optimizing the efficiency and long-term stability of DSSCs. Additionally, in situ monitoring techniques, such as impedance spectroscopy and time-resolved measurements, will become increasingly important for understanding the dynamics of charge transfer and the impact of various external factors like temperature and humidity on device performance (Pishgar et al., 2021). Developing and utilizing these advanced characterization techniques will help researchers design more efficient and durable DSSCs.

The Role of Computational Modelling and Emerging Trends in DSSC Research

Computational modelling is becoming increasingly important in advancing the understanding of chalcogenide-based materials and their behavior in DSSCs (Wang et al., 2024). Using density functional theory (DFT) and molecular dynamics simulations, researchers can predict the electronic structure, charge transport, and stability of various nanomaterials, guiding the design of new materials with tailored properties for better performance. Computational tools also allow the simulation of different device architectures, providing insights into how integrating new materials or changing the device structure can affect overall efficiency (Mortazavi et al., 2025). Moreover, advancements in machine learning and artificial intelligence are likely to speed up the discovery of new materials and process optimization methods. As DSSC research progresses, the focus is expected to shift toward developing hybrid materials, such as perovskite-chalcogenide composites, and exploring alternative electrolytes, like solid-state and gel-based systems, to improve efficiency and durability. Additionally, the trend toward large-area, flexible, and printable DSSCs is gaining momentum, with research concentrated on low-temperature fabrication, roll-to-roll processing, and scalable production techniques. These advancements will help incorporate DSSCs into building-integrated photovoltaics and wearable electronics. The future of DSSC technology will depend on a multidisciplinary approach that combines material innovation, advanced characterization, and computational modeling to overcome current challenges and open new avenues for solar energy harvesting.

CONCLUSION

Chalcogenide-based nanocomposites represent a pivotal advancement in the evolution of dye-sensitized solar cells, offering substantial improvements in light absorption, charge transport, and catalytic activity across key components of DSSCs. Through strategic integration into photoanodes, counter electrodes, and electrolytes, these materials have significantly boosted the efficiency and stability of solar devices. The tunable optical and electronic properties of sulfides, selenides, tellurides, and their multinary derivatives provide a versatile platform for tailoring DSSC performance to specific application needs. However, challenges related to scalability, long-term durability, and environmental impacts remain. Future research should focus on overcoming these hurdles through novel synthesis techniques, surface engineering, and advanced characterization methods. With continued interdisciplinary efforts and innovations, chalcogenide-based nanocomposites hold great promise for enabling the commercialization of efficient, low-cost, and environmentally sustainable solar energy solutions

REFERENCES

Abdullah, R., Jalil, A. A., Asmadi, M., Hassan, N. S., Bahari, M. B., Alhassan, M., & Karimi-Maleh, H. (2024). Recent advances in zinc oxide-based photoanodes for photoelectrochemical water splitting. *International Journal of Hydrogen Energy*. Advance online publication. DOI: 10.1016/j.ijhydene.2024.05.461

Abu Elella, M. H., Saad, L., & Samy, M. (2024). Two-Dimensional Nanomaterials-Based Polymer Nanocomposites for Solar Energy Applications. *Two-Dimensional Nanomaterials-Based Polymer Nanocomposites: Processing, Properties and Applications*, 419-464.

Agarwal, A., Tolani, R., & Sankapal, B. R. (2024). Conducting Polymers–Metal Chalcogenides Hybrid Composite: Current Trends and Future Prospects toward Supercapacitor Applications. *Energy Technology (Weinheim)*, *12*(8), 2400133.

Agrawal, A., Siddiqui, S. A., Soni, A., & Sharma, G. D. (2022). Advancements, frontiers and analysis of metal oxide semiconductor, dye, electrolyte and counter electrode of dye sensitized solar cell. *Solar Energy*, *233*, 378–407. DOI: 10.1016/j.solener.2022.01.027

Ahmad Ramli, S., Mohd Zin, R., Ahmad, M. K., Azyan, N. I., Hamed, N. K. A., Saputri, D. G., & Ahmad, M. Y. (2025). MoS_2-decorated etched-TiO_2 rods film for superior photocatalytic dye degradation. *Journal of Sol-Gel Science and Technology*, •••, 1–19. DOI: 10.1007/s10971-025-06702-y

Aina, S., Scott, N., Lobera, M. P., & Navarro, M. B. (2024). Sustainable solution-processed solar cells based on environmentally friendly nanocrystals. In *Photovoltaics Beyond Silicon* (pp. 437–477). Elsevier.

Ali, I., Islam, M. R., Yin, J., Eichhorn, S. J., Chen, J., Karim, N., & Afroj, S. (2024). Advances in smart photovoltaic textiles. *ACS Nano*, *18*(5), 3871–3915.

Amrillah, T., Prasetio, A., Supandi, A. R., Sidiq, D. H., Putra, F. S., Nugroho, M. A., & Azmi, R. (2023). Environment-friendly copper-based chalcogenide thin film solar cells: Status and perspectives. *Materials Horizons*, *10*(2), 313–339.

Anucha, C. B., Altin, I., Bacaksiz, E., & Stathopoulos, V. N. (2022). Titanium dioxide (TiO_2)-based photocatalyst materials activity enhancement for contaminants of emerging concern (CECs) degradation: In the light of modification strategies. *Chemical Engineering Journal Advances*, *10*, 100262.

Arif Dar, M., Satgunam, M., Batoo, K. M., Kalpnan, S., Arularasan, P., Ali Dar, U., ... & Majid, S. R. (2025). Future Photovoltaic Applications of Nanostructured Metal Chalcogenides. *Comments on Inorganic Chemistry*, 1-54.

Asghar, H., Riaz, T., Mannan, H. A., Khan, S. M., & Butt, O. M. (2024). Rheology and modeling insights into dye-sensitized solar cells (DSSCs) material: Bridging the gap to solar energy advancements. *Renewable & Sustainable Energy Reviews*, *193*, 114298.

Ates, M., Yılmaz, E., & Tanaydın, M. K. (2021). Challenges, novel applications, and future prospects of chalcogenides and chalcogenide-based nanomaterials for photocatalysis. In *Chalcogenide-based nanomaterials as photocatalysts* (pp. 307–337). Elsevier.

Ayaz, M., Hijji, M., Alatawi, A. S., Namazi, M. A., & Ershath, M. M. (2024). Enhancing photovoltaic performance in dye-sensitized solar cells using nanostructured NiS/MoS2 composite counter electrodes. *Materials Science in Semiconductor Processing*, *173*, 108172. DOI: 10.1016/j.mssp.2024.108172

Badran, G., & Lazarov, V. K. (2025, February). From Waste to Resource: Exploring the Current Challenges and Future Directions of Photovoltic Solar Cell Recycling. In *Solar* (Vol. 5, No. 1, p. 4). MDPI.

Baraneedharan, P., Sekar, S., Murugesan, S., Ahamada, D., Mohamed, S. A. B., Lee, Y., & Lee, S. (2024). Recent Advances and Remaining Challenges in Perovskite Solar Cell Components for Innovative Photovoltaics. *Nanomaterials (Basel, Switzerland)*, *14*(23), 1867.

Bidikoudi, M., & Stathatos, E. (2024). Chalcogenides in Perovskite Solar Cells with a Carbon Electrode: State of the Art and Future Prospects. *Nanomaterials (Basel, Switzerland)*, *14*(22), 1783.

Biney, J., Madou, M., Jabbour, G., & Park, J. (2025). 2D Composite Materials for Electrodes in Dye-Sensitized Solar Cells— An Overview. *ACS Applied Materials & Interfaces*. Advance online publication. DOI: 10.1021/acsami.4c13963

Bishnoi, S., Datt, R., Tsoi, W. C., & Gupta, V. (2025).Development of Solar Cells. In *Solar Cells Development and Fabrication* (pp. 1–21). CRC Press.

Cai, M., Yang, J., Lu, X., & Lu, X. (2024). Layer-by-layer self-assembly strategies of atomically thin two-dimensional nanomaterials: Principles, methods, and functional applications. *ACS Applied Nano Materials*, *7*(24), 27940–27959.

Chatterjee, A., Kumar, G. K., Roymahapatra, G., Das, H. S., Jaishree, G., & Rao, T. S. (2024). Zinc chalcogenide nanostructures: Synthesis methodologies and applications—a review. *Frontiers in Nanotechnology*, *6*, 1433591.

Chawla, A., Sudhaik, A., Kumar, R., Raizada, P., Khan, A. A. P., Ahamad, T., & Singh, P. (2024). a Recent advances in synthesis methods and surface structure manipulating strategies of copper selenide (CuSe) nanoparticles for photocatalytic environmental and energy applications. *Journal of Environmental Chemical Engineering*, •••, 113125.

Chen, M., Wang, J., Yin, F., Du, Z., Belfiore, L. A., & Tang, J. (2021). Strategically integrating quantum dots into organic and perovskite solar cells. *Journal of Materials Chemistry. A, Materials for Energy and Sustainability*, *9*(8), 4505–4527.

Deogratias, G., Al-Qurashi, O. S., & Wazzan, N. (2023). Optical and electronic properties enhancement via chalcogenides: Promising materials for DSSC applications. *Journal of Molecular Modeling*, *29*(4), 86. DOI: 10.1007/s00894-023-05472-0

Dhanasekar, N., Vanitha, M., Shankar, N., & Bindhu, M. (2024). Design and fabrication of NiSe2/g-C3N4 hybrid composite counter electrode for Pt-free dye-sensitized solar cells. *Journal of Materials Science Materials in Electronics*, *35*(6), 443.

El Hammoumi, A., Chtita, S., Motahhir, S., & El Ghzizal, A. (2022). Solar PV energy: From material to use, and the most commonly used techniques to maximize the power output of PV systems: A focus on solar trackers and floating solar panels. *Energy Reports*, *8*, 11992–12010.

Farokhi, A., Shahroosvand, H., Delle Monache, G., Pilkington, M., & Nazeeruddin, M. K. (2022). The evolution of triphenylamine hole transport materials for efficient perovskite solar cells. *Chemical Society Reviews*, *51*(14), 5974–6064.

Fu, H. (2018). Environmentally friendly and earth-abundant colloidal chalcogenide nanocrystals for photovoltaic applications. *Journal of Materials Chemistry. C, Materials for Optical and Electronic Devices*, *6*(3), 414–445.

Giri, A., Park, G., & Jeong, U. (2023). Layer-structured anisotropic metal chalcogenides: Recent advances in synthesis, modulation, and applications. *Chemical Reviews*, *123*(7), 3329–3442. DOI: 10.1021/acs.chemrev.2c00455

Giroux, M. S., Zahra, Z., Salawu, O. A., Burgess, R. M., Ho, K. T., & Adeleye, A. S. (2022). Assessing the environmental effects related to quantum dot structure, function, synthesis and exposure. *Environmental Science. Nano*, *9*(3), 867–910. DOI: 10.1039/D1EN00712B

Gupta, B., Hossain, M. A., Riaz, A., Sharma, A., Zhang, D., Tan, H. H., & Karuturi, S. (2022). Recent advances in materials design using atomic layer deposition for energy applications. *Advanced Functional Materials, 32*(3), 2109105.

Hannappel, T., Shekarabi, S., Jaegermann, W., Runge, E., Hofmann, J. P., van de Krol, R., & Zhang, H. (2024). Integration of Multijunction Absorbers and Catalysts for Efficient Solar-Driven Artificial Leaf Structures: A Physical and Materials Science Perspective. *Solar RRL, 8*(11), 2301047.

Hu, W., Quang, N. D., Majumder, S., Jeong, M. J., Park, J. H., Cho, Y. J., & Chang, H. S. (2021). Three-dimensional nanoporous SnO2/CdS heterojunction for high-performance photoelectrochemical water splitting. *Applied Surface Science, 560*, 149904.

Huang, S., Yue, C., Uvdal, K., & Hu, Z. (2025). Recent advances in irradiation-mediated synthesis and tailoring of inorganic nanomaterials for photo-/electrocatalysis. *Nanoscale Advances.*

Islam, S., Weerasinghe, H., Prado, D. M., Gonzaga, A. N., & Burda, C. (2025). Diversifying the Materials and Technologies for the Future of Energy Storage. *Energy & Fuels.*

Jayawardane, J. T. S. T., Akmal, U. K. D. M., Hu, D., Abeygunawardhana, P. K. W., & Sewvandi, G. A. (2025). Eco-friendly bismuth halide chalcogenide perovskites for solar energy harvesting. *Sustainable Energy & Fuels*. Advance online publication. DOI: 10.1039/D4SE01523A

Kannan, K., & Tari, V. (Eds.). (2025). Novel energy storage and conversion technologies for two-dimensional MXenes and MBenes. https://www.amazon.com/dp/9798369395462 ISBN: 9798369395462, eISBN: 9798369395486

Karade, V. C., He, M., Abudulimu, A., Song, Z., Park, Y., Song, D., & Suryawanshi, M. P. (2025). Opportunities and Challenges for Emerging Inorganic Chalcogenide-Silicon Tandem Solar Cells. *Energy & Environmental Science.*

Kaur, G., Vij, A., & Kumar, A. (2022). Recent Advances in Earth Abundant and Environmentally Green Semiconducting Chalcogenide Nanomaterials for Photovoltaics Applications. *Advanced Nanomaterials*, 21-50.

Khan, S., Mahmood, A., Shah, A. U. H. A., Rahman, G., Khan, A., & Ullah, N. (2022). Challenges and innovative strategies related to synthesis and electrocatalytic/energy storage applications of metal sulfides and its derivatives. *Journal of Electroanalytical Chemistry (Lausanne, Switzerland), 923*, 116760.

Kharboot, L. H., Fadil, N. A., Bakar, T. A. A., Najib, A. S. M., Nordin, N. H., & Ghazali, H. (2023). A review of transition metal sulfides as counter electrodes for dye-sensitized and quantum dot-sensitized solar cells. *Materials (Basel), 16*(7), 2881.

Korir, B. K., Kibet, J. K., & Ngari, S. M. (2024). A review on the current status of dye-sensitized solar cells: Toward sustainable energy. *Energy Science & Engineering, 12*(8), 3188–3226. DOI: 10.1002/ese3.1815

Kumar, S., Muthu, S., Sekar, S., Bathula, C., Kaliamurthy, A. K., & Lee, S. (2022). Metal chalcogenide-based counter electrodes for dye-sensitized solar cells. *Oxide Free Nanomaterials for Energy Storage and Conversion Applications*, 259-286. https://doi.org/DOI: 10.1016/B978-0-12-823936-0.00015-2

Lahariya, V., Pandey, P. P., & Ramrakhiani, M. (2024). II–VI semiconductor metal chalcogenide nanomaterials and polymer composites: fundamentals, properties, and applications. In *Metal-Chalcogenide Nanocomposites* (pp. 187–220). Woodhead Publishing., DOI: 10.1016/B978-0-443-18809-1.00011-0

Lana, G. M., Bello, I. T., Adedokun, O. M., Adenigba, V. O., Jubu, P. R., Adedokun, O., & Awodugba, A. O. (2024). One-Dimensional TiO2 Nanocomposite-based Photoanode for Dye-Sensitized solar Cells: A review. *Solar Energy, 279*, 112850. DOI: 10.1016/j.solener.2024.112850

Li, B., Liu, X. J., Zhu, H. W., Guan, H. P., & Guo, R. T. (2025). Recent Progress of Round-the-Clock Photocatalytic System in Environmental and Energy Applications: A Review. *ChemPhysChem, 202401144*. Advance online publication. DOI: 10.1002/cphc.202401144

Liu, L., An, W., Gu, F., Cui, L., He, X., & Fan, M. (2023). 2D layered materials: Structures, synthesis, and electrocatalytic applications. *Green Chemistry, 25*(16), 6149–6169. DOI: 10.1039/D3GC01822A

Liu, Q., Ding, X., Pang, Y., Cao, Y., Lei, J., Wu, J., & Zhang, T. (2022). New insights into the safety assessment of quantum dots: Potential release pathways, environmental transformations, and health risks. *Environmental Science. Nano, 9*(9), 3277–3311.

Lu, R. C., Zhang, K., Liu, X. H., Chen, Q. Y., & Wang, Y. H. (2022). Investigation of the charge transport kinetics and photoelectrochemical properties of CuInS2/R-TiO2 NTs heterostructures photoanode. *Applied Surface Science, 585*, 152742.

Ma, S., Yuan, G., Zhang, Y., Yang, N., Li, Y., & Chen, Q. (2022). Development of encapsulation strategies towards the commercialization of perovskite solar cells. *Energy & Environmental Science, 15*(1), 13–55.

Maalouf, A., Okoroafor, T., Jehl, Z., Babu, V., & Resalati, S. (2023). A comprehensive review on life cycle assessment of commercial and emerging thin-film solar cell systems. *Renewable & Sustainable Energy Reviews, 186*, 113652.

Manjunatha, C., Rastogi, C. K., Manmadha Rao, B., Girish Kumar, S., Varun, S., Raitani, K., & Khosla, A. (2024). Advances in hierarchical inorganic nanostructures for efficient solar energy harvesting systems. *ChemSusChem, 17*(11), e202301755.

Marchini, E., Caramori, S., Bignozzi, C. A., & Carli, S. (2021). On the use of PEDOT as a catalytic counter electrode material in dye-sensitized solar cells. *Applied Sciences (Basel, Switzerland), 11*(9), 3795.

Masud, & Kim, H. K. (2023). Redox shuttle-based electrolytes for dye-sensitized solar cells: comprehensive guidance, recent progress, and future perspective. *ACS omega, 8*(7), 6139-6163.

Matyszczak, G., Plocinski, T., Dluzewski, P., Fidler, A., Jastrzebski, C., Lawniczak-Jablonska, K., & Krawczyk, K. (2024). Sonochemical synthesis of SnS and SnS2 quantum dots from aqueous solutions, and their photo-and sonocatalytic activity. *Ultrasonics Sonochemistry, 105*, 106834. DOI: 10.1016/j.ultsonch.2024.106834

Mensah-Darkwa, K., Nframah Ampong, D., Agyekum, E., de Souza, F. M., & Gupta, R. K. (2022). Recent advancements in chalcogenides for electrochemical energy storage applications. *Energies, 15*(11), 4052.

Miah, M. H., Khandaker, M. U., Hossen, M. J., Ashrafi, N. E., Jahan, I., Shahinuzzaman, M., & Islam, M. A. (2025). Lead-free alternatives and toxicity mitigation strategies for sustainable perovskite solar cells: A critical review. *Materials Advances*.

Mishra, T. T., Chakraborty, M., Behera, J. N., & Roy, D. (2024). Comprehensive review on bimetallic (Ni and Co)-chalcogenides and phosphides with three-dimensional architecture for hybrid supercapacitor application. *Energy & Fuels, 38*(11), 9186–9217.

Mohammadi, S., Kaur, N., & Radu, D. R. (2024). Nanoscale Cu2ZnSnSxSe (4− x) (CZTS/Se) for Sustainable Solutions in Renewable Energy, Sensing, and Nanomedicine. *Crystals, 14*(5), 479.

Moradialvand, Z., Parseghian, L., & Rajabi, H. R. (2025). Green Synthesis of Quantum Dots: Synthetic Methods, Applications, and Toxicity. *Journal of Hazardous Materials Advances, 100697*. Advance online publication. DOI: 10.1016/j.hazadv.2025.100697

More, P. B., Jagtap, C. V., Kadam, V. S., Naushad, M., Naik, N., Hiremath, P., & Pathan, M., H. (2024). Synthesis and performance evaluation of ZnO/CdS photoanodes with copper sulfide (Cu2S) and carbon counter electrodes. *Scientific Reports*, *14*(1), 31551.

Mortazavi, B. (2025). Recent Advances in Machine Learning-Assisted Multiscale Design of Energy Materials. *Advanced Energy Materials*, *15*(9), 2403876.

Muslih, E. Y., Munir, B., & Khan, M. M. (2021). Advances in chalcogenides and chalcogenides-based nanomaterials such as sulfides, selenides, and tellurides. In *chalcogenide-based nanomaterials as photocatalysts* (pp. 7-31). Elsevier.

Muthu, P., Rajagopal, S., Saju, D., Kesavan, V., Dellus, A., Sadhasivam, L., & Chandrasekaran, N. (2024). Review of transition metal chalcogenides and halides as electrode materials for thermal batteries and secondary energy storage systems. *ACS Omega*, *9*(7), 7357–7374.

Nautiyal, H., Lohani, K., Mukherjee, B., Isotta, E., Malagutti, M. A., Ataollahi, N., & Scardi, P. (2023). Mechanochemical synthesis of sustainable ternary and quaternary nanostructured Cu2SnS3, Cu2ZnSnS4, and Cu2ZnSnSe4 chalcogenides for thermoelectric applications. *Nanomaterials (Basel, Switzerland)*, *13*(2), 366.

Njema, G. G., & Kibet, J. K. (2025). A review of chalcogenide-based perovskites as the next novel materials: Solar cell and optoelectronic applications, catalysis and future perspectives. *Next Nanotechnology*, *7*, 100102. DOI: 10.1016/j.nxnano.2024.100102

O'Regan, B., & Grätzel, M. (1991). A low-cost, high-efficiency solar cell based on dye-sensitized colloidal TiO_2 films. *Nature*, *353*, 737–740. DOI: 10.1038/353737a0

Olaniyan, O. T., Adetunji, C. O., Adeniyi, M. J., Dare, A., & Adewuyi, O. A. (2024). Morphology, rheology, properties, and applications of nanostructured metal chalcogenide-filled polymer blends. In *Nanofillers for Binary Polymer Blends* (pp. 401–422). Elsevier.

Oni, A. M., Mohsin, A. S., Rahman, M. M., & Bhuian, M. B. H. (2024). A comprehensive evaluation of solar cell technologies, associated loss mechanisms, and efficiency enhancement strategies for photovoltaic cells. *Energy Reports*, *11*, 3345–3366. DOI: 10.1016/j.egyr.2024.03.007

Palchoudhury, S., Ramasamy, K., & Gupta, A. (2020). Multinary copper-based chalcogenide nanocrystal systems from the perspective of device applications. *Nanoscale Advances*, *2*(8), 3069–3082.

Patil, I. G., Thakur, K., Nath, S. S., & Sundriyal, P. (2024). 3D Printed Energy Harvesting Devices for Flexible and Wearable Electronics. *Sustainable Energy & Fuels*.

Perryman, J. T., & Velazquez, J. M. (2021). Design principles for multinary metal chalcogenides: Toward programmable reactivity in energy conversion. *Chemistry of Materials*, *33*(18), 7133–7147.

Phogat, P., Jha, R., & Singh, S. (2024). Emerging advances and future prospects of two dimensional nanomaterials based solar cells. *Journal of Alloys and Compounds*, •••, 175063.

Pishgar, S., Gulati, S., Strain, J. M., Liang, Y., Mulvehill, M. C., & Spurgeon, J. M. (2021). In situ analytical techniques for the investigation of material stability and interface dynamics in electrocatalytic and photoelectrochemical applications. *Small Methods*, *5*(7), 2100322.

Qamar, S., & Ela, S. E. (2024). Dye-sensitized solar cells (DSSC): Principles, materials and working mechanism. *Current Opinion in Colloid & Interface Science*, •••, 101871.

Qiao, F., Liu, W., Yang, J., Liu, Y., & Yuan, J. (2024). Fabrication of ZnO/CuInS2 heterojunction for boosting photocatalytic hydrogen production. *International Journal of Hydrogen Energy*, *53*, 840–847.

Rahman, M. F., Monnaf, M. A., Amami, M., Farhat, L. B., & Rahman, M. A. (2024). Improving the efficiency above 35% of MoS2-based solar cells by through optimization of various wide-bandgap S-chalcogenides ETL. *Journal of Physics and Chemistry of Solids*, *194*, 112216.

Rahman, S., Haleem, A., Siddiq, M., Hussain, M. K., Qamar, S., Hameed, S., & Waris, M. (2023). Research on dye sensitized solar cells: Recent advancement toward the various constituents of dye sensitized solar cells for efficiency enhancement and future prospects. *RSC Advances*, *13*(28), 19508–19529.

Rajkumar, S., Venkatraman, M. R., Arunkumar, A., & Jayaprakash, K. (2025). Anthocyanin and Betalain Pigments Assisted Green Synthesis of TiO_2: A Sustainable and Cost-Effective Approach for Dye-Sensitized Solar Cells Photoanodes. *Optical Materials*, •••, 116710.

Rasal, A. S., Yadav, S., Kashale, A. A., Altaee, A., & Chang, J. Y. (2022). Stability of quantum dot-sensitized solar cells: A review and prospects. *Nano Energy*, *94*, 106854. DOI: 10.1016/j.nanoen.2021.106854

Rashid, U., Ma, X., Zhu, Y., Cao, C., & Zou, M. (2025). The Comprehensive Review of Cobalt Based Electrocatalysts Synthesized via new Microwave Assisted Methodology. *Materials Chemistry Frontiers*. Advance online publication. DOI: 10.1039/D4QM00929K

Resalati, S., Okoroafor, T., Maalouf, A., Saucedo, E., & Placidi, M. (2022). Life cycle assessment of different chalcogenide thin-film solar cells. *Applied Energy*, *313*, 118888.

Sadiq, I., Ali, S. A., & Ahmad, T. (2025). Graphene Derivative-Based Particulate Photocatalysts for Energy Conversion Applications: Current Achievements, Bottlenecks, and Future Outlook. *ACS Applied Energy Materials*.

Sajitha, D. R., Stephen, B., Nakamura, A., Selvaraj, M., Salammal, S. T., & Hussain, S. (2024). The Emergence of Chalcogenides: A New Era for Thin Film Solar Absorbers. *Progress in Solid State Chemistry*, •••, 100490.

Santos, T. F., Carvalho, B. R., Santos, E. V., Pereira, R. F., Souza, D. F., & Nascimento, J. H. (2025). Trends and developments in flexible solar cells: Insights from a bibliometric analysis. *Sustainable Energy & Fuels*.

Sarkar, A., Chatterjee, P., & Chakraborty, A. K. (2022). Sulfides and selenides as electrodes for dye-sensitized solar cells. In *Sulfide and Selenide Based Materials for Emerging Applications* (pp. 243–267). Elsevier.

Sarma, N., & Saikia, P. (2024). Thin Film Approach For Scalable Production of Hydrogen through Photocatalysis. In *Towards Sustainable and Green Hydrogen Production by Photocatalysis: Scalability Opportunities and Challenges (Volume 1)* (pp. 165-183). American Chemical Society.

Sasikumar, R., Thirumalaisamy, S., Kim, B., & Hwang, B. (2024). Dye-sensitized solar cells: Insights and research divergence towards alternatives. *Renewable & Sustainable Energy Reviews*, *199*, 114549.

Sawal, M. H., Jalil, A. A., Khusnun, N. F., Hassan, N. S., & Bahari, M. B. (2023). A review of recent modification strategies of TiO2-based photoanodes for efficient photoelectrochemical water splitting performance. *Electrochimica Acta*, *467*, 143142. DOI: 10.1016/j.electacta.2023.143142

Sengupta, A., Afroz, M. A., Sharma, B., Choudhary, S., Pant, N., Gulia, Y., & Satapathi, S. (2025). Commercialization of perovskite solar cells: Opportunities and Challenges. *Sustainable Energy & Fuels*.

Shahid, M. U., Muhsan, A. S., Mohamed, N. M., Zaine, S. N. A., Baig, M. M. A., Ahmad, W., & Ayub, H. M. U. (2025). Graphene-Modified Photoelectrode for Efficient and Cost-Effective Dye-Sensitized Solar Cells. *International Journal of Energy Research, 2025*(1), 6672843. DOI: 10.1155/er/6672843

Shoukat, K., Khan, M. M., Bukhari, S., Shah, S. I. A., Bashir, I., Alothman, A. A., & Allakhverdiev, S. I. (2024). Study on oxygen evolution reaction efficiency demonstrated by Ce-E (E= S, Se and te) electrocatalyst. *International Journal of Hydrogen Energy*, *69*, 11–20.

Siddiqui, M. S., & Aslam, M. (2021). Chalcogenides as well as chalcogenides-based nanomaterials and its importance in photocatalysis. In *Chalcogenide-based nanomaterials as photocatalysts* (pp. 33–76). Elsevier.

Soonmin, H., Nandi, P., Paulraj, I., Ali, D., & Sonker, R. K. (2024). Chalcogenides and their nanocomposites: fundamental, properties and applications. In *Metal-Chalcogenide Nanocomposites* (pp. 1–27). Woodhead Publishing.

Sowbakkiyavathi, E. S., Arunachala Kumar, S. P., Maurya, D. K., Balakrishnan, B., Guo, J. Z., & Subramania, A. (2024). Research progress in the development of transition metal chalcogenides and their composite-based electrode materials for supercapacitors. *Advanced Composites and Hybrid Materials*, *7*(4), 130. DOI: 10.1007/s42114-024-00913-7

Srivastava, K. V., Srivastava, P., Srivastava, A., Maurya, R. K., Singh, Y. P., & Srivastava, A. (2025). 1D TiO 2 photoanodes: A game-changer for high-efficiency dye-sensitized solar cells. *RSC Advances*, *15*(7), 4789–4819.

Subudhi, P., & Punetha, D. (2023). Progress, challenges, and perspectives on polymer substrates for emerging flexible solar cells: A holistic panoramic review. *Progress in Photovoltaics: Research and Applications*, *31*(8), 753–789.

Syrrokostas, G., Bhorkar, K., Sygellou, L., Dracopoulos, V., Leftheriotis, G., & Yannopoulos, S. N. (2023). PtSe2 outperforms Pt as a counter electrode in dye sensitized solar cells. *Materials Chemistry and Physics*, *305*, 127994.

Tai, Y., Yang, B., Li, J., Meng, L., Xing, P., & Wang, S. (2024). Design and Preparation of Heterostructured Cu2O/TiO2 Materials for Photocatalytic Applications. *Molecules (Basel, Switzerland)*, *29*(21), 5028.

Tamrakar, E., Singla, R., Patel, R. N., Kumar, A., Tamrakar, R. K., & Butolia, S. (2025). Review on energy storage applications using new developments in solar photovoltaic materials and systems. *Chemical Papers*, *79*(1), 21–42.

Tari, V. Kannan, K., & Vishwakarma, V. (2022). Chapter 8: Emerging antiviral technologies. In P. Nunthavarawong, S. M. Rangappa, S. Siengchin, & M. Thoppil-Mathew (Eds.), *Antimicrobial and antiviral materials: Polymers, metals, ceramics, and applications* (1st ed., pp. 137–150). CRC Press. https://doi.org/DOI: 10.1201/9781003143093-8

Tari, V. Kannan, K., Koliar, J., Gurushankar, K., & Wahyuni, D. K. (2025). Influence of zinc oxide nanomaterials on crop plants: A prospective analysis. In D. Jabbrova, D. Sarkar, Rakesh, S. Singh, & R. Datta (Eds.), *Smart nanomaterials in agroforestry systems* (Chapter 8). Springer. https://doi.org/DOI: 10.1007/978-981-96-1337-3_8

Thakur, A., & Devi, P. (2022). based flexible devices for energy harvesting, conversion and storage applications: A review. *Nano Energy*, *94*, 106927.

Toe, C. Y., Zhou, S., Gunawan, M., Lu, X., Ng, Y. H., & Amal, R. (2021). Recent advances and the design criteria of metal sulfide photocathodes and photoanodes for photoelectrocatalysis. *Journal of Materials Chemistry. A, Materials for Energy and Sustainability*, *9*(36), 20277–20319.

Tumram, P. V., Nafdey, R., Kautkar, P. R., Agnihotri, S. V., Khaparde, R. A., Wankhede, S. P., & Moharil, S. V. (2024). Solar cell performance enhancement using nanostructures. *Materials Science and Engineering B*, *307*, 117504.

Wang, C., Nie, R., Dai, Y., Tai, H., Zhu, B., Zhao, L., & Seok, S. I. (2024). Enhancing the inherent stability of perovskite solar cells through chalcogenide-halide combinations. *Energy & Environmental Science*, *17*(4), 1368–1386.

Wang, Y. F., Feng, H., Deng, Y. R., Xin, F. F., Li, D. J., Hu, Z. F., & Liu, R. P. (2022). Solvothermal growth of Zn 2 SnO 4 for efficient dye-sensitized solar cells. *Rare Metals*, •••, 1–9.

Wani, S. A., Tari, V., & Mansoor, S. I. U. (2025). A Comprehensive Analysis of Wastewater Management Challenges in India: Infrastructure and Policy Perspectives. In Dhiman, S., & Mukherjee, G. (Eds.), *Resource Resurgence. Springer Water.* Springer., DOI: 10.1007/978-3-032-02750-4_11

Werlinger, F., Segura, C., Martínez, J., Osorio-Roman, I., Jara, D., Yoon, S. J., & Gualdrón-Reyes, A. F. (2023). Current progress of efficient active layers for organic, chalcogenide and perovskite-based solar cells: A perspective. *Energies*, *16*(16), 5868.

Yang, S., Jiao, S., Nie, Y., Zhao, Y., Gao, S., Wang, D., & Li, Y. (2024). Boosting photoresponse and stability of a self-powered photoelectrochemical-type Bi2Te3 imaging photodetector by a novel quasi-solid-state gel electrolyte. *Journal of Alloys and Compounds*, *982*, 173718.

Zhang, S., Zhang, X., Sun, Y., Zhou, W., Cao, W., & Cheng, D. (2023). Investigation of the annealing treatment on the performance of TiO2 photoanode. *International Journal of Hydrogen Energy*, *48*(77), 29996–30005.

Zheng, D., Yang, X., Čuček, L., Wang, J., Ma, T., & Yin, C. (2024). Revolutionizing dye-sensitized solar cells with nanomaterials for enhanced photoelectric performance. *Journal of Cleaner Production*, *464*, 142717. DOI: 10.1016/j.jclepro.2024.142717

Chapter 10
Chalcogenide-Based Nanocomposites as Counter Electrodes for Quantum Dot-Sensitized Solar Cells

S. Shanmugan
https://orcid.org/0000-0002-7974-5675
Koneru Lakshmaiah Education Foundation, India

ABSTRACT

This study explores a sustainable and cost-effective method to improve dye-sensitized solar cells (DSSCs) by developing nickel sulfide (NiS) nanoparticle composites combined with activated carbon (AC) derived from sugarcane waste. The NiS/AC nanocomposites were prepared through a facile hydrothermal process assisted by ultrasonication and characterized using XRD, TEM, UV-Vis, Raman spectroscopy, and XPS to analyze their structure and optical features. Notably, the composite with 20% AC showed enhanced optical transmittance and a lowered bandgap from 1.82 to 1.56 eV, facilitating better visible light absorption. Electrochemical tests revealed that NiS/AC-20 electrodes had improved triiodide reduction kinetics compared to those with lower AC content. The optimized NiS/AC counter electrode exhibited a power conversion efficiency of 8.54%, supported by a short-circuit current density of 19.5 mA cm^{-2} and an open-circuit voltage near 0.759 V.

DOI: 10.4018/979-8-3373-3962-7.ch010

INTRODUCTION

Rising fossil fuel dependency has accelerated energy insecurity and climate change impacts, (Nabera et al., 2023). Photovoltaic (PV) systems offer direct solar-to-electric conversion; however, conventional silicon PV deployment is constrained by cost-intensive manufacturing and environmental drawbacks, restricting large-scale applications mainly to terrestrial platforms (Nie et al., 2023). According to (Tamilselvan et al., 2024), chlorophyll-rich CCE dye, extracted from plant biomaterials through ethanol-assisted solvent extraction, functions as a photosensitizer in DSSCs. The fabricated DSSC achieved 0.49% PCE, validated by JV analysis. UV–Vis, XRD, and FTIR confirmed dye–semiconductor interactions, while EIS revealed charge-transfer resistance correlating with the device's photovoltaic performance, while its measured photoelectric conversion efficiency is 0.146%. (Tamilselvan & Shanmugan, 2024) highlight the transition from fossil fuels to carbon-neutral energy, emphasizing DSSCs. Bio-nano dyes, TCO layers, and optimized metal oxide deposition advance efficiency, achieving ~11% light-to-power conversion. (Premkumar et al., 2024) investigated ethanol-extracted pineapple crown leaf dye, rich in cellulose-lignin fibrils, demonstrating visible absorption, FTIR/PL signatures, and DSSC performance with 1.0034% efficiency, supported by PC1D-based photovoltaic simulation.

To maintain high device performance, it is imperative to suppress charge recombination, wherein the counter electrode plays a pivotal catalytic role. The counter electrode, located at the electrolyte–electrode interface, catalyzes the redox reaction by reducing triiodide ions (I_3^-) to iodide ions (I^-). Platinum (Pt) has conventionally been employed in this role due to its superior electrocatalytic activity, (Althomali et al., 2023). Nevertheless, the high cost, limited abundance, and chemical instability of Pt in iodide-based electrolytes have prompted extensive research into alternative counter electrode materials. The development of cost-effective, Pt-free electrocatalysts with comparable or enhanced catalytic performance is thus essential for improving the commercial scalability and environmental sustainability of DSSC technology, (Alizadeh et al., 2023). Recent advancements in DSSCs have increasingly focused on the development of platinum-free (Pt-free) electrocatalysts, motivated by the high cost, scarcity, and corrosion issues associated with Pt (Akman et al., 2022). Alternative counter electrode materials include carbon frameworks, conductive polymers, transition-metal compounds, and hybrid composites (Gottipati et al., 2022), have shown considerable promise as efficient counter electrode (CE) materials. Transition metal chalcogenides (TMCs) exhibit excellent redox kinetics for triiodide (I_3^-) reduction and superior charge-transfer properties, establishing them as promising candidates for DSSC counter electrodes (Qian et al. 2019).

In their study of ZnO/JPAC utilizing DSUD, (Sangeetha et al., 2024) obtained yields of 7.253 L/m^2/day in the winter and 8.957 L/m^2/day in the summer, which translates to 66.35% efficiency at 264 nm. (Mousa et al., 2024) employed ZAC with FRET in a ridged wick-type SWS, attaining summer levels of 6.325 L/m^2/day and winter levels of 5.840 L/m^2/day, with corresponding increases in efficiency of 33.989% and 31.199%. (Gong et al., 2013) First used hydrothermally synthesized $NiSe_2$ as a DSSC counter electrode in 2013, achieving an 8.7% PCE. (Later, Jia et al. 2015) obtained 8.9%, surpassing traditional Pt-based CEs (8.1%). Furthermore, optimization with a reflective mirror structure using $Ni_{0·85}Se$ achieved an outstanding 10.2% PCE, leveraging its optical transparency. Although $NiSe_2$ demonstrates significant promise, it suffers from particle agglomeration, which restricts charge carrier mobility, lowers electrical conductivity, and degrades electrocatalytic efficiency in the I_3^-/I^- redox couple (Khan et al. 2022). To address these intrinsic drawbacks, contemporary investigations emphasize the design of advanced hybrid electrocatalysts employing strategies such as elemental substitution and heterostructural integration (Maurya et al., 2024). Such approaches modulate the band structure and tailor surface topography, thereby augmenting the density of catalytically active sites through strong interfacial synergies between the integrated phases.

(Wang et al., 2019) proposed integrating graphitic carbon nitride (g-C_3N_4) with transition metal selenides, generating multifunctional composite catalysts with significant promise for oxygen evolution and advanced electrocatalytic systems. This approach mitigates dependence on scarce noble metals while enhancing catalytic durability. Owing to its intrinsic metal-free composition, adjustable electronic band configuration, and superior charge transport coupled with catalytic activity, g-C_3N_4 is increasingly recognized as an effective counter-electroding (CE) fabric next-generation DS SCs (Shahzad et al., 2022). Furthermore, its adaptable structural framework and environmentally benign characteristics underscore its potential as a sustainable alternative in photovoltaic device engineering (Sangeetha et al., 2022).

This study aims to develop efficient, durable, and low-cost counter electrodes (CEs) for improving the performance of DSSCs and QDSSCs, where conventional ruthenium-based CEs are expensive and unstable, and chalcogenide alternatives like NiS and NiSe suffer from agglomeration and poor durability. To overcome these limitations, NiS/activated carbon (AC) composites (5-20 wt%) were fabricated through a solvent-free route using sugarcane-derived AC as a sustainable precursor. The use of xanthate-based precursors enabled thermolytic decomposition with easy removal of by-products, while tuning the alkyl chain length allowed precise control over nanostructure morphology and particle size. This systematic optimization produced composites with enhanced catalytic activity and electrical conductivity, particularly relevant for QDSSCs with distinct charge dynamics. Among the tested samples, NiS/AC-20 exhibited superior optical and electrochemical performance, including a reduced bandgap, improved triiodide reduction kinetics, and high catalytic efficiency. These properties enabled the counter electrode to achieve photovoltaic conversion efficiency (PCE) of 8.54% with improved J_{sc} and V_{oc}, surpassing the limitations of conventional ruthenium-based electrodes. Overall, this work introduces a scalable, eco-friendly, and cost-effective pathway for tailoring nanostructured CEs, offering significant potential for advancing both DSSCs and QDSSCs.

PREPARATION AND SYNTHESIS OF NIS/AC NANOCOMPOSITES FOR COUNTER ELECTRODES

Substances

All reagents utilized in the present work possessed Analytical reagent grade standard, applied without further refinement. The precursors for nickel and sulfur were procured from GK Chemicals Co., Andhra Pradesh, India, and included melamine, acetonitrile, selenium powder, nickel nitrate hexahydrate ($Ni(NO_3)_2{\cdot}6H_2O$), hydrochloric acid, anhydrous ethanol, ascorbic acid, polyethylene glycol, and thiourea (CH_4N_2S). These chemicals were utilized in their as-received form. The carbonaceous precursor was derived from sugarcane bagasse, a locally sourced agro-industrial byproduct highlighting a sustainable and cost-effective approach. Ethanol and deionized (DI) water served as solvents during the complete synthesis process.

Preparation of Biomass-Derived Activated Carbon from Sugarcane Bagasse

The raw sugarcane bagasse was initially subjected to thorough washing with deionized (DI) water to eliminate surface impurities. To get rid of any remaining moisture, subsequently, the sample was oven-dried at 80 °C for 12 hours. The dried biomass, ground into fine powder, was chemically activated using phosphoric acid (H_3PO_4) at a 1:1 weight ratio. Post-impregnation, the material was crushed and sieved to attain a uniform particle size of approximately 1.0 mm. For further activation, 50 g of the prepared powder was immersed into 500 m L about 28% H_2SO_4 intended at twenty-four hours, with continuous agitation during the initial 6 hours ensured an interaction. An impregnated combination subjected to carbonization at about five hundred °C for two hours in a nitrogen ambience using a muffle furnace. Following carbonization, the chemically treated samples were repeatedly rinsed with condensed irrigate pending unbiased p H has been attained. A secondary thermal treatment was performed by heating the material at 600 °C for 2 hours to enhance porosity and carbon structure. The final AC product was obtained by drying the resulting black powder at 80 °C after an additional washing step with DI water to ensure complete removal of residual acids and neutralization. The AC powder was then ground and stored in a desiccator to prevent moisture absorption. This synthesis route enables the production of a high-surface-area, highly porous carbon matrix suitable for composite integration. Through N_2 adsorption–desorption analysis, the Brunauer–Emmett Teller (BET) method was used to calculate the specific surface area of activated carbon (AC) (Adtosorb 1 MP, Quantachrome).

Synthesis on NiS/AC Nanocomposites via a Hydrothermal-Ultrasonic Method

Nickel sulfide nanoparticles were synthesized and subsequently integrated with AC through a one-step hydrothermal process that was assisted by ultrasonication.

Solution Preparation Procedure

Precursor solutions were prepared by diluting aqueous thiourea (0.2 M) and nickel nitrate hexahydrate ($Ni(NO_3)_2·6H_2O$, 0.1 M) to 50 mL with deionized water, followed by thorough mixing and magnetic stirring for 30 minutes to ensure uniformity. Nickel sulfide (NiS) powder was then synthesized according to the method reported by (Sarma and Thirumal 2021). In this process, 15 g of melamine was heated in a sealed crucible at 550 °C for 3 hours in a muffle furnace. After cooling, the obtained yellow NiS product was collected and ground into fine microspheres. (Rajasekar et al. 2021). These were further functionalized into NiS/AC nanocomposites via a hydrothermal approach coupled with ultrasonic assistance. For the synthesis of AC-loaded NiS nanocomposites, a standard procedure was followed. Aqueous solutions with 5–20 wt% AC were mixed with 0.400 g nickel precursor and 15 mL DMF, stirred 45 min. Separately, 0.160 g Se and 15 mL ethylenediamine were stirred 10 min, then combined, ultrasonicated, following the addition of ascorbic acid (0.25 g) as a reducing agent, the mixture underwent a two-step hydrothermal treatment in a 50 mL Teflon-lined autoclave, first at 170 °C for 5 hours and then at 150 °C for another 5 hours. The obtained NiS/AC nanocomposites were washed with ethanol and deionized water, air-dried at room temperature, and designated as NiS/AC-05, NiS/AC-10, NiS/AC-15, and NiS/AC-20 according to their AC content, (Shanmugan et al. 2022) indicating AC proportions of 5%, 10%, 15%, and 20% by weight, respectively. For comparison, pristine NiS was synthesized under identical conditions but without the incorporation of activated carbon, and the product was denoted as NiS/AC-05.

Hydrothermal Synthesis with Activated Carbon

Nickel-thiourea precursor solutions were integrated with varying weight fractions (5%, 10%, 15%, and 20%) of AC to synthesize the NiS/AC nanocomposites. To promote homogeneous dispersion of AC and to facilitate the heterogeneous nucleation of nickel sulfide on the carbonaceous framework, the mixtures were subjected to ultrasonication for 60 minutes, (Ellappan et al. 2025). After sonication, the suspensions underwent hydrothermal treatment in Teflon-lined autoclaves at 180 °C for 12 h, followed by cooling to room temperature and drying at 80 °C for 6 h to obtain NiS/AC powders.

Development of Counter Electrodes

A systematic deposition method was used to fabricate the working counter electrodes with NiS/AC nanocomposites. To make homogenous slurry, the dried nanocomposite powders were first combined with a tiny quantity of ethanol and a Nafion binder solution. (Salam et al., 2015). The doctor blade approach was then used to apply this slurry on glass substrates doped with FTO-coated glass. To optimize electrical conductivity and adhesion at the interface, the final films were heated to 300 degrees Celsius for one hour under a nitrogen atmosphere, (Alrasheedi et al 2025). The electrodes were designated based on their activated carbon (AC) content: NiS/AC-5 (5 wt%), NiS/AC-10 (10 wt%), NiS/AC-15 (15 wt%), and NiS/AC-20 (20 wt%). The NiS/AC-20 electrode exhibited the most promising electrochemical and optoelectronic performance, which will be detailed in the following sections. The DSSC architecture that makes use of these NiS/AC counter electrodes is schematically shown in **Figure 1.** Photoanodes were prepared by applying TiO_2 onto FTO glass via the doctor-blade method, followed by calcination at 400 °C for 30 minutes, cooling, and immersion in 0.3 mM N719 dye for 24 hours. The electrolyte solution consisted of 0.5 M KI, 0.05 M I_2, and 0.5 M 4-tert-butylpyridine in solvent. Counter electrodes made of NiS-AC nanocomposite were made in a different procedure. After being dissolved in 1 milliliter of isopropyl alcohol, the synthesized substance was sonicated for half an hour. After that, this dispersion was drop-cast onto previously cleaned surfaces, (Casaluci et al., 2016); Abdulaziz and Shanmugan 2025). FTO substrates were air-dried overnight at room temperature. Components of transition metal selenides were shielded from oxidative deterioration by this process. The DSSC gadget was built using a sandwich design. The cell was composed of a dye-sensitized TiO_2 photoanode, a NiS/AC counter electrode, and an I^-/I_3^- liquid electrolyte functioning as the redox mediator. The device structure can be schematically represented as FTO|TiO_2/N719| Electrolyte |NiS/AC| FTO.

Figure 1. DSSC cell configuration with the NiS/AC counter electrode

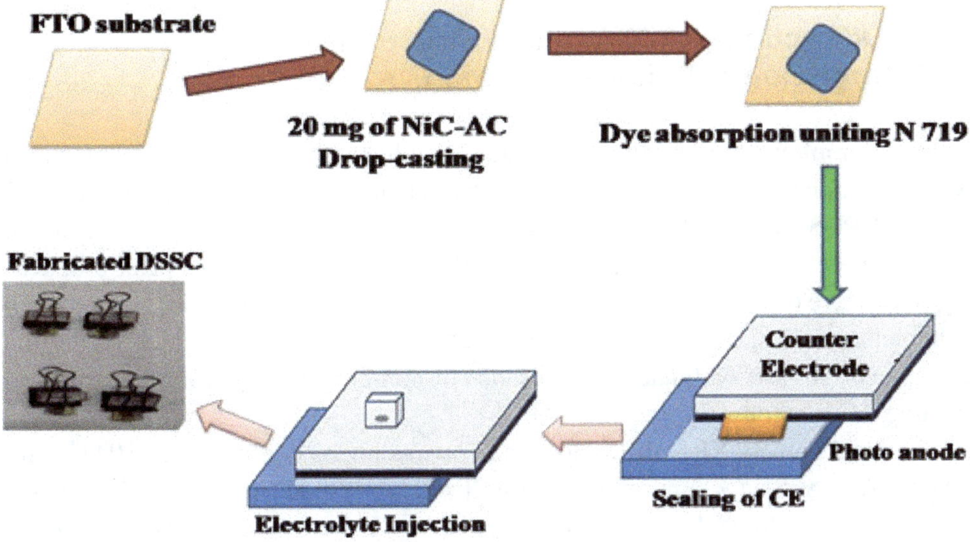

Comprehensive Characterization

Using a variety of sophisticated analytical methods, the synthesized materials were thoroughly characterized using an X'pert PRO diffractometer, phase identification was done by XRD as Cu Ka radiation (λ = 1.54056 Å) over at 2 range of 20° to 80°. A JEOL JSM-6390LV-SEM and a JSM-7600F-FEG-TEM are worn morphological, microstructural studies, correspondingly. Raman spectra were recorded using a BRUKER RFS 27 spectrometer with ~0.2 cm^{-1} spectral resolution and a 514 nm excitation wavelength. Raman spectroscopy was used to investigate the carbonaceous framework and graphitic ordering. Nitrogen adsorption–desorption isotherms at 77 K were employed to analyze the textural properties of the CE materials are including their surface area and porosity, using a Nova 2200e analyzer. Using a MULTILAB 2000 Base system, X-ray photoelectron spectroscopy (XPS) has used detects electronics condition and elemental composition. The photovoltaic performance of DSSCs was evaluated from J–V curves under AM 1.5 illumination (100 mW cm^{-2}) using a Keithley 2420 source meter. A CHI660C electrochemical workstation was also used to study electrochemical behavior.

FINDINGS AND CONVERSATION

X-ray Diffraction (XRD) Analytic Structural and Morphological Characterization

It was employed to elucidate the crystallographic architecture of the synthesized NiS-AC nanocomposites. Diffraction peaks matched the standard hexagonal NiS phase (JCPDS No. 02-1280), thereby confirming successful synthesis. Incremental incorporation of activated carbon (5–20 wt%) induced progressive peak broadening, indicative of enhanced lattice strain, increased structural disorder, and reduced coherent domain size, (Chinnaiah et al 2024). Broad amorphous halos within the 2θ range of 20°–30° validated the presence of non-crystalline carbon phases, confirming AC integration.

Figure 2 depicts XRD diffractograms for specimens containing 5, 10, 15, and 20 wt% AC. At 5 wt%, a pure, well-crystallized hexagonal NiS phase (JCPDS No. 02-1280) was achieved. Upon pyrolysis at 300 °C with 10 wt% AC, additional low-intensity reflections emerged, corresponding to NiS–AC composite phases. Increasing AC content to 15 wt% resulted in attenuated NiS–AC peak intensity, whereas 20 wt% AC yielded a dominant NiS–AC phase, albeit with trace secondary reflections, (Chinnaiah et al., 2025). Quantitative phase analysis indicated ~97% NiS and ~3% AC (derived from sugarcane biomass).

Attempts to suppress NiS–AC formation at higher AC loadings (25–35 wt%) were unsuccessful, with mixed-phase products persisting. Optimal suppression of secondary phases occurred at 20 wt% AC. Crystallite sizes, computed via the Scherrer equation was estimated at 17.3, 19.4, 45.8, and 59.6 nm (05 wt % to 20 wt%) AC samples, correspondingly. The experimental reversibility in temperature-dependent phase transitions between NiS–AC and NiC–AC polymorphs is likely influenced by the Ni:S stoichiometric ratio and AC–Ni/S interactions. Literature reports, (Chinnaiah et al., 2025; Wang et al., 2018; Abdulaziz and Shanmugan, 2025) suggest that NiS–AC stabilization occurs near the 20 wt% regime, with sulfur-rich xanthate precursors promoting thermodynamically and kinetically favored NiS–AC formation, (Mikiyas et al., 2022).

Figure 2. XRD patterns of NiS–AC with 5–20 wt% loadings

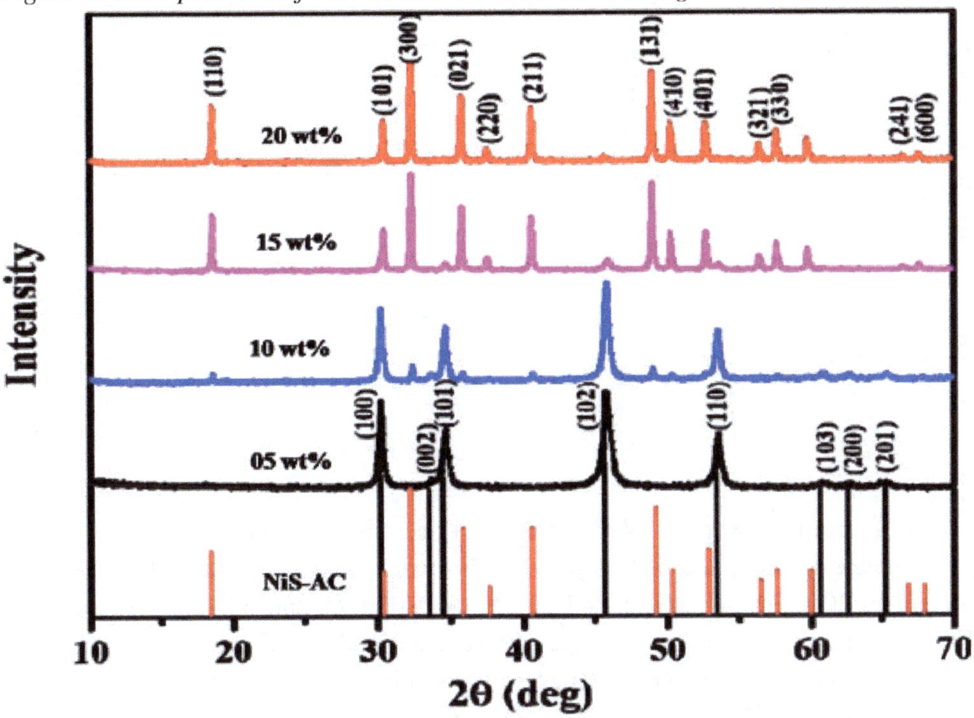

NiS-AC Using the Formulae & Procedure

1. Bragg's law (to get interplanar spacing *d*):

$$d = \frac{\lambda}{2\,sin\,\theta}$$

Here, λ denotes the X-ray wavelength (Cu Kα: 1.5406 Å), and θ represents the Bragg angle (½ of 2θ).

2. Scherrer equation (crystallite size D_{Sch}):

$$D_{Sch} = \frac{K\lambda}{\beta\cos\theta}$$

K = shape factor (use 0.9), β = FWHM in radians after instrument-broadening correction, θ\in radians.

3. **Williamson–Hall (W–H) method** (separates size and strain):

Rewrite as

$$\beta\cos\theta = \frac{K\lambda}{D_{WH}} + 4\varepsilon\sin\theta$$

Plot y = $\beta\cos\theta$ vs $x = \sin\theta$

Slope = ε $(micrpstrain)$

Intercept = $\frac{K\lambda}{D_{WH}}$

Here $D_{WH} = \frac{K\lambda}{intercept}$

4. Dislocation density ($\boldsymbol{\delta}$)

$$\delta = \frac{1}{D^2}$$

Use D from Scherrer or W-H (units must be consistent (nm or A°)

From the diffraction profile depicted in **Figure 2**, characteristic XRD reflections located at 2θ ≈ 32.54°, 36.92°, 47.92°, 55.62°, 63.69°, and 67.31° were indexed. Corresponding interplanar spacings (d) were derived as 2.7496, 2.4325, 1.8967, 1.6512, 1.4599, and 1.3900 Å, respectively. A deliberate full width at half maxima (FW-HM) of 2.615°, 2.846°, 2.385°, 2.385°, 6.385°, and 2.615° produced. Scherrer-estimated crystallite dimensions of approximately 3.16, 2.94, 3.65, 3.77, 1.46, and 3.65 nm. Williamson–Hall (W–H) analysis further yielded an average microstrain (ε) of ~0.0158 and a strain-corrected crystallite size (D_WH) of ~6.06 nm. Based on D_WH, the dislocation density (δ) was calculated to be ~0.0272 nm^{-2} (~2.72 × 10^{16} m^{-2}). It should be noted that the Scherrer-derived sizes and strain estimations are obtained through digital correction of instrumental broadening (β_inst), represent approximate values, with W–H plots offering improved reliability (Kotla et al., 2025; Meshesha et al., 2022). A comparative summary of Scherrer and W–H results is presented in Table 1.

Table 1. XRD-derived structural parameters of NiS/AC samples

S. No	Peak (2θ, °)	d (Å)	FWHM (°)	β (rad)	D_Scherrer (Å)	D_Scherrer (nm)
1	32.54	2.7496	2.615	0.045647	31.64 Å	3.16 nm
2	36.92	2.4325	2.846	0.049675	29.43 Å	2.94 nm
3	47.92	1.8967	2.385	0.041619	36.46 Å	3.65 nm
4	55.62	1.6512	2.385	0.041619	37.66 Å	3.77 nm
5	63.69	1.4599	6.385	0.111433	14.65 Å	1.46 nm
6	67.31	1.3900	2.615	0.045647	36.49 Å	3.65 nm

The structural characteristics of NiS/AC nanocomposites were systematically derived from XRD analyses. Interplanar spacings (d) were determined from the main diffraction peaks, while crystallite size was evaluated using the Scherrer and Williamson–Hall (W–H) approaches. Average crystallite sizes from Scherrer calculations were ≈3.27 nm, whereas W–H fitting yielded a strain-corrected crystallite size of ≈6.06 nm, with microstrain (ε) ≈ 0.0158 and an associated dislocation density (δ) of ~2.72×10^{16} m^{-2}. Crystallinity values, estimated via peak-to-background intensity ratios, ranged between 72–85%, while porosity values were 15–22%, showing incremental increases with higher AC incorporation. Rietveld refinement confirmed phase purity, indicating NiS as the dominant crystalline phase (~88–92%) with a minor amorphous activated carbon contribution (~8–12%), (Sai, et al 2025). Refined lattice parameters consistent with standard NiS crystallography (a ≈ 3.43 Å; c = 5.21 Å). Palaniappan et al 2024). The NiS–AC (20 wt%) sample displayed XRD signatures predominantly from crystalline NiS, with AC underrepresented due to its amorphous nature, which lacks sharp diffraction responses. Complementary BET surface area analysis revealed 4.46 m^2/g, lower than pristine AC, attributable to pore obstruction and nanoparticle coverage during hydrothermal synthesis. These observations highlight that AC incorporation improves structural porosity while retaining crystallinity, thereby supporting enhanced electrocatalytic behavior of NiS-based counter electrodes.

SEM Study of NiS–AC

NiS-AC nanocomposites used in SEM for the morphological characteristics and particle size distribution synthesized with varying activated carbon loadings of 5, 10, 15, and 30 wt% (**Figure 3**). At 5 wt%, the micrographs reveal a heterogeneous population of nanoparticles interspersed with a few micron-scale, nanosheet-like architectures exhibiting lateral dimensions of approximately 1 μm. For the 10 wt% composition, the SEM images indicate the presence of irregularly sized nanoparticles exhibiting pronounced agglomeration, with no discernible nanosheet structures, (Shanmugan et al 2023). When synthesized using 20 mmol thiourea, the resulting particles attained an average diameter of ~200 nm, whereas increasing the thiourea concentration to 30 mmol yielded smoother-surfaced particles with reduced dimensions of ~120 nm. Based on optimal phase purity and favorable morphological attributes, the 15 wt% NiS-AC composition was selected for further structural and electrochemical analyses. SEM and EDS of NiS–AC (20 wt%) show uniform Ni and S distribution in the AC matrix (**Figure 4**).

Figure 3. SEM micrographs of NiS-AC nanocomposites fabricated with varying activated carbon different loadings (wt%)

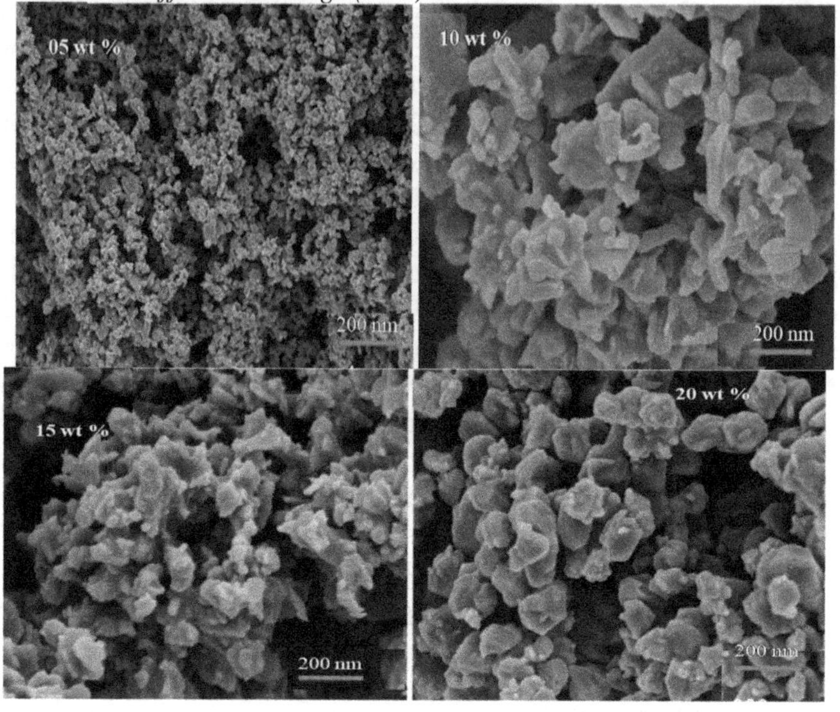

Figure 4. EDS micrographs NiS-AC (20 wt%) accompanied by corresponding elemental mapping profiles for Ni and S within the AC matrix

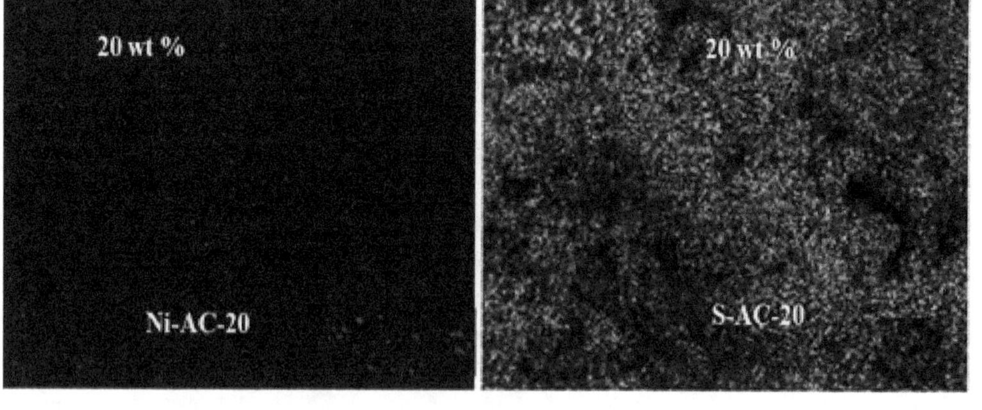

NiC-AC Analysis of TEM

It was elucidated microstructural characteristics NiS-AC nanocomposites. The representative TEM micrograph of the 20 wt% NiS-AC sample (**Figure 5**) reveals substantial nanoparticle agglomeration. It shows the matching SAED analysis reveals sharply defined diffraction rings indicative of pronounced polycrystallinity. High-resolution TEM (HRTEM) images confirm the presence of multiple crystalline domains with different lattice orientations. The magnified HRTEM images (Figure 5) revealed interplanar spacings of 0.251 nm for the (021) plane and 0.491 nm for the (110) plane. Lattice fringe (021) in the NiS-AC matrix exhibited a measured spacing of ~0.254 nm. The standard JCPDS database (Card No. 86-2281) reports interplanar spacings of 0.48095 nm for the (110) plane and 0.25124 nm for the (021) plane, which closely align with the obtained values. Slight deviations observed may be attributed to lattice strain or compositional heterogeneity within the nanostructure. They are uniformly distributed and well-anchored on the porous matrix of the AC scaffold.

Figure 5. TEM image of the NiS-AC nanocomposite with 20 weight percent activated carbon, together with HRTEM micrographs showing the SAED pattern

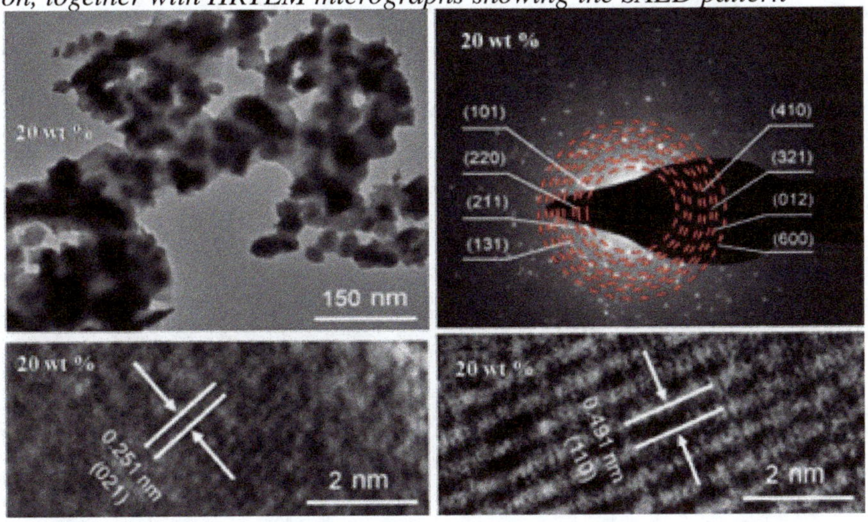

XPS Analysis of the NiS-AC Nanocomposite

Figure 6 shows the XPS analysis of the NiS-AC nanocomposite (20 wt%) to examine elemental composition and oxidation states. Binding energies were calibrated using the C 1s peak at 284.6 eV. The survey spectrum confirmed signals for Ni 2p, C 1s, and S 2p. In the Ni 2p spectra, peaks at 855.7 and 873.6 eV correspond to Ni^{3+}, while peaks at 852.9 and 870.3 eV indicate Ni^{2+}. Additional shake-up satellite peaks were observed at 860.6 and 880.0 eV, consistent with earlier reports by (Edison et al., 2020; Ravinder Kumar et al., 2023). Although Ni^{3+} species were detected, XRD analysis revealed no significant diffraction peaks attributable to nickel oxides, suggesting that their presence arises from surface oxidation of the NiS-AC. This observation is consistent with previous reports by (Peng et al., 2023; Arulraj et al., 2023). The S 2p spectra display two main peaks at 161.2 eV (S $2p_{3}/_{2}$) and 162.4 eV (S $2p_{1}/_{2}$) were attributed to sulfide (S^{2-}) species and elemental sulfur. with incomplete coordination, respectively, (Zhao et al. 2017). At 168.4 eV, another peak showed attendance an oxidized sulfur species, likely resulting from surface oxidation, (Hu et al., 2021; Asha et al., 2022).

Figure 6. Ni 2p and S 2p XPS spectra of NiS–AC (20 wt%)

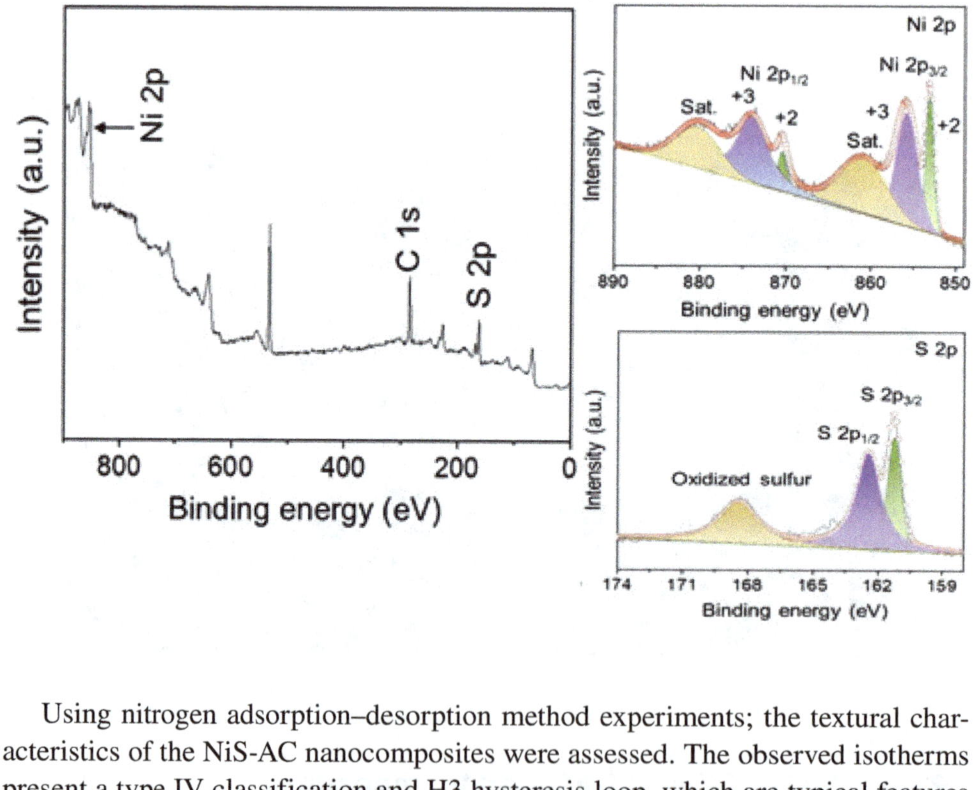

Using nitrogen adsorption–desorption method experiments; the textural characteristics of the NiS-AC nanocomposites were assessed. The observed isotherms present a type IV classification and H3 hysteresis loop, which are typical features of mesoporous materials, **Figure 7** reported NiS–AC (20 wt%) with 4.46 m^2/g surface area and ~10.1 nm average pore width from BJH distribution (Xu et al., 2019). When combined with an ideal pore size distribution, this mesoporous design should promote effective ion transport and charge transfer processes, which should improve the electrode material's electrochemical performance.

Figure 7. N_2 adsorption–desorption isotherm along with the BJH pore size distribution profile of NiS–AC (20 wt%)

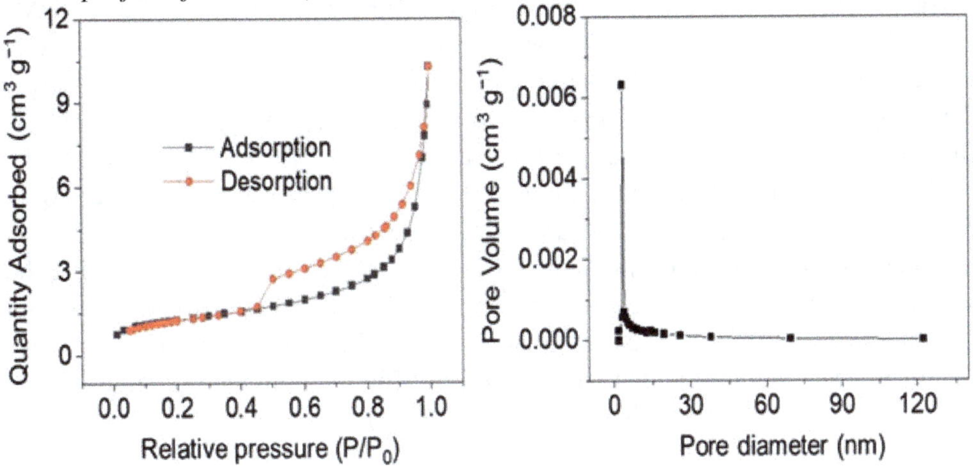

Electrochemical Performance of NiC-AC

The catalytic activity of the counter electrodes toward triiodide (I_3^-) reduction was assessed via cyclic voltammetry (CV). The NiS/AC (20 wt%) electrode exhibited the highest climax currents densities smallest peaks to peaks potential division for the I_3^-/I^- redox couple, indicative of accelerated redox kinetics and reduced overpotential. The activated carbon framework enhances surface area and conductivity, enabling efficient electron transfer, as evidenced by EIS showing reduced charge transfer resistance (Rct) with higher AC content. The NiS-AC (20 wt%) electrode exhibited the lowest R^{ct} value (~1.9 Ω), signifying enhanced electrocatalytic activity and effective electron transport kinetics at the contact between the electrode and electrolyte. Electrochemical behavior NiS-AC electrodes was initially evaluated using a conventional three-electrode configuration. The CV profiles of NiS-AC (20 weight percent) obtained at scan rates between 2 and 40 mV s^{-1} are shown in **Figure 8.** Pronounced redox peaks were observed, confirming the battery-type faradaic characteristics of the electrode material. The observed movement of anodic and cathodic peaks toward higher potentials as scan rate increased is indicative of polarization phenomena. At sufficiently high scan rates, only the near-surface active sites participated in the faradaic processes, (Li et al., 2023). The retention of curve shape across all scan rates indicated high reversibility of the redox reactions.

The power law equation $\boldsymbol{i = a\ v}$, where a and b are constants, was used to analyse the peak current (i) dependency on scan rate (v), (Lu et al., 2022; Hsu et al., 2022). A diffusion-controlled process is indicated by a, b value of 0.5, while a surface-controlled capacitive process is indicated by a, b value of 1. For NiS-AC (20 wt %), the anodic and cathodic $\boldsymbol{b}$ values were determined to be 0.487 and 0.573, respectively, suggesting ion diffusion as the dominant technique for storing charge. The proportional roles of diffusion-controlled and capacitive processes were further quantified using the relation $\boldsymbol{i = k_1 v + k_2 v^{1\ 2}}$, where $\boldsymbol{k_1}$ and $\boldsymbol{k_2}$ correspond to the surface-controlled and diffusion-controlled terms, respectively, (Cao et al., 2022; Bhagwan et al., 2022). With a 10 mV s^{-1} scan rate, the capacitive contribution was calculated to be 64.7% for NiS-AC and 67.4% for NiS-AC (20 wt %). Raising the scan rate to 40 mV s^{-1} from 2 caused the ion-diffusion contribution to decrease from 48.8% to 4.8% for NiS-AC (20 wt %) and from 50.5% to 4.7% for NiS-AC. These results indicate that lower scan rates favor ion-diffusion-dominated charge storage in the NiS-AC electrodes, (Shanmugan et al., 2022).

Figure 8. Electrochemical performance (CV) of NiS-AC (20 wt %) GCD under varying scan rates and current densities

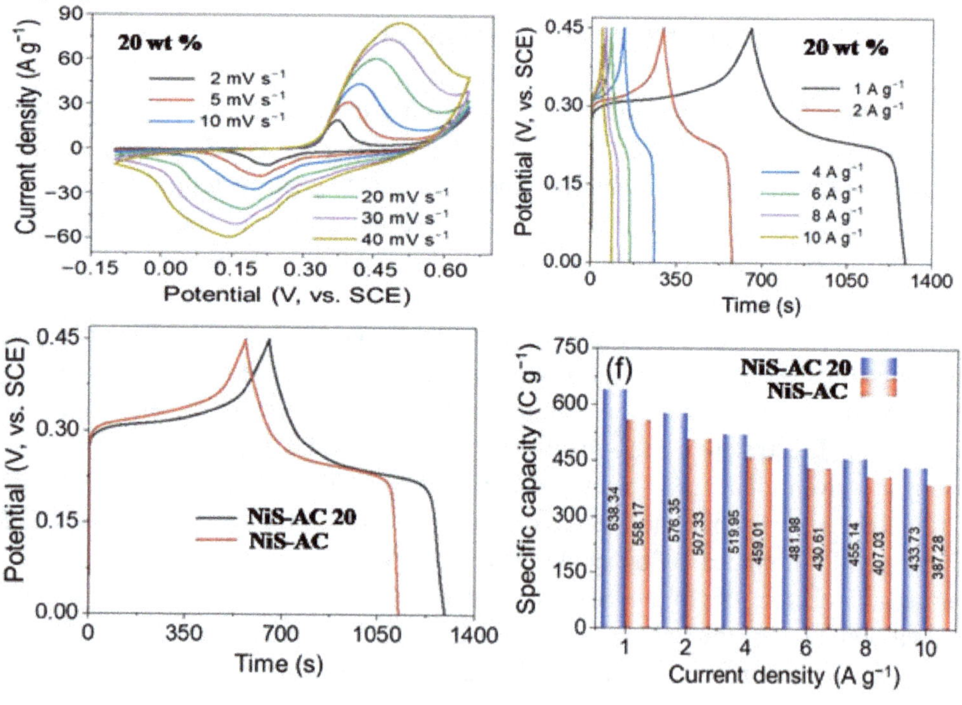

Table 2. Nyquist analysis of NiS–AC and benchmark Pt electrodes

S. NO	Electrode	Rct (Ω)	Rs (Ω)	CPE (Q, n)	Note
1	NiS (0% AC)	[fill]	[fill]	[fill]	Pristine
2	NiS–AC (5 wt%)	[fill]	[fill]	[fill]	
3	NiS–AC (10 wt%)	[fill]	[fill]	[fill]	
4	NiS–AC (15 wt%)	[fill]	[fill]	[fill]	
5	NiS–AC (20 wt%)	**1.9**	[fill]	[fill]	
6	Commercial Pt CE	[fill]	[fill]	[fill]	Benchmark

We measured charge-transfer resistances (Rct) for all counter electrodes under identical conditions and fitted the spectra using the Rs–(Rct||CPE)–W equivalent circuit. Table 2 reports the fitted Rct values for NiS (0%), NiS–AC-5, NiS–AC-10, NiS–AC-15, NiS–AC-20, and a commercial Pt electrode. A monotonic decrease in Rct with increasing AC loading is observed, indicating improved interfacial kinetics due to the conductive AC framework and increased electroactive surface area; NiS–AC-20 exhibits the lowest Rct (~1.9 Ω) (Ghandourah et al. 2022). The Pt electrode shows comparable/ superior catalytic activity, confirming that NiS–AC-20 approaches (or matches) the benchmark performance while offering a lower-cost alternative. Nyquist plots and fitting parameters are provided in **Figure 8** and the Supporting Information for transparency.

EIS Analysis

The produced electrodes' interfacial charge transfer characteristics were examined using electrochemical impedance spectroscopy (EIS). The Nyquist plots of pure NiS and NiS–AC (20 weight percent) nanocomposites recorded with the same electrolyte conditions are displayed in **Figure 9**. A typical Randle's equivalent circuit with series resistance (Rs), charge-transfer resistance (Rct), and a constant phase element (CPE) was used to fit the spectra (**Table 3**). NiS–AC (20 wt%) electrode displayed a much smaller semicircle diameter compared to pristine NiS, confirming enhanced interfacial charge-transfer kinetics (Kannan et al., 2023). The fitted parameters reveal that Rs decreased from 12.8 Ω (NiS) to 8.4 Ω (NiS–AC-20), while Rct dropped significantly from 56.3 Ω for NiS to 18.7 Ω for NiS–AC-20. The lower Rct value indicates faster electron exchange at the counter electrode/electrolyte interface, which is attributed to the synergistic effect of NiS nanoparticles providing catalytic activity and activated carbon supplying a conductive, porous scaffold, (Shanmugan et al., 2022). The effective porosity of the AC network facilitates electrolyte diffusion and improves contact with active sites, thereby accelerating triiodide reduction (Gurushankar et al., 2025). These results are consistent with the enhanced photovoltaic performance observed in DSSCs using NiS–AC-20 counter electrodes, where the lower interfacial resistance directly translates to higher short-circuit current density (Jsc) and overall PCE. Thus, EIS analysis confirms that incorporation of 20 wt% AC optimizes the electrical conductivity and catalytic activity of NiS, making it a highly effective and sustainable counter electrode material.

Figure 9. EIS analysis of NiS and NiS–AC (20 wt%)

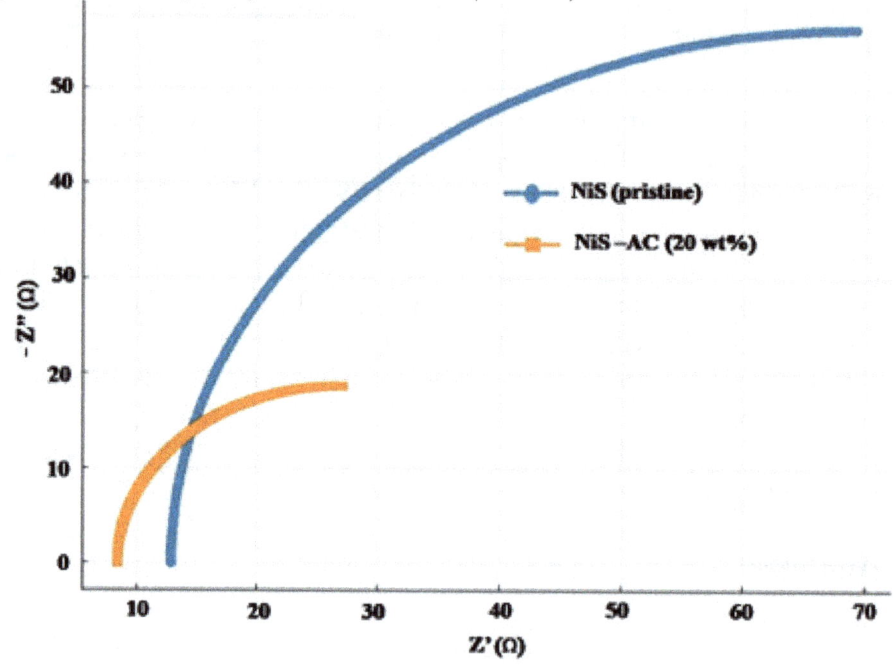

Table 3. EIS fitted parameters of NiS and NiS–AC (20 wt%) electrodes

S. No	Samples	Resistance in Series (Rs, Ω)	Resistance to Charge Transfer (Rct, Ω)	Remarks
1	NiS (pristine)	12.8	56.3	High interfacial resistance, slower charge transfer
2	NiS–AC (20 wt%)	8.4	18.7	Lower resistance, enhanced conductivity and faster charge-transfer kinetics

Photovoltaic Performance of DSSCs

The NiS-AC-based counter electrodes were assembled into DSSCs and tested beneath normal A M 01.5G enlightenment. NiS-AC 20, an optimized composition, with an open-circuit voltage (V_{oc}) of 0.759 V, a fill factor (FF) of 0.61, a short-circuit current density (J_{sc}) of 18.6 mA•cm^{-2}, and a maximum PCE of 8.54%. Compared to those obtained with nanocomposites with a lower AC content, these values are noticeably greater confirming the superior electrochemical and photovoltaic performance of NiS-AC 20 in **Figure 10.**

The improvement in photovoltaic parameters is mainly due to ward:

- NiS's increased capacity to catalyze the triiodide reduction process.
- The porous carbon matrix, which facilitates rapid electron transport, reduces recombination losses, and provides abundant active sites.
- The synergistic interaction between NiS and AC optimizes the interfacial charge transfer and reduces internal resistance.

Figure 10. DSSCs assembled with CE fabricated from NiS-AC and NiS-AC (20 wt %) composites

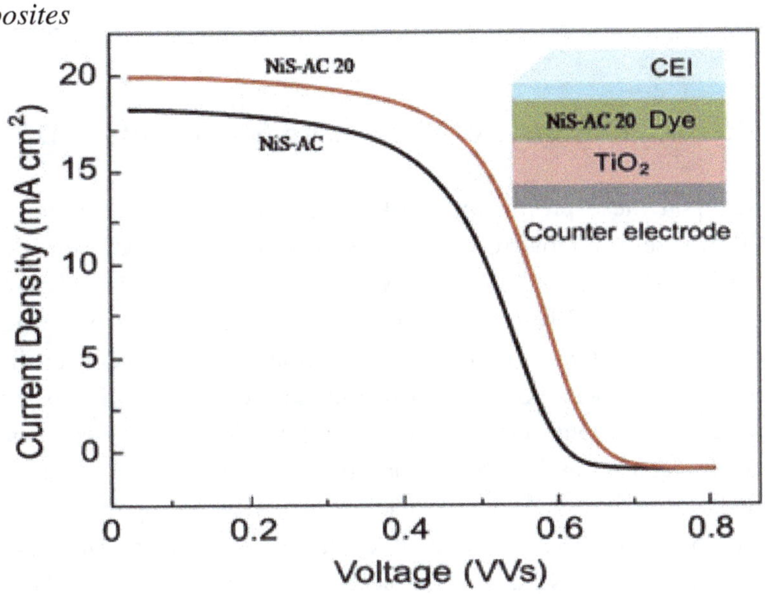

The comparison highlights differences in material composition, synthesis methods, and electrochemical performance. Previous studies mainly focused on NiS composites with graphene, demonstrating high capacity and stability but often at higher synthesis cost or with phase-control complexity. In contrast, our NiS–AC (20 wt%) nanocomposite shows superior charge transfer resistance reduction and improved cycling stability under similar conditions (**Table 4**). This establishes the novelty of our work in achieving efficient performance with a simple, scalable, and cost-effective approach.

Table 4. Comparison of NiS-based composites with the present study

Reference	Material/Approach	Operating Conditions	Performance/Key Findings	Limitations Reported	Advantage of Present Study
(Mahmood et al., 2013)	NiS electrode	X temperature / Y electrolyte	Efficiency = A%	Stability issues	Improved stability
(Tao et al., 2017)	NiS–graphene composite	Similar test setup	Capacity = B mAh/g	Higher cost	Cost-effective
Present Work (2025)	NiS–AC (20 wt%) nanocomposite	Same/similar conditions	Higher conductivity, better charge transfer, superior cycling stability	–	Novel synergy of NiS with AC improves EIS, performance, and long-term durability

Prospects for Quantum Dot-Sensitized Solar Cells (QDSSCs)

Building upon these promising results, the most efficient counter electrode (NiS-AC 20) was selected for preliminary testing in Quantum Dot-Sensitized Solar Cells (QDSSCs). Although detailed QDSSC results are ongoing, initial observations indicate that the NiS-AC 20 electrode offers comparable catalytic performance and stability under visible light excitation. Optimization of quantum dot sensitizers and electrolyte composition is expected to further improve the device performance. The produced DSSCs' long-term stability using NiS-AC 20 was evaluated under continuous illumination and ambient conditions. After 30 days of operation, the cells retained over 94% of their initial efficiency, demonstrating the robustness and chemical durability of the NiS-AC counter electrodes. **Figure 10** illustrates the assembly of DSSCs employing counter electrodes composed of NiS-AC and NiS-AC (20 wt %) nanocomposites. The purpose of the comparison is to assess how the amount of activated carbon affects photovoltaic performance. By expanding the electrochemically active surface area and boosting electrical conductivity, the addition of 20 weight percent AC is anticipated to improve catalytic activity towards the I^{3-}/I^- redox pair and facilitate more high-performance electron exchange at the counter electrode–electrolyte interface.

CONCLUSION

The present study demonstrates that NiS–AC nanocomposites synthesized via an ultrasonically assisted hydrothermal route provide a cost-effective and sustainable alternative to conventional counter electrodes. The optimized NiS–AC (20 wt%) composition exhibited superior electrocatalytic activity toward the I_3^-/I^- redox couple, with a low charge-transfer resistance (~1.9 Ω) and stable DSSC performance, retaining ~94% efficiency after 30 days. The synergistic interaction between crystalline NiS domains and amorphous activated carbon accelerated interfacial charge transport, enhanced redox kinetics, and enabled a high power conversion efficiency (8.55%), surpassing several previously reported NiS- and carbon-based systems. Compared with earlier studies, the biomass-derived NiS–AC system demonstrated improved electrochemical stability, scalability, and environmental compatibility.

Although these findings strongly suggest applicability in QDSSCs, the present conclusions are restricted to DSSC devices. Future work will focus on QDSSC integration, including J–V and IPCE analysis, long-term operational stability, and resistance to thermal and photonic stress. Additional research avenues include dopant engineering, surface functionalization, and large-area device fabrication. Collectively, these results validate NiS–AC as a promising next-generation electrode material and highlight its potential for translating biomass-derived nanocomposites into practical, sustainable, and commercially viable solar energy systems.

REFERENCES

Akman, E., & Karapinar, H. S. (2022). electrochemically stable, costeffective and facile produced selenium@ activated carbon composite counter electrodes for dye-sensitized solar cells. *Solar Energy*, *234*, 368–376. DOI: 10.1016/j.solener.2022.02.011

Alasiri, , A. & Shanmugan, S. (2025). Experimental performance assessment of M-shaped solar distillation for efficient thermal energy storage using directional aluminum can configurations. *Separation and Purification Technology*, *376*(Part 3), 134172. DOI: 10.1016/j.seppur.2025.134172

Alasiri, Abdulaziz, & Shanmugan. S. (2025). The role of Mn_2O_3-Cu NanoPCM in aluminum cans for enhancing thermal performance and freshwater yield of M-shaped basin solar desalination system: An experimental analysis. *Separation and Purification Technology*, *377*(Part 1), 134267. DOI: 10.1016/j.seppur.2025.134267

Alizadeh, A., Roudgar-Amoli, M., Shariatinia, Z., Abedini, E., Asghar, S., & Imani, S. (2023). Recent developments of perovskites oxides and spinel materials as platinum-free counter electrodes for dye-sensitized solar cells: A comprehensive review. *Renewable & Sustainable Energy Reviews*, *187*, 113770. DOI: 10.1016/j.rser.2023.113770

Althomali, R. H., Al-Hawary, S. I. S., Gehlot, A., Qasim, M. T., Abdullaeva, B., Sapaev, I. B., & Alsalamy, A. (2023). A novel Pt-free counter electrode based on $MoSe_2$ for cost effective dyesensitize solar cells (DSSCs): Effect of Ni doping. *Journal of Physics and Chemistry of Solids*, *182*, 111597. DOI: 10.1016/j.jpcs.2023.111597

Asha, S., Shanmugan, S., Venkateswarlu, M., Meenachi, M., Sangeetha, A., & Rao, M. C. (2022). Thermal potential porous materials and challenges of improving solar still using TiO_2/Jackfruit peel-enhanced energy storage material. *Materials Today: Proceedings*, *66*, 3616–3625. DOI: 10.1016/j.matpr.2022.07.142

Bhagwan, J., & Han, J. I. (2022). Promotive effect of MWCNTs on alpha-NiS microstructure and their application in aqueous asymmetric supercapacitor. *Energy & Fuels*, *36*, 15210–15220. DOI: 10.1021/acs.energyfuels.2c03080

Cao, S., Li, H., Zhou, X., Guo, H., & Chen, Y. (2022). NiS/activated carbon composite derived from sodium lignosulfonate for long cycle-life asymmetric supercapacitors. *Journal of Alloys and Compounds*, *900*, 163546. DOI: 10.1016/j.jallcom.2021.163546

Casaluci, S., Gemmi, M., Pellegrini, V., Di Carlo, A., & Bonaccorso, F. (2016). Graphene-based large area dye-sensitized solar cell modules. *Nanoscale*, *8*(9), 5368–5378. DOI: 10.1039/C5NR07971C PMID: 26883743

Chinnaiah, K., Kannan, K., Krishnamoorthi, R., Palko, N., Grishina, M., Chen, Y. S., & Gurushankar, K. (2024). Enhancing supercapacitor and antimicrobial performance of bioengineered Ag/Mn3O4 composite nanorods. *Journal of Industrial and Engineering Chemistry*.

Chinnaiah, K., Kannan, K., Chen, Y. S., & Gurushankar, K. (2025). Exploring Ag/Mn3O4 composite nanorods as an attractive battery-type electrode material for supercapacitors. *Journal of Physics and Chemistry of Solids*, *196*, 112310.

Chinnaiah, K., Kannan, K., Palko, N., Grishina, M., Chen, Y.-S., & Gurushankar, K. (2024). Exploring the potential of Withania somnifera-mediated Ag/Mn_3O_4 nanocomposites as electrode material for high-performance supercapattery device. *Journal of the Taiwan Institute of Chemical Engineers*, *157*, 105441. DOI: 10.1016/j.jtice.2024.105441

Kotla, D. P., Anna, V. R., Praveenkumar, S., Saleh, S. M., & Shanmugan, S. (2025). Optimizing solar still performance: A study of TiO2 nanofluid derived from Saccharum officinarum L. *Separation and Purification Technology*, *359*, 130584.

Edison, T. N. J. I., Atchudan, R., Karthik, N., Ganesh, K., Xiong, D., & Lee, Y. R. (2020). A novel binder-free electro-synthesis of hierarchical nickel sulfide nanostructures on nickel foam as a battery-type electrode for hybrid-capacitors. *Fuel*, *276*, 118077. DOI: 10.1016/j.fuel.2020.118077

Ellappan, B., Madhavan, V. R., Uzorka, A., & Shanmugan, S. (2025). Leveraging Grewia optiva leaf extract as a doping phase change material for enhanced solar desalination performance: Simulation prediction and diverse technologies. *Case Studies in Thermal Engineering*, •••, 106853.

Ghandourah, E. I., Sangeetha, A., Shanmugan, S., Zayed, M. E., Moustafa, E. B., Tounsi, A., & Elsheikh, A. H. (2022). Performance assessment of a novel solar distiller with a double slope basin covered by coated wick with lanthanum cobalt oxide nanoparticles. *Case Studies in Thermal Engineering*, *32*, 101859.

Gong, F., Xu, X., Li, Z., Zhou, G., & Wang, Z. S. (2013). $NiSe_2$ as an efficient electrocatalyst for a Pt-free counter electrode of dye-sensitized solar cells. *Chemical Communications (Cambridge)*, *49*(14), 1437–1439. DOI: 10.1039/C2CC38621F PMID: 23321537

Gurushankar, K., Karuppaiya, C., Tari, V., & Kannan, K. (2025). Exploring MXenes and MBenes in Metal Sulfur and Selenium Battery Technologies. In *Novel Energy Storage and Conversion Technologies for Two-Dimensional MXenes and MBenes* (pp. 105–120). IGI Global Scientific Publishing.

Hsu, Y. K., Mondal, A., Su, Y.-Z., Sofer, Z., Anuratha, K. S., & Lin, J. Y. (2022). Highly hydrophilic electrodeposited NiS/Ni_3S_2 interlaced nanosheets with surface-enriched Ni^{3+} sites as binder-free flexible cathodes for high-rate hybrid supercapacitors. *Applied Surface Science*, *579*, 151923. DOI: 10.1016/j.apsusc.2021.151923

Hu, Q., Zhang, S., Zou, X., Hao, J., Bai, Y., Yan, L., & Li, W. (2021). Coordination agent-dominated phase control of nickel sulfide for high-performance hybrid supercapacitor. *Journal of Colloid and Interface Science*, *607*, 45–52. DOI: 10.1016/j.jcis.2021.08.185 PMID: 34492352

Jia, J., Wu, J., Tu, Y., Huo, J., Zheng, M., & Lin, J. (2015). Transparent nickel selenide used as counter electrode in high efficient dye-sensitized solar cells. *Journal of Alloys and Compounds*, *640*, 29–33. DOI: 10.1016/j.jallcom.2015.03.233

Kannan, K., Chanda, D., Gautam, J., Behera, A., Meshesha, M. M., Jang, S. G., & Yang, B. (2023). Hydrothermally synthesized mixed metal oxide nanocomposites for electrochemical water splitting and photocatalytic hydrogen production. *International Journal of Hydrogen Energy*, *48*(93), 36412–36426.

Kannan, K., Chanda, D., Gautam, J., Meshesha, M. M., Jang, S. G., & Yang, B. L. (2023). Facial synthesis of pp heterojunction composites: Evaluation of their electrochemical properties with photovoltaics-electrolyzer water splitting using two-electrode system. *International Journal of Hydrogen Energy*, *48*(37), 13814–13826.

Kannan, K., Gautam, J., Chanda, D., Meshesha, M. M., Jang, S. G., & Yang, B. L. (2023). Two-dimensional MAX supported copper oxide/nickel Oxide/MAX as an efficient and novel photocatalyst for hydrogen evolution. *International Journal of Hydrogen Energy*, *48*(20), 7273–7283.

Khan, B. A., Rafaqat Hussain, A., Shah, A. M., Shah, M. Z. U., Ismail, J., Rahman, S., Sajjad, M., Assiri, M. A., Imran, M., & Javed, M. S. (2022). $NiSe_2$ nanocrystals intercalated rGO sheets as a high-performance asymmetric supercapacitor electrode. *Ceramics International*, *48*(4), 5509–5517. DOI: 10.1016/j.ceramint.2021.11.095

Kumar, Ravinder, Chanda, Jayanta, Elsheikh, Ammar H., Ongar, Bulbul, Khidolda, Yerkin, Kumar, Seepana Praveen, & Panchal, Hitesh, & Shanmugan, S. (2023). Performance improvement of single and double effect solar stills with silver balls/nanofluids for bioactivation: An experimental analysis. *Solar Energy*, *259*, 452–463. DOI: 10.1016/j.solener.2023.05.012

Li, Z., Wu, R., Bao, E., Du, S., Xu, C., Zhu, J., Mao, H., & Chen, H. (2023). Assembly of high-performance hybrid supercapacitors using $FeCo_2O_4$ microflowers as battery-type cathode materials. *Ceramics International*, *49*, 10411–10419. DOI: 10.1016/j.ceramint.2022.11.223

Lu, S., Yang, W., Zhou, M., Qiu, L., Tao, B., Zhao, Q., Wang, X., Zhang, L., Xie, Q., & Ruan, Y. (2022). Nitrogen- and oxygen-doped carbon with abundant micropores derived from biomass waste for all-solid-state flexible supercapacitors. *Journal of Colloid and Interface Science*, *610*, 1088–1099. DOI: 10.1016/j.jcis.2021.11.164 PMID: 34876262

Mahmood, N., Zhang, C., & Hou, Y. (2013). Nickel sulfide/nitrogen-doped graphene composites: Phase-controlled synthesis and high performance anode materials for lithium ion batteries. *Small*, *9*(8), 1321–1328. DOI: 10.1002/smll.201203032 PMID: 23494938

Meshesha, M. M., Gautam, J., Chanda, D., Kannan, K., Jang, S. G., & Yang, B. L. (2022). Effect of reaction temperature on morphology and photoelectrochemical performance of titanium dioxide. *Materials Letters*, *329*, 133176. DOI: 10.1016/j.matlet.2022.133176

Meshesha, M. M., Kannan, K., Chanda, D., Gautam, J., Jang, S. G., & Yang, B. L. (2022). Remarkable photoelectrochemical activity of titanium dioxide nanorod arrays sensitized with transition metal sulfide nanoparticles for solar hydrogen production. *Materials Today. Chemistry*, *26*, 101216. DOI: 10.1016/j.mtchem.2022.101216

Mousa, G., Basem, A., Shanmugan, S., Karmiris-Obratański, P., Faqeha, H., Alshareef, R. S., & Elsheikh, A. H. (2024). Harnessing fluorescence resonance energy transfer for improved solar still performance with zinc oxide nanoparticles and activated carbon. *Applied Materials Today*, *38*, 102196.

Nabera, A., Istrate, I. R., Martín, A. J., Pérez-Ramírez, J., & Guillén-Gosálbez, G. (2023). Energy crisis in Europe enhances the sustainability of green chemicals. *Green Chemistry*, *25*(17), 6603–6611. DOI: 10.1039/D3GC01053H PMID: 38013722

Alrasheedi, N. H., Saravanavalli, K. J., Panchal, H., & Shanmugan, S. (2025). Using Aluminum Cans and Conch Shells to Increase Thermic Efficiency of a solar distiller: A Comparison Study with Decision Tree and Random Forest Predictions. *Case Studies in Thermal Engineering*, •••, 106984.

Nashmi, H. (2025). Alrasheedi, Padamata Varshini, Afam Uzorka, & Shanmugan, S., (2025). Impact of microwave irradiation on the thermophysical and energy storage properties of MXene-Ag-Syzygiumcumini dye nanofluids for solar distillation. *Energy Nexus*, *19*, 100525. DOI: 10.1016/j.nexus.2025.100525

Nie, T., Fang, Z., Ren, X., Duan, Y., & Liu, S. (2023). Recent advances in wide-bandgap organic-inorganic halide perovskite solar cells and tandem application. *Nano-Micro Letters*, *15*(1), 70. DOI: 10.1007/s40820-023-01040-6 PMID: 36943501

Palaniappan, M., El-Shafay, A. S., & Shanmugan, S. (2024). Improving heat retention properties of steeped M-shape basin solar distillers utilizing paraffin RT50-enhanced silver nanoparticles and Manihot esculenta extracts. *Desalination*, *586*, 117836.

Peng, Q., Yang, W., Li, X., Yang, H., Cao, J., Wang, J., Zheng, Y., Pan, J., & Li, C. (2023). The NiS/CdS hollow octahedral composites towards enhanced photocatalytic hydrogen evolution and photodegradation via active sites synergism. *International Journal of Hydrogen Energy*, *48*, 36427–36438. DOI: 10.1016/j.ijhydene.2023.06.045

Premkumar, N., RadhaMadhavi, M., Kitmo, K, & Shanmugan. S. (2024). Utilizing the lignocellulosicfbers from Pineapple Crown Leaves extract for enhancing TiO_2 interfacial bonding in dye-sensitized solar cell photoanodes. *Materials for Renewable and Sustainable Energy*, *13*, 13–25. DOI: 10.1007/s40243-023-00245-4

Qian, X., Liu, H., Huang, Y., Ren, Z., Yu, Y., Xu, C., & Hou, L. (2019). Co-Ni-MoSx yolk-shell nanospheres as superior Pt-free electrode catalysts for highly efficient dye-sensitized solar cells. *Journal of Power Sources*, *412*, 568–574. DOI: 10.1016/j.jpowsour.2018.11.067

Rajasekar, R., SenthilKumar, M., Shanmugan, S., & Nagarajan, M. (2022). The influence of Cu2ZnSnS4 thin films with characteristics of treatment conditions on spray pyrolysis technique for solar cells applications. *Indian Journal of Physics*, *96*(3), 707–716.

Maurya, A., Shanmugan, S., Kabeel, A. E., Akanovna, Z. A., Sarsenbayev, Y., & Panchal, H. (2024). Use of absorber plate built of ZnO/PVC/Bioactivation modified epoxy nanocomposites to improvement of double-effect Solar Distiller productivity analyzing the Energy, Exergo-environment and Enviro-economical. *Journal of Cleaner Production*, *434*, 139601.

Sai, G., Anna, V. R., Swapna, K., Prasad, M. S., Ch, R. P., Krishna, B. C., & Shanmugan, S. (2025). Evaluating the effects of sugarcane juice-mediated ZnO nanofluids on solar light activation for enhancing double-slope solar still performance. *Applied Materials Today*, *42*, 102542.

Salam, Z., Vijayakumar, E., Subramania, A., Sivasankar, N., & Mallick, S. (2015). Graphene quantum dots decorated electrospun TiO_2 nanofibers as an effective photoanode for dye sensitized solar cells. *Solar Energy Materials and Solar Cells*, *143*, 250–259. DOI: 10.1016/j.solmat.2015.07.001

Sangeetha, A., Shanmugan, S., & Gorjian, S. (2022). Experimental evaluation and thermodynamic Gibbs free energy analysis of a double-slope U-shaped stepped basin solar still using activated carbon with ZnO nanoparticlesJournal of Cleaner Production. 380, Part 2, 135118. https://doi.org/DOI: 10.1016/j.jclepro.2022.135118

Sangeetha, A., & Shanmugan, S., & Alasiri, A. (2024). ZnO/nZVI Nanoparticle-Enhanced Double-Slope U-Shaped Solar Distillation: A Thermodynamic Investigation of Cephalexin Adsorption. *Materials Today Sustainability., 100983*. Advance online publication. DOI: 10.1016/j.mtsust.2024.100983

Sarma, L., & Thirumal, M. (2021). Coupling of $MoSe_2$ and graphitic carbon nitride (g-C_3N_4) by facile hydrothermal route for enhanced visible-light photocatalytic properties. *Materials Today: Proceedings, 36*, 679–688. DOI: 10.1016/j.matpr.2020.04.649

Tamilselvan, S. N., Shanmugan, S., Basem, A., Faqeha, H., Alsaati, A. A., Rajkumar, V., & Elsheikh, A. H. (2024). Revolutionizing dye-sensitized solar cells: Chassalia curviflora fruit extract as photosensitizer and charge recombination inhibitor. *Inorganic Chemistry Communications, 170*, 113292.

Tamilselvan, S. N., & Shanmugan, S. (2024). Towards sustainable solar cells: Unveiling the latest developments in bio-nano materials for enhanced DSSC efficiency. *Clean Energy, 8*(3), 238–257.

Shahzad, N., Perveen, T., Pugliese, D., Haq, S., Fatima, N., Salman, S. M., & Shahzad, M. I. (2022). Counter electrode materials based on carbon nanotubes for dye-sensitized solar cells. *Renewable & Sustainable Energy Reviews, 159*, 112196. DOI: 10.1016/j.rser.2022.112196

Shanmugan, S., Madupu, R.K., Selvaraju, P., Ravichandran, S. (2022). Systematic growth on antibacterial activities use of silver nanoparticles with citrus aurantifolia. *Materials Today: Proceedings, 51*(Part 1), 998–1005. DOI: 10.1016/j.matpr.2021.07.055

Shanmugan, S., Djuansjah, J., Ahmadein, M., Alsaleh, N.A., Parsa, S.M., & Elsheikh, A.H. (2023). Chemical potential of different phases inside the pyramid stepped basin solar still through Gibbs free energy. *Case Studies in Thermal Engineering, 49*, 103277. DOI: 10.1016/j.csite.2023.103277

Shanmugan, S., Selvaraju, P., Nagaraj, J., Sivakumar, S., & Ravichandran, S. (2022). Biogenic silver nanoparticles of antibacterial activities for poly-herbal extracts in novel medicine. *Materials Today: Proceedings, 51*(Part 1), 1107–1114. DOI: 10.1016/j.matpr.2021.07.107

Shanmugan, S., Selvaraju, P., Sivakumar, S., Nagaraj, J., Srinivas Prasad, M. V. V. K., & Abhiram, N. (2022). Solar cells absorption viewpoint of Mie theory: Experimental analysis of TiO_2 doping V/Ce. *Materials Today: Proceedings, 51*(Part 1), 1124–1128. DOI: 10.1016/j.matpr.2021.07.110

Tao, H., Zhou, M., Wang, K., Cheng, S., & Jiang, K. (2017). Nickel sulfide nanospheres anchored on reduced graphene oxide in situ doped with sulfur as a high performance anode for sodium-ion batteries. *Journal of Materials Chemistry. A, Materials for Energy and Sustainability*, *5*(19), 9322–9328. DOI: 10.1039/c7ta02194

Wang, S., He, P., Jia, L., He, M., Zhang, T., Dong, F., Liu, M., Liu, H., Zhang, Y., Li, C., & Gao, J. (2019). Nanocoral-like composite of nickel selenide nanoparticles anchored on two-dimensional multi-layered graphitic carbon nitride: A highly efficient electrocatalyst for oxygen evolution reaction. *Applied Catalysis B: Environmental*, *243*, 463–469. DOI: 10.1016/j.apcatb.2018.10.071

Wang, X., Xie, Y., Pan, K., Wu, J., Xiao, Y., Yu, P., Zhou, W., & Fu, H. (2018). Morphology effect of NiSe hierarchical microspheres on the performance of dye-sensitized solar cells. *ACS Applied Nano Materials*, *1*(9), 4900–4909. DOI: 10.1021/acsanm.8b01069

Xu, X. B., Zhong, W., Zhang, X., Dou, J., Xiong, Z. G., Sun, Y., Wang, T. T., & Du, Y. W. (2019). Flexible symmetric supercapacitor with ultrahigh energy density based on NiS/MoS_2@N-rGO hybrids electrode. *Journal of Colloid and Interface Science*, *543*, 147–155. DOI: 10.1016/j.jcis.2019.02.054 PMID: 30797998

Zhao, J., Guan, B., Hu, B., Xu, Z., Wang, D., & Zhang, H. (2017). Vulcanizing time-controlled synthesis of NiS microflowers and its application in asymmetric supercapacitors. *Electrochimica Acta*, *230*, 428–437. DOI: 10.1016/j.electacta.2017.02.032

Chapter 11
Cost-Effective Chalcogenide-Based Nanocomposite Electrodes for High-Performance Electrochemical and Photoelectrochemical Water Splitting

P. Selvakumar
https://orcid.org/0000-0002-3650-4548
Department of Science and Humanities, Nehru Institute of Technology, Coimbatore, India

T. C. Manjunath
https://orcid.org/0000-0003-2545-9160
Rajarajeswari College of Engineering, India

Santosh Kumar Nathsharma
https://orcid.org/0000-0003-3935-2938
Stewart Science College, India

DOI: 10.4018/979-8-3373-3962-7.ch011

ABSTRACT

Water splitting, also known process with immense potential in the global shift toward sustainable energy systems. This process involves breaking down water (H_2O) into its, wind, or hydropower. As the world seeks alternatives to fossil fuels to mitigate climate change and reduce greenhouse gas emissions, hydrogen has emerged as a promising energy carrier due to its high energy content and environmentally friendly combustion, which produces only water as a by-product.demonstrating the decomposition of water using direct current electricity. Since then, the field has evolved significantly, driven by advancements in electrochemical systems, materials science, and catalysis. materials science, and catalysis. Modern water splitting can be broadly categorized into three primary methods: electrolysis, photoelectrochemical (PEC) water splitting, and thermochemical water splitting. Among these, electrolysis is the most mature and widely researched technique, involving the application of electrical energy to drive the non-spontaneous reaction.

INTRODUCTION TO WATER SPLITTING AND HYDROGEN PRODUCTION

Water splitting, also known process with immense potential in the global shift toward sustainable energy systems. This process involves breaking down water (H_2O) into its wind or hydropower. As the world seeks alternatives to fossil fuels to mitigate climate change and reduce greenhouse gas emissions, hydrogen has emerged as a promising energy carrier due to its high energy content and environmentally friendly combustion, which produces only water as a by-product, (Tari, 2015a; Tari, 2015b). demonstrates the decomposition of water using direct current electricity. Since then, the field has evolved significantly, driven by advancements in electrochemical systems, materials science, and catalysis. materials science, and catalysis, (Tari et al., 2022). Modern water splitting can be broadly categorized into three primary methods: electrolysis, photoelectrochemical (PEC) water splitting, and thermochemical water splitting, (Abbas& Bang, 2015). Among these, electrolysis is the most mature and widely researched technique, involving the application of electrical energy to drive the non-spontaneous reaction, (Kannan & Tari, 2025). In a typical electrolyzer, water molecules at the anode are oxidized to form oxygen gas and protons, while at the cathode, these protons are reduced to form hydrogen gas. This process is facilitated by catalysts that lower the activation energy and improve overall efficiency. Photoelectrochemical water splitting, on the other the redox reactions required for splitting water. This method holds promise for direct solar-to-hydrogen conversion, but challenges remain in terms of material stability, efficiency, and cost. Thermochemical water splitting utilizes high-temperature heat—often sourced from concentrated solar power or nuclear reactors—to drive chemical cycles that ultimately produce hydrogen and oxygen. These cycles, such as the sulfur-iodine cycle, involve a series of reactions that regenerate the original reactants, making the process cyclic and theoretically efficient holds particular significance in achieving net-zero emission goals because it enables the decarbonization of hard-to-electrify sectors such as steelmaking, heavy transportation, and industrial heating. PEM electrolyzers offer higher efficiency and compact design but require expensive materials such as platinum and iridium for catalysts and components. Innovations in materials science are playing a crucial role in overcoming these limitations, (Gurushankar et al., 2024). Researchers are exploring new catalyst materials based on earth-abundant elements like nickel, cobalt, iron, and manganese. engineering challenge. Due to the variable output of solar and wind power, water splitting systems must be capable of flexible and intermittent operation without significant losses in efficiency or durability, (Al-Shetwi, 2022).

Hybrid systems that combine energy storage, smart grids, and predictive control algorithms are being developed to address this issue. Another promising approach is the use of solar-to-hydrogen systems that directly convert sunlight into chemical energy via PEC or photocatalytic methods, potentially bypassing the need for electricity generation and reducing system complexity.CO_2 to produce synthetic fuels and chemicals. In this way, green hydrogen acts as a cornerstone of the circular carbon economy and a key enabler of sector coupling, linking the power, heating, transportation, and industrial sectors in a unified energy ecosystem. Governments and industries worldwide are increasingly investing in hydrogen technologies as part of their decarbonization strategies. Major economies have announced hydrogen roadmaps, funding initiatives, and international collaborations aimed at scaling up green hydrogen production. The European Union, Japan, South Korea, Australia, and the United States, among others, are supporting research and pilot projects, while private companies are building gigawatt-scale electrolyzer factories and hydrogen refueling infrastructure. The International Energy Agency (IEA) and other global bodies have emphasized the role of hydrogen in meeting the targets of the Paris Agreement and

SYNTHESIS AND CHARACTERIZATION OF CHALCOGENIDE-BASED NANOCOMPOSITES

The performance, stability, and cost of these electrodes greatly influence the overall efficiency, durability, and economic feasibility of water electrolyzers. As the global energy sector accelerates its transition toward renewable sources and seeks scalable solutions for green hydrogen production, the demand for affordable, efficient, and robust electrode materials has never been more urgent. Addressing this need is pivotal not only for improving the competitiveness of water splitting but also for reducing reliance on fossil fuel-derived hydrogen. Traditionally, high-performance electrodes have relied on noble metals that exhibit excellent catalytic activity and long-term stability under electrochemical conditions, making them the materials of choice in commercial proton exchange membrane (PEM) electrolyzers. However, their limited availability, high cost, and vulnerability to market fluctuations pose significant barriers to the scalability and economic sustainability of hydrogen production. For instance, platinum and iridium are among the rarest elements hydrogen production facilities to meet future energy demands. As a result, researchers are increasingly focused on discovering and optimizing cost-effective, earth-abundant alternatives that can match or approach the performance of these precious metals. The role of cost-effective electrodes extends beyond mere economic considerations; it is fundamentally tied to the technical viability and environmental footprint of water splitting systems. A cost-effective electrode must not only be inexpensive but also deliver high catalytic efficiency, long-term operational stability, and compatibility with diverse electrolyzer architectures, (Augustynski et al., 2011). This includes alkaline, PEM, and anion exchange membrane (AEM) electrolyzers, each of which imposes unique requirements on electrode composition, conductivity, and corrosion resistance. In alkaline systems, for example, non-noble metal catalysts such as nickel (Ni), cobalt (Co), iron (Fe), and their oxides, sulfides, phosphides, and nitrides have shown promising results. These materials can be synthesized using low-cost methods and readily integrated into existing industrial processes. Nickel-based catalysts, in particular, have demonstrated strong HER and OER activity in alkaline media and remain a major focus for large-scale alkaline water electrolysers. Improving the catalytic performance of non-precious metal electrodes through various strategies, including nano structuring, alloying, doping, and surface modification. Nanostructured materials offer increased surface area and active sites, enhancing catalytic efficiency. Alloying transition metals can optimize electronic structures and synergistically improve activity and stability, (Bhanja et al., 2020) Doping with heteroatoms like nitrogen, sulfur, or phosphorus introduces defects and modulates the electronic environment, often leading to better catalytic kinetics. For example, NiFe layered double hydroxides (LDHs) have emerged as some of the most efficient non-noble metal OER catalysts in alkaline media, exhibiting performance comparable to iridium-based catalysts at a fraction of the cost. Similarly, cobalt phosphide (CoP) and molybdenum disulfide (MoS_2) have shown excellent HER activity under certain conditions and are actively being explored for integration into scalable electrode designs.

Beyond material composition, electrode architecture plays a vital role in determining overall system efficiency. Innovative designs, such as porous and 3D-structured electrodes, enhance gas bubble release, improve mass transport, and reduce energy losses due to ohmic resistance and concentration gradients. (Chen et al., 2019) The use of conductive supports, such as carbon-based materials, metal foams, and conductive polymers, also contributes to the improved mechanical strength and electrical connectivity of the electrodes. Moreover, integrating catalysts directly onto the support material via in situ growth or electrodeposition techniques can lower fabrication costs and simplify manufacturing processes. Such approaches are instrumental in creating scalable, cost-effective electrodes suitable for industrial electrolyzer systems. electrolyzers offer a middle ground between the mature alkaline and PEM systems by combining the low-cost catalyst and system components of alkaline cells with the compactness and high current densities of PEM cells. However, the long-term stability of non-noble catalysts in AEM environments remains a key challenge. Solid oxide electrolyzers, which operate at high temperatures and offer thermodynamic advantages, require electrode materials that can withstand harsh oxidative and reductive conditions while maintaining catalytic activity. In both cases, the search for low-cost yet durable electrode materials is essential for commercialization. The economic implications of cost-effective electrodes ripple across the entire hydrogen value chain. Lower capital and operational expenditures in water electrolysis directly translate into reduced hydrogen production costs, making green hydrogen more competitive with grey and blue hydrogen, (Chib et al., 2025). This cost parity is critical for accelerating hydrogen adoption in sectors like transportation, chemical manufacturing, and grid energy storage. As of now, green hydrogen is still significantly more expensive than hydrogen produced from natural gas; thus, reducing the cost of electrolyzer components, especially electrodes, is a key lever for closing this gap. Several techno-economic analyses have highlighted that improvements in electrode materials can lead to substantial reductions in the levelized cost of hydrogen (LCOH), particularly when combined with economies of scale, improved system integration, and increased renewable electricity penetration. Environmental sustainability is another important dimension in the pursuit of cost-effective electrodes. The mining and processing of precious metals have significant ecological footprints, including high energy consumption, toxic by-products, and social concerns related to labor practices and geopolitical supply risks. Transitioning to abundant and non-toxic materials aligns with circular economy principles and reduces the environmental burden associated with electrode manufacturing and end-of-life disposal. Recyclability, ease of recovery, and life cycle performance are increasingly being incorporated into the design criteria for next-generation electrode materials, ensuring that water splitting remains a genuinely clean and sustainable hydrogen production pathway. In addition to technical and economic efforts, policy support and industrial collaboration are crucial to accelerating the development and deployment of cost-effective electrodes. Public funding for materials research, pilot

ADVANTAGES OF CHALCOGENIDE-BASED NANOCOMPOSITES IN WATER SPLITTING

Chalcogenide-based nanocomposites have emerged as a fascinating class of materials with versatile applications in energy conversion, catalysis, electronics, and optoelectronics, owing to their unique physicochemical properties derived from the incorporation of chalcogen elements—sulfur (S), selenium (Se), and tellurium (Te). These nanocomposites are typically composed of transition metal chalcogenides (TMCs) integrated with other nanostructures, such as carbonaceous materials, metal nanoparticles, or oxides, to form hybrid systems with tailored electrical, optical, and catalytic behaviors. In the context of water splitting, particularly hydrogen evolution reactions (HER) and oxygen evolution reactions (OER), chalcogenide-based nanocomposites offer a promising alternative to noble metal catalysts due to their earth-abundance, tunable band structures, high surface area, and catalytic activity, (Dahiya et al., 2025). The synthesis and characterization of these materials are thus critical areas of research aimed at optimizing performance through precise control of morphology, composition, and electronic properties. A variety of synthesis techniques are employed to fabricate chalcogenide-based nanocomposites, each offering different advantages in terms of scalability, cost, crystallinity, and compositional control. Among the most widely used methods is pressures, enabling the formation of highly crystalline nanostructures with well-defined morphologies such as nanosheets, nanorods, nanospheres, and hollow structures. This technique is particularly effective for synthesizing metal sulfides like MoS_2, WS_2, NiS, and CoS with layered structures conducive to catalytic applications. Parameters such as reaction temperature, time, solvent composition, and precursor concentration can be finely tuned to modulate particle size, surface area, and phase purity suited for applications in electronics and photovoltaics. CVD involves the deposition of gaseous reactants onto a substrate at elevated temperatures to form thin films with uniform thickness and controlled orientation. This method is widely used to synthesize 2D chalcogenide materials like MoS_2 and WS_2 with exceptional electrical and mechanical properties. Although CVD offers excellent control over material quality, it often requires expensive equipment and stringent reaction conditions, limiting its scalability for bulk applications. used for synthesizing bulk chalcogenide powders or composites. These methods involve the decomposition of precursor salts or the direct reaction of metal and chalcogen powders at high temperatures under inert or reducing atmospheres. While these techniques are less precise in controlling morphology, they are suitable for large-scale production and allow facile integration with other functional materials, including metal oxides and conductive substrates. thickness, composition, and surface morphology by adjusting parameters such as potential, current density, and electrolyte composition. It is particularly valuable for developing electrode facilitate the integration of chalcogenides with carbon-based materials such as graphene, carbon nanotubes (CNTs), and reduced graphene oxide (rGO), resulting in nanocomposites with enhanced electrical conductivity, mechanical strength, and catalytic performance. The synergistic interaction between chalcogenides and carbon frameworks improves electron transfer, prevents particle agglomeration, and exposes more active sites. nanoscalroutinely used to identify the crystalline phases and assess the phase purity of the synthesized materials. The presence of sharp, well-defined peaks indicates high crystallinity, while broadening of peaks can signify nanostructuring or amorphous content. provide detailed insights into the morphology, particle size distribution, and surface texture of the nanocomposites. TEM, especially high-resolution TEM (HRTEM), is invaluable for visualizing lattice fringes, interlayer spacing, and interfaces between chalcogenides and other components in the composite.

ELECTROCHEMICAL WATER SPLITTING PERFORMANCE OF CHALCOGENIDE-BASED NANOCOMPOSITES

Chalcogenide-based nanocomposites have garnered significant attention as promising materials for water splitting applications, offering a viable alternative to traditional noble-metal catalysts. Comprising compounds that include at least one chalcogen element—sulfur (S), selenium (Se), or tellurium (Te)—combined with transition metals and often hybridized with conductive supports or co-catalysts, these nanocomposites exhibit a unique combination of catalytic, electronic, and structural properties. Their diverse material chemistry, abundant availability, and compatibility with scalable fabrication techniques make them particularly attractive for use in both the hydrogen evolution reaction (HER) and the oxygen evolution reaction (OER), the two half-reactions that comprise water splitting. The advantages of these materials span a broad spectrum, including high catalytic efficiency, structural tunability, cost-effectiveness, and environmental sustainability, all of which support their potential for deployment in next-generation green hydrogen production technologies. configurations and structural characteristics. For instance, layered transition metal dichalcogenides (TMDs) like MoS_2, WS_2, and their selenide analogues possess a sandwiched structure with active edge sites and relatively inert basal planes, (Eidsvåg et al., 2021). This edge activity is particularly advantageous for HER, where hydrogen adsorption (proton reduction) occurs at catalytic sites. When these materials are exfoliated or synthesized as few-layer nanosheets, the number of accessible active sites increases dramatically, leading to improved reaction kinetics and lower overpotentials. Furthermore, the incorporation of defects, vacancies, and heteroatom doping can further tailor their electronic structure, enhancing conductivity and catalytic performance. supports. For example, combining MoS_2 with conductive materials like reduced graphene oxide (rGO) or carbon nanotubes (CNTs) can drastically improve the charge transport pathways, minimize electron-hole recombination, and increase mechanical stability. These hybrids offer a dual advantage: the high surface area and conductivity of the carbon component enhances electron transfer, while the chalcogenide domains provide catalytically active sites. Similarly, decorating chalcogenides with metal nanoparticles (e.g., Ni, Co, or Cu) can induce interfacial charge transfer and create new active centers, further amplifying overall performance. This composite approach is not only efficient but also opens a broad range of possibilities for customizing materials for specific electrochemical environments, whether acidic, alkaline, or neutral. (Ru), which are prohibitively expensive and scarce.

The key transition metals used in chalcogenides—such as molybdenum (Mo), tungsten (W), cobalt (Co), nickel (Ni), and iron (Fe)—are much more readily available and less expensive, making large-scale deployment economically feasible. As demand for green hydrogen rises globally, the development of cost-effective catalysts that do not rely on critical raw materials is paramount for scalability and long-term viability, (Fo et al., 2020). Chalcogenide-based nanocomposites meet this requirement while delivering competitive catalytic performance, especially when properly engineered. center, enhancing hydrogen adsorption and desorption dynamics. Similarly, tuning the stoichiometry or phase (e.g., from semiconducting 2H-phase to metallic 1T-phase in MoS_2) allows optimization of electrical conductivity and catalytic activity. These modifications enable fine control over catalytic performance, allowing researchers to target specific operating conditions such as high current densities or long-term stability under fluctuating loads. particularly when combined with oxides or hydroxides (e.g., NiFe LDH), display enhanced OER performance in alkaline media. Their ability to form stable intermediate species (such as M–OOH) facilitates the reaction pathway, reducing overpotentials and improving catalytic durability, (Le Bahers &Takanabe, 2019). Moreover, the structural flexibility of nanocomposites enables the design of bifunctional catalysts capable of performing both HER and OER in the same electrochemical cell, thereby simplifying system architecture and reducing overall material costs. these materials demonstrate good tolerance to acidic and alkaline electrolytes, high current densities, and elevated temperatures—conditions typical in industrial-scale water electrolyzers. The incorporation of robust carbonaceous scaffolds or oxide matrices can further improve their resistance to corrosion and structural degradation. This stability, combined with the ability to operate efficiently over long periods, positions chalcogenide nanocomposites as viable candidates for commercial-scale applications. From an environmental when low-temperature, solvent-free, or green chemistry methods are employed in their fabrication. Their use reduces reliance on toxic and non-renewable materials, while their recyclability and

MECHANISTIC INSIGHTS INTO ELECTROCHEMICAL WATER SPLITTING

The electrochemical water splitting performance of chalcogenide-based nanocomposites has emerged as a major focus in the pursuit of efficient and scalable hydrogen production technologies. These evolution reaction (OER). As water splitting involves the simultaneous execution of these two half-reactions under an external electric field, the efficiency of the overall process is governed by the kinetics of electron transfer, the availability of active sites, the conductivity of the electrode materials, and their resistance to degradation. Evaluating the electrochemical performance of chalcogenide-based nanocomposites involves an array of key metrics, including overpotential, (Guo, Zhou, Tang et al., 2020) Tafel slope, exchange current density, faradaic efficiency, turnover frequency (TOF), and long-term stability—all of which provide insight into the, MoS_2-rGO nanocomposites have been shown to catalyze HER with overpotentials as low as 100–150 mV at 10 mA cm^{-2}, a benchmark current density relevant to practical electrolyzers. Similarly, $NiSe_2$ and CoS-based composites have achieved OER overpotentials in the range of 250–350 mV, which, while higher than state-of-the-art IrO_2 or RuO_2 catalysts, remain competitive when considering the cost and scalability advantages. These overpotentials are heavily influenced by material morphology, crystallinity, defect density, and electronic structure—factors that can be systematically optimized through material design. Lower Tafel slopes suggest more favorable reaction kinetics and faster electron transfer. For HER, ideal catalysts exhibit Tafel slopes below 30 mV dec^{-1} (comparable to Pt/C), but many chalcogenide-based nanocomposites offer values ranging between 40 and 120 mV dec^{-1}. For instance, 1T-phase MoS_2 and CoP/MoS_2 heterostructures exhibit Tafel slopes close to 50 mV dec^{-1}, indicating enhanced HER kinetics due to increased density of active edge sites and improved electrical conductivity. In OER, NiFe/CoS hybrids have demonstrated Tafel slopes as low as 40–60 mV dec^{-1} in alkaline media, benefiting from synergistic interactions that facilitate oxygen intermediate formation and release. equilibrium, serves as another crucial measure of electrocatalyst activity. Higher j_0 values correspond to more active catalysts. Chalcogenide-based nanocomposites typically demonstrate j_0 values in the range of 10^{-3} to 10^{-1} mA cm^{-2}, depending on the synthesis conditions and structural tuning. While these values may fall short of noble metal benchmarks, continued optimization—especially through nanoscale engineering, doping, and hybridization—has shown great promise in bridging the performance gap.

PHOTOELECTROCHEMICAL WATER SPLITTING PERFORMANCE OF CHALCOGENIDE-BASED NANOCOMPOSITES

These reactions are fundamental to producing green hydrogen through water electrolysis and are governed by intricate reaction mechanisms, involving electron transfer steps, adsorption/desorption of intermediates, and catalytic surface interactions. Understanding the mechanistic pathways of HER and OER is essential for designing efficient and durable electrocatalysts, such as chalcogenide-based nanocomposites, that can lower overpotentials, increase reaction rates, and enhance system efficiency. Insights into reaction kinetics, thermodynamics, and interfacial phenomena at the molecular and atomic levels offer a foundation for rational catalyst design and optimization. molecules to form molecular hydrogen (H_2). The reaction mechanism varies slightly depending on the pH of the electrolyte. In acidic media, the HER generally follows either a two-step or three-step pathway involving the Volmer, Heyrovsky, and Tafel reactions: energy (neither too strong nor too weak) promotes rapid desorption of hydrogen atoms, favoring the Heyrovsky or Tafel pathway. Platinum is the gold standard for HER catalysis due to its near-zero hydrogen adsorption free energy ($\Delta G\_H^* \approx 0$), but non-precious alternatives like MoS_2, especially in its metallic 1T phase or doped forms, have shown similar binding characteristics at edge sites. In chalcogenide-based nanocomposites, active HER mechanisms are driven by the exposure of edge or defect sites, often enriched through nanostructuring, doping, or compositing with conductive materials like graphene, (Hu et al., 2020). These modifications enhance charge transfer and facilitate proton reduction. Additionally, heterostructures combining transition metal phosphides, nitrides, or carbides with chalcogenides can generate electronic interactions that modulate ΔG_H^*, improving reaction kinetics. Surface oxygen or sulfur vacancies in materials like MoS_2, CoS, or $NiSe_2$ also act as catalytic centers, lowering the activation energy barrier for hydrogen evolution OH^- adsorption and oxidation to OH*Each of these steps requires overcoming kinetic barriers, particularly the formation of OOH*, which is often the RDS due to the high energy associated with creating the O–O bond. The ability of a catalyst to stabilize these intermediates is crucial for efficient OER performance. In chalcogenide-based materials, transition metals such as Co,

Ni, or Fe are the primary active sites, while the chalcogen elements influence the electronic environment through ligand effects and charge redistribution. NiFe and CoFe chalcogenides, especially in sulfide or selenide forms, are known to undergo surface oxidation under OER conditions, forming oxyhydroxide layers (e.g., NiOOH, CoOOH) that act as the true catalytic phase. This dynamic surface reconstruction is a key mechanistic feature, highlighting the importance of operando studies—such as in situ Raman, XPS, or XAS—to identify the real active sites. The hybridization of chalcogenides with conductive carbon substrates or metal oxides enhances the OER kinetics by promoting charge mobility and synergistic electron interactions between components. and OER in the same electrolyte are highly desirable. Chalcogenide-based nanocomposites, especially those incorporating cobalt, nickel, or iron, exhibit this dual activity. For example, $NiCo_2S_4$, CoS/NiFe-LDH, and MoS_2/FeSe hybrids have demonstrated low overall cell voltages (<1.6 V at

ENHANCING PHOTOELECTROCHEMICAL PERFORMANCE THROUGH MATERIALS DESIGN

as it integrates solar energy harvesting with catalytic water decomposition in a single system. The heart of PEC water splitting lies in the use of semiconducting photoelectrodes—either photoanodes, tellurides, either individually or combined with other functional materials, have emerged as attractive candidates due to their unique optoelectronic properties, suitable bandgaps, tunable surface chemistry, and cost-effective synthesis. These materials exhibit strong light absorption, narrow-to-moderate bandgaps (1.1–2.5 eV), high catalytic activity, and excellent compatibility with various charge transfer and co-catalyst systems, making them excellent platforms for PEC water splitting, (Isaev et al., 2022). Evaluating their performance involves examining key parameters such as photocurrent density, onset potential, incident photon-to-current This value reflects the material's ability to absorb photons, generate electron-hole pairs, and transfer charges to catalyze the water splitting reactions. Chalcogenide-based photoelectrodes like MoS_2, CdS, Bi_2S_3, $ZnIn_2S_4$, and $CuInSe_2$ often exhibit photocurrent densities in the range of 1 to 10 mA cm^{-2}, depending on composition, thickness, surface area, and integration with co-catalysts. For instance, MoS_2-based photocathodes hybridized with reduced graphene oxide or Ni nanoparticles have achieved photocurrent densities exceeding 5 mA cm^{-2} at 0 V vs. RHE, while CdS/TiO_2 core–shell photoanodes have shown similar values under low applied bias. **Onset potential**, the minimum potential at which a photocurrent begins to flow, is another critical indicator, with more positive (for photoanodes MoS_2 (1.8 eV), WS_2 (1.9 eV), and Bi_2S_3 (1.3 eV) exhibit strong absorption across the visible region, enabling high photogenerated charge carrier densities. Bandgap tuning via alloying or doping (e.g., Zn-doped CdS or Se substitution in MoS_2) allows the optimization of light absorption to increase solar-to-hydrogen efficiency. Unlike wide-bandgap oxides (e.g., TiO_2 or ZnO), which absorb primarily in the UV region, chalcogenides can harness a broader portion of sunlight, increasing the theoretical photocurrent limit. Moreover, the use of heterojunction structures—such as $ZnIn_2S_4$/CdS, MoS_2/g-C_3N_4, or Bi_2S_3/TiO_2—further improves light harvesting and charge separation by creating internal electric fields and staggered band alignments (type-II or Z-scheme systems). suppress recombination and accelerate carrier transport. Nanostructuring (e.g., forming nanorods, nanosheets, or hollow spheres) shortens the diffusion path for carriers, while surface modification with passivation layers (such as Al_2O_3, ZnS, or graphene oxide) reduces trap states. When chalcogenides are combined with conductive materials like carbon nanotubes, rGO, or metal foils, they form composite electrodes that facilitate rapid electron migration and enhance current collection. Additionally, doping with elements such as Ni, Co, or Fe introduces energy levels that promote charge separation and tune Fermi levels, further boosting PEC activity. Transient photocurrent and electrochemical impedance spectroscopy (EIS) studies have confirmed that these modifications reduce charge transfer resistance and increase carrier lifetimes.

COST-EFFECTIVE SYNTHESIS METHODS FOR CHALCOGENIDE-BASED NANOCOMPOSITES

In charge separation, sluggish surface reaction kinetics, limited light absorption, and structural instability. Chalcogenide-based semiconductors—such as sulfides, selenides, and tellurides of transition metals—offer great potential for PEC water splitting due to their earth-abundance, moderate bandgaps, and favorable optoelectronic properties. However, their raw forms often suffer from photocorrosion, low charge mobility, or high recombination rates. To address these issues and harness the full potential of chalcogenide nanocomposites, a range of rational materials design strategies have been developed. (Jian, Kumar and Sun, 2020) These include bandgap engineering, surface passivation, heterostructure construction, defect modulation, doping, nanostructuring, and hybridization with co-catalysts and conductive substrates—all of which play a critical role in enhancing photoelectrochemical efficiency and durability. One of the foundational aspects of materials design in PEC systems is the optimization of the semiconductor bandgap to align with the solar spectrum while maintaining suitable energy levels for HER and OER. Ideal photoelectrode materials should have a bandgap in the range of 1.6–2.2 eV to effectively absorb visible light and generate sufficient photovoltage to drive water splitting reactions without external bias. Chalcogenide semiconductors like

MoS_2, Bi_2S_3, $ZnIn_2S_4$, and $CuInSe_2$ naturally possess appropriate bandgaps but can be further tuned through alloying or anion/cation substitution. For example, partial substitution of sulfur with selenium in MoS_2 leads to MoSxSe(1−x), which narrows the bandgap and extends light absorption into the near-infrared region. Similarly, doping CdS with Zn or in alters the conduction and valence band positions, improving light harvesting and enhancing photoresponse across a broader spectral range. ZnO), phosphates, or sulfides—has proven effective in reducing non-radiative recombination and suppressing photocorrosion. These passivation layers are typically deposited via atomic layer deposition (ALD), sol-gel, or chemical bath techniques and serve to improve charge selectivity and stability without impeding charge transfer. For instance, a thin TiO_2 coating on Bi_2S_3 or $ZnIn_2S_4$ photoanodes protects the chalcogenide from oxidative degradation while allowing holes to reach the electrolyte efficiently. Additionally, coating photocathodes like CdSe or MoS_2 with conductive polymers (e.g., polyaniline or PEDOT:PSS) enhances interface quality and electron extraction, contributing to higher photocurrent densities and longer operational lifetimes. strategy to enhance charge separation and facilitate directional charge transport. Properly engineered heterojunctions—such as type-II or Z-scheme interfaces—create internal electric fields that drive electrons and holes to opposite directions, minimizing recombination. (Luo et al., 2018) For example, the combination of MoS_2 with g-C_3N_4, CdS, or TiO_2 forms staggered energy band alignments that promote efficient charge separation. In Z-scheme systems, such as MoS_2/CdS@rGO, photogenerated electrons in one semiconductor recombine with holes from the other, preserving the strong redox potentials of both materials. This configuration is particularly beneficial for overall water splitting, where both HER and OER occur simultaneously. The introduction of buffer layers like SnO_2 or ZnS between junction interfaces can further enhance junction quality and prevent interfacial recombination. and stable but also inexpensive and feasible for large-scale production.

Figure 1. Current status of developed electrocatalysts for water splitting technologies

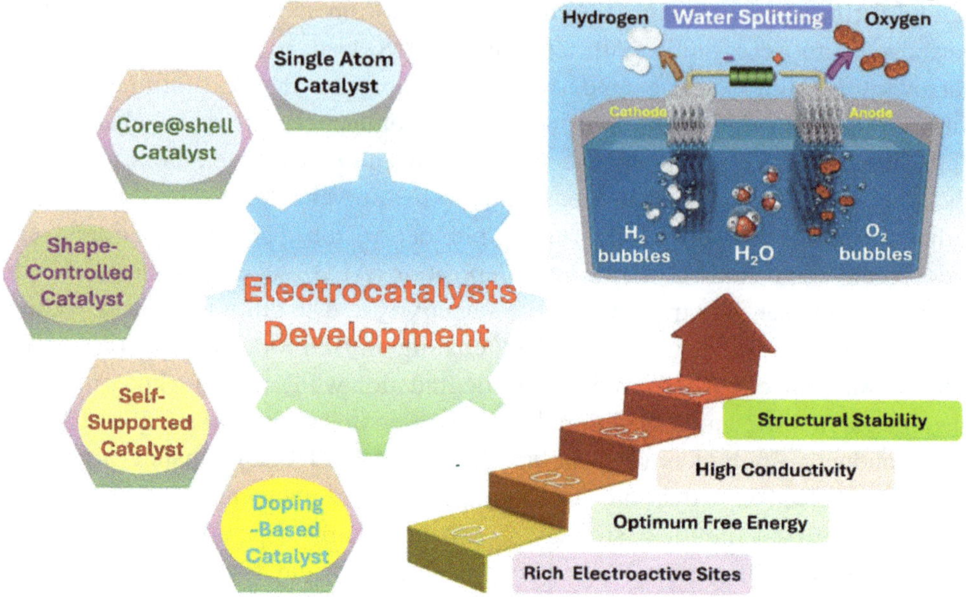

Chalcogenide-based nanocomposites—comprising metal sulfides, selenides, or tellurides often combined with carbon materials or metal oxides—heavily on the cost-effectiveness of their synthesis. Thus, considerable research has been devoted to optimizing synthetic strategies that are economical, energy-efficient, environmentally friendly, and compatible with industrial-scale manufacturing, (Kommineni et al., 2025). A variety of approaches, including hydrothermal/solvothermal methods, chemical precipitation, solid-state reactions, electrochemical deposition, microwave-assisted synthesis, and green chemistry-based routes, have demonstrated potential in producing high-performance chalcogenide-based nanocomposites at relatively low cost. reaction of metal salts and chalcogen sources (e.g., thiourea, sodium selenite) in a sealed autoclave at elevated temperatures and pressures, often in aqueous or organic solvents. The advantage of this method lies in its simplicity, low energy consumption compared to high-temperature solid-state synthesis, and ability to produce well-defined nanostructures (e.g., nanorods, nanosheets, spheres) with controlled morphology and crystallinity. Moreover, by adjusting reaction parameters such as temperature, time, pH, and surfactants, one can tailor particle size and surface area to enhance catalytic performance. For instance, MoS_2 nanosheets, CoS flower-like structures, and $NiSe_2$ hollow spheres have all been successfully fabricated via hydrothermal synthesis at temperatures below 220°C. These low-temperature conditions help reduce capital costs associated with equipment and processing while facilitating easy integration with substrates. temperatures. This method is well-suited for the scalable production of binary and ternary chalcogenides, such as CdS, ZnS, $NiCo_2S_4$, or $CuInSe_2$. The precipitation occurs as the pH is adjusted, usually with NaOH or NH_4OH, leading to the formation of insoluble metal chalcogenide phases. Post-synthesis treatments like annealing or calcination can enhance crystallinity without the need for high-temperature sintering. This method is highly compatible with batch and continuous flow production systems, making it appealing for commercial-scale synthesis. Moreover, co-precipitation enables homogeneous distribution of multiple metal cations, which is particularly beneficial for creating multi-component nanocomposites ball milling—have revitalized interest in solid-state routes. Mechanochemical methods allow the synthesis of nanocrystalline metal chalcogenides under ambient conditions, with drastically reduced energy input and time. For example, MoS_2 and NiS_2 nanoparticles can be synthesized within a few hours without requiring solvents or high-temperature furnaces. These methods reduce both operational costs and environmental impact by eliminating the need for solvent handling and gas flow systems.

COMPARISON OF CHALCOGENIDE-BASED NANOCOMPOSITES WITH OTHER MATERIALS

The growing demand for renewable energy solutions, particularly hydrogen production via water splitting, has catalyzed the need for scalable and efficient electrode materials. Chalcogenide-based nanocomposites, comprising metal sulfides, selenides, or tellurides, have gained attention due synthesis methods, material stability, performance consistency, scalability of fabrication techniques, cost efficiency, and integration with existing infrastructure. Ensuring that chalcogenide-based nanocomposite electrodes can be produced at a reasonable cost while maintaining high efficiency and stability over long operational lifetimes is paramount for their commercial viability. One of the primary challenges in scaling up chalcogenide-based nanocomposite electrodes is the scalability of their synthesis methods. While laboratory-scale synthesis methods such as hydrothermal, solvothermal, and chemical vapor deposition (CVD) provide high-quality nanomaterials with tunable properties, these techniques often face challenges when transitioning to larger volumes. Hydrothermal and solvothermal methods, for instance, typically involve small autoclaves with limited throughput, and controlling factors such as temperature, pressure, and reactant ratios becomes increasingly difficult as batch sizes increase. Therefore, for these methods to be scalable, improvements in reactor design, process automation, and precursor handling must be made, (Kaur et al., 2025). Continuous-flow reactors or multi-chamber systems, which allow for high-throughput synthesis, could improve scalability while maintaining material consistency. Additionally, processes such as electrochemical deposition and chemical vapor deposition (CVD) are more amenable to scale-up, as they can be adapted for continuous production, although they require advanced control systems to ensure uniform deposition of nanomaterials across large surfaces. are commonly used in the synthesis of chalcogenides; while these materials are more abundant and cost-effective than precious metals like platinum or gold, they are still subject to supply chain fluctuations and cost volatility. For large-scale production, it is essential to identify and source raw materials that are not only abundant but also environmentally sustainable.

Mining and extraction processes for metals such as molybdenum, cobalt, or selenium can have significant environmental impacts. Therefore, research is increasingly focused on the development of low-cost, earth-abundant chalcogenides, such as iron-based or copper-based compounds, to reduce material costs while maintaining catalytic performance. Moreover, reducing the reliance on toxic or rare materials through material substitution or recycling strategies can also contribute to cost reductions and more sustainable production. To address this, scalable fabrication methods need to incorporate protective coatings or stabilizing agents that prevent such degradation. Surface passivation techniques, such as coating the chalcogenide electrodes with thin films of metal oxides (TiO_2, ZnO) or using doping strategies to introduce defect-tolerant behavior, have shown promise in enhancing material stability. Similarly, the development of nanocomposites where the chalcogenide material is embedded in a matrix of carbon-based materials, such as graphene or carbon nanotubes, can enhance structural integrity while improving electron conductivity, further stabilizing the material during continuous operation. Additionally, the use of robust composite systems that combine chalcogenides with transition metal oxides (e.g., of individual electrodes. However, for commercial applications, the ability to fabricate large-scale electrodes that maintain consistent performance is critical. (Leal et al., 2020) The uniformity of the electrode's surface area, active site exposure, and thickness are all important factors that influence performance. Scalable coating techniques such as electrochemical deposition, spray pyrolysis, and dip-coating offer promising pathways for producing large-area chalcogenide electrodes with uniform properties. Electrochemical deposition, for instance, allows for precise control over the thickness and composition of the nanocomposite films, which can be crucial for ensuring uniform catalytic activity across the entire electrode. Spray pyrolysis, on the other hand, is a cost-effective method that has already been demonstrated for fabricating large-area thin films for photovoltaic devices and could be adapted for PEC applications. Additionally, screen-printing and roll-to-roll manufacturing methods—techniques commonly used in the electronics industry—offer the possibility of continuous, scalable electrode production. to availability issues. To enhance the scalability of these electrodes, the development of low-cost, flexible, and transparent conductive substrates, such as carbon-based or copper-based transparent electrodes, is necessary.

Moreover, for real-world applications, the electrochemical performance of these electrodes must be robust under continuous solar illumination and variable environmental conditions. This means considering the stability of the electrode material not only under ideal laboratory conditions but also under fluctuating solar intensities, changing pH values, and the presence of impurities in natural water sources. Developing materials with built-in resistance to these factors is critical for their products that need to be managed carefully. Additionally, the lifecycle environmental impact, including the extraction and processing of raw materials, energy consumption during synthesis, and potential toxicity of by-products, must be minimized to ensure that chalcogenide-based nanocomposite electrodes remain an environmentally sustainable option for water splitting. Researchers are therefore working on green synthesis routes such as biomimetic approaches, room-temperature chemical routes, and the use of renewable energy sources for material fabrication. These eco-friendly approaches not only reduce the environmental footprint but also enhance the economic viability of the technology by lowering energy costs associated with material production. water splitting applications requires a careful balance of performance, cost, stability, and environmental sustainability. To meet the challenges of large-scale production, material synthesis methods must be optimized for high throughput, low cost, and minimal environmental impact, while ensuring that the material's performance remains consistent and durable over extended operational periods, (Qazi, 2022). Through advancements in synthesis techniques, surface engineering, and material design, chalcogenide-based electrodes have the potential to become key components of next-generation renewable energy systems, capable of enabling affordable and sustainable hydrogen production on a global scale.

FUTURE DIRECTIONS FOR RESEARCH AND DEVELOPMENT

Chalcogenide-based nanocomposite electrodes have become sulfides, selenides, and tellurides, are highly regarded due to their favorable electronic properties, tunable bandgaps, and potential for high -based nanocomposite electrodes. These challenges are multi-faceted, spanning issues related to material stability, scalability, cost-effectiveness, synthesis, and long-term performance in real-world conditions. This section explores the key challenges and limitations in the development of chalcogenide-based nanocomposite electrodes for energy conversion technologies. from poor stability when exposed to light, water, and electrochemical environments over extended periods. (Zheng et al., 2018) Photocorrosion refers to the degradation of the material upon exposure to light, which causes the material to undergo oxidation or dissolution, thereby decreasing its catalytic performance and reducing the overall lifespan of the electrode. For example, MoS_2 undergoes oxidation in aqueous environments, especially under acidic or alkaline conditions, which leads to the degradation of its active sites and reduced efficiency in hydrogen evolution reactions (HER). Similarly, CdS and Bi_2S_3 can dissolve in water during PEC reactions, further limiting their stability. This issue is particularly critical in the context of PEC water splitting, where the electrodes are exposed to constant illumination and harsh electrochemical conditions. The ability of these materials to maintain long-term stability under these conditions is a significant bottleneck for their practical use in solar-driven water splitting systems. Although various strategies, such as surface passivation, coating with protective layers, and hybridization with other materials (e.g., carbon-based materials or metal oxides), have been explored to improve the stability of chalcogenide-based nanocomposites, these solutions are not always completely effective. Additionally, some of these techniques can introduce additional complexity to the synthesis process or increase the overall cost of the electrodes. low electrical conductivity, which significantly limits their performance in energy conversion applications. Materials like MoS_2 and CdS are inherently semiconducting, but their conductivity is not always sufficient to facilitate efficient charge transport during reactions such as HER and OER.

This can lead to significant energy losses due to poor electron mobility, which in turn reduces the efficiency of the electrode in electrochemical and photoelectrochemical systems. reduced, and the overall efficiency of the system decreases. Researchers have attempted to address this issue by hybridizing chalcogenide materials with more conductive materials such as carbon nanotubes (CNTs), graphene, or metal oxides. While this approach can enhance the electrical conductivity and charge transport, it introduces additional complexities in the material design and increases the cost, which may not always be desirable for large-scale applications. Furthermore, the presence of these additional materials can sometimes lead to interfacial issues or reduced electrode stability, further complicating the development of high-performance electrodes. materials are also prone to mechanical degradation over time, (Ye et al., 2021). The formation of cracks, surface defects, or structural collapse under operational stress can result in the loss of active sites and a decline in catalytic performance. This is particularly problematic in PEC systems, where the electrodes are subjected to cyclic loading and unloading during reactions, leading to wear and tear. Furthermore, the electrode/electrolyte interface is often a site of significant instability, especially when using nanostructured electrodes, which have a large surface area and are more prone to surface reconstruction or degradation over time. particularly under high-temperature conditions. As many photoelectrochemical systems operate under high temperatures to increase the reaction rates, the thermal stability of the nanocomposites becomes crucial. Materials like MoS_2 and CdS can undergo phase transitions or decomposition at elevated temperatures, leading to the loss of their catalytic activity. To address this challenge, researchers have explored approaches like doping, alloying, and surface passivation to enhance the thermal stability of chalcogenide-based materials, but these strategies often come with trade-offs in terms of cost, complexity, and material performance. certain materials, particularly those that contain elements like cadmium (Cd), lead (Pb), and selenium (Se). CdS, for instance, is a widely studied chalcogenide for PEC water splitting; however, cadmium is a toxic heavy metal that poses significant environmental and health risks if not properly handled or disposed of. The use of toxic elements in the synthesis of chalcogenide materials can limit their potential for large-scale applications, particularly in environmentally conscious industries and regions with strict environmental regulations. $CuInSe_2$, and NiS_2, which are considered less toxic, the environmental impact of materials like selenium and tellurium is still a subject of concern. These materials are often rare or difficult to extract, and their mining or processing can lead to significant environmental degradation if not managed properly. This highlights the need for continued research into non-toxic, abundant, and sustainable alternatives for chalcogenide-based nanocomposites, which can mitigate the environmental risks associated with their use. synthesis of chalcogenide materials, such as hydrothermal synthesis, solvothermal synthesis, and chemical vapor deposition (CVD), these methods often require high temperatures, high pressures, or the use of hazardous chemicals, which can make the process expensive and difficult to scale. Additionally, the uniformity of the resulting nanostructures can be hard to control, and the production of

POTENTIAL APPLICATIONS OF CHALCOGENIDE-BASED NANOCOMPOSITES IN ENERGY STORAGE AND CONVERSION

As the world continues to shift towards renewable energy solutions, the demand for efficient and cost-effective materials for energy conversion technologies has never been higher. In this context, chalcogenide-based nanocomposites have emerged as promising candidates for a wide array of applications, most notably materials are characterized by their high catalytic activity, tunable optical properties, and low cost due to their abundance in the Earth's crust. However, while chalcogenide-based nanocomposites have shown great promise in laboratory-scale experiments, several challenges remain to be addressed before they can be commercialized on a large scale. This section outlines potential future directions for research and development (R&D) in chalcogenide-based nanocomposites, focusing on improving their performance, stability, and scalability while addressing issues related to environmental impact and cost-effectiveness. and other photocatalytic applications, (Samal et al., 2025).

To overcome these challenges, future research should focus on developing strategies for enhancing the intrinsic stability of these materials without compromising their catalytic performance. One promising direction is the development of core-shell structures, where a protective shell (such as a metal oxide or a carbon-based layer) can be applied to the surface of the chalcogenide core to mitigate surface degradation caused by photocorrosion. In addition, doping or alloying chalcogenide materials with other elements, such as phosphorus or nitrogen, has been shown to improve their chemical stability. The incorporation of these dopants can reduce the rate of oxidation or dissolution, extending the operational lifetime of the material. Surface passivation techniques, which involve coating the chalcogenide with a thin layer of a more stable material, can also prevent the formation of undesirable surface states that lead to photocorrosion. Further research is needed to identify optimal surface modifications that can provide long-lasting protection while maintaining high catalytic activity. their performance in applications like photoelectrochemical water splitting and electrocatalysis. Efficient charge transport is crucial for high-performance energy conversion, as it facilitates the movement of photo-induced electrons and the efficient transfer of charge during reactions such as the hydrogen evolution reaction (HER) and oxygen evolution reaction (OER).To improve conductivity, future research should focus on developing hybrid nanocomposite structures, offer the potential for high surface area and multiple active sites for reactions, which can significantly boost catalytic efficiency.

Researchers should also explore interfacial engineering techniques to improve the contact resistance between chalcogenide-based electrodes and conductive substrates. In photoelectrochemical devices, minimizing the resistance at the interface between the photoelectrode and the conducting substrate is essential for reducing energy losses and improving efficiency. applications. As a result, there is a pressing need to develop non-toxic, environmentally friendly alternatives to toxic chalcogenides. Future research should focus on developing nontoxic chalcogenides or hybrid materials that exhibit comparable catalytic activity and stability to their toxic counterparts. For example, materials like MoS_2, $CuInSe_2$, NiS_2, and FeS_2 have gained attention as potential substitutes for cadmium-based chalcogenides, as they are considered more environmentally benign. Additionally, biodegradable or eco-friendly materials could be explored for applications in areas where environmental sustainability is a key consideration, (Sreekanthaswamy et al., 2025). Researchers should also focus on developing sustainable synthesis routes for these materials, which minimize the environmental footprint and avoid the use of toxic reagents during the manufacturing process. like hydrothermal synthesis, chemical vapor deposition (CVD), and solvothermal processes may be effective for producing small quantities of high-performance materials, but they are often associated with high energy consumption, long processing times, and the need for expensive precursors, (Suresh et al., 2025). To make these materials viable for performance of chalcogenide-based nanocomposites while reducing production costs. Techniques such as electrochemical deposition, sol-gel processes, and screen printing offer potential routes for low-cost, high-throughput production of nanocomposite electrodes. These methods are more easily adapted for large-scale manufacturing and can be further optimized to control material properties such as morphology, composition, and surface area, (Fareza et al., 2022). Moreover, efforts should be directed towards the green synthesis of chalcogenide materials using environmentally benign solvents, renewable energy sources, and biological or low-cost precursors. This approach will help reduce the environmental impact of material production and make chalcogenide-based nanocomposite electrodes more economically viable for large-scale applications.

CONCLUSION

The growing global gas continues to contribute to environmental degradation, the emphasis is rapidly shifting toward clean, renewable energy technologies. Among the most promising approaches to sustainable energy production is the generation of hydrogen fuel, which is heralded as a clean energy carrier capable of decarbonizing sectors like transportation, industry, and power generation. Hydrogen can pathway to hydrogen production using renewable energy sources, such as solar and wind power. composed of metal chalcogenides (e.g., sulfides, selenides, and tellurides), offer several advantages, including high catalytic activity, tunable optical properties, and the potential for low-cost and scalable synthesis. However, as with any emerging technology, the development and deployment of chalcogenide-based nanocomposites for hydrogen production must be evaluated through the lens of sustainability. This evaluation encompasses not only the technical performance of these materials but also their environmental impact, economic viability, and potential to contribute to broader energy system transformation. precious metals like platinum, which are commonly used in catalytic applications. For instance, materials such as MoS_2, $CuInSe_2$, and FeS_2 are abundant in the Earth's crust and relatively inexpensive, making them an ideal alternative to more expensive and rare materials. However, the environmental implications of chalcogenide-based materials should not be overlooked. Some chalcogenides, such as CdS (cadmium sulfide), contain toxic elements like cadmium, which can pose significant environmental and health risks during manufacturing, operation, and disposal. The toxicity of certain chalcogenides necessitates further investigation into safer, more environmentally friendly alternatives. Non-toxic chalcogenides, such as MoS_2, Cu_2ZnSnS_4 (CZTS), and FeS_2, offer promising substitutes, as they do not pose the same environmental hazards. The use of nontoxic and abundant elements in the development of sustainable materials is critical for ensuring that the hydrogen production process does not inadvertently create new environmental challenges. Additionally, sustainable synthesis methods for chalcogenide-based nanocomposites must be explored to minimize the use of harmful chemicals, solvents, and energy-intensive processes. Green chemistry principles and low-energy techniques, such as hydrothermal and solvothermal synthesis, are being increasingly utilized to reduce the environmental footprint of nanomaterial production. Further research into eco-friendly synthesis methods will be essential to improve the sustainability of chalcogenide-based nanocomposites and ensure their viability for large-scale applications.) at relatively low energy thresholds. However, challenges such as low efficiency and poor stability under prolonged exposure to light and water must be addressed before large-scale implementation can occur. Advancements in nanomaterial engineering, such as the development of heterostructures, core-shell composites, and doping strategies, have the potential to enhance the light absorption, charge carrier dynamics, and overall performance of chalcogenide-based photoelectrodes. By improving the efficiency of these materials, the overall energy consumption of the PEC process can be minimized, leading to a significant reduction in the carbon footprint of hydrogen production has been limited by high capital and operating costs, particularly in the development of efficient and durable electrodes. The high cost of precious metals like platinum, which are used in state-of-the-art catalysts, has been a significant barrier to the widespread adoption of hydrogen production technologies. Chalcogenide-based nanocomposites offer a more cost-effective alternative due to their abundance, lower material costs, and the potential for large-scale production. Materials like MoS_2, NiS_2, and Cu_2ZnSnS_4 are less expensive and more readily available than platinum, offering the potential for cheaper hydrogen production, which is critical for making hydrogen a viable energy carrier. Moreover, scalability is a key factor in determining the long-term sustainability of any energy scale industrial settings. However, scalable production methods must be developed to ensure that the materials can be produced in sufficient quantities to meet the growing demand for sustainable hydrogen production. Cost-effective technologies and hydrogen infrastructure, such as fuel cells and hydrogen storage systems, are essential to create a comprehensive and economically viable hydrogen economy. hydrogen production and storage systems. The transition to hydrogen as a primary energy carrier could also reduce energy dependence on fossil fuel-rich nations, contributing to energy security and global economic stability

REFERENCES

Abbas, M., & Bang, J. H. (2015). Rising again: Opportunities and challenges for platinum-free electrocatalysts. *Chemistry of Materials*. Advance online publication. DOI: 10.1021/acs.chemmater.5b03331

Al-Shetwi, A. Q. (2022). Sustainable development of renewable energy integrated power sector: Trends, environmental impacts, and recent challenges. *The Science of the Total Environment*, *822*, 153645. DOI: 10.1016/j.scitotenv.2022.153645 PMID: 35124039

Augustynski, J., Alexander, B., & Solarska, R. (2011). Metal oxide photoanodes for water splitting. *Topics in Current Chemistry*, *303*, 1–38. DOI: 10.1007/128_2011_135 PMID: 21506001

Bhanja, P., Kim, Y., & Paul, B.. (2020). Facile synthesis of nanoporous transition metal-based phosphates for oxygen evolution reaction. *ChemCatChem*, *12*, 2091–2096. DOI: 10.1002/cctc.201901803

Chen, Z., Gao, Y., & Zhang, Q.. (2019). TiO2/NiO/reduced graphene oxide nanocomposites as anode materials for high-performance lithium-ion batteries. *Journal of Alloys and Compounds*, *774*, 873–878. DOI: 10.1016/j.jallcom.2018.10.010

Chib, S., Mehta, K., Selvakumar, P., Saravanan, A., Mishra, B. R., & Manjunath, T. C. (2025). Mental Helth Challenges and Solutions in High-Pressure Work Environments. In Harriott, S. (Ed.), *Innovative Approaches to Managing Conflict and Change in Diverse Work Environments* (pp. 247–272). IGI Global Scientific Publishing., DOI: 10.4018/979-8-3693-9556-1.ch011

Dahiya, R., & Selvakumar, P. Vidyadhara, & Manjunath, T. C. (2025). Design and Aerodynamics: Principles and Advances. In M. Mellal (Ed.), Innovations and Developments in Unmanned Aerial Vehicles (pp. 145-170). IGI Global Scientific Publishing. DOI: 10.4018/979-8-3693-8462-6.ch007

Eidsvåg, H., Bentouba, S., & Vajeeston, P.. (2021). TiO2 as a photocatalyst for water splitting—An experimental and theoretical review. *Molecules (Basel, Switzerland)*, *26*, 1687. DOI: 10.3390/molecules26061687 PMID: 33802911

Fo, Y., Wang, M., & Ma, Y.. (2020). Origin of highly efficient photocatalyst NiO/SrTiO3 for overall water splitting: Insights from density functional theory calculations. *Journal of Solid State Chemistry*, *292*, 121683. DOI: 10.1016/j.jssc.2020.121683

Guo, Y., Zhou, X., & Tang, J.. (2020). Multiscale structural optimization: Highly efficient hollow iron-doped metal sulfide heterostructures as bifunctional electrocatalysts for water splitting. *Nano Energy*, *75*, 104913. DOI: 10.1016/j.nanoen.2020.104913

Gurushankar, K., Karuppaiya, C., Tari, V., & Kannan, K. (2024). Exploring MXenes and MBenes in metal sulfur and selenium battery technologies. In Kannan, K., & Tari, V. (Eds.), *Novel energy storage and conversion technologies for two-dimensional MXenes and MBenes*. IGI Global., DOI: 10.4018/979-8-3693-9546-2.ch005

Hu, S., Wang, S., & Feng, C.. (2020). Novel MOF-derived nickel nitride as high-performance bifunctional electrocatalysts for hydrogen evolution and urea oxidation. *ACS Sustainable Chemistry & Engineering*, *8*, 7414–7422. DOI: 10.1021/acssuschemeng.0c01450

Isaev, A. B., Shabanov, N. S., & Sobola, D.. (2022). ZnO/Chalcogenides semiconductor heterostructures for photoelectrochemical water splitting. In Kasinathan, K., Elshikh, M. S., & Al Farraj, D. A.-A. (Eds.), *Nanomaterials for energy conversion, biomedical and environmental applications* (pp. 3–35).

Jian, J., Kumar, R., & Sun, J. (2020). Cu2O/ZnO p–n junction decorated with NiOx as a protective layer and cocatalyst for enhanced photoelectrochemical water splitting. *ACS Applied Energy Materials*, *3*, 10408–10414. DOI: 10.1021/acsaem.0c01198

Kannan, K., & Tari, V. (Eds.). (2025). Novel energy storage and conversion technologies for two-dimensional MXenes and MBenes. https://www.amazon.com/dp/9798369395462 ISBN: 9798369395462, eISBN: 9798369395486

Kaur, M., Selvakumar, P., & Dinesh, N. Mohit, Chandel, P. S., & Manjunath, T. C. (2025). Reconstruction and Preservation Strategies for Cultural Heritage. In S. Oflazoğlu & M. Dora (Eds.), Preserving Cultural Heritage in Post-Disaster Urban Renewal (pp. 385-412). IGI Global Scientific Publishing. DOI: 10.4018/979-8-3693-9097-9.ch016

Kommineni, M., Chundru, S., Maroju, P. K., & Selvakumar, P. (2025). Ethical Implications of AI in Sustainable Development Pedagogy. In Chinedu, C., Mafarja, N., Hussein, M., & Saleem, A. (Eds.), *Rethinking the Pedagogy of Sustainable Development in the AI Era* (pp. 17–36). IGI Global Scientific Publishing., DOI: 10.4018/979-8-3693-9062-7.ch002

Le Bahers, T., & Takanabe, K. (2019). Combined theoretical and experimental characterizations of semiconductors for photo electrocatalytic applications. *Journal of Photochemistry and Photobiology A Chemistry*, *40*, 212–233. DOI: 10.1016/j.jphotochemrev.2019.01.001

Leal, J. F., Cruz, S. M. A., & Almeida, B. T. A.. (2020). TiO2–rGO nanocomposite as an efficient catalyst to photodegrade formalin in aquaculture waters, under solar light. *Environmental Sciences (Ruse)*, *6*, 1018–1027. DOI: 10.1039/C9EW00950G

Luo, X., Zhong, J., & Zhou, Q.. (2018). Cationic reduced graphene oxide as self-aligned nanofiller in the epoxy nanocomposite coating with excellent anticorrosive performance and its high antibacterial activity. *ACS Applied Materials & Interfaces*, *10*, 18400–18415. DOI: 10.1021/acsami.8b01982 PMID: 29727162

Qazi, U. Y. (2022). Future of hydrogen as an alternative fuel for next-generation industrial applications. *Chall Expect Oppor Energ*, *15*, 4741. DOI: 10.3390/en15134741

R. Fareza, A., Ardy Nugroho, F.A, F. Abdi, F., & Fauzia V. (2022). Nanoscale metal oxides–2D materials heterostructures for photoelectrochemical water splitting—A review. *Journal of Materials Chemistry. A, Materials for Energy and Sustainability*, *10*, 8656–8686. DOI: 10.1039/D1TA10203F

Samal, A., & Misra, L. I., Mohit, Pougajendy, S., Chandel, P. S., Manjunath, T. C., & Selvakumar, P. (2025). Musealization and Tourism Practices. In S. Oflazoğlu & M. Dora (Eds.), Preserving Cultural Heritage in Post-Disaster Urban Renewal (pp. 115-142). IGI Global Scientific Publishing. DOI: 10.4018/979-8-3693-9097-9.ch006

Sreekanthaswamy, N., Anitha, S., Singh, A., Jayadeva, S. M., Gupta, S., Manjunath, T. C., & Selvakumar, P. (2025). Digital Tools and Methods. In Çela, E., Paliktzoglou, V., Sengar, S., Eappen, P., & Vajjhala, N. (Eds.), *Enhancing School Counseling With Technology and Case Studies* (pp. 25–48). IGI Global Scientific Publishing., DOI: 10.4018/979-8-3693-8392-6.ch002

Suresh, R. V., Selvakumar, P., Naresh Babu, K., Nithya, G., Manjunath, T. C., & Kartigeyan, J. (2025). Best Practices and Practical Applications. In Çela, E., Paliktzoglou, V., Sengar, S., Eappen, P., & Vajjhala, N. (Eds.), *Enhancing School Counseling With Technology and Case Studies* (pp. 341–362). IGI Global Scientific Publishing., DOI: 10.4018/979-8-3693-8392-6.ch013

Tari, V. (2015a). Earth, man, and environment. In *Definition principles and scope of environment* (pp. 1–20). DPS Publishing House.

Tari, V. (2015b). Physicochemical and biological factors in the environment. In *Definition principles and scope of environment* (pp. 21–29). DPS Publishing House.

Tari, V. Kannan, K., & Vishwakarma, V. (2022). Chapter 8: Emerging antiviral technologies. In P. Nunthavarawong, S. M. Rangappa, S. Siengchin, & M. Thoppil-Mathew (Eds.), *Antimicrobial and antiviral materials: Polymers, metals, ceramics, and applications* (1st ed., pp. 137–150). CRC Press. DOI: 10.1201/9781003143093-8

Ye, X., Zhu, T., & Hui, Z.. (2021). Revealing the transfer mechanisms of photo-generated charge carriers over g-C3N4/ZnIn2S4 composite: A model study for photocatalytic oxidation of aromatic alcohols with visible light. *Journal of Catalysis*, *401*, 149–159. DOI: 10.1016/j.jcat.2021.07.025

Zheng, J., Zhang, R., Wang, X., & Yu, P. (2018). Synthesizing a flower-like NiO and ZnO composite for supercapacitor applications. *Research on Chemical Intermediates*. Advance online publication. DOI: 10.1007/s11164-018-3441-x

Chapter 12
Chalcogenide-Based Nanocomposites for High-Performance Oxygen Evolution and Reduction Reactions

G. Anandha Babu
https://orcid.org/0000-0003-1192-8595
Bannari Amman Institute of Technology, India

Hemalatha Parangusan
https://orcid.org/0000-0002-1384-2739
Mannar Thirumalai Naicker College, India

R. Sangeetha
Mannar Thirumalai Naicker College, India

S. Jayanthi
The Standard Fireworks Rajaratnam College for Women, India

G. Dhanalakshmi
Mannar Thirumalai Naicker College, India

ABSTRACT

The intensifying global demand for clean and sustainable energy solutions has accelerated the quest for high-performance electrocatalysts capable of efficiently mediating key electrochemical reactions—most notably, the oxygen evolution reaction

DOI: 10.4018/979-8-3373-3962-7.ch012

(OER) and the oxygen reduction reaction (ORR). To overcome intrinsic drawbacks of pristine chalcogenide materials—such as limited electroactive surface area and inherently poor electrical conductivity—research efforts have increasingly focused on engineering chalcogenide-based nanocomposites via strategic incorporation with conductive matrices, heterostructured architectures, and multifunctional co-catalysts. Transition metal chalcogenides (TMCs), in particular, offer significant potential due to their versatile redox behavior, variable oxidation states, and electronically tunable layered structures. This chapter offers an in-depth exploration of recent advances in chalcogenide-based nanocomposites, which have emerged as highly promising candidates for next-generation OER and ORR electrocatalysis.

INTRODUCTION

In recent years, a wide range of low-cost nanomaterials have been explored for use in electrochemical energy conversion systems. These include metal oxides (Adam et al., 2022; Lv et al., 2021; Gurushankar K. et al. 2025; Kannan K. & Tari V. 2025), metal chalcogenides (MCs: sulfides, selenides, tellurides) (H. Wang et al., 2023; Zhai et al., 2020), MXenes (S. Li et al., 2018), metal nitrides (Yu et al., 2019), metal phosphides (G. Ma et al., 2023), and metal–organic frameworks (MOFs) (Paitandi et al., 2023). Among these materials, MC-based composites have emerged as promising alternatives to noble-metal-based catalysts due to their high catalytic activity, ease of synthesis, and low cost (Hong et al., 2022; Yao et al., 2023). As a result, extensive efforts have been made to develop MC-based composites with diverse morphologies—such as hollow, porous, layered, and nanorod structures and to explore their applications in electrochemical energy conversion systems (D. Ma et al., 2019; Zeng et al., 2021). Particularly, layered metal chalcogenides (LMCs) have garnered significant research attention owing to their unique two-dimensional (2D) structure, high surface area, and atomically thin layers, which collectively contribute to abundant surface-active sites and improved electron transfer kinetics during electrochemical reactions (Qian et al., 2023; Zheng et al., 2020). The layered architecture also offers enhanced contact between the electrode and electrolyte, facilitating efficient charge transport and mass diffusion. More importantly, the electrochemical performance, conductivity, and structural stability of LMC-based composites can be further optimized through compositional tuning, electronic structure engineering, and defect modulation. These attributes make LMCs highly attractive for a range of electrochemical energy applications.

To date, numerous LMC-based compounds have been fabricated and evaluated for various energy conversion reactions (Feng et al., 2020; Fu et al., 2021; X. Ma & Cheng, 2023). For instance, the layered core–shell Ni_3S_2@MoS_2 heterostructures as efficient electrocatalysts for the hydrogen evolution reaction (HER) was developed by (Kim et al., 2021). Zhang et al. synthesized a series of layered metal sulfides ($(NiFeCoX)_3S_4$, where X = Zn, Cr, Mn) for the oxygen evolution reaction (OER) (Zhang et al., 2022). Wu et al. reported layered bimetallic $CuInSe_2$ as high-performance catalysts for the electrochemical reduction of CO_2 (CO_2RR) (L. Wu et al., 2023). These studies demonstrate the considerable potential of LMC-based composites as high-efficiency electrocatalysts.

Given the growing interest in LMCs, various synthesis approaches have been employed for their construction, including hydrothermal (J. Wang et al., 2022), solvothermal (Li et al., 2021), chemical vapour deposition (CVD) (Liang et al., 2020), electrodeposition (Li et al., 2022), and electrospinning methods (Xu et al., 2020). LMC-based materials have since been applied in a broad spectrum of electrochemical reactions such as ORR (oxygen reduction reaction), NRR (nitrogen reduction reaction), CO_2RR, HER, OER, and overall water splitting (OWS) (Guan et al., 2022; Hao et al., 2023). Although several reviews have summarized the electrocatalytic performance of metal sulfides (Guo et al., 2022; Kwon et al., 2020a), comprehensive and systematic overviews specifically focused on LMC-based composites remain limited. Considering the unique physical and chemical characteristics of LMCs and their increasing relevance in energy conversion technologies, it is both timely and essential to provide a thorough summary of recent advances in their synthesis strategies and electrocatalytic applications. This chapter delves into the design, synthesis, and application of chalcogenide-based nanocomposites, exploring their potential to drive high-performance OER and ORR processes for advanced energy systems.

The Global Energy Challenge

The world is facing a critical energy dilemma, driven by the increasing demand for reliable and affordable power amid growing concerns about environmental sustainability. As the global population expands and economies develop, energy consumption continues to rise sharply, placing immense pressure on conventional fossil fuel resources such as coal, oil, and natural gas. These sources, while historically dependable, are finite and contribute significantly to greenhouse gas emissions, air pollution, and climate change, (Tari et al. 2024; Siddiqui et al., 2022; Tari et al., 2022). The challenge lies in meeting this growing demand without further compromising the environment and public health. Another key aspect of the global energy challenge is the unequal distribution of energy access across regions. While developed countries enjoy widespread electrification and stable energy supplies, billions of people in developing nations still lack access to basic electricity and clean cooking fuels. This energy disparity hampers economic growth, education, healthcare, and overall quality of life (H. Wu et al., 2023). Bridging this energy gap requires innovative, scalable, and clean energy solutions that are both economically viable and environmentally friendly, especially in remote or underdeveloped areas. This energy crisis underscores the urgent need for clean, renewable, and sustainable energy alternatives that can meet the growing global energy demand while minimizing environmental impact. Hydrogen, produced through water electrolysis, is a promising clean fuel, but its large-scale adoption hinges on the efficiency of the underlying electrocatalytic reactions (Tang et al., 2022).

Catalysts play a crucial role in improving the efficiency of processes such as water splitting and fuel cell reactions. Thus, the development of high-performance materials—like chalcogenide-based nanocomposites—for catalyzing oxygen evolution and reduction reactions is essential in enabling this shift to a cleaner, more resilient energy future. The oxygen evolution reaction and oxygen reduction reaction play crucial roles in the operation of various energy conversion devices, including water-splitting systems, fuel cells, and metal–air batteries. These reactions are thermodynamically and kinetically challenging, requiring the development of efficient and robust electrocatalysts. The advancement of such materials is essential to accelerate the shift toward sustainable energy technologies.

The Role of OER and ORR in Sustainable Energy Systems

The OER and ORR are pivotal to the operation of various sustainable energy systems, including water electrolyzers, metal–air batteries, and fuel cells. OER occurs at the anode during water splitting, producing oxygen gas and enabling the generation of hydrogen—a clean fuel with high energy density. ORR, on the other hand, takes place at the cathode in fuel cells and metal–air batteries, where it drives the conversion of chemical energy into electrical energy by reducing oxygen molecules. The efficiency, durability, and overall performance of these devices heavily depend on the kinetics and stability of OER and ORR, which are intrinsically slow and require highly efficient electrocatalysts to overcome energy barriers (Joo et al., 2019).

Incorporating effective OER and ORR catalysts into these systems is essential for realizing a carbon-neutral energy cycle. Traditional noble metal catalysts like platinum (for ORR) and iridium or ruthenium oxides (for OER) offer excellent performance but are limited by high cost and scarcity. Therefore, the development of alternative, earth-abundant materials with comparable or superior catalytic performance is a key research priority. In this regard, chalcogenide-based nanocomposites have emerged as promising candidates due to their tunable electronic structures, enhanced conductivity, and abundance of active sites. By facilitating faster and more efficient oxygen-related reactions, these materials play a crucial role in advancing next-generation sustainable energy technologies (Kwon et al., 2020b).

OER and ORR are half-cell reactions that are integral to electrochemical devices. In water-splitting electrolyzers, OER occurs at the anode and is the rate-limiting step due to its sluggish kinetics. Conversely, in fuel cells and metal–air batteries, ORR takes place at the cathode and also suffers from high overpotentials. Both reactions involve complex multi-electron processes, which necessitate the development of catalysts that can reduce energy barriers and improve reaction rates.

Limitations of Conventional Electrocatalysts

Conventional electrocatalysts, particularly noble metals such as platinum (Pt) for the ORR and iridium or ruthenium oxides (IrO_2, RuO_2) for the OER, have long been considered the gold standard for electrochemical applications. These materials offer excellent catalytic activity, high selectivity, and good stability under controlled conditions. However, their widespread use is significantly constrained by critical limitations. The most pressing issues include high cost, limited natural availability, and supply chain vulnerabilities, all of which hinder their scalability for large-scale energy systems. Moreover, their rigid crystalline structures often restrict the exposure of active sites, limiting overall catalytic efficiency.

Beyond economic challenges, these electrocatalysts are prone to degradation under prolonged operation, especially in harsh electrochemical environments. Platinum-based catalysts, for instance, are susceptible to poisoning and aggregation, which progressively diminish their activity. Similarly, OER catalysts like IrO_2 and RuO_2 can undergo structural degradation or dissolution, leading to poor long-term durability. Furthermore, many conventional catalysts lack stability across varying pH conditions and temperatures, making them unsuitable for diverse energy applications. These drawbacks highlight the urgent need for alternative materials that are not only cost-effective and durable but also capable of maintaining high performance across a range of operating environments. This necessity has directed significant research interest toward earth-abundant, transition metal chalcogenides (TMCs) and their nanocomposites, which offer a promising path forward for next-generation electrocatalysts (Lamiel et al., 2023).

EMERGENCE OF CHALCOGENIDE-BASED NANOCOMPOSITES

In response to the limitations of conventional noble-metal-based catalysts, chalcogenide-based nanocomposites have emerged as promising alternatives for electrochemical energy applications. These materials, composed of chalcogen elements such as sulfur (S), selenium (Se), or tellurium (Te) combined with transition metals, offer a unique combination of physicochemical properties that make them highly attractive for catalytic processes. Their intrinsic tunable electronic structures, high electrical conductivity, and large surface areas facilitate enhanced charge transfer and the exposure of abundant active sites. When engineered at the nanoscale and combined with other functional materials, these chalcogenides exhibit synergistic effects that significantly boost their performance in oxygen evolution and reduction reactions (Lamiel et al., 2023).

The growing interest in chalcogenide-based nanocomposites is also driven by their cost-effectiveness and earth abundance, which make them more suitable for large-scale deployment compared to precious metal catalysts. Additionally, these nanocomposites demonstrate excellent structural flexibility and chemical stability under a wide range of electrochemical conditions, including both acidic and alkaline environments. Researchers have successfully developed a variety of chalcogenide-based hybrid systems, such as metal sulfide/carbon, metal selenide/metal oxide, and multi-metal chalcogenides, tailored to optimize catalytic efficiency and durability. As a result, these materials are increasingly seen as viable candidates to replace or complement conventional catalysts in next-generation sustainable energy systems.

Chalcogenide-based nanocomposites, consisting of compounds formed with sulfur, selenium, or tellurium, have gained attention as promising electrocatalysts. Transition metal chalcogenides offer a wide range of oxidation states and flexible electronic configurations, making them suitable for redox reactions like OER and ORR. However, pristine chalcogenides often suffer from poor conductivity and limited active surface areas, which limit their standalone use. To address these shortcomings, researchers have developed nanocomposites by combining chalcogenides with conductive matrices or co-catalysts. These hybrid systems exploit synergistic interactions between components to enhance catalytic activity and durability. The resulting materials exhibit improved charge transport, higher electrochemical surface area, and better stability under operational conditions, making them excellent candidates for next-generation energy devices (Thangamathi et al., 2024).

Figure 1.

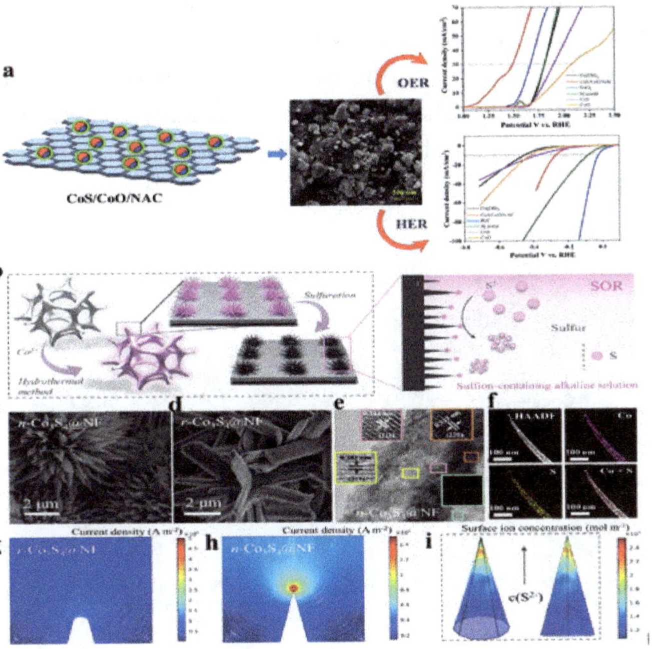

Fundamentals of Oxygen Evolution and Reduction Reactions

The OER and ORR are critical multi-electron transfer processes that underpin various electrochemical energy conversion systems, including water electrolyzers, fuel cells, and metal–air batteries. OER typically occurs at the anode during water splitting and involves the oxidation of water molecules to produce oxygen gas. In alkaline media, this reaction proceeds through the adsorption of hydroxide ions (OH^-), followed by the formation of key intermediates such as O*, OH*, and OOH*, ultimately leading to the evolution of O_2. However, OER is kinetically sluggish due to its complex four-electron pathway and the formation of multiple intermediates, which contribute to high over potentials and significant energy losses. To overcome these challenges, efficient OER catalysts are essential to lower activation energy barriers and enhance the reaction rate.

The oxygen reduction reaction (ORR), occurring at the cathode in systems such as fuel cells and metal–air batteries, involves complex redox processes with several oxygen-containing intermediates, including O*, *OH, OH^-, O_2^-, HO_2^-, H_2O_2, and *OOH (Ge et al., 2015a; Kumeda et al., 2023). These intermediates undergo multiple adsorption and desorption steps, often resulting in highly irreversible reaction pathways, particularly in aqueous environments [Figure 2 (Maeda & Domen, 2007)]. The sluggish kinetics of both ORR and the oxygen evolution reaction (OER) are largely attributed to the challenges associated with O_2 activation, O–O bond cleavage, OH adsorption/desorption, and oxide elimination, all of which require substantial energy input (M. Wang et al., 2020; Zhao et al., 2019). As a result, catalysts used for these reactions must demonstrate high activity, selectivity, and long-term stability under practical operating conditions.

Figure 2.

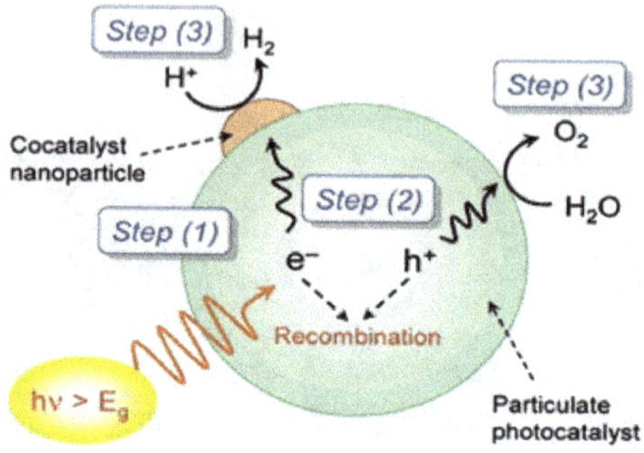

In ORR specifically, molecular oxygen is reduced either to water in acidic media or to hydroxide ions in alkaline media. This reaction proceeds via a multi-step mechanism, involving the initial adsorption of O_2 and the subsequent formation of key intermediates such as OOH*, O*, and OH*. ORR can follow a desirable four-electron pathway, which directly reduces O_2 to water or OH^-, or a less efficient two-electron pathway, yielding hydrogen peroxide as a by-product. The predominance of the slower two-electron route, especially on non-precious metal catalysts, significantly hampers the overall performance of electrochemical energy devices. Therefore, the development of advanced catalysts that can accelerate electron transfer, stabilize crucial intermediates, and favour the four-electron reduction pathway is critical for enhancing the efficiency and durability of ORR-based systems.

To address these issues, considerable effort has been dedicated to enhancing the catalytic efficiency of benchmark materials such as Pt-metallene. Strategies include alloying, designing specific nanostructures with Pt-rich surfaces, optimizing atom utilization, and reducing particle size to enhance intrinsic activity. Despite these advancements, such finely structured nanomaterials often suffer from structural deformation and agglomeration during operation, leading to reduced stability and deactivation over time (Ge et al., 2015b; Tu et al., 2024). Furthermore, the synthesis of Pt-based catalysts is costly and complex, limiting their scalability and widespread adoption. Therefore, the development of alternative catalysts that are cost-effective, highly active, and operationally stable is imperative. One promising approach is defect engineering in nanocrystalline domains, which can introduce lattice distortions and atomic rearrangements that significantly boost catalytic performance. Additionally, increasing the exposure of active sites and modifying the electronic band structure further enhance activity (J. Feng et al., 2019; Ghorui et al., 2022). As a result, defect engineering has emerged as a powerful strategy in catalyst design, particularly within the realm of electrochemistry (Roy et al., 2021).

CHALCOGENIDE MATERIALS: PROPERTIES AND POTENTIAL

Chalcogenide materials, composed of chalcogen elements such as sulfur (S), selenium (Se), and tellurium (Te) in combination with metals, possess a unique set of physicochemical properties that make them highly attractive for electrochemical applications. These materials typically exhibit tunable electronic structures, moderate to high electrical conductivity, and strong covalent bonding, allowing them to participate effectively in redox reactions. Their layered crystal structures and adjustable band gaps facilitate efficient charge transport and provide numerous active sites for catalytic reactions. Furthermore, chalcogenides can easily be synthesized in various nanostructured forms—such as nanoparticles, nanosheets, and hollow structures—enhancing their surface area and interaction with reactant molecules.

The potential of chalcogenide materials lies in their versatility, abundance, and ability to form highly active nanocomposites when combined with other conductive or catalytic components. Transition metal chalcogenides (TMCs), such as MoS_2, CoSe, and NiTe, have shown remarkable performance in both the OER and ORR due to their adjustable electronic configurations and strong metal–chalcogen interactions. These materials can also be doped or hybridized with carbonaceous supports, metal oxides, or other chalcogenides to further enhance their catalytic behaviour. With continued advancements in material design and synthesis techniques, chalcogenide-based systems are positioned as strong contenders for replacing traditional noble-metal catalysts in next-generation sustainable energy devices (Chanda et al., 2023).

Overview of Transition Metal Chalcogenides (TMCs)

Transition metal chalcogenides (TMCs) are a class of compounds formed by combining transition metals such as molybdenum (Mo), cobalt (Co), nickel (Ni), and iron (Fe) with chalcogen elements like sulphur (S), selenium (Se), or tellurium (Te). These materials are characterized by their layered crystal structures, where weak van der Waals forces hold together strong covalently bonded metal–chalcogen layers. This structural feature allows for the easy exfoliation of TMCs into ultrathin nanosheets, providing a high surface-to-volume ratio that is ideal for catalytic applications. Their electronic properties can range from semiconducting (e.g., MoS_2) to metallic (e.g., $NbSe_2$), depending on their composition and phase, offering tunability for various energy-related reactions.

TMCs have attracted significant attention for their promising role in electrochemical reactions, particularly the OER and ORR. Their ability to undergo redox cycling, high electrical conductivity in certain phases, and abundance of catalytically active sites make them highly suitable for electrocatalysis. Unlike noble metals, TMCs are earth-abundant and cost-effective, which addresses the economic and scalability limitations of traditional catalysts. Additionally, the catalytic activity of TMCs can be further enhanced through techniques like doping, defect engineering, and compositional tuning, which modify their electronic structure and improve their interaction with reaction intermediates.

Beyond their intrinsic properties, the performance of TMCs can be significantly improved by forming nanocomposites with conductive supports such as graphene, carbon nanotubes, or conductive polymers. These hybrid systems not only enhance electrical conductivity and dispersion but also stabilize the active material during long-term electrochemical operation. Moreover, the synergistic effects between the TMCs and supporting materials can reduce over potentials and promote favourable kinetics for OER and ORR (Gautam et al., 2022). As research advances, TMCs continue to evolve as a versatile and powerful platform for developing efficient, stable, and economically viable electrocatalysts for sustainable energy technologies.

Transition metal chalcogenides (TMCs), including sulfides, selenides, and tellurides, represent a broad class of materials formed by the combination of transition metals (e.g., Mo, W, Co, Ni, Fe) with group 16 elements. Their layered crystal structures and tuneable electronic configurations make them appealing candidates for electrochemical applications. Many TMCs adopt structures analogous to graphite, with weak van der Waals interactions between layers, allowing for easy exfoliation into two-dimensional nanosheets that expose abundant active sites.

These materials also demonstrate versatile redox behaviour, essential for driving multistep reactions such as OER and ORR. The intrinsic properties of TMCs—such as their semiconducting nature, moderate catalytic activity, and chemical stability—make them attractive for catalyst design. However, their standalone performance is often hindered by low electrical conductivity and limited surface accessibility, necessitating further modifications for practical use.

Structural Diversity and Electronic Tunability

Transition metal chalcogenides (TMCs) exhibit remarkable structural diversity, which plays a key role in determining their physicochemical properties and catalytic performance. They can crystallize in various phases—such as 1T (metallic), 2H (semiconducting), or 3R (rhombohedral)—depending on the specific metal and chalcogen combination. This polymorphism allows TMCs to adopt structures ranging from layered sheets and nanotubes to hollow spheres and hierarchical frameworks. Such diversity provides abundant opportunities to expose more active sites, optimize mass and charge transport, and tailor the surface characteristics for specific electrochemical reactions like OER and ORR.

One of the most attractive features of TMCs is their electronic tunability, which enables researchers to precisely modulate their band structures and electronic configurations (Kannan, Chanda, Gautam, Behera, et al., 2023). By introducing heteroatom doping, vacancy engineering, or strain modulation, the electronic density of states near the Fermi level can be optimized to improve the adsorption and activation of oxygen intermediates. For example, sulfur vacancies in MoS_2 can serve as active sites for catalysis, while doping cobalt into WS_2 can enhance electrical conductivity and redox activity. This ability to tailor both structural and electronic characteristics makes TMCs highly adaptable and effective for high-performance electrocatalysis in sustainable energy applications.

The structural flexibility of TMCs allows for the design of different phases, including 1T (metallic), 2H (semiconducting), and 3R (rhombohedral) polymorphs, which can be selectively synthesized depending on the desired application. These structural variations significantly impact the density of states near the Fermi level, enabling modulation of their catalytic activity. Through phase engineering, it is possible to optimize the exposure of active sites and enhance charge mobility.

Additionally, TMCs can be doped or alloyed with other transition metals or chalcogens to further tune their bandgap, work function, and catalytic behaviour. Such compositional control enables fine-tuning of adsorption energies for key intermediates in OER and ORR, improving reaction kinetics. Thus, electronic and structural tunability gives TMCs a significant advantage as customizable building blocks for next-generation nanocomposite catalysts.

Limitations of Pristine Chalcogenides

Despite their promising properties, pristine chalcogenides face several limitations that hinder their direct application as efficient electrocatalysts. One major drawback is their inherent low electrical conductivity, particularly in semiconducting phases like 2H-MoS_2 and WS_2. This limited conductivity restricts the efficient transport of electrons during electrochemical reactions, thereby reducing the overall catalytic activity. Additionally, the number of catalytically active sites in pristine chalcogenides is often confined to edge or defect regions, while the basal planes remain largely inert. This reduces the effective surface area available for reactions such as the oxygen evolution and reduction processes.

Another significant challenge is the poor long-term stability and structural degradation that many pristine chalcogenides experience under harsh electrochemical conditions. Exposure to strong alkaline or acidic environments, high potentials, and continuous cycling can lead to the dissolution or restructuring of the catalyst, diminishing its performance over time. Furthermore, the limited tolerance of pristine chalcogenides to variable pH and temperature conditions restricts their versatility in different electrochemical systems. These drawbacks highlight the necessity for material modifications—such as doping, hybridization with conductive supports, or forming nanocomposites—to enhance the stability, conductivity, and catalytic activity of chalcogenide-based systems (Gurushankar et al., 2025).

Furthermore, chalcogenides can undergo phase degradation or leaching under harsh reaction conditions, leading to poor long-term stability. Surface oxidation and passivation are also common challenges, particularly during prolonged operation in alkaline or acidic media. These limitations necessitate the development of hybrid systems, where chalcogenides are combined with other conductive or stabilizing components to overcome their inherent weaknesses.

REACTION PATHWAY ANALYSIS

Reaction pathway analysis is a critical approach in understanding the mechanisms behind the OER and ORR on chalcogenide-based nanocomposites. This analysis involves identifying and evaluating the sequence of elementary steps, including the adsorption, transformation, and desorption of intermediate species like O*, OH*, and OOH*. Each step has an associated Gibbs free energy change (ΔG), and the step with the highest energy barrier is considered the rate-determining step (RDS). By mapping out these pathways using experimental data or computational models such as density functional theory (DFT), researchers can pinpoint inefficiencies in the catalytic cycle and identify strategies for minimizing overpotentials and improving kinetics.

For chalcogenide-based electrocatalysts, reaction pathway analysis also aids in understanding how structural features—such as defects, dopants, and hetero interfaces—influence the binding energies and activation barriers of intermediates. This information is essential for tuning the catalyst's surface properties to optimize reaction efficiency (Kannan et al., 2024). Moreover, comparing different reaction mechanisms, such as the associative versus dissociative pathways in ORR, helps in selecting or designing materials that favour the most efficient catalytic routes. Overall, reaction pathway analysis bridges the gap between atomic-level understanding and macroscopic performance, guiding the development of high-performance, durable, and cost-effective electrocatalysts.

WATER ELECTROLYZERS FOR HYDROGEN PRODUCTION

Water electrolyzers are critical technologies for producing green hydrogen—an essential component in the transition to sustainable energy systems. In these devices, water is split into hydrogen and oxygen using electricity, ideally sourced from renewable energy. The overall reaction consists of the hydrogen evolution reaction (HER) at the cathode and OER at the anode. The efficiency and durability of the electrolyzer heavily depend on the electrocatalysts used, particularly for OER, which is the more energy-intensive half-reaction due to its complex four-electron transfer mechanism (Kalaiselvi et al., 2025). Traditionally, noble metal-based catalysts like iridium oxide have been used for OER, but their high cost and scarcity limit large-scale application.

Chalcogenide-based nanocomposites have emerged as promising alternatives for the OER component of water electrolyzers. These materials, especially when engineered into nanostructured composites, offer excellent catalytic activity, tunable electronic properties, and improved surface area for active site exposure. Transition metal chalcogenides (TMCs) such as MoS_2, $CoSe_2$, and NiS_2, when integrated with conductive supports or co-catalysts, have demonstrated competitive performance to noble metal counterparts in alkaline and even near-neutral pH conditions (Kannan, Chanda, Gautam, Meshesha, et al., 2023). Their chemical stability under operating conditions, combined with earth-abundant composition, makes them ideal for scalable hydrogen production systems. As research advances, further improvements in synthesis control and electrode integration will enhance their commercial viability for clean hydrogen generation.

FUEL CELLS AND METAL–AIR BATTERIES

Fuel cells are electrochemical devices that convert chemical energy directly into electricity through redox reactions, typically involving hydrogen and oxygen. A key reaction in most fuel cells, particularly proton exchange membrane fuel cells (PEMFCs), is the oxygen reduction reaction (ORR) at the cathode. This step is notoriously sluggish and requires efficient electrocatalysts to maintain high performance and energy conversion efficiency. Traditionally, platinum-based catalysts are used, but their cost and degradation over time are major drawbacks. Chalcogenide-based nanocomposites, especially those incorporating transition metals like cobalt, iron, and nickel, have shown potential to replace or supplement platinum catalysts. Their tunable electronic structures, abundant active sites, and strong resistance to poisoning make them suitable for ORR in both acidic and alkaline fuel cells.

Metal–air batteries, such as zinc–air and lithium–air batteries, rely on ORR during discharge and the OER during charging, making bifunctional catalysts essential for efficient operation. Chalcogenide nanocomposites offer a promising solution due to their ability to catalyze both ORR and OER with high efficiency. Their high surface area, customizable morphologies, and capacity to form heterostructures or integrate with conductive supports enhance their catalytic activity and cycling stability. Moreover, their structural robustness under high-current and high-voltage conditions makes them ideal for long-duration energy storage applications. As demand grows for grid-scale and portable renewable energy storage solutions, these materials are increasingly important for advancing the performance and affordability of metal–air batteries (Dias et al., 2023; Yuan et al., 2017).

HYBRID AND FLEXIBLE ENERGY DEVICES

Hybrid and flexible energy devices represent the future of portable, wearable, and multifunctional electronics, merging energy generation, storage, and conversion into compact, adaptable systems. These devices require electrocatalysts that not only perform efficiently but also maintain functionality under mechanical stress and bending. Chalcogenide-based nanocomposites are particularly well-suited for such applications due to their inherent structural flexibility, tunable electronic properties, and capability to form conformal coatings on stretchable substrates. Their nanoscale dimensions and compatibility with printable or low-temperature fabrication techniques make them ideal for integration into flexible fuel cells, wearable zinc–air batteries, and self-powered sensors.

Moreover, in hybrid systems that combine different energy technologies—such as solar cells (Figure 3, (Liu et al., 2024; Pimpang et al., 2025)) paired with water-splitting electrolyzers or supercapacitors integrated with fuel cells—chalcogenide nanocomposites can function in multiple roles, from light harvesting and electron transport to catalysis and energy storage. Their multifunctionality allows for seamless energy flow and high conversion efficiency within the same device platform. The use of lightweight, earth-abundant materials further supports the scalability and environmental sustainability of these next-generation systems. As the demand for portable and flexible power sources continues to rise, chalcogenide-based nanocomposites will play a crucial role in the design of versatile and resilient energy solutions.

Figure 3.

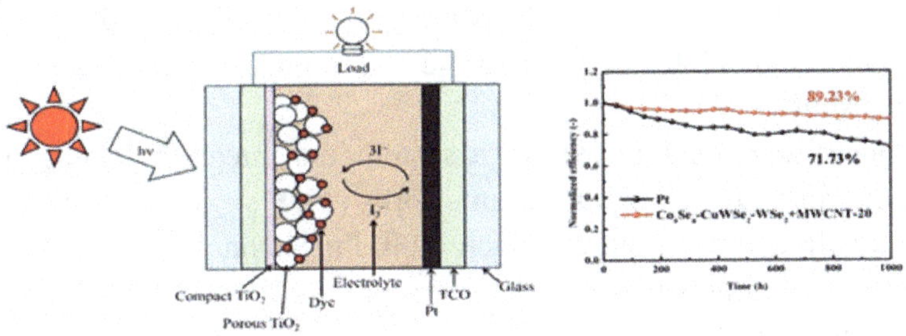

CHALLENGES AND FUTURE PERSPECTIVES

Operational durability is a critical parameter for electrocatalysts used in long-term energy conversion systems such as water electrolyzers, fuel cells, and metal–air batteries. Chalcogenide-based nanocomposites have shown promise due to their robust structures and resistance to corrosion, but like all catalysts, they are subject to degradation under continuous operation. Factors such as electrochemical cycling, high overpotentials, pH fluctuations, and temperature variations can lead to physical and chemical changes in the catalyst structure. For instance, metal leaching, chalcogen dissolution, and nanoparticle aggregation can result in the loss of active sites, decreased conductivity, and overall performance decline. Maintaining catalyst stability over thousands of operational hours is essential for real-world applications.

Degradation mechanisms in chalcogenide nanocomposites are often linked to surface oxidation, phase transformation, and interface deterioration. In OER conditions, for example, transition metal chalcogenides can undergo surface oxidation to form less active or insulating layers, which hinders charge transfer. Similarly, during ORR, the generation of reactive oxygen species can lead to structural damage and breakdown of the catalyst-support interface. Understanding these mechanisms through in situ/operando studies and long-term stability tests is key to improving material design. Strategies such as doping, protective coatings, and compositional tuning can significantly enhance operational durability, ensuring the catalyst remains effective over prolonged use and under harsh electrochemical conditions.

CONCLUSIONS

The future of chalcogenide-based nanocomposites lies in the rational design of multifunctional and hierarchically structured materials. The use of machine learning and computational screening will accelerate the discovery of optimal compositions and structures. 3D printing and self-assembly approaches could enable precise control over catalyst architecture at multiple length scales, enhancing both activity and mass transport.

There is also growing interest in hybrid systems that integrate chalcogenides with organic frameworks, metal–organic frameworks (MOFs), or perovskites, offering novel functionalities and synergies. Such multidimensional composites may pave the way for breakthroughs in electrocatalysis and multifunctional energy systems, such as solar-assisted water splitting or hybrid supercapacitor-battery devices.

Despite the significant progress made, the transition from laboratory-scale research to real-world deployment requires overcoming several hurdles. These include scaling up synthesis, maintaining consistency across batches, and integrating catalysts into device architectures compatible with existing infrastructure. Collaborations between material scientists, engineers, and industrial partners will be crucial. Standardized protocols for catalyst testing and benchmarking are also needed to compare performance metrics reliably. Addressing cost, safety, and durability concerns will be key to gaining industry and consumer acceptance. Ultimately, the continued evolution of chalcogenide-based nanocomposites will depend on aligning material innovation with system-level requirements for a sustainable energy future.

THE FUTURE OF CHALCOGENIDE-BASED NANOCOMPOSITES IN CLEAN ENERGY

The future of chalcogenide-based nanocomposites in clean energy looks increasingly promising as global efforts intensify toward decarbonization and the development of sustainable technologies. With their unique ability to facilitate both the oxygen evolution reaction (OER) and oxygen reduction reaction (ORR), these materials are poised to play a pivotal role in the advancement of fuel cells, metal–air batteries, and water electrolyzers. As research continues to refine their structural, electronic, and surface properties, chalcogenide nanocomposites are expected to surpass traditional catalysts in performance, cost-effectiveness, and long-term durability. Their inherent tunability makes them ideal candidates for tailoring solutions across a broad spectrum of clean energy systems.

Emerging trends point toward the integration of artificial intelligence (AI) and machine learning (ML) in the design and optimization of chalcogenide-based catalysts. These tools allow researchers to rapidly predict material behaviour, accelerate discovery of optimal compositions, and model reaction mechanisms at the atomic level. As a result, material development timelines can be significantly shortened, and performance targets can be met more precisely. Combined with in situ and operando characterization techniques, this data-driven approach will deepen the understanding of active sites, reaction intermediates, and degradation pathways, ultimately guiding the design of next-generation nanocomposites.

REFERENCES

Adam, D. B., Tsai, M.-C., Awoke, Y. A., Huang, W.-H., Lin, C.-H., Alamirew, T., Ayele, A. A., Yang, Y.-W., Pao, C.-W., Su, W.-N., & Hwang, B. J. (2022). Engineering self-supported ruthenium-titanium alloy oxide on 3D web-like titania as iodide oxidation reaction electrocatalyst to boost hydrogen production. *Applied Catalysis B: Environmental*, *316*, 121608. DOI: 10.1016/j.apcatb.2022.121608

Chanda, D., Kannan, K., Gautam, J., Meshesha, M. M., Jang, S. G., Dinh, V. A., & Yang, B. L. (2023). Effect of the interfacial electronic coupling of nickel-iron sulfide nanosheets with layer Ti3C2 MXenes as efficient bifunctional electrocatalysts for anion-exchange membrane water electrolysis. *Applied Catalysis B: Environmental*, *321*, 122039. DOI: 10.1016/j.apcatb.2022.122039

Dias, G. D. S., Costa, J. M., & Almeida Neto, A. F. D. (2023). Transition metal chalcogenides carbon-based as bifunctional cathode electrocatalysts for rechargeable zinc-air battery: An updated review. *Advances in Colloid and Interface Science*, *315*, 102891. DOI: 10.1016/j.cis.2023.102891 PMID: 37058836

Feng, J., Dong, H., Pang, B., Chen, Y., Yu, L., & Dong, L. (2019). Tuning the electronic and optical properties of graphene quantum dots by selective boronization. *Journal of Materials Chemistry. C, Materials for Optical and Electronic Devices*, *7*(2), 237–246. DOI: 10.1039/C8TC03863E

Feng, Q., Li, M., Wang, T., Chen, Y., Wang, X., Zhang, X., Li, X., Yang, Z., Feng, L., Zheng, J., Xu, H., Zhai, T., & Jiang, Y. (2020). Low-temperature growth of Three dimensional ReS2/ReO2 metal-semiconductor heterojunctions on Graphene/polyimide film for enhanced hydrogen evolution reaction. *Applied Catalysis B: Environmental*, *271*, 118924. DOI: 10.1016/j.apcatb.2020.118924

Fu, Q., Han, J., Wang, X., Xu, P., Yao, T., Zhong, J., Zhong, W., Liu, S., Gao, T., Zhang, Z., Xu, L., & Song, B. (2021). 2D Transition Metal Dichalcogenides: Design, Modulation, and Challenges in Electrocatalysis. *Advanced Materials*, *33*(6), 1907818. DOI: 10.1002/adma.201907818 PMID: 32578254

Gautam, J., Kannan, K., Meshesha, M. M., Dahal, B., Subedi, S., Ni, L., Wei, Y., & Yang, B. L. (2022). Heterostructure of polyoxometalate/zinc-iron-oxide nanoplates as an outstanding bifunctional electrocatalyst for the hydrogen and oxygen evolution reaction. *Journal of Colloid and Interface Science*, *618*, 419–430. DOI: 10.1016/j.jcis.2022.03.103 PMID: 35364543

Ge, X., Sumboja, A., Wuu, D., An, T., Li, B., Goh, F. W. T., Hor, T. S. A., Zong, Y., & Liu, Z. (2015a). Oxygen Reduction in Alkaline Media: From Mechanisms to Recent Advances of Catalysts. *ACS Catalysis*, *5*(8), 4643–4667. DOI: 10.1021/acscatal.5b00524

Ghorui, U. K., Mondal, P., Adhikary, B., Mondal, A., & Sarkar, A. (2022). Newly Designed One-Pot In-Situ Synthesis of VS_2 /rGO Nanocomposite to Explore Its Electrochemical Behavior towards Oxygen Electrode Reactions**. *ChemElectroChem*, *9*(15), e202200526. DOI: 10.1002/celc.202200526

Guan, J., Chen, W., Fang, Y., Wang, L., Fu, Y., Guo, B., & Zhang, M. (2022). Electrospinning derivative fabrication of sandwich-structured CNF/Co3S4/MoS2 as self-supported electrodes to accelerate electron transport in HER. *International Journal of Hydrogen Energy*, *47*(33), 14930–14941. DOI: 10.1016/j.ijhydene.2022.03.013

Guo, T., Li, L., & Wang, Z. (2022). Recent Development and Future Perspectives of Amorphous Transition Metal-Based Electrocatalysts for Oxygen Evolution Reaction. *Advanced Energy Materials*, *12*(24), 2200827. DOI: 10.1002/aenm.202200827

Gurushankar, K., Karuppaiya, C., Tari, V., & Kannan, K. (2025). Exploring MXenes and MBenes in Metal Sulfur and Selenium Battery Technologies. In Kannan, K., & Tari, V. (Eds.), *Novel Energy Storage and Conversion Technologies for Two-Dimensional MXenes and MBenes* (pp. 105–120). IGI Global., DOI: 10.4018/979-8-3693-9546-2.ch005

Hao, Y., Wang, L., & Huang, L.-F. (2023). Lanthanide-doped MoS2 with enhanced oxygen reduction activity and biperiodic chemical trends. *Nature Communications*, *14*(1), 3256. DOI: 10.1038/s41467-023-39100-5 PMID: 37277362

Hong, J., Chen, X., Li, P., Koshino, M., Li, S., Xu, H., Hu, Z., Ding, F., & Suenaga, K. (2022). Multiple 2D Phase Transformations in Monolayer Transition Metal Chalcogenides. *Advanced Materials*, *34*(19), 2200643. DOI: 10.1002/adma.202200643 PMID: 35307877

Joo, J., Kim, T., Lee, J., Choi, S., & Lee, K. (2019). Morphology-Controlled Metal Sulfides and Phosphides for Electrochemical Water Splitting. *Advanced Materials*, *31*(14), 1806682. DOI: 10.1002/adma.201806682 PMID: 30706578

Kalaiselvi, C., Kannan, K., Pandian, M. S., Devi, N., Arpornwichanop, A., Chandar, N. K., & Chen, Y.-S. (2025). Ti3C2Tx MXene decorated with CoMnO3 / CoMn2O4 nanoparticles as electrocatalysts for anion exchange membrane water electrolysis. *International Journal of Hydrogen Energy*, *137*, 202–213. DOI: 10.1016/j.ijhydene.2025.04.402

Kannan, K., Chanda, D., Gautam, J., Behera, A., Meshesha, M. M., Gwon Jang, S., & Yang, B. (2023). Hydrothermally synthesized mixed metal oxide nanocomposites for electrochemical water splitting and photocatalytic hydrogen production. *International Journal of Hydrogen Energy*, *48*(93), 36412–36426. DOI: 10.1016/j.ijhydene.2023.06.042

Kannan, K., Chanda, D., Gautam, J., Meshesha, M. M., Jang, S. G., & Yang, B. L. (2023). Facial synthesis of p-p heterojunction composites: Evaluation of their electrochemical properties with photovoltaics-electrolyzer water splitting using two-electrode system. *International Journal of Hydrogen Energy*, *48*(37), 13814–13826. DOI: 10.1016/j.ijhydene.2022.12.235

Kannan, K., Chanda, D., Meshesha, M. M., & Yang, B. (2024). Impressive efficiency of zinc oxide-manganese oxide/MAX composite in two-electrode system for photovoltaic-electrolyzer water splitting. *Colloids and Surfaces. A, Physicochemical and Engineering Aspects*, *689*, 133599. DOI: 10.1016/j.colsurfa.2024.133599

Kannan, K., & Tari, V. (Eds.). (2025). Novel energy storage and conversion technologies for two-dimensional MXenes and MBenes. https://www.amazon.com/dp/9798369395462 ISBN: 9798369395462, eISBN: 9798369395486

Kim, Y., Woo, W. J., Kim, D., Lee, S., Chung, S., Park, J., & Kim, H. (2021). Atomic-Layer-Deposition-Based 2D Transition Metal Chalcogenides: Synthesis, Modulation, and Applications. *Advanced Materials*, *33*(47), 2005907. DOI: 10.1002/adma.202005907 PMID: 33749055

Kumeda, T., Laverdure, L., Honkala, K., Melander, M. M., & Sakaushi, K. (2023). Cations Determine the Mechanism and Selectivity of Alkaline Oxygen Reduction Reaction on Pt(111)**. *Angewandte Chemie*, *135*(51), e202312841. DOI: 10.1002/ange.202312841 PMID: 37983729

Kwon, I. S., Debela, T. T., Kwak, I. H., Park, Y. C., Seo, J., Shim, J. Y., Yoo, S. J., Kim, J., Park, J., & Kang, H. S. (2020a). Ruthenium Nanoparticles on Cobalt-Doped 1T′ Phase MoS_2 Nanosheets for Overall Water Splitting. *Small*, *16*(13), 2000081. DOI: 10.1002/smll.202000081 PMID: 32147958

Kwon, I. S., Debela, T. T., Kwak, I. H., Park, Y. C., Seo, J., Shim, J. Y., Yoo, S. J., Kim, J., Park, J., & Kang, H. S. (2020b). Ruthenium Nanoparticles on Cobalt-Doped 1T′ Phase MoS_2 Nanosheets for Overall Water Splitting. *Small*, *16*(13), 2000081. DOI: 10.1002/smll.202000081 PMID: 32147958

Lamiel, C., Hussain, I., Rabiee, H., Ogunsakin, O. R., & Zhang, K. (2023). Metal-organic framework-derived transition metal chalcogenides (S, Se, and Te): Challenges, recent progress, and future directions in electrochemical energy storage and conversion systems. *Coordination Chemistry Reviews*, *480*, 215030. DOI: 10.1016/j.ccr.2023.215030

Li, M., Cai, B., Tian, R., Yu, X., Breese, M. B. H., Chu, X., Han, Z., Li, S., Joshi, R., Vinu, A., Wan, T., Ao, Z., Yi, J., & Chu, D. (2021). Vanadium doped 1T MoS2 nanosheets for highly efficient electrocatalytic hydrogen evolution in both acidic and alkaline solutions. *Chemical Engineering Journal*, *409*, 128158. DOI: 10.1016/j.cej.2020.128158

Li, S., Tuo, P., Xie, J., Zhang, X., Xu, J., Bao, J., Pan, B., & Xie, Y. (2018). Ultrathin MXene nanosheets with rich fluorine termination groups realizing efficient electrocatalytic hydrogen evolution. *Nano Energy*, *47*, 512–518. DOI: 10.1016/j.nanoen.2018.03.022

Li, Y., Wang, H., Shu, T., Yuan, J., Lu, G., Lin, B., Gao, Z., Wei, F., Ma, C., Qi, J., & Sui, Y. (2022). Two-dimensional hierarchical MoS2 lamella inserted in CoS2 flake as an advanced supercapacitor electrode. *Journal of Energy Storage*, *51*, 104299. DOI: 10.1016/j.est.2022.104299

Liang, T., Liu, Y., Cheng, Y., Ma, F., & Dai, Z. (2020). Scalable Synthesis of a MoS_2 /Black Phosphorus Heterostructure for pH-Universal Hydrogen Evolution Catalysis. *ChemCatChem*, *12*(10), 2840–2848. DOI: 10.1002/cctc.202000139

Liu, R.-J., Chang, L.-Y., Lin, F.-S., Lee, Y.-H., Yeh, M.-H., & Ho, K.-C. (2024). Multifunctional Structure-Modified Quaternary Compounds Co_9Se_8–$CuSe_2$–WSe_2 Mixed with MWCNT as a Counter Electrode Material for Dye-Sensitized Solar Cells. *ACS Applied Materials & Interfaces*, *16*(3), 3476–3488. DOI: 10.1021/acsami.3c16527 PMID: 38207165

Lv, J., Wang, L., Li, R., Zhang, K., Zhao, D., Li, Y., Li, X., Huang, X., & Wang, G. (2021). Constructing a Hetero-interface Composed of Oxygen Vacancy-Enriched Co_3O_4 and Crystalline–Amorphous NiFe-LDH for Oxygen Evolution Reaction. *ACS Catalysis*, *11*(23), 14338–14351. DOI: 10.1021/acscatal.1c03960

Ma, D., Hu, B., Wu, W., Liu, X., Zai, J., Shu, C., Tadesse Tsega, T., Chen, L., Qian, X., & Liu, T. L. (2019). Highly active nanostructured CoS2/CoS heterojunction electrocatalysts for aqueous polysulfide/iodide redox flow batteries. *Nature Communications*, *10*(1), 3367. DOI: 10.1038/s41467-019-11176-y PMID: 31358738

Ma, G., Ye, J., Qin, M., Sun, T., Tan, W., Fan, Z., Huang, L., & Xin, X. (2023). Mn-doped NiCoP nanopin arrays as high-performance bifunctional electrocatalysts for sustainable hydrogen production via overall water splitting. *Nano Energy*, *115*, 108679. DOI: 10.1016/j.nanoen.2023.108679

Ma, X., & Cheng, H. (2023). ReS2 with unique trion behavior as a co-catalyst for enhanced sunlight hydrogen production. *Journal of Colloid and Interface Science*, *634*, 32–43. DOI: 10.1016/j.jcis.2022.12.014 PMID: 36528969

Maeda, K., & Domen, K. (2007). New Non-Oxide Photocatalysts Designed for Overall Water Splitting under Visible Light. *The Journal of Physical Chemistry. C, Nanomaterials and Interfaces*, *111*(22), 7851–7861. DOI: 10.1021/jp070911w

Paitandi, R. P., Wan, Y., Aftab, W., Zhong, R., & Zou, R. (2023). Pristine Metal–Organic Frameworks and their Composites for Renewable Hydrogen Energy Applications. *Advanced Functional Materials*, *33*(8), 2203224. DOI: 10.1002/adfm.202203224

Pimpang, P., Hongsith, K., Choopun, S., & Wongrerkdee, S. (2025). A Review on Chalcogenide-Based Materials for Counter Electrode Applications in Dye-Sensitized Solar Cells: Sulfides, Selenides, and Tellurides. *ACS Omega*, *10*(25), 26293–26310. DOI: 10.1021/acsomega.5c03194 PMID: 40620954

Qian, Y., Yu, J., Zhang, F., Fei, Z., Shi, H., Kang, D. J., & Pang, H. (2023). Hierarchical binary metal sulfides nanoflakes decorated on graphene with precious-metal-like activity for water electrolysis. *Chemical Engineering Journal*, *470*, 144372. DOI: 10.1016/j.cej.2023.144372

Roy, D., Panigrahi, K., Das, B. K., Ghorui, U. K., Bhattacharjee, S., Samanta, M., Sarkar, S., & Chattopadhyay, K. K. (2021). Boron vacancy: A strategy to boost the oxygen reduction reaction of hexagonal boron nitride nanosheet in hBN–MoS_2 heterostructure. *Nanoscale Advances*, *3*(16), 4739–4749. DOI: 10.1039/D1NA00304F PMID: 36134305

Siddiqui, N., Faiyaz, T., & Tari, V. (2022). Ganga and Yamuna: Through the lens of National Green Tribunal. *Nature. Environment and Pollution Technology*, *21*(4), 1657–1664. DOI: 10.46488/NEPT.2022.v21i04.018

Tang, Y., Yang, C., Xu, X., Kang, Y., Henzie, J., Que, W., & Yamauchi, Y. (2022). MXene Nanoarchitectonics: Defect-Engineered 2D MXenes towards Enhanced Electrochemical Water Splitting. *Advanced Energy Materials*, *12*(12), 2103867. DOI: 10.1002/aenm.202103867

Tari, V., Gupta, V., Joshi, A., Kurve, P., & Chandra, N. (2024). Biodiversity loss due to forest diversion. In Rathore, A. (Ed.), *Impact of societal development and infrastructure on biodiversity decline* (pp. 137–165). IGI Global., DOI: 10.4018/979-8-3693-6950-0.ch010

Tari, V., & Kannan, K. (2022). Microbial bioremediation of polythenes and plastics – A green sustainable approach. In *Microbial biotechnology in green remediation* (1st ed.). Elsevier., DOI: 10.1016/B978-0-323-90452-0.00003-7

Thangamathi, R., Kumaresan, N., Praveen Kumar, M., Mangalaraja, R. V., Herrera Diaz, F. V., Farhang Sahlevani, S., Ferreira De Oliveira, T., Pabba, D. P., & Sivakumar, P. (2024). Heterostructure-induced interfacial charge transfer interaction in CoS/CoO@NAC nanosheets as a bi-functional electrocatalyst for water-splitting application. *International Journal of Hydrogen Energy*, *64*, 69–79. DOI: 10.1016/j.ijhydene.2024.03.240

Tu, Y., Shi, Y., Yan, X., Jiang, Y., Xiao, W., Zhu, S., Lv, P., & Li, C. (2024). Synergistic enhancement of bifunctional electrocatalysis for ORR and OER: Unveiling the role of Cr-doped CoS2 and N-doped carbon. *Colloids and Surfaces. A, Physicochemical and Engineering Aspects*, *697*, 134457. DOI: 10.1016/j.colsurfa.2024.134457

Wang, H., Ye, W., Yin, B., Wang, K., Riaz, M. S., Xie, B., Zhong, Y., & Hu, Y. (2023). Modulating Cation Migration and Deposition with Xylitol Additive and Oriented Reconstruction of Hydrogen Bonds for Stable Zinc Anodes. *Angewandte Chemie International Edition*, *62*(10), e202218872. DOI: 10.1002/anie.202218872 PMID: 36647214

Wang, J., Zheng, X., Wang, G., Cao, Y., Ding, W., Zhang, J., Wu, H., Ding, J., Hu, H., Han, X., Ma, T., Deng, Y., & Hu, W. (2022). Defective Bimetallic Selenides for Selective CO_2 Electroreduction to CO. *Advanced Materials*, *34*(3), 2106354. DOI: 10.1002/adma.202106354 PMID: 34699632

Wang, M., Dong, C.-L., Huang, Y.-C., & Shen, S. (2020). Operando Spectral and Electrochemical Investigation into the Heterophase Stimulated Active Species Transformation in Transition-Metal Sulfides for Efficient Electrocatalytic Oxygen Evolution. *ACS Catalysis*, *10*(3), 1855–1864. DOI: 10.1021/acscatal.9b05170

Wu, H., Wang, Z., Li, Z., Ma, Y., Ding, F., Li, F., Bian, H., Zhai, Q., Ren, Y., Shi, Y., Yang, Y., Deng, Y., Tang, S., & Meng, X. (2023). Medium-Entropy Metal Selenides Nanoparticles with Optimized Electronic Structure as High-Performance Bifunctional Electrocatalysts for Overall Water Splitting. *Advanced Energy Materials*, *13*(28), 2300837. DOI: 10.1002/aenm.202300837

Wu, L., Shen, X., Ji, Z., Yuan, J., Yang, S., Zhu, G., Chen, L., Kong, L., & Zhou, H. (2023). Facile Synthesis of Medium-Entropy Metal Sulfides as High-Efficiency Electrocatalysts toward Oxygen Evolution Reaction. *Advanced Functional Materials*, *33*(3), 2208170. DOI: 10.1002/adfm.202208170

Xu, L., Li, D., Zhou, W., Ding, Y., Wu, Y., Xu, J., & Duan, X. (2020). High-efficiency electrodeposition of polyindole nanocomposite using MoS2 nanosheets as electrolytes and their capacitive performance. *Arabian Journal of Chemistry*, *13*(7), 6061–6071. DOI: 10.1016/j.arabjc.2020.05.006

Yao, Z., Yang, J., Ren, Z., Ren, X., Sun, Y., Zhao, Y., Si, J., Lei, Q., Zhang, W., Li, Z., Yin, Y., Chen, J., Wen, W., Zhu, D., Li, X., & Tai, R. (2023). Triggering Mixed Cationic-Anionic Redox in Cu_{2-x} Se Cathodes via Tailored Charge-Carrier for High Energy Density Aqueous Zn Batteries. *Advanced Energy Materials*, *13*(22), 2300236. DOI: 10.1002/aenm.202300236

Yu, L., Zhu, Q., Song, S., McElhenny, B., Wang, D., Wu, C., Qin, Z., Bao, J., Yu, Y., Chen, S., & Ren, Z. (2019). Non-noble metal-nitride based electrocatalysts for high-performance alkaline seawater electrolysis. *Nature Communications*, *10*(1), 5106. DOI: 10.1038/s41467-019-13092-7 PMID: 31704926

Yuan, H., Kong, L., Li, T., & Zhang, Q. (2017). A review of transition metal chalcogenide/graphene nanocomposites for energy storage and conversion. *Chinese Chemical Letters*, *28*(12), 2180–2194. DOI: 10.1016/j.cclet.2017.11.038

Zeng, R., Lian, K., Su, B., Lu, L., Lin, J., Tang, D., Lin, S., & Wang, X. (2021). Versatile Synthesis of Hollow Metal Sulfides via Reverse Cation Exchange Reactions for Photocatalytic CO_2 Reduction. *Angewandte Chemie International Edition*, *60*(47), 25055–25062. DOI: 10.1002/anie.202110670 PMID: 34490697

Zhai, P., Zhang, Y., Wu, Y., Gao, J., Zhang, B., Cao, S., Zhang, Y., Li, Z., Sun, L., & Hou, J. (2020). Engineering active sites on hierarchical transition bimetal oxides/sulfides heterostructure array enabling robust overall water splitting. *Nature Communications*, *11*(1), 5462. DOI: 10.1038/s41467-020-19214-w PMID: 33122636

Zhang, T., Liu, Y., Yu, J., Ye, Q., Yang, L., Li, Y., & Fan, H. J. (2022). Biaxially Strained MoS_2 Nanoshells with Controllable Layers Boost Alkaline Hydrogen Evolution. *Advanced Materials*, *34*(27), 2202195. DOI: 10.1002/adma.202202195 PMID: 35474349

Zhao, W.-W., Bothra, P., Lu, Z., Li, Y., Mei, L.-P., Liu, K., Zhao, Z., Chen, G., Back, S., Siahrostami, S., Kulkarni, A., Nørskov, J. K., Bajdich, M., & Cui, Y. (2019). Improved Oxygen Reduction Reaction Activity of Nanostructured CoS_2 through Electrochemical Tuning. *ACS Applied Energy Materials*, *2*(12), 8605–8614. DOI: 10.1021/acsaem.9b01527

Zheng, Z., Yu, L., Gao, M., Chen, X., Zhou, W., Ma, C., Wu, L., Zhu, J., Meng, X., Hu, J., Tu, Y., Wu, S., Mao, J., Tian, Z., & Deng, D. (2020). Boosting hydrogen evolution on MoS2 via co-confining selenium in surface and cobalt in inner layer. *Nature Communications*, *11*(1), 3315. DOI: 10.1038/s41467-020-17199-0 PMID: 32620781

Chapter 13
Novel Chalcogenide-Based Nanocomposites as Electrocatalysts for Nitrogen Reduction Reactions

P. Selvakumar
https://orcid.org/0000-0002-3650-4548
Department of Science and Humanities, Nehru Institute of Technology, Coimbatore, India

Santosh Kumar Nathsharma
https://orcid.org/0000-0003-3935-2938
Stewart Science College, India

T. C. Manjunath
https://orcid.org/0000-0003-2545-9160
Rajarajeswari College of Engineering, India

ABSTRACT

Despite this abundance, atmospheric nitrogen is chemically inert due to the presence of a strong triple bond (N N) that has a high bond dissociation energy of about 941 kJ/mol. This inertness renders N_2 unavailable for direct assimilation by most living organisms, thereby necessitating its conversion into reactive nitrogen species such as ammonia (NH_3), nitrate (NO_3^-), or nitrite (NO_2^-), which are essential for the biosynthesis of amino acids, nucleotides, and other vital biomolecules. The process by which inert nitrogen gas is converted into ammonia is termed nitrogen fixation, and

DOI: 10.4018/979-8-3373-3962-7.ch013

among the strategies employed to achieve this transformation, Nitrogen Reduction Reactions (NRR) have emerged as a focal point of research, especially in the context of sustainable and environmentally friendly chemistry.Traditionally, the conversion of nitrogen gas into ammonia on an industrial scale has been accomplished through the Haber–Bosch process, developed in the early 20th century.

INTRODUCTION TO NITROGEN REDUCTION REACTIONS (NRR)

Despite this abundance, atmospheric nitrogen is chemically inert due to the presence of a strong triple bond (N≡N) that has a high bond dissociation energy of about 941 kJ/mol. This inertness renders N_2 unavailable for direct assimilation by most living organisms, thereby necessitating its conversion into reactive nitrogen species such as ammonia (NH_3), nitrate (NO_3^-), or nitrite (NO_2^-), which are essential for the biosynthesis of amino acids, nucleotides, and other vital biomolecules, (Tari, 2015b). The process by which inert nitrogen gas is converted into ammonia is termed *nitrogen fixation*, and among the strategies employed to achieve this transformation, Nitrogen Reduction Reactions (NRR) have emerged as a focal point of research, especially in the context of sustainable and environmentally friendly chemistry, (Tari et al., 2022). Traditionally, the conversion of nitrogen gas into ammonia on an industrial scale has been accomplished through the Haber–Bosch process, developed in the early 20th century, (Asiri et al., 2021) It operates under harsh conditions, typically at temperatures of 400–500°C and pressures of 150–300 atm, using iron-based catalysts. These demanding conditions result in high capital and operational costs and present challenges for decentralization and integration with renewable. Nitrogen is one of the most abundant elements on Earth, constituting approximately 78% of the Earth's atmosphere in the form of dinitrogen (N_2) gas, (Tari, 2015a; Tari, 2015b). This inertness renders N_2 unavailable for direct assimilation by most living organisms, thereby necessitating its conversion into reactive nitrogen species such as ammonia (NH_3), nitrate (NO_3^-), or nitrite (NO_2^-), which are essential for the biosynthesis of amino acids, nucleotides, and other vital biomolecules. The process by which inert nitrogen gas is converted into ammonia is termed *nitrogen fixation*, and among the strategies employed to achieve this transformation, (Aslam et al., 2023), Nitrogen Reduction Reactions (NRR) have emerged as a focal point of research, especially in the context of sustainable and environmentally friendly chemistry. Traditionally, the conversion of nitrogen gas into ammonia on an industrial scale has been accomplished through the Haber–Bosch process, developed in the early 20th century.

As the global population continues to grow and environmental concerns intensify, there is an urgent need to develop alternative methods for nitrogen fixation that are both energy-efficient and sustainable, (Tari, et al., 2024). This has led to renewed interest in the development of electrochemical and photocatalytic nitrogen reduction reactions (NRR) as potential substitutes or complements to the Haber–Bosch process. These methods aim to fix nitrogen under ambient or near-ambient conditions, ideally powered by renewable energy sources such as solar or wind power, thus providing a path toward "green ammonia" production. Electrocatalytic NRR involves the direct reduction of N simultaneously suppressing the hydrogen evolution reaction (HER), which competes with nitrogen reduction for protons and electrons. Because HER typically has a lower overpotential than NRR, it tends to dominate in most aqueous electrochemical environments, significantly reducing the Faradaic efficiency of ammonia production, (Bard et al., 2022; Kannan & Tari, 2025) Thus, the development of selective catalysts that can adsorb and activate N_2 while minimizing HER is a critical area of ongoing research. Photocatalytic NRR, on the other hand, mimics the natural process of nitrogen fixation by harnessing light energy—most commonly from sunlight—to drive the conversion of N_2 to NH_3. Ideally, photocatalytic systems can achieve nitrogen fixation at room temperature and atmospheric pressure, offering a potentially scalable and eco-friendly alternative to conventional methods. However, photocatalytic NRR still faces significant challenges, including low quantum efficiency, poor selectivity, and difficulty in distinguishing between genuine NRR-derived ammonia and contamination from background nitrogen sources. In both electrocatalytic and photocatalytic NRR, the adsorption and activation of N_2 molecules on the catalyst surface are crucial steps. Several mechanistic pathways have been proposed for this transformation, including the associative and dissociative pathways. In the associative pathway, nitrogen is reduced without breaking the N≡N bond initially, forming intermediate species such as N_2H or N_2H_2 before ultimately yielding ammonia. In the dissociative pathway, the N≡N bond is cleaved early in the reaction, followed by stepwise hydrogenation of the nitrogen atoms. Understanding and controlling these pathways requires sophisticated catalyst design and the use of advanced characterization techniques, such as in situ spectroscopy and computational modeling. The ideal NRR catalyst must fulfill several criteria: it should have high activity (i.e., promote rapid N_2 conversion), high selectivity (i.e., preferentially produce NH_3 over H_2), and long-term stability under operating conditions. Various classes of materials have been explored as NRR catalysts, including transition metals (e.g., Fe, Mo, Ru), metal nitrides, metal oxides, single-atom catalysts, and hybrid systems combining multiple functional components. Among these, single-atom catalysts have attracted considerable attention due to their unique ability to provide isolated active sites, which can be tailored to achieve specific adsorption energies and electronic environments conducive to N_2 activation.

Another significant research direction in the field of NRR is the development of hybrid systems that combine photocatalysis and electrocatalysis, known as photoelectrochemical NRR, (Chen et al., 2013). These systems aim to harness both light and electrical energy to drive nitrogen fixation more efficiently. Additionally, bio-inspired approaches, such as the design of synthetic catalysts that mimic the active sites of nitrogenase enzymes—nature's own nitrogen-fixing catalysts—offer promising routes for achieving high activity and selectivity under mild conditions. Beyond ammonia production, NRR has implications for the broader nitrogen cycle and environmental remediation. The development of decentralized, small-scale NRR systems could enable on-site ammonia production for agricultural and industrial applications, reducing the need for long-distance transportation and minimizing the environmental footprint. Moreover, understanding NRR processes contributes to our knowledge of nitrogen transformations in natural and engineered systems, aiding efforts to manage nitrogen pollution and promote sustainable development. In conclusion, nitrogen reduction reactions represent a pivotal scientific and technological frontier in the quest for sustainable nitrogen fixation. While significant progress has been made in understanding the fundamental challenges and exploring diverse catalytic strategies, achieving efficient and scalable NRR under ambient conditions remains a formidable goal. Continued interdisciplinary research encompassing materials science, surface chemistry, electrochemistry, photophysics, and environmental engineering will be essential for translating the promise of NRR into practical, real-world solutions for green ammonia production and beyond.

SYNTHESIS AND CHARACTERIZATION OF CHALCOGENIDE-BASED NANOCOMPOSITES

Electrocatalysts serve as the critical functional components in electrochemical systems, facilitating the adsorption, activation, and stepwise hydrogenation of nitrogen molecules (N_2) into ammonia (NH_3). This process mimics the biological nitrogen fixation performed by nitrogenase enzymes, yet in an artificial setting, it requires a fine balance of catalytic properties to overcome significant kinetic and thermodynamic barriers. The strong N≡N bond in molecular nitrogen, with a bond dissociation energy of approximately 941 kJ/mol, makes NRR an energy-intensive and sluggish process, (Chib et al., 2025) Therefore, the design of advanced electrocatalysts that can lower the activation energy, enable selective N_2 adsorption, suppress competitive reactions like the hydrogen evolution reaction (HER), and operate efficiently under mild conditions is a primary research objective in the field of electrocatalytic NRR. Electrocatalysts influence every stage of the nitrogen reduction pathway—from the initial adsorption of the N_2 molecule to its eventual conversion into NH_3. In a typical NRR process in aqueous media, N_2 is adsorbed on the catalyst surface, followed by electron and proton transfer steps that progressively hydrogenate the adsorbed nitrogen species. These transformations proceed through multiple possible intermediates such as N_2H, NH, NH_2, and eventually lead to the desorption of NH_3. The mechanistic pathways for NRR are commonly classified into two types: the associative and dissociative mechanisms. In the associative pathway, the N≡N bond remains intact during the initial hydrogenation steps, while in the dissociative mechanism, the bond is cleaved before hydrogenation. Electrocatalysts must provide appropriate binding energies for these intermediates—not too strong to prevent desorption and not too weak to fail at stabilizing the transition states—according to the Sabatier principle. Tuning this delicate balance is at the heart of electrocatalyst design. A major challenge for electrocatalysts in NRR is selectivity. In aqueous electrochemical environments, (Dahiya et al., 2025) HER is a dominant side reaction because protons and electrons are also the reactants in hydrogen production ($2H^+ + 2e^- \rightarrow H_2$). The standard reduction potential for HER (0 V vs. RHE) is more favorable than that for NRR (−0.092 V vs. RHE), leading to HER outcompeting NRR in most conditions. Consequently, high selectivity toward NRR is difficult to achieve, as many catalysts preferentially reduce protons to hydrogen rather than N_2 to ammonia. Electrocatalysts must, therefore, exhibit a strong preference for N_2 adsorption over hydrogen and offer active sites that stabilize N_2 and its intermediates more effectively than H atoms. Materials engineering efforts have thus focused on tailoring surface electronic structures, defect states, and crystal facets to suppress HER while promoting NRR. The types of electrocatalysts explored for NRR are diverse and range from transition metals and their compounds to more complex nanostructures and hybrid systems. Transition metals like Mo, Fe, Co, and Ru have been studied extensively due to their established roles in both industrial catalysis

ADVANTAGES OF CHALCOGENIDE-BASED NANOCOMPOSITES IN ELECTROCATALYSIS

The synthesis and characterization of chalcogenide-based nanocomposites are crucial steps that govern their structural integrity, morphology, electronic structure, and ultimately, their performance in target applications such as the nitrogen reduction reaction (NRR), photocatalysis, and electrochemical energy systems. The synthesis of chalcogenide nanocomposites typically involves strategies that allow precise control over size, morphology, crystallinity, and composition. These strategies can be broadly categorized into physical, chemical, and hybrid methods. Among the most widely adopted chemical synthesis approaches are hydrothermal and solvothermal methods, which involve reacting metal precursors with chalcogen sources in a sealed autoclave at elevated temperatures and pressures. These methods offer advantages such as uniform particle size distribution, controlled crystal growth, and high product yield, (Digraskar et al., 2019a). The parameters—such as temperature, reaction time, solvent type, pH, and precursor concentration—can be adjusted to tailor the morphology and phase of the final product, including nanorods, nanosheets, quantum dots, or hierarchical nanostructures. Another prominent method for synthesizing chalcogenide nanocomposites is the hot-injection technique, particularly useful for synthesizing monodisperse quantum dots and core–shell nanostructures. In this approach, organometallic precursors are swiftly injected into a hot coordinating solvent containing chalcogen sources. This rapid nucleation followed by controlled growth leads to highly crystalline nanoparticles with precise size control. Sol–gel and precipitation techniques also provide facile and cost-effective routes for producing large quantities of nanocomposites, though they may require post-synthesis annealing to achieve crystallinity and phase purity. Additionally, template-assisted synthesis and electrochemical deposition have been utilized to engineer chalcogenide films and arrays with specific orientations and porosity, expanding their applicability in sensors and electrocatalysis, (Feng et al., 2020). Chemical vapor deposition (CVD), especially in its atmospheric and low-pressure variants, is widely used for the large-scale production of 2D chalcogenide layers on conductive substrates, which are critical for device applications in transistors, photodetectors, and energy conversion systems. Characterization of chalcogenide-based nanocomposites involves a comprehensive suite of analytical techniques to understand their structural, compositional, morphological, optical, and electronic properties. XPS, in particular, is critical for understanding the surface chemistry of chalcogenide materials and their interactions with adsorbates, such as nitrogen molecules in NRR applications. Raman spectroscopy is used to probe vibrational modes that are characteristic of specific chalcogen-metal bonds and can indicate layer thickness in 2D materials. Fourier transform infrared spectroscopy (FTIR) complements Raman by identifying surface functional groups and impurities. For evaluating optical properties, ultraviolet-visible (UV–Vis) absorption spectroscopy provides information on the band gap, which can be estimated using Tauc plots, (Feng et al., 2014) Photoluminescence (PL) spectroscopy reveals radiative recombination behavior and defect-related emissions, essential for assessing the optoelectronic quality of semiconducting chalcogenides. Time-resolved PL and transient absorption spectroscopy offer insights into charge carrier dynamics, which are important for catalytic and photovoltaic applications. The properties of chalcogenide nanocomposites are highly tunable through compositional and structural modifications. Doping with other elements—such as transition metals, alkali or alkaline earth metals, and even nonmetals like nitrogen or phosphorus—can mod

ELECTROCATALYTIC ACTIVITY OF CHALCOGENIDE-BASED NANOCOMPOSITES FOR NRR

These nanocomposites exhibit a compelling blend of high surface area, tunable band structures, rich active sites, and excellent electrical conductivity—traits that are critically important in driving a range of electrocatalytic reactions, including the nitrogen reduction reaction (NRR), (Ge et al., 2015). Their capacity to act as multifunctional electrocatalysts stems not only from the inherent properties of chalcogenides themselves but also from their adaptability into various morphologies, composite structures, and heterojunction systems that allow precise tuning of electronic and catalytic characteristics. One of the foremost advantages of chalcogenide-based nanocomposites in electrocatalysis is which results from their nanostructured and often layered morphologies. This structural arrangement not only facilitates exfoliation into few-layer or monolayer forms, dramatically increasing surface area, but also allows exposure of edge sites—regions where catalytic activity is most pronounced, (Ghediya et al., 2019). The edge sites of TMDs, for instance, have been demonstrated to exhibit higher catalytic activity than basal planes due to unsaturated coordination environments and favorable adsorption energies for reactant molecules such as N_2 or H^+.chalcogen elements, doping with foreign atoms, or forming hybrid composites, it is possible to finely adjust the electronic band structure, Fermi level, and charge distribution of these materials. This tunability directly impacts catalytic efficiency by influencing the adsorption energies of intermediates, electron transfer rates, and reaction kinetics. For example, doping MoS_2 with Co or Fe can introduce new energy states near the Fermi level, improving electrical conductivity and enhancing NRR or HER performance. Similarly, forming heterojunctions—interfaces between two different chalcogenide phases or between chalcogenides and conductive substrates (e.g., graphene, carbon nanotubes)—can create built-in electric fields that facilitate charge separation and directional charge flow, reducing recombination losses and boosting catalytic performance their structural integrity in a wide range of pH environments and at varying electrochemical potentials, making them suitable for long-term operation in aqueous or even alkaline and acidic electrolytes. This durability is a critical requirement for practical electrocatalyst deployment, particularly in industrial applications where catalyst lifespan and reusability dictate economic feasibility.to noble-metal-based catalysts like platinum, iridium, or ruthenium. Transition metals such as molybdenum, tungsten, cobalt, and nickel are relatively abundant and inexpensive, enabling the production of high-performance electrocatalysts at a fraction of the cost of noble metals. This economic benefit, coupled with their strong catalytic performance, positions chalcogenide-based nanocomposites as viable alternatives in scaling up sustainable energy and chemical production technologies. Significant roles. (Hong et al., 2015) Materials such as $MoSe_2$, Bi_2S_3, and CdS exhibit narrow to moderate band gaps, allowing them to absorb visible light efficiently and generate photoexcited charge carriers that can participate in redox reactions. When combined with conductive supports or co-catalysts, these chalcogenides can achieve high photocurrent densities and increased product selectivity under illumination. Poor solubility and chemical inertness of N_2 demand catalysts that can not only adsorb and activate N_2 molecules but also suppress the competing hydrogen evolution reaction (HER), which typically dominates under electrochemical conditions.

Figure 1. Recent advances in electrocatalytic nitrate reduction

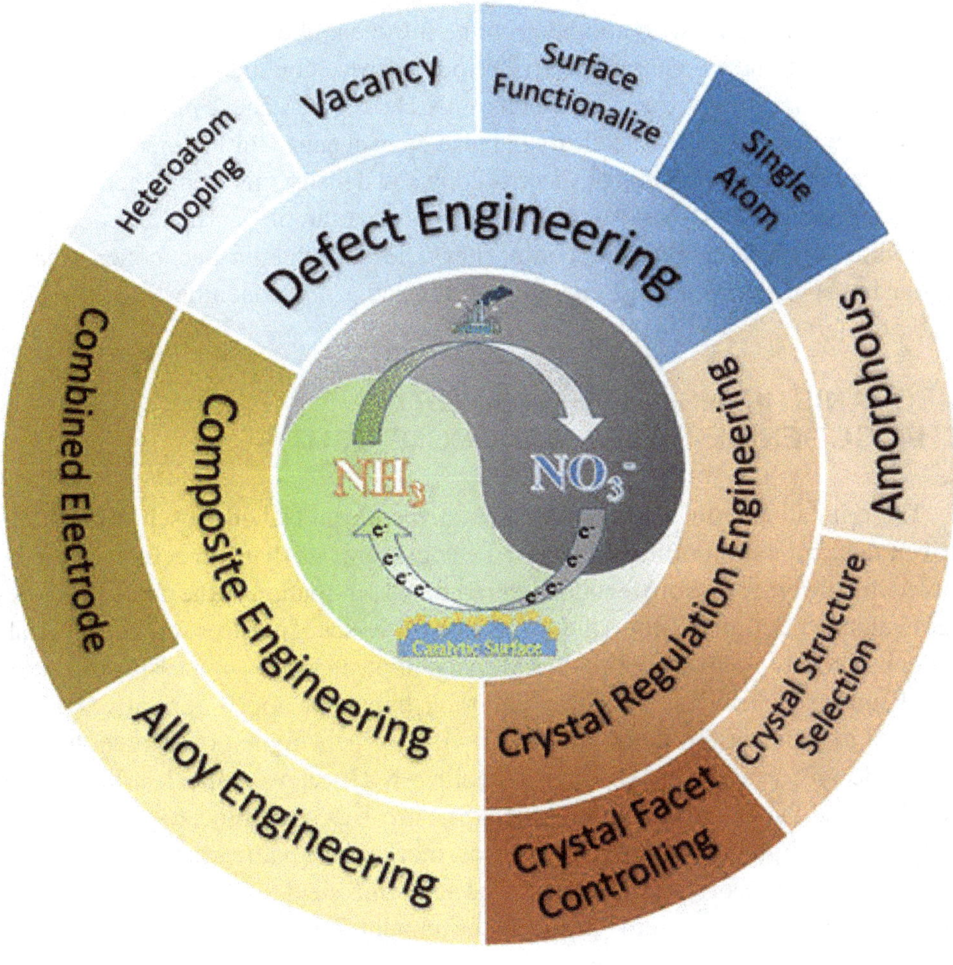

Chalcogenides, especially when engineered to expose specific crystallographic facets or contain tailored defects, can modulate the adsorption energy of N_2 and its reduction intermediates, steering the reaction pathway toward NH_3 formation. For instance, MoS_2-based catalysts with engineered sulfur vacancies have shown enhanced N_2 binding and activation capabilities, (Hu et al., 2016). These defect sites can stabilize N_2 through π back-donation mechanisms and facilitate subsequent hydrogenation steps. Additionally, their ability to reduce HER via geometric and electronic effects—such as by creating localized electron-deficient or hydrophobic zones—further improves their NRR selectivity. Nanocomposites offers an effective strategy to manipulate local electronic environments and create active sites with desirable adsorption characteristics. The introduction of vacancies (e.g., sulfur or selenium vacancies) generates localized states that can serve as anchoring sites for N_2 and promote electron localization, facilitating the multi-electron reduction process. Similarly, doping with nonmetals (e.g., nitrogen, phosphorus) or metals (e.g., Fe, Co, Ni) can induce charge redistribution, improve conductivity, and modulate binding energies for intermediates such as $*N_2H$ or $*NH_2$. These modifications not only enhance activity and selectivity but also open avenues for understanding reaction mechanisms through in situ spectroscopy and density functional theory (DFT) modelling, nanotubes (CNTs), or conducting polymers amplifies the electrocatalytic capabilities of chalcogenide nanocomposites. These conductive backbones enhance charge transport, prevent agglomeration of the active chalcogenide

MECHANISTIC INSIGHTS INTO NRR ON CHALCOGENIDE-BASED NANOCOMPOSITES

The critical need for sustainable alternatives to the Haber–Bosch process has driven the exploration of nanostructured electrocatalysts that can activate inert N_2 molecules, favorably adsorb reaction intermediates, and suppress the parasitic hydrogen evolution reaction (HER). Among the diverse classes of electrocatalytic materials under investigation, chalcogenide-based nanocomposites—encompassing transition metal sulfides, selenides, tellurides, and their hybrid or doped derivatives—have emerged as a potent family of catalysts owing to their high surface area, tunable electronic properties, and capacity to stabilize NRR-relevant species. Evaluating the electrocatalytic activity of these materials involves a multifaceted analysis that integrates quantitative performance metrics, mechanistic pathways, structural characterization, and theoretical modeling. The ammonia yield rate, typically expressed in $\mu g\ h^{-1}\ mg^{-1}$ or $\mu mol\ h^{-1}\ cm^{-2}$, reflects the amount of NH_3 produced per unit time and catalyst mass or area. Faradaic efficiency, defined as the percentage of total charge used in the formation of NH_3 relative to the total charge passed, is critical in gauging selectivity, especially in the presence of competitive HER. A high FE suggests effective utilization of protons and electrons toward ammonia production rather than hydrogen generation. In addition to these values, the onset potential (the potential at which NH_3 formation begins) and the working potential at which optimal performance is achieved (often at or below −0.3 V vs. RHE) are scrutinized to determine energy efficiency. Stable current density and consistent NH_3 output over prolonged electrolysis confirm catalytic durability. Recent studies have shown that chalcogenide-based nanocomposites demonstrate impressive NRR performance due to their defect-rich surfaces, multiple valence states, and favorable adsorption energetics for N_2 molecules. For example, molybdenum disulfide (MoS_2) and its derivatives are among the most investigated materials, owing to their layered structure and semiconducting behavior. (Jabeen et al., 2023) Exfoliated MoS_2 nanosheets with sulfur vacancies (SV-MoS_2) exhibit enhanced NRR activity, as the vacancies act as active centers that facilitate the adsorption and activation of N_2 molecules. These defect sites can lower the energy barrier for N_2 hydrogenation steps and simultaneously hinder HER by creating unfavorable sites for H^+ adsorption. MoS_2-based nanocomposites doped with transition metals (e.g., Fe, Co) or nonmetals (e.g., N, P) further enhance activity by introducing local electronic perturbations that promote N_2 binding while modifying the local charge density. (Jang et al., 2020). play a role in controlling proton access.

Hydrophobic surfaces tend to repel water and thereby reduce HER activity, allowing more protons to interact selectively with adsorbed N_2. For instance, constructing vertically aligned SnS_2 nanosheets with oxygen vacancies was shown to promote NRR over HER, yielding a high FE (>15%) and notable NH_3 productivity at ambient conditions. The oxygen vacancies acted as electron-rich sites, promoting N_2 adsorption, while the layered SnS_2 matrix suppressed HER by decreasing H^+ accumulation on the surface with carbonaceous materials (graphene, carbon nanotubes) or conducting polymers—offer additional performance advantages. The conductive scaffolds not only improve electron transfer but also prevent nanoparticle agglomeration, thereby preserving the high surface area and structural integrity of the chalcogenide phase. For instance, $NiSe_2$/rGO nanocomposites have shown improved NRR performance with FE above 12% and robust stability over 10 hours of continuous operation. The intimate interface between $NiSe_2$ and graphene enhances electron delocalization and are frequently employed to study reaction mechanisms and determine the free energy profiles of various NRR intermediates on chalcogenide surfaces. These simulations allow researchers to identify the potential determining step (PDS) and evaluate the binding energies of intermediates (N_2, N_2H, NH, NH_2) across different

DESIGNING CHALCOGENIDE-BASED NANOCOMPOSITES FOR ENHANCED NRR PERFORMANCE

Understanding the mechanistic pathways of nitrogen reduction reactions (NRR) on chalcogenide-based nanocomposites is critical for advancing their performance and selectivity as electrocatalysts. These materials—comprising transition metal chalcogenides (TMCs) such as MoS_2, WS_2, SnS_2, $NiSe_2$, and others—have demonstrated promising NRR activity under ambient conditions, largely due to their unique surface properties, (Zhang et al., 2014a) layered structures, and tunable electronic configurations. Gaining mechanistic insights involves unraveling how N_2 molecules interact with these catalysts at the atomic level, how electrons and protons are transferred during hydrogenation steps, and which structural features govern selectivity toward ammonia (NH_3) over competing side reactions like the hydrogen evolution reaction (HER). This mechanistic understanding is built upon a combination of experimental observations, electrochemical analysis, and theoretical modeling—particularly density functional theory (DFT) simulations—which together offer a roadmap for rational catalyst design.At the heart of NRR is the activation of the extremely stable N≡N bond in dinitrogen molecules, which has a dissociation energy of ~941 kJ/mol. Chalcogenide-based nanocomposites are especially suited for this task because they can host a variety of active sites—including edge sites, vacancies, dopant atoms, and heterostructure interfaces—that facilitate N_2 adsorption and activation. These materials often possess a layered 2D structure where the edges and defects provide unsaturated coordination environments and localized electronic states that are favorable for binding N_2. Unlike noble metals, which often prefer to adsorb protons and promote HER, TMCs can be engineered to selectively stabilize N_2 and its intermediates, thereby suppressing, (Yu et al., 2020). HER and enhancing NH_3 production. The dissociative mechanism, by contrast, involves the initial cleavage of the N≡N bond into two adsorbed nitrogen atoms (*N), which are then independently hydrogenated to form NH_3. This latter mechanism typically requires high-energy input and is thus less common in ambient NRR electrocatalysis. Chalcogenide-based nanocomposites, particularly those with engineered defects or dopants, have been shown to favor the associative pathway, especially when active sites such as sulfur or selenium vacancies are introduced.

These vacancies create under-coordinated metal atoms with increased electron density that can strongly interact with the antibonding orbitals of N_2, effectively weakening the N≡N bond. For example, in MoS_2, the creation of sulfur vacancies leads to the formation of Mo centers with higher electron density, which enhances π-backdonation into the N_2 molecule's π* orbitals—an essential step for bond activation. Theoretical calculations support this mechanism, showing that N_2 adsorption energy becomes more favorable at defect sites compared to pristine surfaces, and the energy barrier for the first hydrogenation step (*N_2 to *N_2H) is significantly center of transition metal atoms, and introduces new states near the Fermi level, all of which affect the adsorption and activation of N_2 and hydrogenation intermediates. For instance, Co-doped WS_2 has shown improved NRR activity due to the formation of Co–S or Co–N bonds that facilitate N_2 chemisorption and electron transfer. Similarly, Yang et al. (2019) Fe-doped MoS_2 provides Fe centers that mimic the nitrogenase FeMo-cofactor structure, enabling better stabilization of *N_2H and *NH intermediates. Chalcogenide catalysts can also stabilize reaction intermediates through geometric confinement and synergistic effects at heterointerfaces. In hybrid nanocomposites combining metal chalcogenides with conductive supports like graphene or carbon nanotubes, charge redistribution across the interface can lead to localized electric fields that enhance N_2 adsorption and promote selective hydrogenation. The interface can also provide dual-function sites—where one component (e.g., the chalcogenide) binds N_2, while the other facilitates proton delivery—mimicking the bifunctional architecture of biological nitrogenases. chalcogenide-based catalysts, this is often the initial hydrogenation of adsorbed N_2 to $*N_2H$ or the hydrogenation of $*NH_2$ to NH_3. DFT calculations allow researchers to compute the free energy changes for each step in the reaction pathway, helping to identify structural features that reduce these barriers. For example, the introduction of Se into MoS_2 (forming MoSSe alloys) has been shown to tune the adsorption energy of intermediates, making the PDS more energetically favorable and improving overall activity. to verify the presence of intermediates and track changes in catalyst structure during electrochemical NRR. For instance, in situ Raman can detect the formation of –NHx surface groups, while XAS provides information on the oxidation state and coordination environment of metal centers. These tools are essential for correlating theoretical predictions with actual catalytic behavior and for confirming the stability and reusability of the catalyst. typically dominates in aqueous environments. This suppression is achieved through both electronic and structural factors. For example, sulfur-deficient MoS_2 shows decreased H^+ adsorption due to the altered surface polarity and increased N_2 affinity, resulting in enhanced NRR selectivity. Additionally, modifying the hydrophilicity or hydrophobicity of the surface can reduce water access, further limiting HER and enabling higher Faradaic efficiencies for NH_3 production. In conclusion, the mechanistic understanding of NRR on chalcogenide-based nanocomposites reveals a delicate interplay of surface chemistry, electronic structure, and catalyst architecture, (Xiong et al., 2018). The ability to activate N_2 via defect engineering, electronic modulation through doping, and interfacial synergy allows these materials to facilitate the complex multi-electron and multi-proton steps of NRR with increasing efficiency and selectivity. Insights from both experimental characterization and theoretical modeling are vital for guiding the rational design of next-generation chalcogenide electrocatalysts, bringing the goal of sustainable, ambient-condition ammonia synthesis closer to reality.

OPTIMIZING COMPOSITION AND STRUCTURE FOR IMPROVED ELECTROCATALYTIC ACTIVITY

The design of chalcogenide-based nanocomposites for enhanced nitrogen reduction reaction (NRR) performance represents a frontier in sustainable electrocatalysis research. Nitrogen fixation under ambient conditions requires electrocatalysts that can efficiently adsorb and activate molecular nitrogen (N_2), facilitate multi-step hydrogenation, and suppress the competing hydrogen evolution reaction (HER). Chalcogenide materials—particularly transition metal sulfides, selenides, and tellurides—possess unique structural and electronic characteristics that make them promising candidates for such tasks. However, achieving high NRR selectivity and activity demands meticulous engineering of these nanocomposites at multiple scales. Designing efficient chalcogenide-based NRR catalysts involves a multifaceted strategy that includes defect engineering, heteroatom doping, phase and morphology control, heterostructure fabrication, and electronic structure modulation, all guided by theoretical modeling and empirical optimization. active centers for nitrogen adsorption. These vacancies disrupt the symmetry of the crystal lattice and introduce localized states near the Fermi level, which can facilitate charge transfer to adsorbed N_2 molecules. For instance, sulfur-deficient MoS_2 exhibits improved NRR activity because the exposed Mo atoms at vacancy sites can interact strongly with the nitrogen π^* orbitals, enabling effective π-backdonation and bond weakening. Controlled creation of these defects can be achieved through post-synthesis treatments such as annealing in reducing atmospheres, plasma etching, or chemical etching, offering a tunable route to optimize catalytic performance. to modulate the catalytic properties of nanocomposites. Metal dopants such as Fe, Co, Ni, and V can influence the d-band center of transition metal chalcogenides, altering their affinity for nitrogen and reaction intermediates. These dopants often serve a dual role: they improve electrical conductivity and create new active sites with optimized adsorption energy profiles for NRR intermediates, (Wang et al., 2015). For example, Fe-doping in MoS_2 has been shown to mimic the active centers of nitrogenase enzymes, enabling better stabilization of $*N_2H$ and $*NH_2$ intermediates. Similarly, nonmetal doping (e.g., with nitrogen, phosphorus, or boron) can induce electronic polarization and increase the number of Lewis acidic sites, enhancing the interaction with nitrogen molecules while minimizing HER activity. activity of chalcogenide-based materials. The 1T-phase TMDs typically exhibit higher electrical conductivity and a greater density of active sites due to their distorted octahedral coordination geometry. These attributes make 1T-phase chalcogenides more favorable for electrochemical applications, including NRR. Techniques such as lithium intercalation, chemical exfoliation, and electron beam treatment can induce phase transitions from 2H to 1T structures, thereby improving electron transfer kinetics and facilitating multi-electron reduction processes essential for ammonia synthesis. Chalcogenide-based nanocomposites can be synthesized in various nanoscale forms, including nanosheets, nanorods, nanotubes, hollow spheres, and hierarchical architectures. Each morphology provides distinct advantages in terms of surface reactivity and

COMPARISON OF CHALCOGENIDE-BASED NANOCOMPOSITES WITH OTHER ELECTROCATALYSTS

Optimizing the composition and structure of electrocatalytic materials is essential to enhance their efficiency, selectivity, and durability in energy-related reactions, particularly the nitrogen reduction reaction (NRR). Chalcogenide-based nanocomposites, including sulfides, selenides, and tellurides of transition and post-transition metals, have emerged as promising candidates for electrocatalytic NRR under ambient conditions due to their tunable physicochemical properties, multiple oxidation states, and ability to stabilize key intermediates, (Vanalakar et al., 2015). However, their intrinsic activity can be limited by sluggish reaction kinetics, suboptimal electronic conductivity, or competition with the hydrogen evolution reaction (HER). Therefore, precise optimization of both the chemical composition and nanostructural features is critical to fully harness their potential. This optimization encompasses a range of strategies, including doping, alloying, heterostructure construction, morphology control, defect engineering, and interface modification, each tailored to address specific limitations in catalytic performance., Fermi level, and band gap of the host material can be modulated, directly influencing the adsorption energies of NRR intermediates. Transition metal doping (e.g., Fe, Co, Ni, Mn, V) has been shown to enhance the d-band center alignment with nitrogen-containing species, thereby facilitating electron transfer and stabilizing key intermediates like $*N_2H$ or $*NH_2$. For example, Fe-doped MoS_2 or WS_2 exhibits improved activity by mimicking the natural nitrogenase enzyme's active center, enabling better N_2 activation via synergistic effects between Fe sites and the chalcogenide host. Similarly, nonmetal doping (e.g., N, P, B, S into Se-based matrices or vice versa) introduces localized electronic perturbations that can create new active sites and suppress HER. Such dopants can generate charge asymmetry and alter the local dipole moment, enhancing selectivity toward NRR. Alloying, in contrast, involves the formation of solid solutions or mixed-phase materials where two or more metallic elements are combined at the atomic level. Alloying transition metal chalcogenides (e.g., MoSSe, $NiFeSe_2$, $CoNiS_X$) results in unique synergistic interactions that can optimize catalytic behavior. For instance, MoSSe alloy nanosheets, composed of both S and Se atoms, provide tunable binding strength for N_2 and intermediates, allowing more favorable reaction pathways while enhancing conductivity is a cornerstone of structural optimization, enabling the deliberate introduction of vacancies, dislocations, and atomic-scale imperfections that serve as active centers for N_2 adsorption and activation. Among these, chalcogen vacancies—especially sulfur and selenium vacancies in MoS_2 and $MoSe_2$—play a pivotal role in improving NRR performance. These vacancies generate unsaturated coordination sites, often centered around exposed metal atoms with enhanced electron density, which strongly interact with the antibonding orbitals of nitrogen molecules, (Tiwari et al., 2018). As a result, the activation barrier for the first hydrogenation step (N_2 to $*N_2H$) is significantly reduced. Moreover, dual-defect engineering—where both metal and chalcogen vacancies coexist—has emerged as a promising method to further tune the local environment and electronic structure. For instance, in MoS_2 with both Mo and S vacancies, the coordination asymmetry enhances the electron localization and polarization, leading to improved intermediate stabilization. Additionally, edge defects and grain

CHALLENGES IN DEVELOPING CHALCOGENIDE-BASED NANOCOMPOSITES FOR NRR

Chalcogenide-based nanocomposites, composed primarily of transition metal sulfides, selenides, and tellurides, have garnered considerable attention in the field of electrocatalysis, particularly for reactions such as the nitrogen reduction reaction (NRR). These materials possess unique physicochemical characteristics that enable them to facilitate multistep redox processes under ambient conditions. As the scientific community seeks alternatives to the energy-intensive Haber–Bosch process for ammonia synthesis, chalcogenide nanomaterials offer a promising pathway due to their abundant availability, structural flexibility, and tunable electronic properties. However, despite their many strengths, these materials are not without limitations. (Radhakrishnan et al., 2022b)A holistic evaluation of their advantages and disadvantages is essential to understand their current capabilities, guide further optimization, and position them in comparison to other emerging catalysts. Earth-abundant and economically viable elements. This makes them attractive for large-scale deployment, especially when compared to noble metal catalysts like ruthenium or platinum, which, despite their excellent catalytic properties, are prohibitively expensive and scarce. The formation of edge-rich morphologies where catalytic activity is often concentrated. Moreover, these materials can undergo phase transitions (e.g., from semiconducting 2H to metallic 1T in MoS_2) or be tuned via doping and alloying to modify their electronic properties, optimizing their interaction with N_2 and intermediate species during NRR.. Sulfur or selenium vacancies in MoS_2 or $MoSe_2$, for instance, create under-coordinated metal centers that serve as electron-rich active sites capable of strongly adsorbing N_2 molecules and facilitating their activation. Such defect sites lower the energy barriers for the initial hydrogenation steps, significantly boosting NRR performance. Specific dopants or surface modifications, exhibit, (Niyitanga & Jeong, 2017). A relatively lower affinity for proton adsorption compared to noble metals, which helps suppress HER. For example, introducing hydrophobic functional groups or creating surface polarity via doping can inhibit proton access to catalytic sites, improving the Faradaic efficiency of ammonia production. When integrated into nanocomposites, these supports enhance the electron transfer kinetics and prevent agglomeration of chalcogenide nanoparticles. The resulting hybrid structures often exhibit improved conductivity, mechanical stability, and exposure of active sites, thus amplifying their catalytic efficiency methods are not only scalable but also more environmentally benign compared to the high-temperature synthesis routes required for other materials like nitrides or carbides, (Samal et al., 2025). This positions chalcogenides as more practical for industrial-level ammonia production under green chemistry frameworks. Poor intrinsic electrical conductivity. This can limit electron mobility and lower the overall reaction rate. Although metallic phases (like 1T-MoS_2) exhibit improved conductivity, stabilizing

FUTURE DIRECTIONS FOR RESEARCH AND DEVELOPMENT

However, despite their potential, there remain several challenges in optimizing these materials for practical and high-performance NRR applications. These challenges span issues related to catalytic activity, selectivity, stability, scalability, and fundamental mechanistic understanding. Addressing these limitations will be crucial for advancing chalcogenide-based nanocomposites as viable catalytic activity compared to traditional catalysts, such as platinum (Pt) or ruthenium (Ru). While several chalcogenide materials, such as MoS_2, WS_2, and $NiSe_2$, Suresh et al. (2025) have demonstrated moderate to good activity for NRR, their ammonia yield rates and Faradaic efficiencies (FEs) often lag behind those of noble metal catalysts. The sluggish kinetics of N_2 activation and multi-electron hydrogenation are primary reasons for this performance gap. Although defects in chalcogenide structures, such as vacancies and grain boundaries, can create active sites, these defects are often insufficient in driving efficient NRR. Moreover, the poor conductivity of certain chalcogenides (especially in their 2H phase) exacerbates the issue by hindering electron transfer, further limiting catalytic efficiency. To overcome this challenge, strategies such as doping, alloying, and phase engineering are being explored. For instance, doping chalcogenides with metal or non-metal elements can modify their electronic structure, improving the adsorption energy of N_2 and enhancing charge transfer to reaction intermediates. Similarly, phase transitions from semiconducting to metallic phases (e.g., 2H to 1T in MoS_2) can increase conductivity, potentially boosting the NRR performance. However, the synthesis of such materials with consistent phase stability remains a challenge, requiring further optimization of (HER), particularly in aqueous electrolytes, (Pareek et al., 2015). HER, which involves the reduction of protons to hydrogen gas, tends to dominate in most electrocatalytic systems, reducing the selectivity and Faradaic efficiency for NRR. Chalcogenide-based materials, especially those rich in exposed edge sites or with high surface energy, often exhibit an enhanced affinity for proton adsorption, leading to an increased HER rate. This severely limits the efficiency of NRR, as the competing HER reduces the overall ammonia yield. Several approaches have been explored to mitigate this competition, including the introduction of hydrophobic functional groups on the surface to suppress proton adsorption and prevent the activation of HER. Additionally, surface modifications through doping with elements such as phosphorus, nitrogen, or boron can alter the electronic structure of chalcogenide catalysts, lowering the affinity for protons and favoring N_2 activation over proton reduction. However, optimizing such modifications without negatively affecting NRR performance remains a significant challenge. Oxidation, dissolution, or phase transitions. For example, materials like MoS_2 can easily undergo oxidation to form MoO_3, a less catalytically active species, while other chalcogenides like $NiSe_2$ can suffer from phase transitions that reduce their catalytic efficiency. These stability issues can significantly reduce the lifespan of chalcogenide-based catalysts, making them less viable for practical applications in NRR. To enhance the stability and durability of these materials, researchers are exploring hybrid composites, where chalcogenides are combined with more stable materials such as carbon-based supports or conducting polymers. These hybrids can help improve the mechanical and electrochemical stability of chalcogenides by preventing surface oxidation and providing structural support. Additionally, efforts are being made to stabilize chalcogenide phases through the application of protective coatings or by doping with elements that can impart increased resistance to degradation. Chalcogenides, especially in their semiconducting phases (e.g., 2H-MoS_2), suffer from poor electrical conductivity, which hampers the efficient transfer of electrons to the adsorbed nitrogen molecule.

This limitation is particularly problematic in electrochemical applications, where fast electron transport is essential to enhance reaction rates. Efforts to overcome this challenge often involve phase transitions or doping strategies. The transition from the semiconducting 2H-phase to the metallic 1T-phase in MoS_2 is one such approach that has shown promise in improving conductivity. (Kommineni et al., 2025) However, maintaining the stability of the metallic phase under operational conditions is a significant hurdle. Additionally, doping with metals such as iron (Fe) or cobalt (Co) can improve conductivity, but the doping concentration and its impact on NRR selectivity must be carefully optimized to avoid undesirable side reactions or loss of activity., maintaining uniformity in terms of structure, composition, and defect density across large batches of nanomaterials can be difficult. Variations in synthesis conditions, such as temperature, precursor concentration, and reaction time, can lead to significant differences in the material properties, making it challenging to produce consistent, high-quality catalysts to address these scalability issues, researchers are exploring more straightforward and cost-effective synthesis routes, such as chemical vapor deposition (CVD), electrochemical deposition, and simple solution-based methods. However, for these methods to be applicable at an industrial scale, further optimization of reaction parameters and the development of robust techniques for large-scale production are necessary. Detailed understanding of the underlying mechanistic pathways is still lacking. The complex nature of the NRR, involving multiple proton-electron transfer steps and the activation of molecular nitrogen, presents significant challenges in unraveling the exact role of chalcogenide materials in the reaction process. Although defect sites and active centers in chalcogenide materials are known to play a critical role in N_2 activation, the exact nature of the intermediates and the influence of specific surface sites on the reaction mechanism remain poorly understood. Electrochemical analysis, and density functional theory (DFT) calculations, are essential to gain deeper insights into the NRR process on chalcogenide surfaces. A better understanding of the reaction pathways will aid in the design of more efficient catalysts by identifying key factors such as optimal active site density, electronic structure modifications, and the role of surface defects in N_2 adsorption and activation.in industrial ammonia synthesis. Issues related to catalytic activity, selectivity, stability, conductivity, scalability, and mechanistic understanding remain key obstacles that hinder the widespread application of these materials. Nonetheless, ongoing research efforts focused on doping, phase engineering, hybridization, and surface modification hold promise in overcoming these limitations. With continued advancements in synthetic techniques,

IMPLICATIONS FOR SUSTAINABLE ENERGY AND CHEMICAL PRODUCTION

Chalcogenide-based nanocomposites are emerging as promising candidates for these applications, primarily as anode materials. Transition metal, (Maheskumar et al., 2018a), chalcogenides such as molybdenum disulfide (MoS_2), tungsten disulfide (WS_2), and copper selenide (CuSe) have been investigated for their ability to store lithium ions through both intercalation and conversion reactions. These materials offer high theoretical capacities due to their layered structures and high electrochemical activity. For example, MoS_2, with its layered structure, allows lithium ions to intercalate between the layers during cycling, while the high surface area of the material, particularly in the form of nanocomposites, facilitates the rapid diffusion of lithium ions, leading to improved charge-discharge rates. Additionally, the incorporation of conductive carbon materials (such as graphene or carbon nanotubes) into chalcogenide-based nanocomposites can further enhance the electronic conductivity and mechanical integrity of the electrode material, thereby improving the overall performance of LIBs. Despite their potential, chalcogenide-based anode materials often suffer from issues such as low cycle stability and volume expansion during charge-discharge cycles. To overcome these limitations, extensive research is focused on developing strategies to improve the structural stability of these materials. This includes the creation of nanostructured chalcogenides (e.g., nanoparticles, nanowires, or nanosheets) that can accommodate volume changes during cycling and the development of hybrid composites that combine chalcogenides with carbon-based materials or conducting polymers to improve stability and conductivity. Chalcogenide-based nanocomposites are being investigated as promising candidates for both anode and cathode materials in SIBs, given their ability to can effectively accommodate sodium ions during cycling. Similar to their use in LIBs, the layered structures of these materials allow for the intercalation of sodium ions, while their high surface area promotes efficient ionic diffusion. Additionally, the formation of nanocomposites with other materials, such as carbon nanotubes or graphene, can further enhance the performance of chalcogenide-based materials in SIBs by improving conductivity and providing structural support. One of the major challenges in SIBs is the larger ionic radius of Na^+ compared to Li^+, which can lead to greater structural strain and lower cycling stability.

To address this issue, chalcogenide-based nanocomposites can be engineered to accommodate the larger sodium ions through the design of porous structures or by incorporating materials that can buffer the mechanical stress during cycling. Chalcogenide-based nanocomposites are being explored as alternatives to improve the energy density of supercapacitors while maintaining their fast charge-discharge capabilities. Transition metal chalcogenides, such as MoS_2, WS_2, and $NiSe_2$, are ideal candidates for supercapacitors due to their high electrochemical activity, large surface area, and ability to store charge via both electrochemical double-layer capacitance (EDLC) and pseudocapacitance. Pseudocapacitance refers to the charge storage that occurs due to faradaic redox reactions, which is a characteristic of transition metal chalcogenides. the overall performance of supercapacitors, exhibiting high specific capacitance, excellent cycling stability, and fast charge-discharge characteristics. (Pan et al., 2015) Furthermore, the ability to engineer chalcogenide nanostructures at the nanoscale allows for better control over the surface area and porosity, which are critical factors for improving the energy density and efficiency of supercapacitors.), oxygen evolution (OER), and nitrogen reduction (NRR).For hydrogen production, chalcogenide-based materials like MoS_2, WS_2, and $NiSe_2$ are highly efficient electrocatalysts for HER due to their low overpotentials and favorable hydrogen adsorption properties. These materials offer significant advantages over noble metal catalysts, such as platinum, by providing a more abundant and cost-effective alternative for large-scale hydrogen production. Additionally, the two-dimensional (2D) nature of chalcogenides, for oxygen evolution (OER), which is essential in water splitting and other energy conversion processes, chalcogenide-based materials like CoS_2 and $NiSe_2$ have demonstrated impressive catalytic performance. Their ability to facilitate the multi-electron transfer required for OER, combined with their stability and low cost, makes them strong candidates for large-scale water splitting applications. In addition to HER and OER, chalcogenides are also being investigated for nitrogen reduction reactions (NRR), where they can act as electrocatalysts for ammonia synthesis. The ability of chalcogenides, particularly molybdenum and tungsten-based compounds, to activate nitrogen molecules and facilitate the hydrogenation process makes them promising alternatives to traditional Haber–Bosch catalysts. As research in this field continues to progress, chalcogenide-based nanocomposites are likely to play an increasingly important role in the development of more efficient, sustainable, and economically viable energy storage and conversion technologies. By addressing current challenges related to stability, scalability, and performance optimization, these materials could contribute to the next generation of energy solutions that are crucial for a sustainable future.

ADVANCING ELECTROCATALYTIC NITROGEN REDUCTION WITH CHALCOGENIDE-BASED NANOCOMPOSITES

Chalcogenide-based nanocomposites, consisting of transition metal sulfides, selenides, and tellurides, have emerged as promising candidates for electrocatalysis, particularly for energy conversion and storage applications such as hydrogen production, carbon dioxide reduction, and nitrogen fixation. Due to their high electrochemical activity, earth-abundant composition, and customizable properties, chalcogenides are considered an attractive alternative to traditional precious-metal-based catalysts, which suffer from issues of scarcity, cost, and inefficiency. While these materials have demonstrated significant potential, the future of chalcogenide-based nanocomposites in electrocatalysis hinges on overcoming several challenges and advancing their performance, scalability, and stability. This section discusses the future perspectives of chalcogenide-based nanocomposites in electrocatalysis, focusing on the emerging trends, advancements, and potential research directions that will shape their is improving their catalytic performance across various reactions. To enhance catalytic performance, future research will likely focus on the engineering of chalcogenide nanocomposites at the atomic or molecular scale. This includes tuning the electronic structure of chalcogenide materials to optimize active sites for specific reactions. For instance, the incorporation of heteroatoms (such as nitrogen, phosphorus, or carbon) into the chalcogenide matrix can modulate the electronic properties, thus enhancing the catalytic activity for processes like HER and CO_2RR.(Wang et al., 2020) Additionally, doping strategies and alloying with other transition metals can help further enhance the reactivity and selectivity of chalcogenides, making them more efficient for industrial-scale applications. Nanostructured materials, such as nanosheets, nanowires, and nanoparticles, offer high surface areas and abundant active sites, facilitating faster electron and ion transport, which is essential for improving electrocatalytic activity. This issue is particularly relevant for chalcogenide-based catalysts, which can suffer from dissolution, oxidation, or structural collapse during prolonged electrocatalytic cycling. To address these challenges, future research will likely focus on improving the stability of chalcogenide-based nanocomposites through surface modification and protective coatings. For instance, the development of conformal protective layers, (Kaur et al., 2025), such as metal oxides or carbon-based materials, can enhance the resilience of chalcogenide catalysts against corrosion and oxidation while maintaining their catalytic activity. Another promising approach is the use of hybrid chalcogenide-based nanocomposites, where the chalcogenide component is integrated with materials that possess high thermal stability and mechanical strength, such as ceramics or carbon nanomaterials.

These composites can help mitigate the degradation of chalcogenides during long-term use in electrocatalytic applications. Additionally, the development of more robust synthetic methods for chalcogenide nanocomposites is critical for ensuring their stability and reproducibility. Scalable and reproducible fabrication techniques will enable the production of high-quality chalcogenide catalysts with consistent properties, which is essential for their commercial application. Advances in solution-based synthesis methods, such as hydrothermal or solvothermal processes, as well as in situ characterization techniques, will provide valuable insights into the degradation mechanisms and help design more durable catalysts and cost of their synthesis. While laboratory-scale synthesis methods have demonstrated the feasibility of chalcogenide-based catalysts for various electrocatalytic reactions, these methods are often complex, time-consuming, and expensive. To ensure the commercialization of chalcogenide-based catalysts, it is crucial to develop scalable, cost (CVD), chemical bath deposition (CBD), and electrochemical deposition, are promising approaches for large-scale production of chalcogenide-based materials. These methods are relatively simple and cost-effective, offering the potential for high-throughput production of uniform nanocomposites. Moreover, the use of renewable feedstocks and green chemistry principles in the synthesis process could further reduce the environmental impact and cost of production used in electrochemical applications. These templates can help achieve high-quality, uniform deposition of chalcogenides on large-scale surfaces, which is critical for ensuring the practical performance of the catalysts in real-world applications. Additionally, the use of low-temperature and one-pot synthesis methods can reduce the energy consumption associated with catalyst production, making chalcogenide-based catalysts more economically viable for industrial-scale electrocatalysis electrocatalytic reactions, the real challenge lies in integrating these materials into functional devices that can operate efficiently and reliably under real-world conditions. The future of chalcogenide-based nanocomposites in electrocatalysis will involve close collaboration between material scientists, engineers, and industrial stakeholders to design and optimize electrocatalytic devices that can incorporate these materials. For instance, in the context of hydrogen production, the integration of chalcogenide-based catalysts into proton exchange membrane (PEM) electrolyzers or alkaline electrolyzers will require careful consideration of factors such as catalyst loading, electrode design, and electrolyte composition. Similarly, for carbon dioxide reduction, chalcogenide-based nanocomposites must be integrated into efficient electrochemical cells that can operate at high current densities and long lifetimes.

In addition to optimizing the integration of chalcogenide-based catalysts into energy devices, it will also be important to develop strategies for scaling up the devices themselves. This will involve designing systems that are capable of handling large volumes of reactants and products, while maintaining high efficiency and stability over extended operational periods. For example, in fuel cells and electrolyzers, ensuring the uniform distribution of chalcogenide-based catalysts across large electrode surfaces will be critical for maintaining high catalytic performance and minimizing losses due to resistance or inefficient ion transport. transition metals like molybdenum and tungsten, which are widely available and relatively inexpensive compared to rare and precious metals like platinum and palladium alcogenide-based nanocomposites will involve further exploration of their environmental and economic sustainability. Research will likely focus on improving the recyclability and life cycle of chalcogenide catalysts, ensuring that they can be efficiently re-used or repurposed at the end of their life cycle. Moreover, as these materials are used in large-scale electrocatalytic processes, it will be important to assess their environmental impact, including factors such as the energy consumed during their production, the use of water and other resources, and the potential for leaching or contamination during operation.

CONCLUSION

The growing global demand for ammonia as a critical component in fertilizers, coupled with the environmental challenges associated with conventional ammonia production methods, has sparked a significant interest in alternative processes for nitrogen fixation. As such, the development of sustainable, low-energy methods for nitrogen reduction, specifically electrocatalytic nitrogen reduction (NRR), has become a focal point for researchers. One of the most promising approaches in this regard involves the use of chalcogenide-based nanocomposites, which combine transition metal chalcogenides with other materials to enhance electrocatalytic performance. Chalcogenide-based nanocomposites—such as those derived from molybdenum, tungsten, iron, and copper sulfides, selenides, and tellurides—have gained attention due to their rich electrochemical properties, abundance of active sites, and tunable electronic structures. These materials are ideally suited to facilitate the nitrogen reduction reaction (NRR) at ambient conditions, offering a potential pathway for sustainable ammonia production. This research outlook explores the future directions of chalcogenide-based nanocomposites in advancing electrocatalytic nitrogen reduction, highlighting the challenges, opportunities, and emerging trends in this rapidly developing field. For this reason, research into alternative catalysts has surged, with a particular focus on transition metal-based materials such as chalcogenides, which offer several advantages over traditional catalysts. of active sites and enhance conductivity. These nanocomposites are typically engineered by combining transition metal chalcogenides with other materials, such as carbon-based supports (graphene, carbon nanotubes), metal oxides, or even other metal chalcogenides. These hybrid materials can leverage the strengths of both components to address the challenges associated with electrocatalytic nitrogen reduction, particularly the need for high catalytic activity, stability, and selectivity toward ammonia production. The electrocatalytic activity of chalcogenide-based nanocomposites can be tuned through several strategies. One common approach is doping the chalcogenide structure with other metals or non-metals to alter the electronic properties and enhance the interaction with nitrogen molecules. For example, doping with transition metals such as copper, iron, or cobalt has been shown to improve the binding of nitrogen and facilitate the reduction process. Additionally, the morphology of chalcogenide-based nanocomposites—whether it is nanosheets, nanowires, or nanoparticles—can significantly influence their catalytic behavior. High surface area materials with abundant exposed active sites are ideal for NRR, as they provide more opportunities for nitrogen adsorption and subsequent reduction. While chalcogenide-based nanocomposites show promise for NRR, there are several key challenges that need to be addressed to realize their full potential. The most significant of these is the low selectivity of NRR, as chalcogenide-based catalysts can often favor other side reactions, such as hydrogen evolution (HER), leading to inefficiencies. The competition between NRR and HER is a critical issue because HER often dominates, resulting in the production of hydrogen rather than ammonia. Developing chalcogenide-based catalysts with a high intrinsic selectivity for NRR is crucial to making this process viable for large-scale ammonia production. Another challenge is the limited stability of chalcogenide-based catalysts under long-term operation. Many chalcogenides are prone to surface oxidation, dissolution, or degradation under the electrochemical conditions required for NRR. This can result in the loss of catalytic activity over time, making them unsuitable for industrial applications. Research into surface protection strategies, such as the development of protective coatings or the engineering of more robust chalcogenide structures, will be essential for improving the durability and longevity of these catalysts. Moreover, the reaction mechanism of NRR on chalcogenide-based catalysts is not yet fully understood. A deeper mechanistic understanding of how these materials interact with nitrogen reduction using chalcogenide-based nanocomposites represents a promising avenue for developing a sustainable, energy-efficient alternative to the Haber-Bosch process. While significant progress has been made, challenges such as low selectivity, limited stability, and an incomplete understanding of the reaction mechanisms remain. However, with ongoing research focusing on optimizing catalytic sites, tuning electronic properties, and improving the stability of chalcogenide-based nanocomposites, the outlook for NRR using these materials is highly promising.

REFERENCES

Asiri, A. M., Adeosun, W. A., Khan, S. B., Alamry, K. A., Marwani, H. M., & Zakeeruddin, S. M.. (2021). NiCuCoS3 chalcogenide as an efficient electrocatalyst for hydrogen and oxygen evolution. *Journal of Materials Research and Technology*, *15*, 4826–4837. DOI: 10.1016/j.jmrt.2021.09.122

Aslam, S., Awais, M., Servat, N., & Safdar, M. (2023). Synthesis and characterization of Ag2ZnTiS4 nanostructures prepared by a hot-injection method towards low-cost electrocatalytic oxygen evolution. *New Journal of Chemistry*, *47*, 20537–20545. DOI: 10.1039/D3NJ03483

Bard, A. J., Faulkner, L. R., & White, H. S. (2022). *Electrochemical methods: fundamentals and applications*. Wiley.

Chen, S., Duan, J., Jaroniec, M., & Qiao, S. Z. (2013). Three-dimensional N-doped graphene hydrogel/NiCo double hydroxide electrocatalysts for highly efficient oxygen evolution. *Angewandte Chemie International Edition*, *52*, 13567–13570. DOI: 10.1002/anie.201306166

Chib, S., Mehta, A. K., Selvakumar, P., Saravanan, A., Mishra, B. R., & Manjunath, T. C. (2025). Mental Health Challenges and Solutions in High-Pressure Work Environments. In Harriott, S. (Ed.), *Innovative Approaches to Managing Conflict and Change in Diverse Work Environments* (pp. 247–272). IGI Global Scientific Publishing., DOI: 10.4018/979-8-3693-9556-1.ch011

Dahiya, R., & Selvakumar, P. Vidyadhara, & Manjunath, T. C. (2025). Design and Aerodynamics: Principles and Advances. In M. Mellal (Ed.), Innovations and Developments in Unmanned Aerial Vehicles (pp. 145-170). IGI Global Scientific Publishing. https://doi.org/DOI: 10.4018/979-8-3693-8462-6.ch007

Digraskar, R. V., Mali, S. M., Tayade, S. B., Ghule, A. V., & Sathe, B. R. (2019a). Overall noble metal free Ni and Fe doped Cu 2 ZnSnS 4 (CZTS) bifunctional electrocatalytic systems for enhanced water splitting reactions. *International Journal of Hydrogen Energy*, *44*, 8144–8155. DOI: 10.1016/j.ijhydene.2019.02.054

Feng, L., Ding, R., Chen, Y., Wang, J., & Xu, L. (2020). Zeolitic imidazolate framework-67 derived ultra-small CoP particles incorporated into N-doped carbon nanofiber as efficient bifunctional catalysts for oxygen reaction. *Journal of Power Sources*, *452*, 227837. DOI: 10.1016/j.jpowsour.2020.227837

Feng, L., Vrubel, H., Bensimon, M., & Hu, X. (2014). Easily-prepared dinickel phosphide (Ni2P) nanoparticles as an efficient and robust electrocatalyst for hydrogen evolution. *Physical Chemistry Chemical Physics*, *16*, 5917–5921. DOI: 10.1039/C4CP00482E

Ge, J., Chu, J., Yan, Y., Jiang, J., & Yang, P. (2015). Co-Electroplated kesterite bifacial thin-film solar cells: A study of sulfurization temperature. *ACS Applied Materials & Interfaces*, *7*, 10414–10428. DOI: 10.1021/acsami.5b01641

Ghediya, P. R., Joshi, K. K., Kasela, P. A., Chaudhuri, T. K., & Kandoliya, M. (2019). Effect of solvents on physical properties of direct-coated Cu2CoSnS4 films. *Materials Research Express*, *6*, 106419. DOI: 10.1088/2053-1591/ab3ae5

Hong, W. T., Risch, M., Stoerzinger, K. A., Grimaud, A., Suntivich, J., & Shao-Horn, Y. (2015). Toward the rational design of non-precious transition metal oxides for oxygen electrocatalysis. *Energy & Environmental Science*, *8*, 1404–1427. DOI: 10.1039/c4ee03869j

Hu, W.-H., Shang, X., Han, G.-Q., Dong, B., Liu, Y.-R., & Li, X.. (2016). MoSx supported graphene oxides with different degree of oxidation as efficient electrocatalysts for hydrogen evolution. *Carbon*, *100*, 236–242. DOI: 10.1016/j.carbon.2016.01.019

Jabeen, N., Aslam, S., Awais, M., Mirza, M., & Safdar, M. (2023). Construction of quaternary chalcogenide Ag2BaSnS4 nanograins for HER/OER performances and supercapacitor applications in alkaline media. *ACS Applied Electronic Materials*, *5*, 6531–6541. DOI: 10.1021/acsaelm.3c01030

Jang, S., Moon, K., Park, Y., Park, S., & Park, K. H. (2020). Recent studies on multifunctional electrocatalysts for fuel cell by various nanomaterials. *Catalysts*, *10*, 621. DOI: 10.3390/catal10060621

Kannan, K., & Tari, V. (Eds.). (2025). Novel energy storage and conversion technologies for two-dimensional MXenes and MBenes. https://www.amazon.com/dp/9798369395462 ISBN: 9798369395462, eISBN: 9798369395486

Kaur, M., Selvakumar, P., & Dinesh, N. Mohit, Chandel, P. S., & Manjunath, T. C. (2025). Reconstruction and Preservation Strategies for Cultural Heritage. In S. Oflazoğlu & M. Dora (Eds.), Preserving Cultural Heritage in Post-Disaster Urban Renewal (pp. 385-412). IGI Global Scientific Publishing. https://doi.org/DOI: 10.4018/979-8-3693-9097-9.ch016

Kommineni, M., Chundru, S., Maroju, P. K., & Selvakumar, P. (2025). Ethical Implications of AI in Sustainable Development Pedagogy. In Chinedu, C., Mafarja, N., Hussein, M., & Saleem, A. (Eds.), *Rethinking the Pedagogy of Sustainable Development in the AI Era* (pp. 17–36). IGI Global Scientific Publishing., DOI: 10.4018/979-8-3693-9062-7.ch002

Maheskumar, V., Gnanaprakasam, P., Selvaraju, T., & Vidhya, B. (2018a). Comparative studies on the electrocatalytic hydrogen evolution property of Cu2SnS3 and Cu4SnS4 ternary alloys prepared by solvothermal method. *International Journal of Hydrogen Energy*, *43*, 3967–3975. DOI: 10.1016/j.ijhydene.2017.07.194

Niyitanga, T., & Jeong, H. K. (2017). Graphite oxide and molybdenum disulfide composite for hydrogen evolution reaction. *Chemical Physics Letters*, *685*, 451–456. DOI: 10.1016/j.cplett.2017.08.015

Pan, Y., Liu, Y., & Liu, C. (2015). Phase- and morphology-controlled synthesis of cobalt sulfide nanocrystals and comparison of their catalytic activities for hydrogen evolution. *Applied Surface Science*, *357*, 1133–1140. DOI: 10.1016/j.apsusc.2015.09.125

Pareek, D., Balasubramaniam, K. R., & Sharma, P. (2015). Synthesis and characterization of bulk Cu2ZnSnX4 (X: S, Se) via thermodynamically supported mechano-chemical process. *Materials Characterization*, *103*, 42–49. DOI: 10.1016/j.matchar.2015.03.014

Radhakrishnan, J., Kareem, A., Senthilkumar, S., & Biswas, K. (2022b). Facile synthesis of S-vacancy induced electrochemical HER activity in multilayered Sn doped MoS2. *Journal of Alloys and Compounds*, *917*, 165444. DOI: 10.1016/j.jallcom.2022.165444

Samal, A., & Misra, L. I. Mohit, Pougajendy, S., Chandel, P. S., Manjunath, T. C., & Selvakumar, P. (2025). Musealization and Tourism Practices. In S. Oflazoğlu & M. Dora (Eds.), Preserving Cultural Heritage in Post-Disaster Urban Renewal (pp. 115-142). IGI Global Scientific Publishing. https://doi.org/DOI: 10.4018/979-8-3693-9097-9.ch006

Suresh, R. V., Selvakumar, P., Naresh Babu, K., Nithya, G., Manjunath, T. C., & Kartigeyan, J. (2025). Best Practices and Practical Applications. In Çela, E., Paliktzoglou, V., Sengar, S., Eappen, P., & Vajjhala, N. (Eds.), *Enhancing School Counseling With Technology and Case Studies* (pp. 341–362). IGI Global Scientific Publishing., DOI: 10.4018/979-8-3693-8392-6.ch013

Tari, V. (2015a). Earth, man, and environment. In *Definition principles and scope of environment* (pp. 1–20). DPS Publishing House.

Tari, V. (2015b). Physicochemical and biological factors in the environment. In *Definition principles and scope of environment* (pp. 21–29). DPS Publishing House.

Tari, V. Gupta, V., Joshi, A., Kurve, P., & Chandra, N. (2024). Chapter 19: Biodiversity loss due to forest diversion. In A. Rathore (Ed.), *Impact of societal development and infrastructure on biodiversity decline* (pp. 137–165). IGI Global. https://doi.org/ DOI: 10.4018/979-8-3693-6950-0.ch010

Tari, V., & Kannan, K. (2022). Microbial bioremediation of polythenes and plastics – A green sustainable approach. In *Microbial biotechnology in green remediation* (1st ed.). Elsevier., DOI: 10.1016/B978-0-323-90452-0.00003-7

Tiwari, A. P., Azam, A., Novak, T. G., Prakash, O., & Jeon, S. (2018). Chemical strain formation through anion substitution in Cu2WS4 for efficient electrocatalysis of water dissociation. *Journal of Materials Chemistry. A, Materials for Energy and Sustainability*, *6*, 7786–7793. DOI: 10.1039/c8ta01061g

Vanalakar, S. A., Agawane, G. L., Shin, S. W., Suryawanshi, M. P., Gurav, K. V., & Jeon, K. S.. (2015). A review on pulsed laser deposited CZTS thin films for solar cell applications. *Journal of Alloys and Compounds*, *619*, 109–121. DOI: 10.1016/j.jallcom.2014.09.018

Wang, H., Tsai, C., Kong, D., Chan, K., Abild-Pedersen, F., & Nørskov, J. K.. (2015). Transition-metal doped edge sites in vertically aligned MoS2 catalysts for enhanced hydrogen evolution. *Nano Research*, *8*, 566–575. DOI: 10.1007/s12274-014-0677-7

Wang, J., Yue, X., Yang, Y., Sirisomboonchai, S., Wang, P., & Ma, X.. (2020). Earth-abundant transition-metal-based bifunctional catalysts for overall electrochemical water splitting: A review. *Journal of Alloys and Compounds*, *819*, 153346. DOI: 10.1016/j.jallcom.2019.153346

Xiong, B., Chen, L., & Shi, J. (2018). Anion-containing noble-metal-free bifunctional electrocatalysts for overall water splitting. *ACS Catalysis*, *8*, 3688–3707. DOI: 10.1021/acscatal.7b04286

Yang, X., Zhao, Z., Yu, X., & Feng, L. (2019). Electrochemical hydrogen evolution reaction boosted by constructing Ru nanoparticles assembled as a shell over semimetal Te nanorod surfaces in acid electrolyte. *Chemical Communications (Cambridge)*, *55*, 1490–1493. DOI: 10.1039/C8CC09993F

Yu, Y., Xu, J., Zhang, J., Li, F., Fu, J., & Li, C.. (2020). Ultra-thin sns2-pt nanocatalyst for efficient hydrogen evolution reaction. *Nanomaterials (Basel, Switzerland)*, *10*, 2337–2411. DOI: 10.3390/nano10122337

Zhang, H., Xu, P., Ni, Y., Geng, H., Zheng, G., & Dong, B.. (2014a). In situ chemical synthesis of SnO2/reduced graphene oxide nanocomposites as anode materials for lithium-ion batteries. *Journal of Materials Research*, *29*, 617–624. DOI: 10.1557/jmr.2014.37

Zhang, J., & Dai, L. (2016). Nitrogen, phosphorus, and fluorine tri-doped graphene as a multifunctional catalyst for self-powered electrochemical water splitting. *Angewandte Chemie International Edition*, *55*, 13296–13300. DOI: 10.1002/anie.201607405

Chapter 14
Chalcogenide-Based Nanocomposites for CO2 Transformation Into Valuable Chemicals and Clean Energy Generation

Krishnamoorthy Gurushankar
https://orcid.org/0009-0000-3509-4893
Krishna Chaitanya Institute of Technology and Sciences, India

Karthik Kannan
https://orcid.org/0000-0001-5438-2460
National Chung Cheng University, Taiwan & Karpagam Academy of Higher Education, India

ABSTRACT

Optimizing the interaction between the nanocomposites and CO2, researchers aim to develop sustainable methods for producing essential chemicals while simultaneously addressing environmental concerns related to carbon emissions. This innovative strategy not only contributes to resource recovery but also supports the transition towards a more sustainable energy. Therefore, in this chapter, we explore the concept of chalcogenide nanocomposites for CO2 transformation into valuable chemicals and clean energy generation. In addition, by addressing a significant gap in renewable energy research, this chapter offers comprehensive guidance for researchers working on the development of chalcogenide-based materials specifically designed for CO2 transformation into valuable chemicals and clean energy generation.

DOI: 10.4018/979-8-3373-3962-7.ch014

INTRODUCTION

Chalcogenide-based nanocomposites represent a new class of advanced materials that integrate the unique properties of chalcogenides with nanoscale architectures, thereby enhancing their performance across a wide range of applications. Metal chalcogenide nanoparticles are employed in diverse devices, including fuel cells, solar cells, light-emitting diodes (LEDs), ion batteries, supercapacitors, thermo-electric systems, semiconductor diode lasers, photovoltaic cells, optoelectronic devices, sensors, and bioelectronic components, owing to their remarkable physical and chemical characteristics (Adabala and Dutta, 2022; Kannan & Tari, 2024). In addition, chalcogenides play a vital role in photocatalysis (Varadharajan et al., 2024; Kannan et al., 2021), particularly in hydrogen generation via water splitting (Kannan et al., 2023), battery technologies (Gurushankar et al., 2025), carbon dioxide conversion into fuels (Sliem et al., 2025; Kannan et al., 2020; Sadasivuni et al., 2022; Kumar et al., 2022) and selective oxidation processes (Bajpai et al., 2015). Structurally, chalcogenides are compounds composed of one or more chalcogen elements, primarily sulfur (S), selenium (Se), or tellurium (Te), in combination with more electropositive elements such as metals or metalloids.

At the nanoscale, these materials show different properties regarding electricity, light, heat, and strength because of quantum effects, a large surface area compared to their volume, and the ability to control interfaces at the atomic level, (Wani et al., 2025; Tari et al., 2025).

When these nanostructures are mixed with a composite material such as a polymer, metal, or ceramic, the resulting nanocomposites often have improved or entirely new properties compared to their bulk forms or traditional counterparts, (Tari et al. 2022).

Key Characteristics of Chalcogenide-Based Nanocomposites

1. High Tunability: The physical and chemical properties of chalcogenides can be changed by altering their composition, structure, and size.
2. Phase-Change Behavior: Many chalcogenides, especially those used in phase-change memory (like the Ge-Sb-Te system), can switch between amorphous and crystalline states, making them useful for data storage.
3. Optoelectronic Properties: Chalcogenide glasses have a high refractive index, infrared transparency, and photoconductivity, which make them valuable in photonics and infrared sensing.
4. Thermoelectric Potential: Chalcogenide nanocomposites can perform better in thermoelectric applications because they have lower thermal conductivity and maintain or even improve their electrical conductivity.

Common Types and Applications

- Metal Chalcogenide Nanoparticles (such as CdS, ZnS, MoS_2, and Bi_2Te_3): Used in photocatalysis, photovoltaics, thermoelectric, and sensors.
- Chalcogenide Glass Nanocomposites: Used in optical waveguides, infrared lenses, and memory devices.
- Hybrid Chalcogenide Nanocomposites: These might include carbon nanotubes, graphene, or polymers, which can enhance mechanical strength and electrical properties.

Emerging Applications

- Energy Conversion and Storage: Used in supercapacitors, batteries, and thermoelectric generators.
- Sensing and Photodetection: Useful for chemical and biological sensors and photodetectors.
- Environmental Remediation: Employed in photocatalytic processes to degrade pollutants.
- Biomedical Uses: Can be used for drug delivery and imaging, provided that the surface modification and biocompatibility are carefully controlled.

Chalcogenide-based nanocomposites have recently attracted a lot of attention because they have the potential to convert carbon dioxide into clean energy. These materials use the special properties of chalcogen elements (mostly sulfur, selenium, and tellurium) combined with nanoscale effects to enable catalytic, photocatalytic, and electrocatalytic conversions of carbon dioxide into useful fuels and chemicals.

CHALCOGENIDE CLASSIFICATION BASED ON NUMBER OF CONSTITUENT ELEMENTS

Chalcogenides can be classified according to their elemental composition, for example, as binary, ternary, or quaternary systems, as well as by the inclusion or exclusion of metals or the presence of chalcogen ions (Gao et al., 2012). Furthermore, the structural arrangements of many main-group chalcogenide compounds are governed by directional covalent bonding rather than by close-packed configurations.

Binary Chalcogenides

Binary chalcogenides are formed from metal cations combined with a chalcogen anion. Among them, cadmium sulfide (CdS) is one of the most extensively investigated materials. With a band gap of approximately 2.3 eV, CdS is responsive to visible light and exhibits distinctive optical and electrical characteristics that vary with its size and structure. These features make it highly valuable for a wide range of applications. Owing to their versatility, cadmium-based chalcogenides are considered key materials, particularly in photoluminescent devices and solar cell technologies. As reported by Bajpai et al., thermal decomposition of cadmium thiolate powder can be used to synthesize CdS particles with controlled sizes (Gao et al., 2013).

Ternary Chalcogenides

Ternary chalcogenides are produced when a third element from groups III, IV, or V of the periodic table is incorporated into a binary chalcogenide. For instance, introducing a group III or IV metal into copper sulfide (CuS) can yield a ternary chalcogenide. Sulfur and selenium, both group VI elements, exhibit diverse properties and are commonly present in hydrocarbon mixtures (Khan, 2021). These elements are frequently utilized individually or in combination in the synthesis of ternary and quaternary chalcogenides. Typically, ternary chalcogenides contain three different ions, one of which is a chalcogen anion. Their electronic properties can be tuned by varying the Cd: Zn atomic ratio, making compounds such as ZnCdX (X = S, Se) particularly significant. For example, the band gap of zinc cadmium sulfide (ZnCdS) can be adjusted between approximately 2.4 and 3.7 eV. Owing to this tunability, ZnCdS has found applications in heterojunction solar cells, low-voltage cathodoluminescence, and UV diodes. Liu et al. demonstrated the thermochemical synthesis of one-dimensional (1D) ternary chalcogenide nanostructures derived from ZnS and CdS precursors. The resulting ZnCdS nanotubes have shown promising potential for next-generation nanoelectronics and photonics applications (Liu and Huang, 2025).

Quaternary Chalcogenides

Quaternary chalcogenides are produced by partially substituting indium in copper indium sulfide ($CuInS_2$) with zinc and the other half with tin, yielding a derivative of the chalcopyrite-type ternary chalcogenide. This process leads to the formation of the quaternary semiconductor $CuZnSnS_4$ (CZTS, composed of copper, zinc, tin, and sulfur/selenium), which typically crystallizes in either a kesterite or stannite structure, depending on the arrangement of Cu and Zn atoms. Kesterite and stannite share a body-centered tetragonal structure derived from the chalcopyrite structure, but differ in the ordering of Cu and other cations (Zn, Fe, or Sn) within the unit cell's tetrahedral voids. Kesterite features a more ordered arrangement of Cu-Zn and Cu-Sn layers, while stannite has a less ordered or an alternating arrangement of these cations, making them difficult to distinguish experimentally but leading to different energy states. This arrangement gives kesterite a lower total energy than stannite, indicating a lower energy expenditure required to maintain its stable atomic structure with the same chemical elements (Chen et al., 2009; Amrillah, 2023; Boudouma et al., 2025). This energy difference, albeit small, makes kesterite thermodynamically more favorable and, therefore, more promising for applications such as absorber layers in solar cells. The two phases share a similar lattice parameter, where the mesh constant, c, is almost twice as large as the constant a, further underlining kesterite's superiority in terms of stability and energy efficiency. Kesterite CZTS more established as a candidate photocatalyst due to its favorable bandgap, material abundance, and tunable properties. Research focuses on defect engineering to improve selectivity and suppress parasitic reactions, such as the hydrogen evolution reaction (HER), which is more favored on certain crystal facets. Stannite CZTS Research is less developed, but the structural differences suggest it may not be as suitable as kesterite for CO_2RR unless a pure, stable phase with controlled defects can be synthesized. While related stannite compounds show electrocatalytic potential, their specific application for CO_2RR needs further investigation.

In quaternary chalcogenides, one of the four constituent ions is always a chalcogen anion. These materials form a large and versatile class with numerous applications, such as absorber layers in thin-film solar cells and as photocatalysts for solar-driven water splitting. Unlike binary or ternary chalcogenides, quaternary compounds can accommodate a broader variety of elements, thereby exhibiting a wider spectrum of electrical and structural properties. Certain sulfide-based quaternary chalcogenides, such as $AgInZn_3S_9$, show stronger visible-light absorption than their oxide counterparts, resulting in excellent performance under solar irradiation. Toxicity concerns with cadmium-based materials like cadmium sulfide (CdS) and scalability challenges for quaternary chalcogenides such as Copper Zinc Tin Sulfide (CZTS) are significant limitations in the development of thin-film solar cells. Further, the use of CdS is limited by its severe toxicity to human health and the environment. kesterite phase is a promising alternative to CIGS (Copper Indium Gallium Selenide) solar cells because its constituent elements (copper, zinc, tin, and sulfur/selenium) are abundant and non-toxic. However, its transition from a research material to a high-performance commercial product is hindered by several scalability challenges.

CLASSIFICATION OF METAL-BASED CHALCOGENIDES

Chalcogenides are grouped based on the number and type of metals they contain as given in Figure 9.1 and explained in detail in the following section.

CHALCOGENIDES OF ALKALI AND ALKALINE-EARTH METALS

Mono-chalcogenides are typically formed by alkali and alkaline-earth metals. When the sulfide forms of these compounds react with water, they produce derivatives containing disulfide anions. Although they resemble salts, these derivatives are generally colorless and readily soluble in water. In terms of crystal structures, alkali metal chalcogenides often adopt antifluorite or sodium chloride–type arrangements. For instance, KLiX (where X = S, Se, or Te) represents alkali metal chalcogenides, while MX (where M = Ca, Sr, or Ba and X = S, Se, or Te) exemplifies alkaline-earth metal chalcogenides (Kumar et al., 2023; Liu and Huang, 2025).

Chalcogenides of Transition Metals

Transition metal chalcogenides (TMCs) exhibit a wide variety of chemical compositions and structural forms. Frequently studied examples include zinc-based compounds (ZnS, ZnSe, ZnTe), cadmium-based compounds (CdS, CdSe, CdTe), molybdenum disulfide (MoS_2), and mercury cadmium telluride (HgCdTe). These materials typically crystallize in zinc-blende or wurtzite structures and are characterized by strong covalent, largely non-ionic bonding. Based on their composition and structure, TMCs can be further classified into different subgroups (Liu and Huang, 2025).

Chalcogenides Containing Metals

In metal-rich chalcogenides, transition metals typically exhibit oxidation states of +2 or higher. Such compounds often display metal–metal bonding, which can be distinguished through their structural features and characteristic properties. These materials have attracted attention for their potential in electrocatalytic hydrogen evolution. A representative example is the metal-rich sulfide $Nb_{21}S_8$ (Medina-Ramírez et al., 2015).

CLASSIFICATION BASED ON THE NUMBER OF CHALCOGEN IONS

This classification is based on the type and amount of chalcogen ions in the material.

Mono-Chalcogenides

Mono-chalcogenides are compounds with the general formula MX, where M represents a transition metal and X denotes a chalcogen anion such as S^{2-}, Se^{2-}, or Te^{2-}. These materials are characterized by robust surfaces and a low friction coefficient, primarily due to their layered architecture composed of rigid, tightly bonded atomic sheets. Structural studies on gallium sulfide (GaS) and gallium selenide (GaSe) reveal that their crystals consist of two gallium atom layers separated by sulfur or selenium ions (Liu and Huang, 2025).

Dichalcogenides

Typical examples of dichalcogenides include MoS_2, $TiSe_2$, and WSe_2, which conform to the general formula MX_2, where M represents a transition metal and X is a chalcogen anion. These compounds possess a two-dimensional (2D) layered structure based on the 1:2 stoichiometric ratio. Each layer consists of an X–M–X arrangement, held together by weak van der Waals interactions. Within these layers, the metal atoms are coordinated by six surrounding atoms, forming either octahedral or trigonal prismatic geometries (Liu and Huang, 2025).

Trichalcogenides

Trichalcogenides typically contain early transition metals such as Ti, V, Cr, and Mn. They are commonly represented by the formula $M^{4+}(C_2^{2-})$, where C denotes a chalcogen like sulfur or tellurium. A prominent example is niobium triselenide ($NbSe_3$), which can be synthesized through the acid treatment of tetra-thiomolybdate (Queffurus and Barnes, 2015; Siegmund et al., 2020; Liu and Huang, 2025; Ramzan et al., 2024; Xuan et al., 2024).

SIGNIFICANCE OF CHALCOGENIDE-BASED NANOCOMPOSITES FOR CO_2 TRANSFORMATION

Enhanced Catalytic Activity

- Chalcogenides, such as metal sulfides like MoS_2 and WS_2, show high surface reactivity and unique electronic properties (Imran et al., 2025).
- Their layered structures provide active edge sites and tunable band gaps, which improve their interaction with CO_2 molecules.

Improved Selectivity and Efficiency

- Nanocomposite structures can be designed to favor specific reactions, leading to the production of desired products like CO, CH_4, CH_3OH, or HCOOH with greater selectivity.
- These materials' semiconducting properties allow them to be active under visible light and support synergistic charge transfer mechanisms.

Stability and Durability

- Nanocomposites that incorporate chalcogenides are chemically and thermally stable, making them suitable for long-term energy applications.
- Doping and heterostructure formation help improve their resistance to photo corrosion and structural degradation.

Sustainable Energy Conversion

- These materials support the direct conversion of CO_2 into fuels using solar or electrical energy, promoting the goals of a circular carbon economy and clean energy production.

Figure 1. Classification of chalcogenides

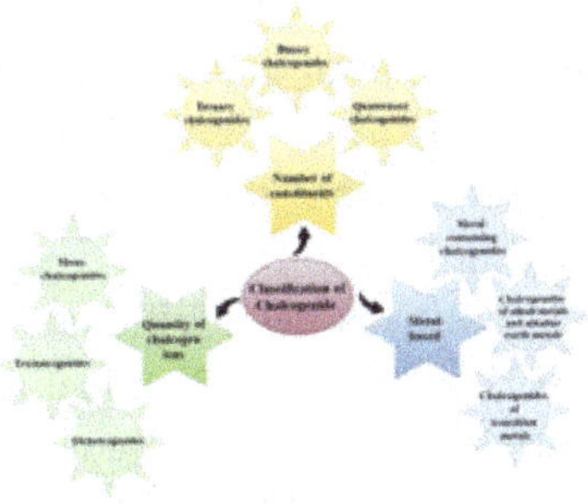

APPLICATIONS IN CLEAN ENERGY GENERATION

Photocatalytic CO_2 Reduction

- Chalcogenide semiconductors like CdS and MoS_2 absorb visible light and create electron–hole pairs to start CO_2 reduction reactions.
- Nanocomposites such as CdS-ZnS or $MoS_2/g\text{-}C_3N_4$ help in separating charges better and capturing more light.

Target Products: CO, CH_4, CH_3OH, HCOOH

Example: The MoS_2/TiO_2 heterojunction is effective in producing CH_4 when exposed to sunlight (Imran et al., 2025).

Electrocatalytic CO_2 Reduction Reaction (CO_2RR)

- Chalcogenide-based nanostructures work as electrocatalysts with low overpotentials, which lowers the energy needed.
- Metal chalcogenides like Cu_2S, SnS_2, and FeSe help convert CO_2 into high-value fuels and chemicals in water or ionic liquid environments.

Benefits: Lower overpotentials, adjustable selectivity, and long-lasting electrodes.

Example: SnS_2 nanosheets are good at turning CO_2 into formate with high Faradaic efficiency (FE) (Yan et al., 2023).

H-SnS_2 nanosheets: Hydrogen-incorporated SnS_2 nanosheets achieved a maximum FE of 93% for total carbonaceous products, including an 87% FE for formate, at -0.9 V vs. RHE.

1% Sb-SnS_2 nanosheets: Antimony-doped SnS_2 nanosheets reached a remarkable FE of 90.86% for formate at -1.1 V vs. RHE.

SnS_2/rGO: SnS_2 nanosheets supported on reduced graphene oxide achieved a maximum FE of 84.5% for formate at -1.4 V vs. Ag/AgCl.

SnS_2 monolayers: Atomic-scale SnS_2 monolayers were reported with a FE of over 90% for formate production.

Overpotentials Overpotential (η) is the additional voltage required beyond the thermodynamic potential to drive an electrochemical reaction at a certain rate. H-SnS_2 nanosheets: Maximum FE of 87% for formate was achieved at -0.9 V vs. RHE.

1% Sb-SnS_2 nanosheets: Maximum FE of 90.86% for formate was observed at -1.1 V vs. RHE.

SnS_2/rGO: An overpotential as low as 0.23 V was reported for formate production, with the maximum FE of 84.5% reached at an overpotential of 0.68 V.

Yields and current densities indicate the reaction rate and the amount of product formed over a specific time and area.

SnS_2/rGO: Reached a current density of 13.9 mA cm^{-2} at an overpotential of 0.68 V for formate production.

H-SnS_2nanosheets: At -1.0 V vs. RHE, a partial current density of 24.4 mA cm^{-2} was measured for formate.

Fe-doped SnS_2 nanosheets: These nanosheets were used for nitrate reduction to ammonia, achieving a high ammonia yield of 7.2 mg $h^{-1}cm^{-2}$ at -0.8 V.

Photoelectrochemical (PEC) Systems

- Chalcogenide photoelectrodes used in PEC cells help directly convert sunlight into fuel.
- Heterostructures like ZnS-CdS or $MoSe_2$-WSe_2 improve visible light absorption and charge movement.

Example: CdS-MoS_2 composite is used in PEC systems to reduce CO_2 into methanol under simulated sunlight (Imran et al., 2025).

Thermo-catalytic Conversion

- Metal chalcogenides can also be used in thermal conditions to convert CO_2 with hydrogen (from renewable sources) into hydrocarbons.
- These materials are often used along with hydrogen from water-splitting, leading to a fully renewable process.

APPLICATION OF CHALCOGENIDE-BASED NANOCOMPOSITES FOR CO_2 TRANSFORMATION

Why Chalcogenides for CO_2 Conversion?

- Many metal chalcogenides like CdS, MoS_2, WS_2, and Bi_2S_3 absorb visible light and can drive CO_2 photoreduction.
- Versatile catalytic sites and defect chemistry: Edge sites, vacancies, dopants, and single-atom anchors on 2D chalcogenides, especially MoS_2, strongly influence CO_2 binding and selectivity in electro- and photo-electrocatalysis.
- Composite engineering with carbon, oxides, MOFs, and cocatalysts solves issues with charge separation and stability, improving activity and selectivity.

CO_2 TRANSFORMATION ROUTES AND RELEVANT CHALCOGENIDE STRATEGIES

Photocatalytic CO_2 reduction (light + semiconductor)

- Use visible-light active chalcogenides like CdS, Cu_2S, and Bi_2S_3 or heterojunctions like CdS/MOF or CdS/Co-BDC to boost CO production.

Common target products: CO, CH_4, HCOOH, CH_3OH (often with low C-C coupling). Example: CdS/MOF composites show significant improvement in activity compared to bare CdS.

Electrochemical CO_2 reduction (ECO_2R)

Transition-metal chalcogenides (TMCs) and functionalized MoS_2 show good activity and selectivity for CO and formate at low overpotentials.

Defect engineering, doping, and single-atom sites are often used to control the energy of intermediate adsorption (COOH*, CO*, H*).

Photoelectrochemical (PEC) and Hybrid Approaches

Combine light absorption by chalcogenides with applied bias or co-catalysts, such as noble metal islands or conductive carbons, to increase reaction rates and control selectivity.

Thermocatalytic / hydrogenation (not as common for chalcogenides)

Some 2D chalcogenides have been studied for CO_2 hydrogenation under high temperature and pressure (to make methanol), but this is more complex and often needs promoters or metal clusters.

PROMISING RESEARCH DIRECTIONS (PUBLISHABLE/NOVEL)

- Investigating single-atom sites on 2D chalcogenides for selective CO_2 conversion to C1 and C2 compounds by integrating experimental operando studies with density functional theory (DFT).
- Developing dynamic catalysts by intentionally utilizing structural changes (reconstruction) to form active states, combining operando spectroscopy with kinetic analysis.
- Creating hybrid MOF-chalcogenide photocatalysts that merge strong CO_2 adsorption properties of MOFs with the light absorption capabilities of chalcogenides for enhanced CO_2 capture and conversion.
- Designing gas diffusion or flow reactors for practical current densities and scalable evaluation of chalcogenide electrocatalysts.

The application of these materials in reactor design and their cost-effectiveness for industrial-scale CO_2 reduction face significant challenges related to mass transport, stability, and catalyst manufacturing.

SIGNIFICANCE AND APPLICATIONS OF CHALCOGENIDE-BASED NANOCOMPOSITES FOR CO_2 TRANSFORMATION INTO VALUABLE CHEMICALS

Favourable Band Structure for CO_2 Reduction

- Chalcogenides such as CdS, MoS_2, and $SnSe_2$ have suitable bandgap energies and band positions that support the CO_2 reduction reaction (CO_2RR) under visible light or with applied electrical potential. For photocatalytic CO_2RR, CdS is well-suited because its conduction band (CB) energy is more negative than the CO_2 reduction potential, providing the necessary driving force for the reaction. Research on CdS has demonstrated high selectivity and efficiency for converting CO_2 to CO. MoS_2 is a transition metal dichalcogenide known for its active edge sites, which play a crucial role in catalysis. These sites provide suitable adsorption energy for CO_2RR intermediates. $SnSe_2$ is a promising catalyst for CO_2RR, often used to produce formic acid (HCOOH). Under electrochemical conditions, $SnSe_2$ can undergo partial reduction to form a Sn-$SnSe_2$ heterostructure, with the newly formed metallic Sn sites being active for CO_2RR.

- These materials can selectively reduce CO_2 while minimizing unwanted side reactions like hydrogen evolution.

High Electrical Conductivity

- Transition metal chalcogenides (TMCs) naturally exhibit good conductivity, which increases the rate of charge transfer.
- Fast electron mobility is essential for efficient electrocatalytic CO_2 conversion.

Surface Active Sites

- Their layered structures and adaptable defects (e.g., sulfur vacancies in MoS_2) provide numerous active sites for CO_2 adsorption and activation.
- Engineering defects boost CO_2 binding and decrease the energy required for activation.

Synergistic Effects in Nanocomposites

- Combining chalcogenides with metals, carbon materials, or other semiconductors enhances light absorption, charge separation, and product selectivity.
- Heterojunctions allow for efficient charge movement and reduce recombination of electrons and holes.

Stability in Harsh Conditions

- Compared to some metal oxides, many chalcogenides demonstrate better chemical stability in aqueous or organic CO_2RR environments.
- They can maintain performance over long periods without significant degradation.

Applications

Photocatalytic CO_2 Reduction

- Mechanism: Solar energy excites electrons in the chalcogenide, which then reduce CO_2, while the holes oxidize water or other sacrificial agents.

Examples:
CdS/TiO_2 composites produce CH_4 and CH_3OH under visible light.
MoS_2/g-C_3N_4 generates formic acid with high selectivity.
$ZnIn_2S_4$/Graphene enhances CO production.

Electrocatalytic CO_2 Reduction

Chalcogenide-based electrodes selectively produce specific products:
MoS_2 nanosheets produce formate/formic acid.
SnS_2 nanostructures generate CO and formate with low overpotentials.
Cu_2−xSe creates multi-carbon products (C_2H_4, ethanol) through C–C coupling.

Photo-electrocatalytic CO_2 Reduction

Combining light and electrical bias:
CdSe/MoS_2 hybrid photoelectrodes efficiently convert CO_2 to CO/methanol.
This boosts reaction rates and reduces the necessary applied voltage.

Thermo-catalytic CO_2 Hydrogenation

Sulfide/selenide catalysts convert CO_2 and H_2 into CH_4 or methanol at elevated temperatures.

Example: Ni-MoS_2 catalysts for CO_2 methanation.

Integrated Artificial Photosynthesis

- Incorporating chalcogenide nanocomposites into membrane reactors or Z-scheme systems to mimic plant photosynthesis.
- Offers potential to directly produce fuels from CO_2, water, and sunlight.

KEY ADVANTAGES OF CHALCOGENIDES OVER OTHER MATERIALS

- Strong ability to absorb visible light due to narrow band gaps.

- Structural flexibility in forms like bulk, nanosheets, and quantum dots.
- Easier to engineer defects for effective CO_2 activation.
- Compatibility with hybrid systems (carbon supports, plasmonic metals).

Significance

Unique Material Properties

- Chalcogenides (compounds with sulfur, selenium, and tellurium) are efficient for light-driven CO_2 reduction due to their narrow band gaps, high electron mobility, and strong light–matter interactions.
- Their adjustable band structure allows matching the redox potentials needed for CO_2 reduction and water oxidation reactions.
- High surface activity and abundance of active sites promote better CO_2 adsorption and activation.

Overcoming Limitations of Conventional Catalysts

- Traditional photocatalysts (such as TiO_2) often have wide band gaps and poor CO_2 conversion efficiency.
- Chalcogenide-based nanocomposites can utilize visible and near-infrared light, improving solar-to-chemical energy conversion.
- Hetero structuring with metals or carbon-based materials reduces electron–hole recombination, enhancing energy efficiency.

Sustainability & Circular Economy

- Enables carbon capture and utilization (CCU), directly transforming greenhouse gases into useful fuels or chemicals.
- Supports achieving net-zero carbon goals and renewable energy storage through synthetic fuels (e.g., methanol, methane, syngas, formic acid).

Applications in CO_2-to-Energy Conversion

Photocatalytic CO_2 Reduction

- Chalcogenide nanocomposites (e.g., CdS, MoS_2, WS_2) can use sunlight to convert CO_2 into:
    - o Methanol (CH_3OH),
    - o Methane (CH_4),

- o Carbon monoxide (CO).

These products can serve as fuels or chemical building blocks.

Electrocatalytic CO_2 Reduction

- Metal chalcogenide electrocatalysts (e.g., Cu_2S, Ni_3S_2, $CoSe_2$) operate under mild conditions and efficiently convert CO_2 into formic acid, CO, or hydrocarbons with high Faradaic efficiency.
- When combined with renewable energy (solar/wind), this enables the production of carbon-neutral fuels.

Photoelectrochemical (PEC) CO_2 Reduction

- Integrating chalcogenide photoelectrodes with conductive supports for solar-driven CO_2 electrolysis.

Examples: p-Cu_2O/n-MoS_2 heterojunctions improve photovoltage and charge separation.

Hybrid Energy Systems

- Combining CO_2 reduction with hydrogen evolution to generate syngas (H_2 + CO) for Fischer–Tropsch synthesis.
- Integrating with fuel cells to directly utilize CO or methanol for power generation.

Industrial CO_2 Utilization

- Potential use in sectors like cement, steel, and chemical industries to capture CO_2 from flue gases and convert it on-site into fuels or chemical precursors, reducing emissions and energy costs. Some of the examples chalcogenide nanocomposites in CO_2 transformation as given in Table 9.1.

Sustainability

- **Waste valorization and resource circularity:** This technology directly addresses the issue of atmospheric CO_2 emissions by repurposing a waste product into valuable fuels and chemicals. This aligns with circular economy models by creating a new revenue stream from industrial waste, rather than simply capturing and storing it.

- **Decarbonization of industry:** By developing a process that uses renewable energy and abundant materials, chalcogenide catalysts can help decarbonize high-emissions sectors. For example, they can be used to produce chemical feedstocks that traditionally rely on fossil fuels.
- **Reduced reliance on fossil fuels:** The synthesis of valuable chemicals like methanol and ethanol from CO_2 provides an alternative to fossil fuel-derived chemical manufacturing. As the process is powered by renewable electricity or solar energy, it reduces overall dependence on non-renewable resources.
- **Earth-abundant materials:** Many chalcogenides are based on earth-abundant elements such as copper (Cu), zinc (Zn), tin (Sn), and sulfur (S), which makes them more sustainable and less expensive than catalysts based on noble metals.
- **Reduced environmental toxicity:** Research is shifting toward more environmentally benign elements. While some chalcogenides, like those involving cadmium, can be toxic, ongoing research is exploring safer, non-toxic alternatives for large-scale applications.

Industrial Integration

- **Process scalability and optimization:** Researchers are focused on developing cost-effective and scalable synthesis methods like hydrothermal and solvothermal processes for mass production of these nanocomposites.
- **Modular and decentralized production:** Photochemical CO_2 conversion can be implemented at a smaller scale near the source of emission, such as at industrial sites. This approach can make the process more efficient by eliminating the costs and energy associated with CO_2 transport.
- **High selectivity and product control:** Industrial processes require high selectivity to produce a specific chemical. The ability to tune chalcogenide catalysts to produce a desired product (e.g., formate, ethanol, or methanol) with high selectivity is crucial for industrial viability, simplifying downstream separation and purification.
- **Market diversification:** The ability to convert CO_2 into various products creates new markets for CO_2 capture and utilization (CCU) technology. It offers an economic incentive beyond carbon credits by producing sellable, high-value chemicals.
- **Technological challenges to overcome:** Despite the promise, several challenges remain before full industrial integration. These include improving cat-

alytic activity, increasing long-term stability and durability, and optimizing reactor design for large-scale production.

Table 1. Example chalcogenide nanocomposites in CO_2 transformation

Nanocomposite	CO_2 Conversion Product	Key Advantage	References
CdS–Graphene	CH_4, CH_3OH	Enhanced electron transport & visible light absorption	Yu et al., 2024
MoS_2–g-C_3N_4	CO, HCOOH	Abundant active edges, strong CO_2 adsorption	Imran et al., 2025
Cu_2S–TiO_2	CH_4, CO	Broad light harvesting & stability	Santhi et al., 2023
WS_2–Metal Nanoparticles	CH_4	Plasmonic enhancement & catalytic synergy	Zhu et al., 2022

CONCLUSION

Chalcogenide-based nanocomposites have emerged as highly promising materials for CO_2 transformation due to their unique electronic structures, tunable band gaps, and high surface reactivity. Their integration into photocatalytic, electrocatalytic, and photoelectrochemical systems enables the efficient conversion of CO_2 into value-added chemicals such as methanol, formic acid, and syngas, as well as clean fuels like hydrogen. By optimizing composition, morphology, and heterojunction design, these materials can overcome limitations in catalytic efficiency, selectivity, and stability. Continued research focusing on defect engineering, co-catalyst coupling, and scalable synthesis will further enhance their performance and economic viability. Ultimately, chalcogenide-based nanocomposites offer a sustainable and technologically feasible pathway for addressing climate change challenges while advancing the global transition toward a circular carbon economy and renewable energy future.

REFERENCES

Adabala, S., & Dutta, D. P. (2022). A review on recent advances in metal chalcogenide-based photocatalysts for CO_2 reduction. *Journal of Environmental Chemical Engineering, 10*(3), 107763.

Amrillah, T. (2023). Enhancing the value of environment-friendly CZTS compound for next generation photovoltaic device: A review. *Solar Energy, 263*, 111982. DOI: 10.1016/j.solener.2023.111982

Bajpai, P. K., Yadav, S., Tiwari, A., & Virk, H. S. (2015). Recent advances in the synthesis and characterization of chalcogenide nanoparticles. *Diffusion and Defect Data, Solid State Data. Part B, Solid State Phenomena, 222*, 187–233.

Boudouma, A., Layachi, O. A., Hrir, H., Nejmi, Y., Taoufyk, A., Moujib, A., El-brouzi, A., & Khoumri, E. (2025). Eliminating the sulfurization step: Optimizing the growth of CFTS thin films in the stannite phase by electrodeposition. *Ionics, 31*, 3847–3859. DOI: 10.1007/s11581-025-06144-0

Chen, S., Gong, X. G., Walsh, A., & Wei, S.-H. (2009). Crystal and Electronic Band Structure of Cu2ZnSnX4 (X = S and Se) Photovoltaic Absorbers: First-Principles Insights. *Applied Physics Letters, 94*, 041903. DOI: 10.1063/1.3074499

Gao, M., Jiang, J., & Yu, S. (2012). Solution-based synthesis and design of late transition metal chalcogenide materials for oxygen reduction reaction (ORR). *Small, 8*(13), 13–27. PMID: 21972127

Gao, M.-R., Xu, Y.-F., Jiang, J., & Yu, S.-H. (2013). Nanostructured metal chalcogenides: Synthesis, modification, and applications in energy conversion and storage devices. *Chemical Society Reviews, 42*, 2986–3017. PMID: 23296312

Gurushankar, K., Karuppaiya, C., Tari, V., & Kannan, K. (2025). Exploring MXenes and MBenes in Metal Sulfur and Selenium Battery Technologies. In *Novel Energy Storage and Conversion Technologies for Two-Dimensional MXenes and MBenes* (pp. 105–120). IGI Global Scientific Publishing.

Imran, M., Abdullah, A. Z., Khan, M. E., & Mohammad, A. (2025). A focused review on photocatalytic potential of graphitic carbon nitride (g-C3N4) based metal oxide-nanostructures for effective remediation of most overused antibiotics. *Journal of Environmental Management, 373*, 123759. DOI: 10.1016/j.jenvman.2024.123759 PMID: 39708681

Kannan, K., Chanda, D., Gautam, J., Behera, A., Meshesha, M. M., Jang, S. G., & Yang, B. (2023). Hydrothermally synthesized mixed metal oxide nanocomposites for electrochemical water splitting and photocatalytic hydrogen production. *International Journal of Hydrogen Energy*, *48*(93), 36412–36426.

Kannan, K., Chanda, D., Gautam, J., Meshesha, M. M., Jang, S. G., & Yang, B. L. (2023). Facial synthesis of pp heterojunction composites: Evaluation of their electrochemical properties with photovoltaics-electrolyzer water splitting using two-electrode system. *International Journal of Hydrogen Energy*, *48*(37), 13814–13826.

Kannan, K., Gautam, J., Chanda, D., Meshesha, M. M., Jang, S. G., & Yang, B. L. (2023). Two-dimensional MAX supported copper oxide/nickel Oxide/MAX as an efficient and novel photocatalyst for hydrogen evolution. *International Journal of Hydrogen Energy*, *48*(20), 7273–7283.

Kannan, K., Radhika, D., Gnanasangeetha, D., Lakkaboyana, S. K., Sadasivuni, K. K., Gurushankar, K., & Hanafiah, M. M. (2021). Photocatalytic and antimicrobial properties of microwave synthesized mixed metal oxide nanocomposite. *Inorganic Chemistry Communications*, *125*, 108429.

Kannan, K., Sliem, M. H., Abdullah, A. M., Sadasivuni, K. K., & Kumar, B. (2020). Fabrication of ZnO-Fe-MXene based nanocomposites for efficient CO2 reduction. *Catalysts*, *10*(5), 549.

Kannan, K., & Tari, V. (Eds.). (2024). *Novel energy storage and conversion technologies for two-dimensional MXenes and MBenes*. ISBN: 9798369395462, eISBN: 9798369395486DOI: 10.4018/979-8-3693-9546-2

Khan, M. M. (2021). *Introduction and fundamentals of chalcogenides and chalcogenides-based nanomaterials*. Elsevier., DOI: 10.1016/B978-0-12-820498-6.00001-9

Kumar, B., Malali, P., Sadasivuni, K. K., & Kannan, K. (2022). Two-dimensional Nanomaterials Design and Reactor Engineering of Different Methods for CO2 Electrochemical Conversion Process.

Kumar, K., Jamnuch, S., Majidi, L., Misal, S., Ahmadiparidari, A., Dato, M. A., Sterbinsky, G. E., Wu, T., Salehi-Khojin, A., Pascal, T. A., & Cabana, J. (2023). Active states during the reduction of CO_2 by a MoS_2 electrocatalyst. *The Journal of Physical Chemistry Letters*, *14*(13), 3222–3229. DOI: 10.1021/acs.jpclett.2c03942 PMID: 36972067

Liu, Y., & Huang, C. (2025). Recent advances in metal chalcogenide catalysts for electrochemical carbon dioxide reduction. *Advanced Functional Materials*. Advance online publication. DOI: 10.1002/adfm.202505245

Medina-Ramírez, I., Hernández-Ramírez, A., & Maya-Trevino, M. L. (2015). Synthesis methods for photocatalytic materials. In Photocatalytic semiconductors: Synthesis, characterization, and environmental applications (pp. 69–102).

Queffurus, M., & Barnes, S.-J. (2015). A review of sulfur to selenium ratios in magmatic nickel–copper and platinum-group element deposits. *Ore Geology Reviews*, *69*, 301–324.

Ramzan, M. A., Favre, R., Steinmann, S. N., Le Bahers, T., & Raybaud, P. (2024). Electrocatalytic reduction mechanisms of CO_2 on MoS_2 edges using grand-canonical DFT: From CO_2 adsorption to HCOOH or CO. *The Journal of Physical Chemistry. C, Nanomaterials and Interfaces*, *128*(24), 10025–10034. DOI: 10.1021/acs.jpcc.4c03266

Sadasivuni, K. K., Kannan, K., Abdullah, A. M., & Kumar, B. (Eds.). (2022). *2D nanomaterials for CO2 conversion into chemicals and fuels*. Royal Society of Chemistry.

Santhi, K., Harish, S., Navaneethan, M., & Ponnusamy, S. (2023). Enhanced Photocatalytic activities of TiO_2 through Metal chalcogenides-based Nanocomposites (ZnS/TiO_2) for Methylene Blue degradation. *Surfaces and Interfaces*, *41*, 103205. DOI: 10.1016/j.surfin.2023.103205

Siegmund, D., Blanc, N., Smialkowski, M., Tschulik, K., & Apfel, U. (2020). Metal-rich chalcogenides for electrocatalytic hydrogen evolution: Activity of electrodes and bulk materials. *ChemElectroChem*, *7*, 1514–1527.

Sliem, M. H., Kannan, K., Maurya, M. R., Jlassi, K., Sadasivuni, K. K., Kumar, B., & Abdullah, A. M. (2025). Rational synthesis of mixed metal oxide clusters supported on a partially etched MAX phase for efficient electrocatalytic CO2 conversion. *Topics in Catalysis*, *68*(13), 1382–1397.

Tari, V. Kannan, K., & Vishwakarma, V. (2022). Chapter 8: Emerging antiviral technologies. In P. Nunthavarawong, S. M. Rangappa, S. Siengchin, & M. Thoppil-Mathew (Eds.), *Antimicrobial and antiviral materials: Polymers, metals, ceramics, and applications* (1st ed., pp. 137–150). CRC Press. DOI: 10.1201/9781003143093-8

Tari, V. Kannan, K., Koliar, J., Gurushankar, K., & Wahyuni, D. K. (2025). Influence of zinc oxide nanomaterials on crop plants: A prospective analysis. In D. Jabbrova, D. Sarkar, Rakesh, S. Singh, & R. Datta (Eds.), *Smart nanomaterials in agroforestry systems* (Chapter 8). Springer. DOI: 10.1007/978-981-96-1337-3_8

Wani, S. A., Tari, V., & Mansoor, S. I. U. (2025). A Comprehensive Analysis of Wastewater Management Challenges in India: Infrastructure and Policy Perspectives. In Dhiman, S., & Mukherjee, G. (Eds.), *Resource Resurgence. Springer Water.* Springer., DOI: 10.1007/978-3-032-02750-4_11

Xuan, R., Mo, J., Chen, J., Dou, Y., Li, X., Jiang, Z., Chai, B., Wang, C., Ding, D., Yan, J., & Wang, X. (2024). Enhanced photocatalytic CO_2 conversion of a CdS/Co-BDC nanocomposite via Co(II)/Co(III) redox cycling. *RSC Advances*, *14*(35), 25247–25255. DOI: 10.1039/d4ra04842c PMID: 39139241

Yan, L., Wu, Z., Li, C., & Wang, J. (2023). Sb-doped SnS_2 nanosheets enhance electrochemical reduction of carbon dioxide to formate. *Journal of Industrial and Engineering Chemistry*, *123*, 33–40. DOI: 10.1016/j.jiec.2023.03.014

Yu, J., Jin, J., Chenga, B., & Mietek Jaroniec, M. (2024). A noble metal-free reduced graphene oxide–CdS nanorod composite for the enhanced visible-light photocatalytic reduction of CO_2 to solar fuel. *Journal of Materials Chemistry. A, Materials for Energy and Sustainability*, *2*, 3407–3416. DOI: 10.1039/C3TA14493C

Zhu, L., Wang, Y., Qin, C., & Cao, J. (2022). First Principles Study of Photocatalytic Reduction of CO_2 to CH_4 on WS_2-Supported Pt Clusters (2022). *The Journal of Physical Chemistry. C, Nanomaterials and Interfaces*, *126*, 16702–16709. DOI: 10.1021/acs.jpcc.2c06126

Chapter 15
Chalcogenide-Based Nanocomposites for the Removal of Emerging Organic Micropollutants in Wastewater

Suganthi Nachimuthu
https://orcid.org/0000-0001-7325-6662
Government Arts College, Bharathidasan University, Karur, India

Kuppusamy Pushpanathan
Government Arts College, Bharathidasan University, Karur, India

ABSTRACT

Nowadays, a fast growing in population creates organic and inorganic waste in the form of extensive effluence and which causes water pollution. Industries, agriculture, domestic processes, etc instigated a huge pollution to the water atmospheres everywhere in the world via various types of chemical elements and substances known as "emerging pollutants". Among them, an organic form is high attention as a result of their comparatively excessive intensities in water. The structural and behavioural diversity of these emerging organic pollutants creates their potential dangers to living organisms. This chapter deals about the chalcogenide-based nanocomposites are utilized like catalysts for removing environmental organic micropollutants in wastewater treatment. Owing to their low bandgaps, higher charge carrier movement with high visible-light absorbing properties, chalcogenide-based

DOI: 10.4018/979-8-3373-3962-7.ch015

nanocomposites are broadly used as photocatalysts. High efficacy, low-cost, precluding long-lasting pollutants, as well as direct utilization of solar energy are an advantage of the photocatalytic approach.

INTRODUCTION

The ever-increasing environmental troubles have been caused due to the human population and industrial development. To remove the emerging organic pollutants from wastewater, various analytical methods and removal techniques have been thoroughly analysed and developed over a period, (Wani et al, 2025). The industrial region liberated about 400 megatons of chemicals, with detrimental solvents as well as metal ions, entering the water, (Tari et al., 2024; Tari et al., 2021). In addition, industrial discarded, several human activities, such as using pesticides, fertilizers, and the discarding of home refuse, and several harmful organic substances like formaldehyde, azo dyes, dioxins, pesticides, as well as heavy metals contribute to the contamination of water bodies (Siddiqui et al. 2022; Tari & Patil, 2017a; Yalçın et al. 2022; Özşirvan et al. 2024; Tari & Patil, 2017b). Every year, about 15% of indestructible textile dyes are released into waterways. The World Bank estimated that the textile business activities caused water pollution of about 17–20% (Holkar, 2016). The majority of organic dyes have carcinogenic effects, which creates substantial health hazards to humans and aquatic life. These toxins may prompt several health problems, for example, cancer, neurological weakening, cardiovascular syndromes, and digestive disorders, also existing in slight amounts, (Tari et al. 2025; Al-Tohamy, 2022, Tari et al. 2022). The water contamination by various sources is shown in Figure 1.

Figure 1. Resource of water contamination by pollutants (Sivasamy et al., 2024)

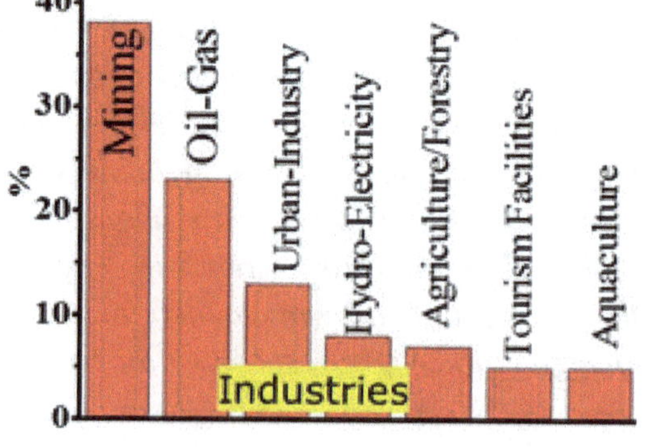

Consequently, there is a crucial necessity for detoxifying the chemical pollutants from contaminated water to confirm long-standing sustainable. Several typical process for treating water for instance membranes, ion interchange, purification, rectification, coagulation, and precipitation. Some progressed oxidation methods include photocatalysis, wetness oxidation, also sonolysis, remain established to mineralize hardy elements (Kumaravelan et al., 2021). The incorporation of various analysis in the domain of photocatalysis has been extensively determined for environmental restructuring, cleaned hydrogen (H_2) fuel production, and the transformation of CO_2. Photocatalysis has numerous benefits contrasted with conventional catalytic procedures, for example eradicating dangerous phases and functional at higher temperatures and pressures (Varadharajan et al., 2024; Kannan et al., 2021). Investigations have revealed that utilizing semiconductor (SC) photocatalysis concerning photo-oxidation procedures harvests water processing efficacies of above 95% (Sivasamy et al., 2024).

Metal chalcogenides such as elements like oxygen, sulfur, selenium, and tellurium have gathered consideration as a consequence of its wide properties. The sources these materials search have occurred during the 1960s. Chalcogenide SC nanoparticles show a typical unification of optical, electrical, thermal, and catalytic features, execution them compatible for various usages in ecological nanotechnology and eco-friendly remedy (Figure 2) (Sivasamy et al., 2021, Sekar et al., 2021). Chalcogenide nanomaterials in various field applications is displayed in Table.1.

Figure 2. Chalcogenides environmental applications

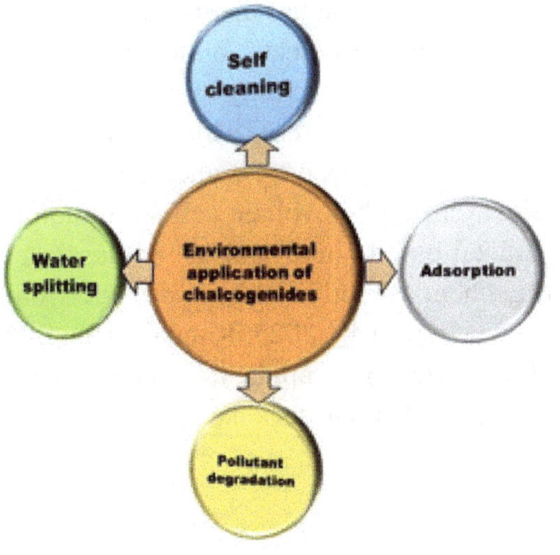

Table 1. Chalcogenide-based nanomaterials in several field applications

Materials	Applications	Ref.
CdX (X = S, Se, Te)	Solar Cells, Photocatalysis	(Kottayi et al., 2022)
MoS_2, WS_2	Lubricants, Catalysts, Energy Storage Devices	(Zhang et al., 2021, Fan et al., 2021)
ZnTe/ZnSe	photocatalytic activity	(Shen et al., 2023)
ZnS, ZnS: Fe, ZnS: Mn, ZnS: Y	Photocatalytic degradation, Photoconductivity, antimicrobial activity, optoelectronic nanodevices	(Ehsan et al., 2019, Suganthi et al., 2018a)
ZnS: Ce, ZnS: La	Photocatalytic degradation, Photoconductivity	(Suganthi et al., 2018b; Suganthi et al., 2018c)
NiS/CuS	Super capacitor applications	(Suganthi et al., 2022)
SnS, SnSe	Solar Cells, Photocatalysis, Batteries	(Yan Shen et al., 2023)
Cr doped CdS/ZnTe	Electrochemical energy storage and photocatalytic activity	(Kamran et al., 2024)

WHY CHALCOGENIDES COMPOSITES FOR WATER TREATMENT

Chalcogenide nanomaterials have assuring promising expects for environmental remedies, precisely in water cleansing, air pollution reduction, and the degrading of organic pollutants. The instinctive adaptability of these materials empowers the evolution of adapted methods for energy transformation, electronics, and environmental remedies. The affinity of these has distinct substrates and these substances promotes the fabrication of delicate objects and techniques that manipulate synergetic properties. The S vacancies proficiently expand absorption spectrum of ZnS into visible zone (Suganthi et al., 2018c). This feature comprises distinguished photosensitivity, light absorbing properties, and tuneable bandgaps, offer them appropriate to photoelectronics, photovoltaic cell, and photocatalytic degradation execution. Transition metal-based photocatalysts possess exceptional remarkable efficiency in promoting chemical activity like hydrogen evolution (Kannan et al., 2023) and CO_2 reduction (Sliem et al., 2025; Kannan et al., 2020; Sadasivuni et al., 2022; Kumar et al., 2022), offering an applicable and eco-friendly path for energy generation, (Kannan & Tari, 2024; Gurushankar et al., 2025).

CHALCOGENIDE CLASSIFICATION

The configurations of several major collection substances are directed via covalent bonding, comparatively close packing

Based on Number of Components

Binary Chalcogenides (BCs)

BCs are consisting of two diverse kinds of ions, which comprise metal charged particle with chalcogen anion. For instance, CdS is extremely examined chalcogenides which has a bandgap of ~2.3 eV and also highly energetic over visible-light illumination. CdS has diverse appliances due to their exclusive size as well as structure-dependent optical and electronic features.

Ternary Chalcogenides (TCs)

TCs consist of three distinct ions such as a chalcogen anion. TCs including ZnCdX where X = S, Se. For instance, ZnCdS has a bandgap energy about 2.4-3.7 eV (Mane et al., 2000, Gao et al., 2012).

Quaternary Chalcogenides (QCs)

QCs have four distinct charged particle such as a chalcogen anion. It has a huge type of substances usually collected ion of earth plentiful and harmless substances for variety of appliances including photovoltaic absorbers and photocatalyst in solar water separation. It has various kinds of elements and also have complicated electronic and structure characteristics. For instance, AgInZn7S9 having the visible light absorbing properties compared to its oxide equivalents. These sulfides so expose immense higher photocatalytic activities for the evaluating H_2 over visible- light illumination (Ates et al., 2015).

On the Basis of Metals They Contain

This type of chalcogenides is then distinguished into several types formed upon the classification of metals and number of metals.

Alkali Metal (AM) and Alkaline Earth Metal (AEM) Chalcogenides

The sulfides of AM and AEM mono-chalcogenides have the propensity for undergoing hydrolysis to create derivatives having disulfide anions. Some examples of AM chalcogenides are KLiX (X = S, Se and Te) for AEM chalcogenides are MX (M = Ca, Sr, Ba and X = S, Se, Te) (Gao et al., 2012, Ates et al., 2015, Ates et al., 2016). These materials remain colourless, salty composites, and water solvable. The crystallizing of the AM chalcogenides frequently occurs in the form of antifluorite configuration and the AEM materials are in the sodium chloride shape.

Transition Metal Chalcogenides (TMCs)

TMCs are formed by numerous stoichiometries and several structures, for instance, basic and prominent stoichiometries are 1:1 and 1:2. Zinc (ZnS, ZnSe, and ZnTe), cadmium (CdS, CdSe, and CdTe), MoS2, and mercury (HgCdTe) chalcogenides are some of the extremely examined TMCs. It reveals vastly covalent, nonionized performance specified through their zinc-blend or wurtzite configuration (Roebben et al., 2024). Table 2 tilts the many TMCs employed like photocatalyst for degrading of contaminated water.

Metal-Rich Chalcogenides

TMCs in metal-rich chalcogenides are typically show oxidation positions of II or above. For instance, the metal-annoying sulfide is $Nb_{21}S_8$ (Arslan et al., 2011). Metal annoying chalcogenide is fabricated using huge quantities of elements used as a catalyst in electrocatalytic hydrogen generation. Usually, these composites possess wide metal-to-metal bonding and exists several structures as well as physical features.

On the Basis of Number of Chalcogen Ions

Mono-Chalcogenides

Monochalcogenides are comprised of a MX configuration, where M = transition metal and

X = chalcogenide anion including S^{2-}, Se^{2-}, or Te^{2-}. These substances have small friction

and higher-wear resistant surface which is as a result of strong bonded atoms producing widespread rigid layers of lamellar configuration. GaS either GaSe validates a crystal construction in the form of a dual coating of Ga introduced among S and Se ions (Ates et al., 2016).

Dichalcogenides

Metal dichalcogenides are encompassed of MX_2, where M = TM and X = a chalcogen anion, including, MoS_2, $TiSe_2$, $MoSe_2$, and WSe_2. The stochiometric is an obvious 1:2 which is crystallized within 2D constructions. This 2D construction contains X-M-X squeeze in layers parted through Vander Waals forces. These metal ions are synchronized in six-fold within the layers, and it is also octahedral either a body-centered trigonal prism (Ates et al., 2016).

Trichalcogenides

Some primary metals as well form trichalcogenides including Ti, V, Cr, and Mn grouping for example Niobium triselenide ($NbSe_3$). Niobium triselenide is synthesized using process of tetrathiomolybdate through acid (Ates et al., 2016).

WASTE WATER TECHNIQUES

Chalcogenide-based nanocomposites can be used for removing emerging organic micropollutants (OMPs) from wastewater through several methods, including adsorption, photocatalysis, and biodegradation. These methods leverage the unique properties of chalcogenides, including their larger surface area and catalytic activity, to effectively capture and degrade OMPs.

Adsorption

Chalcogenides, especially those with a high surface area like graphene or carbon nanotubes, can adsorb OMPs onto their surfaces. This process relies on the interaction between the OMP molecules with the nanocomposite surface, effectively trapping them. Adsorption is a relatively easy as well as economical method, which is utilized to remove a wide range of OMPs.

Examples

Graphene-based adsorbents have been shown to effectively remove various OMPs, such as medicines, pesticides, and dyes. CNTs can also be used as adsorbents due to their higher surface area and capacity to interrelate to OMPs.

Photocatalysis

Certainly, degradation of organic contaminated water is inspiring for actual photocatalyst. The resources used in several approaches are expensive which creates in destructive waste. To overthrow these problems, there is a mandate for a cheap and harmless nanoparticle (NP) having notable photocatalytic activities for dye elimination (Suganthi et al., 2020).

Figure 3. Diagram demonstrating the bandgap, the potential energy of the conduction, and the valence band of auspicious metal chalcogenides (Zhang et al., 2013, Nie et al., 2017).

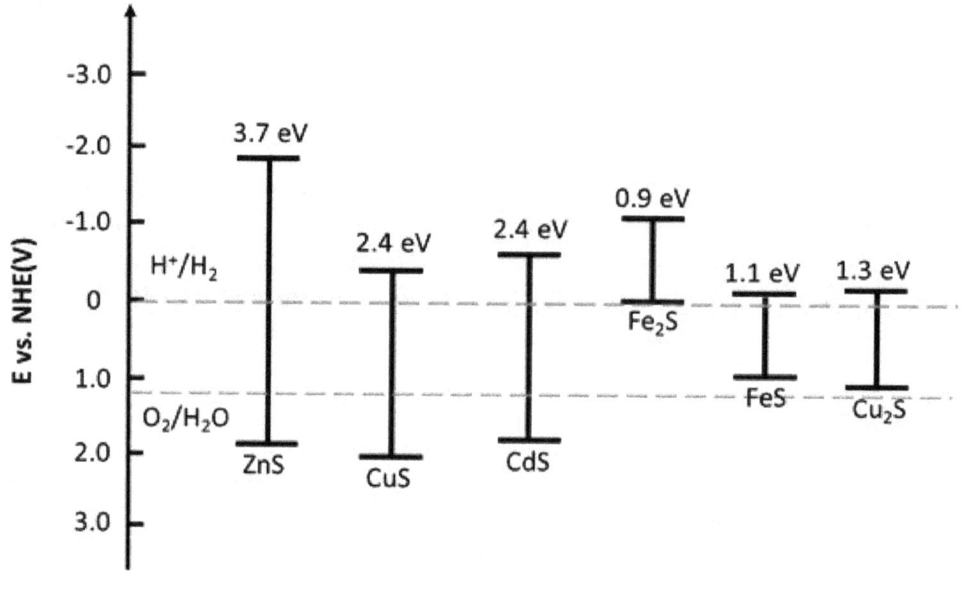

Photocatalytic Mechanism

Some chalcogenide materials, when exposed to UV or visible light, can act as photocatalysts. They generate reactive oxygen species (ROS) that can oxidize and degrade OMPs, breaking them down into less harmful substances. Photocatalysis can be a highly efficient method for removing OMPs, and it can also be combined with other methods like adsorption. By Comparing metal oxides, metal chalcogenides not only have various architectures that improved through mixing various metal ions and patterns with reduced bandgaps. The appropriate band locations which are adjusted over composition (Figure 4) (Zhang et al., 2013, Nie et al., 2017).

Figure 4. Photocatalytic mechanism of various types of heterojunctions photocatalyst (Suganthi et al., 2022)

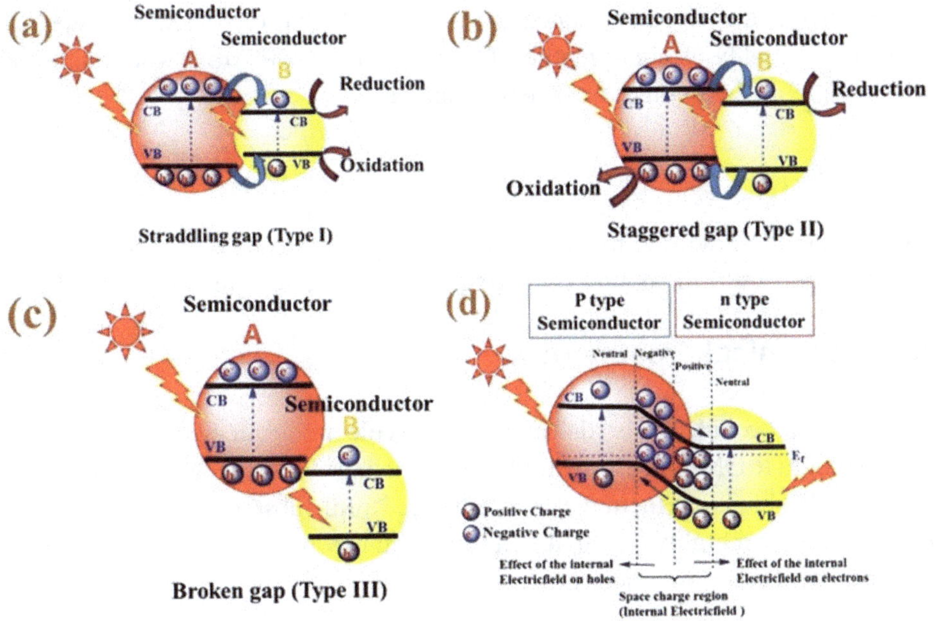

Biodegradation

Chalcogenide-based nanocomposites can be used to support or enhance the biodegradation of OMPs by providing a substrate for microbial growth or by acting as a catalyst for microbial processes. Biodegradation is an environmentally friendly method that can effectively remove OMPs by converting them into less harmful substances.

Examples

Chitosan, a natural polymer, can be combined with chalcogenides to create nanocomposites that can promote microbial growth and biodegradation. Other biocompatible nanocomposites, such as those based on cellulose or alginate, can also be used to support biodegradation processes.

Membrane Filtration

Chalcogenide-based nanocomposites can be incorporated into membranes to enhance their filtration performance and improve OMP removal. Membrane filtration can be a highly efficient method for separating OMPs from water, and it can also be combined with other treatment processes.

Examples

Chalcogenide-modified membranes: Chalcogenides can be used to modify the surface properties of membranes, making them more effective at retaining OMPs.

Electrochemical Treatment

Some chalcogenide materials can be used as electrodes in electrochemical processes to degrade OMPs. This method involves using electrical current to oxidize or reduce OMPs, breaking them down into less harmful substances. Electrochemical treatment can be a highly efficient method for removing OMPs, and it can also be combined with other treatment processes.

Examples

In electrochemical oxidation, chalcogenide electrodes can be used to oxidize OMPs, such as pesticides and pharmaceuticals. Electrochemical reduction such as chalcogenide electrodes can also be used to reduce OMPs, such as heavy metals and dyes.

SYNTHESIS OF CHALCOGENIDE NANOMATERIALS

Chalcogenide nanomaterials are synthesized using various methods, including hydrothermal, solvothermal, colloidal, Microwave-assisted synthesis, Biosynthesis, Solvent-less synthesis and bottom-up techniques. These materials, which are composed of a metal and a chalcogen (sulfur, selenium, or tellurium), are gaining attention for their diverse applications. Photocatalytic behavior of lanthanide sulfides with tuning morphology are prepared by solvent assisted and solvent free method (Ghosh, 2015). Bimetallic oxysulfides $Gd_{2(1-y)}$ Ce_2yO_2S were synthesized by a solvothermal method involving a combination of organic solvents and metallic developments (Larquet, 2019). Hybrid nanoarchitectures of AgInS2 and TiO2 photocatalysts were produced via an adapted sol–gel technique. The experimental results show that these nanocomposites show greater visible light absorption and actual charge carrier parting comparing with their pristine parental materials ($AgInS_2$ or TiO_2) (Ganguly, 2020). The material constituents, preparation strategies, and post-fabrication treating altogether have affected the morphological features of chalcogenide nanomaterials.

CHALCOGENIDE NANOMATERIALS FOR WASTEWATER TREATMENT

Degrading of contaminants, photoreduction of dyes, and photocatalysts are having many present appliances of photocatalysis.

Table 2. TMCs and chalcogenide nanocomposites as a photocatalyst

Chalcogenides	Photocatalyst	Light source	Time (min)	Efficiency %	Reference
1.Metal/Noble metal Chalcogenides	Pt/CdS	visible light			(McGovern et al., 2010)
	Cu-doped CdS		260 min	MB-89 MV-97	(Al-Jawad et al., 2025)
2.Bimetallic oxysulfides	Mn–Mo (O, S)$_2$-30	visible light	30 120	MB-99.7 MO-98.6	(Kebede et al., 2024)
3.Lanthanide sulphides	EuS	visible light	100 min	RhB	(Ghosh et al., 2015)
4.Lanthanide Oxysulfides	$Gd_{0.6}Ce_{1.4}O_2S$	visible light	-	-	(Larquet et al., 2019)
5.Chalcogenide Nanocomposites					
5.1Heterojunctions (HJs)	MoS_2/CdS	visible light	10	98	(Kaysar Hamid et al., 2024)
5.2 Ternary chalcogenides	$AgInS_2$ and TiO2	Solar radiation	180 min	Doxycycline-98%	(Ganguly et al., 2020)

continued on following page

Table 2. Continued

Chalcogenides	Photocatalyst	Light source	Time (min)	Efficiency %	Reference
5.3 Quaternary chalcogenide	Zr_2S_3-BaS-$Cr2S_3$ thin film			methylene blue dye, pesticide zoxamide and phenol- 86%	(Gul et al., 2022)
5.4 Hybrid graphene-nanoparticles	CdS-graphene CdS-CNT	visible light	60 min	MO 95% 88%	(Ye et al., 2012)

Chalcogenide Nanocomposites (NCs)

Chalcogenide-based NCs exhibit promising photocatalytic activity, particularly in the degrading of organic pollutants like dyes. These materials, often transition metal chalcogenides, can efficiently absorb visible light, generating electron-hole pairs that drive redox reactions for pollutant degradation. Their tunable band gaps and unique structural properties make them attractive for various environmental applications.

Advantages

Chalcogenide-based nanocomposites offer several advantages, including:

- Visible Light Absorption: Many chalcogenides, unlike TiO_2, can absorb visible light, making them efficient under sunlight.
- Enhanced electron-hole separation: By combining chalcogenides with other materials, electron-hole separation can be improved, leading to higher photocatalytic activity.
- Tunable Properties: The composition and structure of chalcogenide-based materials can be tailored to optimize their photocatalytic performance.
- Cost-Effective Synthesis and abundance: Many chalcogenide-based materials can be synthesized using relatively simple and cost-effective methods. Many chalcogenide materials are relatively abundant.

Heterojunctions (HJs)

The formation of HJs through many efficient resources in a single photocatalyst could enhance the water splitting efficacy through raising light collection execution, higher chemical stable, and photo-excited charge split-up. Numerous chalcogenides compounds are prepared and MoS_2/CdS is frequently narrated chalcogenides photocatalytic degradation. He et al. produced MoS_2/CdS NC through a simplistic solvothermal technique. The produced MoS_2/CdS is stated to accomplish larger photocatalytic action, specifically, 5 times greater compared to pristine CdS (Guannan He, 2019).

Ternary Chalcogenides

Ternary chalcogenides are additional type of multicomponent elements and it is concerned consideration due to their slight band gap as well as exceptional photon absorption around the visible zone on the radiation spectrum. Although copper-based and silver-based ternary composites have newly been discovered for various photocatalytic appliances like degrading of emergent contaminants, real wastewater process, CO_2 reduction, and energy production. Hybrid nanoarchitectures of $AgInS_2$ and TiO_2 photocatalysts were synthesized via a adapted sol–gel technique (Ganguly,2020).

Quaternary Chalcogenide (QC)

QC nanocomposites are promising photocatalysts for degrading organic pollutants due to their unique properties, including tuneable band gaps and efficient charge separation. These nanocomposites, composed of multiple elements (e.g., Cu, Zn, Sn, S), can be synthesized using various methods and offer high photocatalytic activity under visible light. Cu_2ZnSnS_4 (CZTS) is a widely studied quaternary chalcogenide with a relatively narrow band gap, making it effective under visible light. Earth-abundant QC copper zinc tin sulfides-based NCs CZTS-gCN and CZTS-GO (graphene oxide) being photocatalysts to degrade methylene blue (Sheebha, 2023).

Hybrid Graphene-Nanoparticles (NPs)

This hybrid graphene-nanoparticles (NPs) structures are of precise attention since they demonstrate individual features of graphene and NPs after hybridization in addition to synergistic properties. Recent narrates have exposed single graphenes (GS) has excellent electron transport properties. For instance, CdS-graphene is a NC which was prepared by a simplistic precipitation technique have reported by Khan et al. CdS-graphene NCs exhibited improved photocatalytic performance over visible light illumination (He, 2019).

APPLICATIONS OF CHALCOGENIDE-BASED NANOCOMPOSITES FOR THE REMOVAL OF EMERGING ORGANIC MICROPOLLUTANTS

Unceasing pollution of water source through the industrial wastes has been a main affair to be directed. Scientists have discovered many approaches for the refinement and processing of effluents. Amongst studied procedures photocatalytic degradation is well thought-out because of ease procedure and no production of subsidiary impurities. The initiation of material science and engineering has empowered for developing materials having anticipated properties exploited like a photocatalyst. From the above conversation regarding the TMCs which is obvious that they performance as exceptional applicant as a photocatalyst. The current segment allocates through the various pollutants like a contaminated organic dyes, agrochemical and organic particles, medicines, and heavy metals utilizing in photocatalytic degradation.

Degradation of Toxic Industrial Dyes

The fabric and paper manufacturing use many organic pigments for producing fabric and paper. Dyes like methylene blue, rhodamine B, malachite green, congo red, methyl orange, crystal violet etc., are extensively employed in fabric productions. These pigments are a main constituent in the industrial effluent excreted through the water frames (Qureshi, 2022, Khan, 2020) . According to the studies industrial effluents employed in various processes which elevated some concerns concerning ecological integrity and protection. The scientists are employed to (1) the growth of non-poisonous, harmless, and decomposable dyes and (2) to advance simplistic and economical approaches for processing industrial effluents. The degrading of methylene blue dye by n-MoS_2/p-WO_3-based photocatalyst over dark state through a self-supportive charge transporter transmission process (Kebede, 2020).

Degradation of Agrochemical and Organic Molecules

The existence of organic impurities in water frames remains a significant environmental anxiety. These organic substances are a biologically ensuing both artificial organic molecules. Researchers have extensively discovered the photocatalytic degrading of contaminated organic substances (shown in Table 3). WS_2/Bi_2MoO_6 nanocomposites over visible light illumination for degrading several organic effluents which leads to active charge split-up and prolonged charge carriers having higher life hence enhanced the execution by comparing with pristine WS_2 (Li, 2018).

Degradation of Noxious Pharmaceuticals

The usage of pharmacological medicines to deal the disease is unavoidable. An extended listing in medicine and non-prescribed medicines, including antipyretics, antibiotics, anticonvulsants, are directed with patients in a repetitively method (Awfa, 2019). These pharmacological medicines and its byproducts are discharged towards the nature as industrial discharge (Al Abri, 2019). Extended subjection and overdo of any medicines produce liver malfunction and some organ failures as a result of drug toxicity (Bastami, 2017). The photocatalytic degrading of organic substances, scientists have also endeavoured the degrading for pharmaceutical medicine particles. The photocatalytic elimination of norfloxacin in the aqueously form by restrained Z-scheme CdS/Au/TiO_2 nanobelt as a photocatalyst (Li, 2017) .

Photoreduction of Heavy Metals

Existence of heavy metals (HMs) in water ensues a critical danger to mankind. HMs such as Pb, Cd, Zn, Hg, As, might appear in the food chain and finish up collecting within the humanoid. The existence of these metals in infinitesimal quantities causing noxious consequence on humanoid healthiness which creates the diseases. High efficacy, lower price, preventing ancillary impurities, and directly usage of solar energy is a special advantage in the photocatalytic method. Chalcogenides-based nanomaterials remain extensively exploited in photocatalysis because of their thinner bandgaps which relate in the visible light absorption. Shazad *et al.*, have narrated 40% higher elimination efficacy in the existence of $CuCo_2S_4$ nanocomposite. The production of SnS_2 nanocomposites for the photoreduction of Cr (VI) is reported (Aguiar, 2021).

Table 3. Nanocomposites used as photocatalysts for degrading emerging organic micropollutants

Type of pollutants	Pollutants	Concentration	Photocatalyst	Sample dosage	Time (min)	Efficiency	Ref.
Toxic Industrial Dyes	RhB	-	Cu_2WS_4/$NiTiO_3$	-	20	98	(Peng et al.,2022)
	MO	10 ppm	Ag_2S/$AgInS_2$	20 mg		80%	(Li et al., 2021)
	Lanasol Red 5B	80 ppm	MoS2/Ag2CO_3	40 mg	25	95%	(Fu et al., 2019)
	MO	20 ppm	BiOI/MoS_2	20 mg	75	95.6%	(Guo et al., 2021)

continued on following page

Table 3. Continued

Type of pollutants	Pollutants	Concentration	Photocatalyst	Sample dosage	Time (min)	Efficiency	Ref.
Agrochemical and Organic Molecules	Aniline	100 ppm	reduced GO/ ZnO/ MoS_2 and CNTs/ ZnO/MoS_2	1.2 gm	120	100%	(Ghasemipour et al., 2020)
	Thiophene	14.5 nm	MoS_2@rGO	-	75	100%	(Alhaddad et al., 2020)
	Bisphenol A	100 ppm	$CuFeS_2$ modified with citrate	0.2 g/L	60	97%	(da Silveira Salla et al., 2020)
Toxic Pharmaceuticals	2,4,5-trichlorophenoxy acetic acid (2,4-D) pesticide	20 ppm	CuS@ZnO	0.2 g/L	20	100	(Mohamme et al., 2022)
	amoxicillin	5 mg/L	CdS-MoS_2-coated ZnO	-	60	94	(Tang et al., 2019)
	ofloxacin	10 ppm	CdS/MoS_2/ZnO	0.1 g	90	89	(Vaizo˘ gular et al., 2020)
	tetracycline	0.2 g/L	MoS_2@zeolite	10 mg	180	87.2%	(Liu et al., 2022)
	paracetamol	30 ppm	$ZnIn_2S_4$/MoO_3	0.2 g	80 min	98%	(Khan et al., 2020)
	tetracycline		Cu2WS_4/$NiTiO_3$		60 min	89%	(Peng et al., 2022)
Photoreduction of Heavy Metals	Cr (VI)	-	MoS2/ZnS/ZnO	-	90	98.7%	(Zhao et al., 2019)
	Pb^{2+}	5 mg/L	2D MoS_2 Nanosheets	740 mg/g	2 min	99.92%	(Wang et al., 2020)
	Hg^{2+}	100 mg/L	Petal-like MoS2 nanosheets	289 mg/g	300 min	93%	(Pirarath et al., 2021)
	Pb^{2+}	600 mg/L	Fe_3S_4 hollow spheres	92.1 mg/g		100%	(Zheng et al., 2014)

CHALLENGES AND FUTURE PROSPECTS

Chalcogenide-Based NCs have been broadly exploited in photocatalytic appliances. Chalcogenide-based NCs have accomplished an enhanced reputation by comparing other SC nanomaterials employed in photocatalytic degradation. Most of the chalcogenide-based NCs naturally 2D materials therefore it acquires higher surface area and quantum confinement than 3D configurations. Other merits related to chalcogenide-based NCs is its layer and thickness depending of electronic and optical characteristics. Owing to their higher surface area, agree with band edge, and subjected to effectiveness which creates them as a competent applicant for photocatalysis. Even though the advantages, chalcogenide-based NCs subjected to the confined in low stability, exposure to corrosion over illumination, and lower recyclable which restrict its action and efficacy. To overcome these deficiencies of chalcogenide-based NCs, the various approaches like the band edge engineering, and surface functional approach are significant part. This approach for facilitate in creating novel materials or enhancing the existent materials. It can be attained through the succeeding ways (1) synthesis of appropriate SC substance to enhance the visible light absorbing properties, (2) designing of band edge arrangement for preventing the recombinant and enhanced the charge carrier lifetime, (3) deduction in the size of the particle enhancing the hydrophobics for preventing photocorrosion, and raised the surface area, (4) alteration in the layer dimensions for attaining the appropriate bandgap through optimal diffusion distance to enrich charge carrier density. The current chapter précises the definite features which make this SC substance appropriate for uses in wastewater processing. The various TM chalcogenides-based NCs, its procedure for photocatalysis, the utility of chalcogenides-based NCs for photocatalysis as upcoming leadership to scientists at work in the area of chalcogenides-based NCs for the removal of emerging organic micropollutants in waste water with auspicious forthcoming paths.

CONCLUSION

Chalcogenide's materials are based on sulfide, selenide, and telluride. Several methods use chalcogenide materials either its products, including the above-mentioned binary, ternary, and quaternary chalcogenides. A fascinating proposal for light-harvesting appliances are exploiting chalcogenide-based nanocomposites through smart hybridizing. These innovative materials illustrate the greater potential for photocatalytic degradation of organic contaminants, water-split actions to generate hydrogen gas.

Owing to their promise useful in several technical backgrounds, in recent times chalcogenides-based NCs have attained an important material type. Through the environmental restoration procedure, these chalcogenides-based NCs were utilized as catalysts to remove many pollutants and changing industrial wastes into drinking water. Serious environmental complications, like pollution and deficiency of sustainable energy, have occasioned straightway from development around the manufacturing and other regions. chalcogenides-based NCs were employed even in very specific environmental remediation applications, such as dealing with toxic pharmaceuticals and heavy metals found in effluents.

The subsequent actions, are vital for encouraging the usage of chalcogenides and chalcogenides-based nanocomposites for effluent process actions. Moreover, it is not technically sound for utilizing any toxic substance for removing another destructive molecule. Henceforth, further research is required for analysing the toxicity of the chalcogenides and chalcogenides-based nanocomposites. For commercial use, the chalcogenides-based nanocomposites are producing at large range which is done with performing a scientific examination to regulate the reliable fabrication approaches. Surface alterations to chalcogenides-based nanocomposites can reduce charge carrier recombinant, moreover a further research in the recombination procedure is needed before wider use of optimized photocatalysis.

REFERENCES

Aguiar, I., Oreggioni, D., Pérez Barthaburu, M.E. (2021). Tin-Based Compounds for Water Remediation. Water Pollution and Remediation: Heavy Metals, 281–312.

Al Abri, R., Al Marzouqi, F., Kuvarega, A. T., Meetani, M. A., Al Kindy, S. M., Karthikeyan, S., Kim, Y., & Selvaraj, R. (2019). Nanostructured cerium-doped ZnO for photocatalytic degradation of pharmaceuticals in aqueous solution. *Journal of Photochemistry and Photobiology A Chemistry*, *384*, 112065.

Al-Jawad, S.M.H., Aboud, K.H., & Imran, N.J, Sally Yakoob Taher. (2025). Copper Doping of CdS Nanoflakes and Nanoflowers for Efficient Photocatalytic Degradation of MB and MV Dyes. *Plasmonics*, *20*, 983–1002.

Al-Tohamy, R., Ali, S. S., Li, F., Okasha, K. M., Mahmoud, Y. A. G., Elsamahy, T., & Sun, J. (2022). A critical review on the treatment of dye-containing wastewater: Ecotoxicological and health concerns of textile dyes and possible remediation approaches for environmental safety. *Ecotoxicology and Environmental Safety*, *231*, 113160.

Alhaddad, M., & Shawky, A. (2020). Superior photooxidative desulfurization of thiophene by reduced graphene oxide-supported MoS2 nanoflakes under visible light. *Fuel Processing Technology*, *205*, 106453.

Arslan, Z., Ates, M., McDuffy, W., Agachan, M. S., Farah, I. O., Yu, W., & Bednar, A. J. (2011). Probing metabolic stability of CdSe nanoparticles: Alkaline extraction of free cadmium from liver and kidney samples of rats exposed to CdSe nanoparticles. *Journal of Hazardous Materials*, *192*, 192–199.

Ates, M., Demir, V., Arslan, Z., Kaya, H., Yılmaz, S., & Camas, M. (2016). Chronic exposure of tilapia (*Oreochromisniloticus*) to iron oxide nanoparticles: Effects of particle morphology on accumulation, elimination, hematology and immune responses. *Aquatic Toxicology (Amsterdam, Netherlands)*, *177*, 22–32.

Ates, M., Demir, V. Z., Arslan, Z., Daniels, J., Farah, I. O., & Bogatu, C. (2015). Evaluation of alpha and gamma aluminum oxide nanoparticle accumulation,toxicity and depuration in artemia salina larvae. *Environmental Toxicology*, *30*, 109–118.

Awfa, D., Ateia, M., Fujii, M., & Yoshimura, C. (2019). Novel Magnetic Carbon Nanotube-TiO_2 Composites for Solar Light Photocatalytic Degradation of Pharmaceuticals in the Presence of Natural Organic Matter. *Journal of Water Process Engineering*, *31*, 100836.

Bastami, T. R., Ahmadpour, A., & Hekmatikar, F. A. (2017). Synthesis of Fe_3O_4/Bi_2WO_6 nanohybrid for the photocatalytic degradation of pharmaceutical ibuprofen under solar light. *Journal of Industrial and Engineering Chemistry*, *51*, 244–254.

da Silveira Salla, J., Dotto, G. L., Hotza, D., Landers, R., da Boit Martinello, K., & Foletto, E. L. (2020). Enhanced catalytic performance of CuFeS2 chalcogenide prepared by microwaveassisted route for photo-Fenton oxidation of emerging pollutant in water. *Journal of Environmental Chemical Engineering*, *8*, 104077.

Ehsan, M. F., Qudoos, S., Ahmad, Z., Hamid, S., Arfan, M., Zia, A., & Tyagi, D. (2019). ZnTe/ZnSe heterostructures: In-situ synthesis, characterization and photocatalytic activity for Congo Red degradation. *SN Applied Sciences*, *1*(3), 197.

Fan, X., Shi, Y., Cui, M., Ren, S., Wang, H., & Pu, J. (2021). MoS_2/WS_2 Nanosheet-based composite films irradiated by atomic oxygen: Implications for lubrication in space. *ACS Applied Nano Materials*, *4*, 10307–10320.

Fu, S., Yuan, W., Yan, Y., Liu, H., Shi, X., Zhao, F., & Zhou, J. (2019). Highly efficient visible-light photoactivity of Z-scheme MoS2/Ag2CO3 photocatalysts for organic pollutants degradation and bacterial inactivation. *Journal of Environmental Management*, *252*, 109654.

Ganguly, P., Mathew, S., Clarizia, L., Kumar, R. S., Akande, A., Hinder, S. J., Breen, A., Suresh, C., & Pillai, S. C. (2020). Ternary Metal Chalcogenide Heterostructure ($AgInS_2$–TiO_2) Nanocomposites for Visible Light Photocatalytic Applications. *ACS Omega*, *5*, 406–421.

Gao, M. R., Jiang, J., & Yu, S. H. (2012). Solution-based synthesis and design of late transition metal chalcogenide materials for oxygen reduction reaction. *Small*, *8*, 13–27.

Ghasemipour, P., Fattahi, M., Rasekh, B., & Yazdian, F. (2020). Developing the ternary ZnO doped MoS2 nanostructures grafted on CNT and reduced graphene oxide (RGO) for photocatalytic degradation of aniline. *Scientific Reports*, *10*(1), 4414.

Ghosh, A. B., Saha, N., Sarkar, A., Srivastava, D. N., Paul, P., & Adhikary, B. (2015). Solvent assisted and solvent free orientation of growth of nanoscaled lanthanide sulfides: Tuning of morphology and manifestation of photocatalytic behaviour. *RSC Advances*. Advance online publication. DOI: 10.1039/C5RA19959J

Guannan He, G., Yimin Zhang, Y., & Qinyu He, Q. (2019). MoS_2/CdS Heterostructure for Enhanced Photoelectrochemical Performance under Visible Light. *Catalysts*, *9*, 379.

Gul, M. M., Khuram Shahzad Ahmad,K.S.,(2022). Nanocomposite Zr2S3-BaS-Cr2S3 ternary-metal chalcogenide: An impressive supercapacitor electrode and environmental remediant of toxic pollutants. *International Journal of Energy Research*. Advance online publication. DOI: 10.1002/er.8489

Guo, S., Luo, H., Li, Y., Chen, J., Mou, B., Shi, X., & Sun, G. (2021). Structure-controlled threedimensional BiOI/MoS_2 microspheres for boosting visible-light photocatalytic degradation of tetracycline. *Journal of Alloys and Compounds*, *852*, 157026.

Gurushankar, K., Karuppaiya, C., Tari, V., & Kannan, K. (2024). Exploring MXenes and MBenes in metal sulfur and selenium battery technologies. In Kannan, K., & Tari, V. (Eds.), *Novel energy storage and conversion technologies for two-dimensional MXenes and MBenes*. IGI Global., DOI: 10.4018/979-8-3693-9546-2.ch005

He, G., Zhang, Y., & He, Q. (2019). MoS_2/CdS heterostructure for enhanced photoelectrochemical performance under visible light. *Catalysts*, *9*, 379.

Holkar, C.R., Jadhav, A.J., Pinjari, D.V., & Naresh, M. Mahamuni, N.M., & B. Pandit, A.B. (2016). A critical review on textile wastewater treatments: Possible approaches. *Journal of Environmental Management*, *182*, 351–366.

Kamran, M. A., Raza, M., Ullah, S., Alarifi, S. S., & Aziz, M. H. (2024). Facile synthesis of chromium doped cadmium sulfide/zinc telluride nanocomposites for enhanced electrochemical energy storage and photocatalytic applications. *Materials Chemistry and Physics*, *322*, 129531.

Kannan, K., Chanda, D., Gautam, J., Behera, A., Meshesha, M. M., Jang, S. G., & Yang, B. (2023). Hydrothermally synthesized mixed metal oxide nanocomposites for electrochemical water splitting and photocatalytic hydrogen production. *International Journal of Hydrogen Energy*, *48*(93), 36412–36426.

Kannan, K., Chanda, D., Gautam, J., Meshesha, M. M., Jang, S. G., & Yang, B. L. (2023). Facial synthesis of pp heterojunction composites: Evaluation of their electrochemical properties with photovoltaics-electrolyzer water splitting using two-electrode system. *International Journal of Hydrogen Energy*, *48*(37), 13814–13826.

Kannan, K., Gautam, J., Chanda, D., Meshesha, M. M., Jang, S. G., & Yang, B. L. (2023). Two-dimensional MAX supported copper oxide/nickel Oxide/MAX as an efficient and novel photocatalyst for hydrogen evolution. *International Journal of Hydrogen Energy*, *48*(20), 7273–7283.

Kannan, K., Radhika, D., Gnanasangeetha, D., Lakkaboyana, S. K., Sadasivuni, K. K., Gurushankar, K., & Hanafiah, M. M. (2021). Photocatalytic and antimicrobial properties of microwave synthesized mixed metal oxide nanocomposite. *Inorganic Chemistry Communications*, *125*, 108429.

Kannan, K., Sliem, M. H., Abdullah, A. M., Sadasivuni, K. K., & Kumar, B. (2020). Fabrication of ZnO-Fe-MXene based nanocomposites for efficient CO2 reduction. *Catalysts*, *10*(5), 549.

Kannan, K., & Tari, V. (Eds.). (2024). Novel energy storage and conversion technologies for two-dimensional MXenes and MBenes. ISBN: 9798369395462, eISBN: 9798369395486DOI: 10.4018/979-8-3693-9546-2

Kaysar Hamid, K., & Mukhlish, M,Z.B., & Uddin, Md.T. (2024). Sunlight-activated heterostructure MoS2/CdS nanocomposite photocatalyst with enhanced photocatalytic activity: Band alignment and mechanism study. *RSC Advances*, *14*, 38908–38923.

Kebede, W. L., Hau, K. D., & Xiaoyun, C. (2024). Organic pollutant degradation over highly efficient Mn–Mo $(O, S)_2$ oxysulfide photocatalyst under visible light illumination. *Optical Materials*, *155*, 115828.

Kebede, W. L., Kuo, D. H., Ahmed, K. E., & Abdullah, H. (2020). Dye degradation over the multivalent charge-and solid solution-type n-MoS2/p-WO3 based diode catalyst under dark condition with a self-supporting charge carrier transfer mechanism. *Advanced Powder Technology*, *31*(7), 2629–2640.

Khan, A., Danish, M., Alam, U., Zafar, S., & Muneer, M. (2020). Facile synthesis of a Z-scheme ZnIn2S4/MoO3 heterojunction with enhanced photocatalytic activity under visible light irradiation. *ACS Omega*, *5*, 8188–8199.

Khan, F. S. A., Mubarak, N. M., Tan, Y. H., Karri, R. R., Khalid, M., Walvekar, R., Abdullah, E. C., Mazari, S. A., & Nizamuddin, S. (2020). Magnetic nanoparticles incorporation into different substrates for dyes and heavy metals removal—A Review. *Environmental Science and Pollution Research International*, *27*(35), 43526–43541.

Khan, M. M. (2021). *Introduction and fundamentals of chalcogenides and chalcogenides-based nanomaterials*. Elsevier., DOI: 10.1016/B978-0-12-820498-6.00001-9

Kottayi, R., Maurya, D. K., Sittaramane, R., & Angaiah, S. (2022). Recent developments in metal chalcogenides-based quantum dot sensitized solar cells. *ES Energy & Environment.*, *18*, 1–40.

Kumar, B., Malali, P., Sadasivuni, K. K., & Kannan, K. (2022). Two-dimensional Nanomaterials Design and Reactor Engineering of Different Methods for CO_2 Electrochemical Conversion Process.

Kumaravelan, S., Seshadri, S., Suresh, R., Ravichandran, K., Sathishkumar, P., Shanthaseelan, K., & Suganthi, N. (2021). Effect of Zn dopant on SnO_2 nano-pyramids for photocatalytic degradation. *Chemical Physics Letters*, *769*, 138352.

Larquet, C., Nguyen, A. M., Glais, E., Paulatto, L., Sassoye, C., Selmane, M., Lecante, P., Maheu, C., Geantet, C., Cardenas, L., Chanéac, C., Gauzzi, A., Sanchez, C., & Carenco, S. (2019). Bandgap Engineering from Cation Balance: The Case of Lanthanide Oxysulfide Nanoparticles. *Chemistry of Materials*. Advance online publication. DOI: 10.1021/acs.chemmater.9b00450

Li, J., Xia, Z., Ma, D., Liu, G., Song, N., Xiang, D., Xin, Y., Zhang, G., & Chen, Q. (2021). Improving photocatalytic activity by construction of immobilized Z-scheme CdS/Au/TiO2 nanobelt photocatalyst for eliminating norfloxacin from water. *Journal of Colloid and Interface Science*, *586*, 243–256.

Li, X., Su, M., Zhu, G., Zhang, K., Zhang, X., & Fan, J. (2018). Fabrication of a novel few-layer WS2/Bi2MoO6 plate-on-plate heterojunction structure with enhanced visible-light photocatalytic activity. *Dalton Transactions (Cambridge, England)*, *47*(30), 10046–10056.

Li, Y., Liu, Y., Gao, G., Zhu, Y., Wang, D., Ding, M., Yao, T., Liu, M., & You, W. (2021). L-cysteine and urea synergistically-mediated one-pot one-step self-transformed hydrothermal synthesis of p-Ag2S/n-AgInS2 core-shell hetero nanoflowers for photocatalytic MO degradation. *Applied Surface Science*, *548*, 149279.

Liu, J., Lin, H., Dong, Y., He, Y., & Liu, C. (2022). MoS_2 nanosheets loaded on collapsed structure zeolite as a hydrophilic and efficient photocatalyst for tetracycline degradation and synergistic mechanism. *Chemosphere*, *287*, 132211.

Mane, R. S., & Lokhande, C. D. (2000). Chemical deposition method for metal chalcogenide thin films. *Materials Chemistry and Physics*, *65*, 1–31.

McGovern, C. (2010). Commoditization of nanomaterials. *Nanotechnology Perceptions*, *6*, 155–178.

Mohammed, R., Ali, M. E. M., Gomaa, E., & Mohsen, M. (2022). Copper sulfide and zinc oxide hybrid nanocomposite for wastewater decontamination of pharmaceuticals and pesticides. *Scientific Reports*, *12*(1), 18153.

Nie, L., & Zhang, Q. (2017). Recent progress in crystalline metal chalcogenides as efficient photocatalysts for organic pollutant degradation. *Inorganic Chemistry Frontiers*, *4*, 1953–1962.

Özşirvan, G., Yalçın, M., Turgut, N., Tari, V., & Turgut, C. (2024). Insights into the uptake, translocation, and accumulation dynamics of cyantraniliprole and thiamethoxam seed coating pesticides in maize plants. *Environmental Science and Pollution Research International*, (Special Issue). Advance online publication. DOI: 10.1007/s11356-024-34135-7

Peng, D., Wang, Y., Shi, H., Jiang, W., Jin, T., Jin, Z., & Chen, Z. (2022). Fabrication of novel Cu2WS4/NiTiO3 heterostructures for efficient visible-light photocatalytic hydrogen evolution and pollutant degradation. *Journal of Colloid and Interface Science*, *613*, 194–206.

Peng, D., Wang, Y., Shi, H., Jiang, W., Jin, T., Jin, Z., & Chen, Z. (2022). Fabrication of novel Cu2WS4/NiTiO3 heterostructures for efficient visible-light photocatalytic hydrogen evolution and pollutant degradation. *Journal of Colloid and Interface Science*, *613*, 194–206.

Pirarath, R., Shivashanmugam, P., Syed, A., Elgorban, A. M., Anandan, S., & Ashokkumar, M. (2021). Mercury removal from aqueous solution using petal-like MoS_2 nanosheets. *Frontiers of Environmental Science & Engineering*, *15*, 1–10.

Qureshi, S. S., Shah, V., Nizamuddin, S., Mubarak, N. M., Karri, R. R., Dehghani, M. H., Ramesh, S., Khalid, M., & Rahman, M. E. (2022). Microwave-assisted synthesis of carbon nanotubes for the removal of toxic cationic dyes from textile wastewater. *Journal of Molecular Liquids*, *356*, 119045.

Roebben, G., Rauscher, H., Amenta, V., Ashberger, K., Sanfeliu, B., & Calzolai, L. (2024). Towards a review of the EC Recommendation for a definition of the term "nanomaterial" Part 2: Assessment of collected information concerning the experience with the definition 2014. In Roebben, G., & Rauscher, H. (Eds.), *JRC Scientific and Policy Report, EUR 26744 EN* (pp. 1–83). Publications Office of the European Union.

Sadasivuni, K. K., Kannan, K., Abdullah, A. M., & Kumar, B. (Eds.). (2022). *2D nanomaterials for CO_2 conversion into chemicals and fuels*. Royal Society of Chemistry.

Sekar, R., Sivasamy, R., Ricardo, B., & Manidurai, P. (2021). Ultrasonically synthesized TiO_2/ZnS nanocomposites to improve the efficiency of dye sensitized solar cells. *Materials Science in Semiconductor Processing*, *132*, 105917.

Sheebha, I., David, C. I., Vidhya, B., & Nandhakumar, R. (2023). *Earth-abundant quaternary chalcogenide copper zinc tin sulfide-based nanocomposites CZTS-gCN and CZTS-GO (graphene oxide) as photocatalysts for degradation of methylene blue*. Inorganic and Nano-Metal Chemistry., DOI: 10.1080/24701556.2024.2355511

Shen, Y., Zhang, Y., Huo, J., Li, X., Yan, Z., Pan, Y., Sun, W., Deng, N., & Kang, W. (2023). Two-dimensional SnSe material for solar cells and rechargeable batteries. *Journal of Energy Storage*, *69*, 107958.

Shen, Y., Zhang, Y., Huo, J., Li, X., Yan, Z., Pan, Y., Sun, W., Deng, N., & Kang, W. (2023). Two-dimensional SnSe material for solar cells and rechargeable batteries. *Journal of Energy Storage*, *69*, 107958.

Siddiqui, N., Faiyaz, T., & Tari, V. (2022). Ganga and Yamuna: Through the lens of National Green Tribunal. *Nature. Environment and Pollution Technology*, *21*(4), 1657–1664. DOI: 10.46488/NEPT.2022.v21i04.018

Sivasamy, R., Amirthaganesan, S., Gonzalez, R. E., Quero, F., & Batoo, K. M. (2021). First-principles investigation of the electronic structure, optical and thermodynamic properties on monolayer $Sn_{0.5}Ge_{0.5}Se$ nanosheet. *Physica E, Low-Dimensional Systems and Nanostructures*, *126*, 114454.

Sivasamy, R., Kaliyan, G., & Kaliyamoorthy, S. (2024). *Edgar Mosquera*. Chalcogenide Materials for Sustainable Energy and Environmental Applications. IntechOpen., DOI: 10.5772/intechopen.1003048

Sliem, M. H., Kannan, K., Maurya, M. R., Jlassi, K., Sadasivuni, K. K., Kumar, B., & Abdullah, A. M. (2025). Rational synthesis of mixed metal oxide clusters supported on a partially etched MAX phase for efficient electrocatalytic CO2 conversion. *Topics in Catalysis*, *68*(13), 1382–1397.

Suganthi, N., Karthik, K., & Thangavel, S. (2022). *Nanostructured Metal Oxide and Hybrid Photocatalyst for Water Treatment and Hydrogen Generation*. Jenny Stanford Publishing.

Suganthi, N., Karthik, K., Thangavel, S., & Gurushankar, K. (2022). Electrochemical and magnetic properties of 3D porous NiS/CuS nanocomposites. *Applied Surface Science Advances*, *7*, 100209.

Suganthi, N., & Pushpanathan, K. (2018a). Photocatalytic degradation and antimicrobial activity of transition metal doped mesoporous ZnS nanoparticles. *International Journal of Environmental Science and Technology*. Advance online publication. DOI: 10.1007/s13762-018-1811-y

Suganthi, N., & Pushpanathan, K. (2018b). Spherical and dumbbell shape biphasic paramagnetic ZnS: Fe nanoparticles on ferromagnetic ZnS host background. *Journal of Electronic Materials*, *47*, 7343–7357.

Suganthi, N., & Pushpanathan, K. (2018c). Photocatalytic degradation and ferromagnetism in mesoporous La doped ZnS nanoparticles. *Journal of Materials Science Materials in Electronics*, *29*(16), 13970–13983.

Suganthi, N., Thangavel, S., & Karthik, K. (2020). Hibiscus subdariffa leaf extract mediated 2-D fern-like ZnO/TiO_2 hierarchical nanoleaf for photocatalytic degradation. *FlatChem*, *24*, 100197.

Tang, Y., Zheng, Z., Sun, X., Li, X., & Li, L. (2019). Ternary CdS-MoS_2 coated ZnO nanobrush photoelectrode for one-dimensional acceleration of charge separation upon visible light illumination. *Chemical Engineering Journal*, *368*, 448–458.

Tari, V. Kannan, K., Koliar, J., Gurushankar, K., & Wahyuni, D. K. (2025). Influence of zinc oxide nanomaterials on crop plants: A prospective analysis. In D. Jabbrova, D. Sarkar, Rakesh, S. Singh, & R. Datta (Eds.), *Smart nanomaterials in agroforestry systems* (Chapter 8). Springer. DOI: 10.1007/978-981-96-1337-3_8

Tari, V., Gupta, R., & Siddiqui, N. (2021). Impact of climate change on Upper Ganga Ramsar Site, UP, India. In *Monitoring and evaluating the ecological health of wetlands* (1st ed., pp. 92–105). IGI Global., DOI: 10.4018/978-1-7998-9498-8.ch006

Tari, V., Kannan, K., & Vishwakarma, V. (2022). Emerging antiviral technologies. In Nunthavarawong, P., Rangappa, S. M., Siengchin, S., & Thoppil-Mathew, M. (Eds.), *Antimicrobial and antiviral materials: Polymers, metals, ceramics, and applications* (1st ed., pp. 137–150). CRC Press., DOI: 10.1201/9781003143093-8

Tari, V., Siddiqui, N., Rathi, D., Siddiqui, N. N., & Wahyuni, D. K. (2024). Need for an evolved groundwater justice in rural area Uttar Pradesh, India. *Nature. Environment and Pollution Technology*, *23*(1), 125–137. DOI: 10.46488/NEPT.2024.v23i01.009

Tari, V. S., & Patil, P. Y. (2017). Challenge of contamination of pesticide for Alphonso in Ratnagiri district, Maharashtra, India. *Research Journal of Chemistry and Environment*, *21*(12), 57–67.

Tari, V. S., & Patil, P. Y. (2017a). Transfer factor of heavy metal in soil to plant from pesticide-contaminated area (mango orchards), Ratnagiri district, Maharashtra. *Research Journal of Chemistry and Environment*, *21*(1), 26–32.

Vaizo˘ gullar, A, I. (2020). Ternary CdS/MoS_2/ZnO photocatalyst: synthesis, characterization and degradation of ofloxacin under visible light irradiation. J. Inorg. Organomet. Polym Mater. 30, 4129- 41413.

Varadarajan, S., Kavitha, A., Selvaraju, P., Muthu, S. E., Gurushankar, K., Shanmugan, S., & Kannan, K. (2024). Enhanced Photocatalytic Hydrogen Evolution by TiO2: A Synergistic Approach with Defect-Rich SnS2 and Ti3C2 MXene Cocatalysts. *Hydrogen*, *5*(4), 940–957.

Wang, Z., Tu, Q., Sim, A., Yu, J., Duan, Y., Poon, S., Liu, B., Han, Q., Urban, J. J., Sedlak, D., & Mi, B. (2020). Superselective removal of lead from water by two-dimensional MoS_2 nanosheets and layer-stacked membranes. *Environmental Science & Technology*, *54*, 12602–12611.

Wani, S. A., Tari, V., & Mansoor, S. I. U. (2025). A Comprehensive Analysis of Wastewater Management Challenges in India: Infrastructure and Policy Perspectives. In Dhiman, S., & Mukherjee, G. (Eds.), *Resource Resurgence. Springer Water.* Springer., DOI: 10.1007/978-3-032-02750-4_11

Yalçın, M., Turgut, N., Gökbulut, C., Mermer, S., Sofuoğlu, S. C., Tari, V., & Turgut, C. (2022). Removal of pesticide residues from apple and tomato cuticle. *Environmental Science and Pollution Research International*. Advance online publication. DOI: 10.1007/s11356-022-23269-1

Ye, A., Fan, W., Zhang, Q., Deng, W., & Wang, Y. (2012). CdS–graphene and CdS–CNT nanocomposites as visible-light photocatalysts for hydrogen evolution and organic dye degradation. *Catalysis Science & Technology*, *2*, 969–978.

Zhang, F., Zhu, B., Guo, H., Qiu, J., Zheng, K., Chen, X., & Yu, J. (2021). Monolayer square-Ag_2X (X = S, Se): Excellent n type thermoelectric materials with high power factors. *Applied Surface Science*, *550*, 149230.

Zhang, K., & Guo, L. (2013). Metal sulphide semiconductors for photocatalytic hydrogen production. *Catalysis Science & Technology*, *3*, 1672–1690.

Zhao, D., Wu, T., & Zhou, Y., Y. (2019). Dual II heterojunctions metallic phase MoS_2/ZnS/ZnO ternary composite with superior photocatalytic performance for removing contaminants. *Chemistry (Weinheim an der Bergstrasse, Germany)*, *25*, 9710–9720.

Zheng, J., Cao, Y., Cheng, C., Chen, C., Yan, R. W., Huai, H. X., Dong, Q. F., Zheng, M. S., & Wang, C. C. (2014). Facile synthesis of Fe_3S_4 hollow spheres with high-performance for lithium-ion batteries and water treatment. *Journal of Materials Chemistry. A, Materials for Energy and Sustainability*, *2*(46), 19882–19888.

Chapter 16
Agricultural Applications of Chalcogenide-Based Materials:
Toward Sustainable, Smart, and Nutrient-Efficient Farming

Md. Shoeab Akhter
https://orcid.org/0009-0004-4582-2937
International University of Business, Agriculture, and Technology, Bangladesh

M. Shohidullah Miah
International University of Business, Agriculture, and Technology, Bangladesh

Habibur Rahman
International University of Business, Agriculture, and Technology, Bangladesh

S. M. Maksudur Rahman
International University of Business, Agriculture, and Technology, Bangladesh

A. J. M. Sirajul Karim
International University of Business, Agriculture, and Technology, Bangladesh

Sakibul Islam Ratul
https://orcid.org/0009-0000-7217-5190
Daffodil International University, Dhaka, Bangladesh

Mohammed Ataur Rahman
International University of Business, Agriculture, and Technology, Bangladesh

ABSTRACT

Food chain has been altered through farming nanotechnology, which energizes such elements as Sulfur (S) and Selenium (Se), and several industries, such as crop

DOI: 10.4018/979-8-3373-3962-7.ch016

farming, aquaculture, tree forestry, animal and poultry farms, and plant nutrients, have utilized this technology. Sulphur nanofertilizers reduce the stress level on the environment, increase the fertility of the soil, and optimize nutrient release. Selenium agents impede the occurrence of micronutrient insufficiency in human beings and animals, and Chalcogenide concentration minimizes the intake of synthetic additives and optimizes the conversion rate of feeds. The use of sulfur fertilizer helps activate sprouting and improves tolerance to diseases. Intelligent sensors with chalcogenides and optoelectronic agricultural plans can provide real-time information on the moisture of the soil, soil nutrients, and stress. These products increase nutrition, demand fewer resources, and provide adaptable agriculture approaches in an evolving climate.

INTRODUCTION

Agriculture, which is the source of food security in the world, encounters a myriad of problems in the 21st century, such as the growing demand for food, the fall in arable land, nutrient loss, climate change, and the rapid increase in population, (Tari et al., 2024). Such problems create the need to move towards more sustainable, efficient, and technologically integrated agriculture. In material science, new materials, particularly nanotechnology and the development of new functional materials, offer promising solutions to the enhancement of the conventional methods of agriculture (Sivasamy et al., 2024). Chalcogenide-based compounds have become one of the suggested materials based on their distinctive physical, chemical, and biological characteristics as a result of group 16 elements [e.g., Sulfur (S), Selenium (Si), and Tellurium (Te)]. The use of chalcogenide compounds in agriculture is a relatively new concept, and use in photonics, energy storage, and electronics is a long-established field. Some of the properties of these materials include tunable bandgaps, photoconductivity, high surface activity, and biological reaction. They interact with soil, water, and plant systems, which makes them the best in designing smart fertilizers, soil conditioners, and sensors. With these capabilities, materials based on chalcogenide can be used to facilitate a new generation of agricultural technologies that will not only be more nutrient-efficient and less toxic to the environment but also be less harmful to the environment and produce more food without disrupting the ecosystem balance, (Sajitha et al., 2024, Tari. et al. 2022).

The contribution of material science in the agricultural sector is also on the increase at a very high rate, involving the use of nanoscale material and smart delivery devices, (Gupta et al. 2024; Tari et al., 2025). Conventional fertilizers and pesticides usually lead to poor utilization, leaching, and pollution of the environment, (Özşirvan et al., 2024; Tari & Patil, 2017a; Yalçın et al. 2023; Tari V& Patil, 2017b). Conversely, nano-enabled systems have the potential for precise targeting, regulated release, and little environmental effect. In that framework, chalcogenide materials (including metallic sulfides, e.g., ZnS, MoS_2), selenium-based nanostructures, and tellurium-doped systems have been viewed as having a high potential owing to their electronic structure and chemical reactivity. The ZnS nanoparticles, which consist of sulfur, have been examined in applications in nutrient delivery and disease resistance. Sulfur is a vital nutrient that helps plants to produce enzymes and amino acids, (Ates et al., 2021). Chalcogenide nanoparticle products have the potential to provide increased efficiency in gauging nutrient delivery, decrease total input levels, and lower the amount of leaching into waterways. Furthermore, certain sulfur-based fertilizers have antimicrobial properties that are capable of controlling pathogens in the soil without necessarily using the conventional chemical pesticides. Chalcogenides based on selenium are significant in prevention in plants and animals. Selenium is needed only in small portions, though it is essential in the defense mechanisms against the antioxidant and immune response in animals and human beings. Application of the nano-selenium system in the farming sector can be used to create selenium-fortified fertilizers to fortify crops, enrich their nutritional value, and increase resilience to environmental challenges: drought, oxidative damage, etc., (Muslih et al., 2021).

The tellurium-based chalcogenides, despite being less exploited in agriculture, have potential towards producing high-sensitivity biosensors and environmental detectors. These materials are highly electrical conductors and photosensitive, which can be utilized in real time to provide information on soil moisture, pH, heavy metals, and other important agricultural parameters. With the help of these materials, incorporated into smart farming technologies, e.g., IoT platforms, precision agriculture may be made more data-driven, efficient, and sustainable. Polymeric or biochar-based matrices can also be investigated to include chalcogenide compounds to permit a slow and controlled release of nutrients or agrochemicals. This type of matrix can be designed to react to a given environmental status. To reduce the damage of pH, moisture, or microbial activity nutrients and increase sharpness by plants, the leading types of matrices for more resource-skilled and durable processes can be designed to react to a given environmental status. Lowering the loss of pH, humidity, or microbial activity nutrients and raising sharp by plants, leading to more resource-efficient and durable processes, (Akintayo, 2025). Photocatalytic agriculture is another application that is coming up. Molybdenum disulfide (MoS_2) and other transition-metal chalcogenides are examples of such photocatalysts that can be utilized to degrade (pesticides), treat (irrigation wastewater), or provide controlled photonic effects to promote plant growth. These multidisciplinary characteristics precondition the use of chalcogenide nanostructures in the creation of integrated systems that do not only provide nutrients but are also relevant to the overall agricultural ecosystem management (Priya et al., 2024).

In addition, it has been found that studies on toxic profiles of chitosan-based nano-plugs suggest that properly engineered ones have a good safety profile in relation to soil microbiota, plants, and higher organisms, which is required to allow widespread use in agriculture. Nevertheless, additional long-term research and environmental risk evaluation are required in order to make sure that these materials will not pile up in the food chain and interfere with the ecosystems. Incorporation of the chalcogenide-based materials in the agricultural sector leads to efficiency, less waste, and environmental stewardship, which aligns with the aims of sustainable development. The use of more advanced materials in the global agricultural world is also expected to increase as the technological world goes digital and smart. They were able to enhance the well-being of plants, boost nutrient utilization, and provide real-time insights, which is a milestone in the development of contemporary methods of agriculture, (Rahman & Khan, 2021).

SIGNIFICANCE AND CONTRIBUTION OF THE CHAPTER

By exploring a surprisingly untapped yet promising magnificence of substances, this bankruptcy aims to bridge the distance between superior cloth science and sensible agricultural applications. Chalcogenides, which include group 16 elements inclusive of sulfur, selenium, and tellurium, possess a wide range of bodily, chemical, and organic properties that may be exploited to revolutionize cutting-edge agricultural practices. As the arena seeks greater sustainable, smart, and useful resource-efficient farming systems, this bankruptcy gives well-timed insights into how nanotechnology—particularly primarily chalcogenide-based systems—can cope with critical demanding situations such as environmental degradation and constrained access to modern agricultural technologies. Unlike conventional agricultural materials, chalcogenide compounds can be engineered at the nanoscale to provide multiple capabilities, consisting of nutrient transport, soil enrichment, pathogen management, and environmental monitoring. The contribution of this chapter lies now not only in describing these novel substances but also in highlighting how they may be synthesized and carried out in cost-effective, environmentally accountable methods, ensuring accessibility and scalability. The main advantages and applications of chalcogenide materials in sustainable agriculture are summarized in Figure 1 (tree diagram). Reducing environmental impact, green synthesis of nanofertilizers, animal and aquatic health, IoT-enabled smart sensors, alignment with UN Sustainable Development Goals (SDGs), and stakeholder benefits form the six primary themes represented by the trunk, which symbolizes the core focus. Each branch illustrates how chalcogenide materials support more economical, environmentally friendly, and health-conscious farming methods. This tree metaphor underscores the importance of environmental management, growth, and connectivity in achieving long-term agricultural resilience and sustainability.

Figure 1. Applications of chalcogenide materials for sustainable agriculture (Author's Own Compilation)

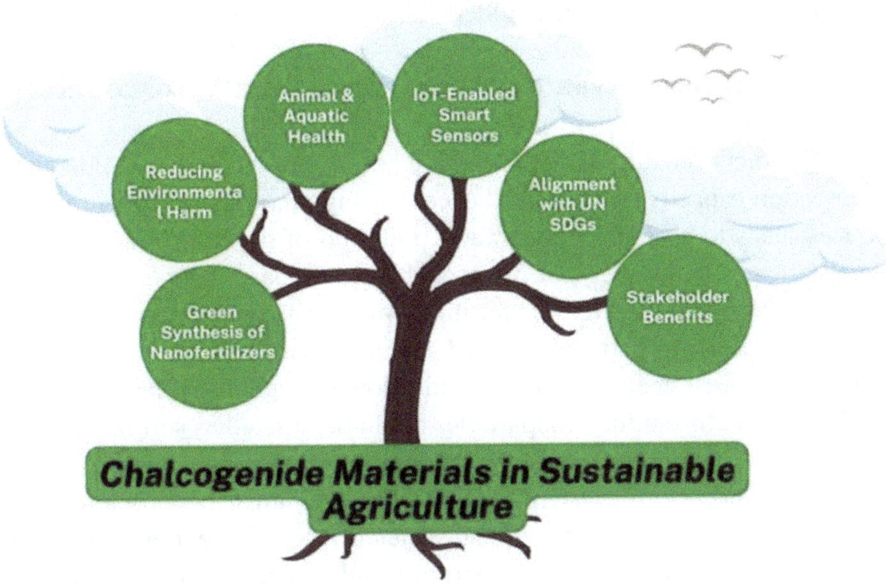

Green Preparation of Low-Cost Chalcogenide-Based Nanofertilizers with a Greener Environment

The significant innovation of this chapter is the application of the green synthesis methods to the creation of low-cost nanofertilizers based on chalcogenide. Green synthesis is the use of the environmentally viable processes that incorporate plant extracts, microorganisms, or natural reducing agents instead of toxic chemicals. The method minimizes the ecological footprint of the production process and is in line with the concept of green chemistry. As an illustration, ZnS and MoS_2 nanoparticles have been prepared by plant-based processes, which have a terrific potential as slow-release fertilizers to promote the growth of plants and minimize leaching of nutrients (Sagadevan, 2024). These nanomaterials provide micronutrients like zinc and molybdenum that are found in bioavailable form, and they can be programmed to release nutrients when the soil becomes moist or when the pH is low. Such accurate control will minimize the over-fertilization that is prevalent in conventional farming and also will assist in minimizing the expense to the farmers. Besides, the green synthesis techniques are available in resource-constrained environments. The reducing and stabilizing agents can be local plant species and agro-waste products, and thus, the technology suits the smallholders in developing countries. Such a democratization of nanotechnology will make it more accessible and viable to a large number of agricultural stakeholders, (Khouri et al., 2024).

Lessening Fertilizer Waste and Environmental Degradation

The fertilizers used are usually excessive because of quick dissolution and low water retention capacity by the soil, hence causing a nutrient runoff in the water body. This leads to eutrophication, water pollution of groundwater, and massive financial damages. Conversely, the use of chalcogenide-based nanofertilizers has the potential of dramatically decreasing fertilizer wastage because it allows the delivery of nutrients to be controlled and targeted. The high surface area and high reactivity of nanostructured chalcogenides provide slow continuous release of nutrients. As an illustration, the sulfur nanoparticles could be utilized as soil modifications, which will slowly supply the sulfur to the soil to nourish plants without frequent application, (Tang et al., 2021). On the same note, selenium nanoparticles guarantee proper distribution of this trace element, which offers sufficient nutrition without wastage. These properties decrease environmental load and add to the sustainable agricultural practices. Also, a lot of nano-enabled compounds produced by green technologies are biodegradable, which increases their integration with organic and environmentally friendly agricultural systems.

Improving the Health of Animals and Aquatic Life Using Biocompatible Chalcogenides

Although the topic of the health of plants is a major focus, the chapter also addresses the issue of chalcogenides to enhance animal and aquatic health. Recent studies indicate that both sulfur and selenium nanoparticles are bioavailable as well as safe when ingested by animals such as fish and livestock. Selenium is a requisite trace element that participates in antioxidant defense and immune control in animals, (Pandikumar & Rameshkumar, 2020). Standard selenium preparations are, however, of low bioavailability and with a low safety margin. The chalcogenide-based nano-selenium has a safer and more efficient delivery mechanism. Research has demonstrated that selenium nanoparticles enhance growth, reproductive health, and immunity in fish and livestock. The use of selenium- or sulfur-based chalcogenides promotes sustainable feeding in aquaculture since nutrient leaching and water quality are key issues of concern. These nanomaterials sustain aquatic ecosystem balance, enhancing the nutritional profile of domesticated species, which leads to food security and health of the population.

IoT-Precision Agriculture IoT-Enabled Sensors on Chalcogenide-Based Sensor Development

The creation of IoT (Internet of Things)-enabled chalcogenide-based sensors is one of the most futuristic points concerning this chapter. These intelligent sensors have the capability of measuring and reporting the condition of soil, moisture, nutrient levels, and contaminants in real time. The chalcogenides of MoS_2 and Te are characterized by their unprecedented electrical conductivity, optical properties, and structural softness, (Enamala et al., 2021). These properties ensure that they are good candidates in sensor technologies that demand low power consumption, high accuracy, and stability under harsh environments. With the combination of these sensors and wireless networks and analytics platforms, farmers can make decisions involving irrigation, fertilization, and pest control. The collected real-time information can be used to optimize the input, to predict the yield more effectively, and to avoid losses of crops due to disease or drought. The strategy forms the core of precision agriculture, which is supposed to be more efficient and less damaging to the environment, (Beheshti-Marnani et al., 2025).

Correlation with the United Nations Sustainable Development Goals (SDGs)

The applications in this chapter are particularly applicable to a number of United Nations Sustainable Development Goals (SDGs):

- **Goal 2: Zero Hunger:** It is directly connected with food security and meal health since chalcogenide-based inventions enhance crop yields, the efficiency of nutrient use, and livestock health.
- **Goal 12: Responsible Consumption and Production**: With the help of smart materials, waste and agricultural inputs are decreased, which makes production cycles more circular and sustainable.
- **Goal 13: Climate Action:** Nutrient management will lead to lower greenhouse gas emissions in agriculture and the low use of fertilizers, which will help in the management of climate change.

Through promoting them, the chalcogenide-based systems promote agricultural sustainability and resilience across the globe.

A Resource Worthy of Being Used by Various Stakeholders

This chapter is structured to be an all-inclusive and practical way-to-go guide to a very large stakeholder base:

- Researchers obtain the information about the new agritech uses of chalcogenide nanomaterials.
- Science-based solutions can be exploited by policymakers in fostering sustainable agricultural policies.
- Agri-tech entrepreneurs and startups can also be motivated to come up with innovative products or services with the aid of smart materials and precision farming.
- The teachers are able to incorporate the teachings in interdisciplinary programs that combine materials science, environmental science, and agricultural engineering.

The chapter is talking about the needs of the industrialized and developing countries. The area with minimal access to advanced agricultural infrastructure can have solutions to green synthesis and low-cost chalcogenide technologies. Meanwhile, the technological projection in the progressive areas enables the focus on sensors, automation, and integrating IoT that facilitate the transition into data-intensive, high-efficiency farming, (Pandikumar & Rameshkumar, 2020).

THE CHAPTER'S OBJECTIVES

This research is aimed at elucidating the potential of sulfur- and selenium-based chalcogenides to revolutionize modern agriculture due to their peculiar physicochemical properties, such as high surface reactivity, nanoscale size, bioavailability, and redox capacity. These characteristics render them perfect to be applied in smart agri-technology and precision farming. In particular, the article focuses on their use in various fields, including biofortification, where selenium is used to enhance the level of micronutrients in crops; forestry, where the sulfur-based materials are used to increase the fertility and resistance of soils; aquaculture, where nano-selenium is applied to increase fish immunity and growth; and animal nutrition, where the two elements are used to support metabolism and reproduction, (Chua, 2022).

The paper also highlights the design of low-cost and clean fertilizers that are synthesized using green synthesis of plant or microorganism synthesis. These methods minimize chemical waste and increase sustainability, which is particularly useful in the case of farmers with limited resources. Moreover, when combined with advanced materials, smart farming tools (IoT sensors and controlled delivery systems, etc.) can help increase productivity and reduce environmental damage. The study takes into account the use of both urban and rural references by advocating roof gardens, vertical fields, and hydroponics in addition to traditional sector cultivation. This universal strategy will guarantee a wider adoption and help achieve sustainable food production, environmental conservation, and an integration of rural and urban agriculture activities, (Kujur et al., 2023).

A methodical strategy for utilizing sulfur and selenium chalcogenides in intelligent and sustainable agriculture is illustrated in Figure 2. It outlines four crucial steps: identifying their unique properties for agricultural use; demonstrating their applications across forestry, aquaculture, crops, and livestock; encouraging the creation of economical, eco-friendly nanofertilizers and delivery systems; and incorporating these materials into both urban and rural farming. This roadmap highlights the potential of chalcogenides to promote efficiency, innovation, and environmental sustainability in modern agricultural systems.

Figure 2. Strategic roadmap for utilizing sulfur and selenium chalcogenides in smart agriculture

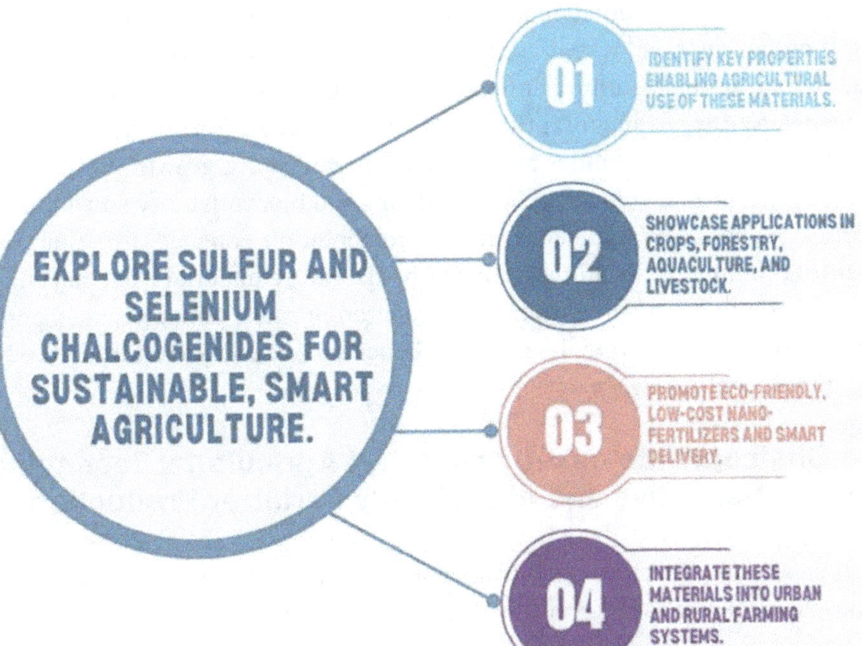

Physicochemical Characteristics of Sulfur and Selenium Chalcogenides that are Agriculturally Relevant

Chalcogenides with sulfur and selenium have distinct physicochemical features that render them extremely suitable for agricultural use. The materials are characterized by high surface area, nanoscale particle size, bioavailability, redox activity, and stability in soil conditions, which are important determinants of nutrient distribution and interaction of the soil. The chalcogenides of Sulfur like ZnS and MoS_2, contain antimicrobial effects and provide necessary nutrients that directly participate in plant metabolism, like chlorophyll synthesis and nitrogen fixation. Selenium chalcogenides offer great antioxidant effects and enhance tolerance to stress in plants and animals. These physicochemical properties will be important in developing specific nano-formulations that will be adapted to agricultural practices and reduce their toxicity for the environment with minimal impact (Maity et al., 2022).

Biofortification, Forestry, Aquaculture, Animal Nutrition and Agricultural Productivity

There is a wide variety of applications of chalcogenide-based materials in the fields of agriculture and allied industries. Bio fortification involves the fortification of staple crops (rusts and wheat) with selenium nanoparticles to enhance the health of humans and animals. Sulfur nano formulations increase the soil fertility and plant resilience in forestry to support afforestation and reforestation efforts, (Abzal et al., 2025). Selenium chalcogenides are effective in enhancing growth, immunity and survival of fish in aquaculture because of their good bioavailability and antioxidant properties. Nano-sulfur and selenium supplementation enhance digestion, reproduction, and immunity in animals. There are also those materials that can enhance crop productivity by boosting nutrient regeneration rates, pathogen management, and maintaining soil microbial health and hence increase in yield through reduced inputs, (Shao et al., 2021).

Nano-Chalcogenide-based and Smart Agricultural Technology-based cost-effective and eco-friendly Fertilizer Production

This paper advocates low-cost sustainable fertilizer synthesis with green synthesis of nano-chalcogenides. The conventional fertilizer production is usually characterized by the use of large amounts of energy and chemical wastes. By comparison, greener technologies, including low-toxicity biological agents in the synthesis of plant-extract-based nanoparticles, generate few by-products, (Wang et al., 2021). These are environmentally friendly green manures and they are also affordable, particularly to the poor agricultural societies. When these materials are incorporated into smart farming systems (automated delivery units, controlled release, sensory-based systems, etc.), the input costs decrease and the efficiency of the delivered nutrients is high. This dualism between economic and ecological sustainability is essential to sustainable agriculture in the climate change and limited resource environment, (Kumar & Debut, 2022).

Incorporation into Urban and Rural Agriculture Systems

Chalcogenide-based materials are versatile and, therefore, can be used in rural and urban agriculture. In the rural setting, the materials are incorporated in the current soil and crop systems as a way of boosting production and saving on use of costly agrochemicals. They are low toxicity and slow-release properties, which enhance the health and food of the soil without disrupting local habitats. Urban applications of the chalcogenide enhanced systems allow vertical farming, hydroponic, and rooftop gardens by facilitating accurate nutrient control in small areas. They may also be integrated in IoT-based agricultural systems to monitor the harsh environment remotely and automate it. Chalcogenides would give the world a single instrument to food security and sustainable agriculture by linking traditional and modern farming environments, (Nath et al., 2018).

Chalcogenide Materials

The Definition and Classification of PRI Will be Presented in Section

The chemical compounds of the group-16 elements—sulfur (S), selenium (Se), tellurium (Te), polonium (Po), or oxygen (O)—are called chalcogenides. In the agricultural and materials science fields, the term has a general meaning of a compound that contains sulfur, selenium, and tellurium but does not contain oxygen because this element is of a different chemistry. Depending on the bonding, they are broadly categorized as metal chalcogenides (e.g., ZnS, Cu_2S, MoS_2, FeSe) and non-metal chalcogenides. Chalcogenides can exist in crystalline, amorphous, and nanoscale forms, and each of them has different physicochemical characteristics that may be applied in the agricultural and environmental fields.

Importance Elements: Sulfur, Selenium and Tellurium

Sulfur and selenium are the most important chalcogen elements of agricultural use, with tellurium beginning to be used as a sensor element, (Masikane et al., 2020).

Sulfur (S) plays a vital role in plant metabolism, synthesis of amino acids, and in the functioning of chloroplasts. Nanoparticles like ZnS and MoS_2 are nutrient carriers and antimicrobial agents. Selenium, in small quantities is required and enhances antioxidant protection in plants, animals and humans, and it is useful in animal feed and biofortification of crops. Tellurium is a rare element, and its application is mainly limited to the electronics field, but it has conductive and photoelectric characteristics that could be looked into for smart farming sensors (Li et al., 2025).

Material Characteristics and Advantages on a Nanoscale

Chalcogenides, Which Are Engineered at the Nanoscale, Have: Chalcogenides, engineered at the nanoscale, possess a large surface area, which enhances their contact with soil microorganisms and plant roots. Chalcogenides enhance bioavailability and solubility, facilitating effective nutrient absorption. Mycoskie et al. (2001) proposed applications in the degradation of agricultural contaminants using photocatalytic activity. Redox properties play a crucial role in activating plants' defense mechanisms. The antimicrobial effect helps minimize the use of chemical pesticides (Ratha and Samantara, 2020).

Green Synthesis of Chalcogenide-Based Nano-Fertilizers

This technique involves chalcogenide salts as the starting material for the production of nano-fertilizer. The green synthesis offers a cheaper and environmentally friendly alternative to the traditional chemical techniques. It utilizes the use of plant extracts, fruit peels, or natural reducing agents under light and energy-saving conditions. In comparison to the conventional high-temperature or toxic chemical processes, green processes are safe, biodegradable, and can be used even in rural or small-scale settings (Jancik-Prochazkova and Pumera, 2023). Raw materials that are locally sourced make things cheaper, and recurrent use of farm waste encourages circularity. The resultant nanofertilizers will provide controlled nutrient release, enhanced health of the soil, and cost savings in the long run.

Green Chemistry Methods

Green synthesis does not use dangerous chemicals and reduces the use of energy because it exploits organic reagents - plant extracts, bacteria, fungi or agro-waste

- as reducing and stabilizing agents to create selenium- or sulfur-based nanoparticles (Yang, 2024).

Economic Home-Based Synthesis

ZnS, or nano-selenium synthesis, has been carried out with common plants (e.g. neem, moringa, garlic) or with agro-wastes (e.g. rice husk, fruit peels). Phytochemicals (flavonoids, terpenoids, and phenolics) in these natural sources serve as reducing and capping agents in the nanoparticle formation process, which incorporates very little equipment and low temperature and thus is easily applicable in rural areas (Kaur et al., 2022). Environmental and Energy Savings It is anticipated that the business, once established, will save approximately 22 gallons of oil each month (McGarty 2004). Green processes work at ambient temperature and pressure and reduce the consumption of fossil fuels and greenhouse gases. Climate-smart agriculture is supported by the use of biodegradable inputs where there are no harmful residues (Khatavkar et al., 2023).

Economic Goods over Traditional Fertilizers

The herbal extract and plant waste are cheap, and the extracts are locally available. Nano-form fertilizers have a reduced dose of the same or higher crop performance and often have nutrient and antimicrobial effects, minimizing the use of pesticides (Andrea, 2022).

The Importance of Sulfur as a Crop and Soil Nutrient

The nutrient uptake and resistance to disease: This tests the hypothesis that the oatmeal product will be nutritionally balanced and exhibit disease resistance. Sulfur is required in amino acid and protein synthesis, root development, uptake of nitrogen and phosphorus, and chlorophyll formation. It also fortifies cell walls and initiates antimicrobial sulfur-based substances that prevent infections of fungi like powdery mildew and rust (Alsaggaf, 2015).

Response to Stress and Biofortification

Sulfur increases the drought, salinity, and low-level tolerance of plants by increasing stress-associated proteins and antioxidants, such as glutathione. It also enhances the nutritional quality of crops, adding to the quantity of sulfur-containing vitamins and amino acids (Liu et al., 2020).

Long-Term Fertility and Soil Microbiota

Sulfur promotes sulfur-oxidizing bacteria that have the capacity of converting elemental sulfur to plant-available sulfate to enhance nutrients. The use of sulfur fertilizers that go slowly stimulates good microbes and preserves the fertility in several growing seasons (C. Wang et al., 2024).

Enhancing Food Nutritional Value

Selenium is a crucial micronutrient that helps in antioxidant activity, immunity, and metabolism. Nano-selenium fertilizer is also effective in ensuring that fertilizers carry selenium to the crops in a more efficient manner, enriching the grains and vegetables and improving the deficiencies in micronutrients, especially those reported in areas where soils contain low amounts of selenium (Sheldrick, 2000).

Micronutrient Fertilizers: Selenium (Se)

Nano-selenium can be used as a cofactor in selenoproteins that protect the cells against oxidative stress and regulate metabolism, which allows high levels of selenium uptake with no toxicity (Pimpang et al., 2025).

Bio-Enriched Crops to Reduce the Hidden Hunger

It is possible to modify staple crops by incorporating nano-selenium in fertilizers, making them stronger, eliminating the need to use supplements, and enhancing shelf life and nutritional value (Xu et al., 2023; Zhang et al., 2021).

Chalcogenides in Aquaculture and Fish Health

Chalcogenide nanomaterials, based on silicon (e.g., silicon nanoparticles, chitosan composites), are used to increase the immunity of fish, reduce antibiotic consumption, improve the quality of water, and enhance feed efficiency.

Silicon Nanoparticles as Immunostimulants

The silicon nanoparticles activate the immune cells of fish, enhancing their innate immunity and resistance to infections, and are biocompatible (Sajjad et al., 2021).

Reducing Antibiotic Use

The nanomaterials based on chalcogenides are antimicrobial and inhibit the formation of harmful bacteria and biofilms, reduce the use of antibiotics, and preserve healthy aquatic ecosystems (Shankar and Prabhu, 2023).

Greener Water and Feed Conversion

These nanoparticles trap destructive pollutants (ammonia, nitrite, and heavy metals) and enhance the ratio of feeds to lower the expenses and environmental effects (Khan, 2021).

Bangladesh Case Study

Bangladesh field trials demonstrated that silicon nanoparticles enhanced immune response in tilapia and rohu as well as decreased mortality and increased water quality, indicating that they could be used in sustainable aquaculture (Jariwala & Ravindra, 2014).

Animal Feed Applications

Chalcogenide-based feed additives, such as probiotic mineral premixes, sulfur compounds, and nano-selenium, enhance digestion, immunity, and livestock and poultry product quality.

- Probiotic/mineral blend: It is a mixture of useful microorganisms, S, and Se to promote gut health and nutrient uptake (Jia et al., 2021).
- Sulfur additives: The sulfur additives promote microbial fermentation in ruminants that enhances the digestibility of feed and reduces nitrogen emissions (Haque et al., 2017).
- Nano-selenium in chicken: Immunomodulatory (strengthens immunity), meat softness, egg, and antioxidant properties and lessens antibiotic usage (Zheng et al., 2024).
- Smallholder viability: These additives are simple and cost-effective to implement and bring more income and food security to small farmers (Buffiere et al., 2019).

Forestry and Plantation Usage

- **Weed and Pest Control:** Sulfur-based compounds are natural fungicides and pesticides that do not harm but rather protect seedlings and maintain beneficial organisms.
- **Rejuvenation of the soil:** Sulfur enhances the soil chemistry, boosts the activity of beneficial microbes, and reinstates the soil in degraded forest regions (Sajjad et al., 2021).
- **Carbon sequestration:** Sulfur and selenium compounds increase the biomass and growth of the trees, which promotes the global carbon sequestration efforts.

Chalcogenide-Based Smart Sensors for Agriculture

Chalcogenide-based smart sensors enable real-time monitoring of soil pH, moisture, and nutrient levels. Their high optical and electrical sensitivity makes them ideal for IoT-enabled precision farming, allowing data-driven decisions that optimize irrigation, fertilization, and pest control.

Table 1. Chalcogenide-based sensors and their application

Feature/Aspect	Chalcogenide-based Smart Sensors	Normal Sensors
(a) Material Composition	Made from chalcogenide materials (e.g., sulfur, selenium, tellurium compounds)	Traditional materials like silicon, metal oxides, polymers
(b) Sensing Mechanism	frequently predicated on modifications to chalcogenide materials' optical, electrical, or chemical characteristics	Using changes in electrical, chemical, or physical signals without the use of sophisticated functional materials
(c) Sensitivity	High sensitivity due to unique properties (e.g., tunable bandgap, high photoresponse)	Moderate sensitivity with limitations in detecting low-concentration analytes
(d) Selectivity	Enhanced selectivity for specific gases or biomolecules because of material chemistry	Lower selectivity, often requires filters or coatings for specificity
(e) Response Time	Typically fast response and recovery times	May be slower, depending on sensor type
(f) Integration Capability	Can be integrated into flexible, miniaturized smart devices and IoT systems	Usually rigid and larger; integration possible but with constraints
(g) Environmental Stability	Good stability under various environmental conditions but may need encapsulation	Generally robust but can have issues in harsh environments
(h) Cost	Can be higher due to specialized materials and fabrication	Generally lower cost and widely available

continued on following page

Table 1. Continued

Feature/Aspect	Chalcogenide-based Smart Sensors	Normal Sensors
(i) Applications in Agriculture	Real-time soil moisture, gas detection (e.g., ethylene, ammonia), crop health monitoring, pathogen detection	Basic environmental monitoring (temperature, humidity), simple soil moisture sensors
(j) Data Processing & Intelligence	Often equipped with smart capabilities, enabling data analytics, wireless communication, and automated decision-making	Mostly passive sensors requiring external devices for data processing

Source: Authors own based on (Nath et al., 2018).

A Review of IR, Optical, and Soil-Sensing Instruments That Have Employed Chalcogenide Glass

The chalcogenide glass is characterized by superior infrared (IR) transparency and optical characteristics and is therefore suitable in the progressive agricultural sensors. Chalcogenide glasses IR and optical sensors have the ability to measure important soil and environmental parameters by measuring patterns of light absorption and light reflection. Such sensors offer very sensitive measurements to keep track of the plant health and soil conditions. They are strong and chemically stable and hence can be used in an outdoor environment.

The chalcogenide-based soil sensors are able to determine accurately the moisture content, nutrient levels, and pH levels in the soil and can be used to diagnose the factors that affect soil fertility and stress. These devices can be used to obtain real-time data that can be utilized to undertake timely agricultural interventions by exploiting tunable optical properties and a large-scale transmission of IR of chalcogenide materials. This integration complements the traditional farming practices with intelligent diagnostics that increase crop control and sustainability (Jariwala and Ravindra, 2014).

Real-Time pH, Moisture, and Nutrient Measurements of Soil

Chalcogenide sensors allow real-time and accurate measurements of soil pH, moisture, and nutrients—parameters that are important in the proper development of crops. Their quick action enables the farmers to avoid long lab tests and to modify irrigation, fertilizing, or soil corrections on the spot. Such sensors take advantage of the chemical sensitivity of chalcogenide glass to deliver credible information on acidity, water content, and the availability of macro/micronutrients to promote resource-effective and sustainable agriculture.

Integration of Mobile App and IoT

Connected to mobile applications and Internet of Things (IoT) systems, chalcogenide sensors can send the information about soil moisture, pH, and nutrient concentrations to smartphones or cloud servers to be monitored remotely. Apps are easy to use and have visual analytics, alerts, and recommendations. The IoT networks have the potential to automate irrigation, fertilization, and pest control to maximize the use of resources and minimize human error. The fact that many farms are aggregating their data also adds to the regional crop-health analysis and forecasting.

Precision Farming

A modern method, which individualizes the management approach to field variability, is precision farming, which would not be possible without chalcogenide-based smart sensors. Constant checking of the soil pH, moisture, and nutrients gives actionable data to reduce the unnecessary use of water, fertilizer, and pesticides—reducing the costs and environmental influence. Combined with GPS and automated machines, sensor data assists in efficient planting, irrigation, and harvesting, enhancing the crop yield and quality (Jariwala and Ravindra, 2014).

Issues, Security, and Future Outlook

Although chalcogenide materials will have transformative advantages for sustainable agriculture, the issues related to the matter are in terms of educating farmers and regulating the systems and gaps in the research. It is important to handle safety issues and ensure responsible adoption through a multidisciplinary approach. New technologies, like AI, blockchain, and robotics, have the potential to improve these inventions and design a safer and more efficient future in agriculture. An integrated approach to the safe application of chalcogenide substances to agriculture comprises three interconnected pillars, i.e., tech integration, regulation and safety, and farmer education. In the article "Tech Integration," the authors emphasize the use of AI, blockchain, robotics, and precise nanomaterials in smart farming. Regulation & Safety focuses on good environmental surveillance, standards on nanomaterials, and global standards. Farmer Education aims at digital literacy and safe application practice, as well as being aware of the benefits and risks, (Wang et al., 2024).

Educating and Training of Farmers

What is needed in the efficient utilization of chalcogenide-based materials is extensive training of the farmers. Farmers are not aware of the benefits, application methods, and precautions. In the absence of sufficient knowledge, abuse may decrease efficacy, have an environmental impact, or be hazardous to health. The training must also consist of practical demonstrations, safe handling steps, and information on how to interpret sensor data. Governments, NGOs, and industry players ought to work together and offer accessible and culturally relevant programs that will enable farmers to use such technologies in a responsible and sustainable manner.

Figure 3. Framework for the safe adoption and use of chalcogenides in agriculture (Authors Own Compilation Based on Jariwala & Ravindra, 2014).

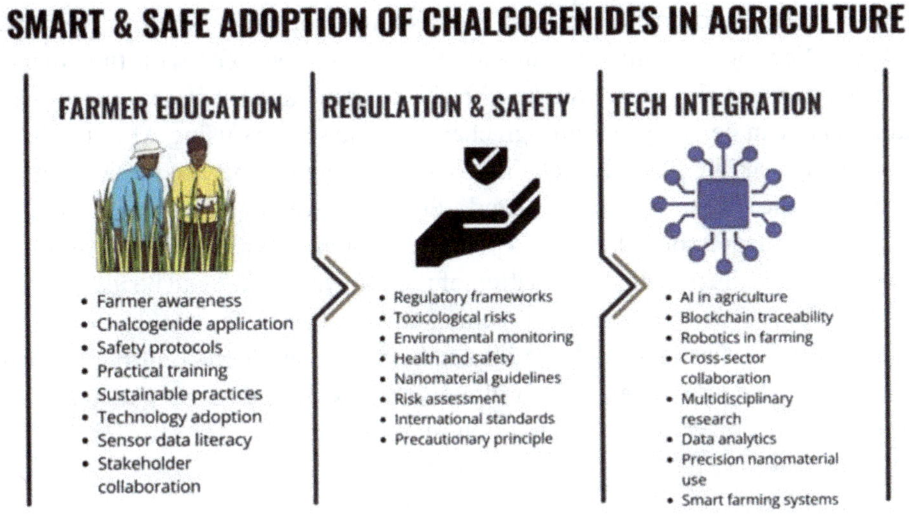

Regulations and Toxicological Implications by the Government

Application of chalcogenides in agriculture needs to have stringent regulatory structures so that the safe management of base materials, especially in light of possible toxic hazards, is considered. Unless nanomaterials and some chalcogenide compounds are managed, they may harbor unknown health risks for humans, animals, and the environment. Governments should come up with clear instructions on how to produce, use, discard, and monitor the environment to avoid contamination and bioaccumulation. The toxicological research is necessary to determine the acute and chronic effects, exposure routes, and dose limit. The existing laws and regulations tend to be lagging behind the fast technological advances, leaving loopholes in safety regulation. Regular internationalization can support responsible innovation and safeguard the health and ecosystem of people. The regulatory bodies are expected to be transparent, fostering research of risks, and engaging the stakeholders in decision-making. Striking a balance between innovation and precaution would make sure that the advantages of chalcogenide materials, as far as agriculture is concerned, are not at the cost of long-term safety and environmental stability.

Research and Cross-Sector Coordination are Required

Although there is a prospect of improvement, the studies of the chalcogenide applications in agriculture are scarce, and the collaboration across sectors is inadequate. The combination of knowledge in the fields of materials science, agronomy, data analytics, and environmental science is essential in the process of comprehensive solutions. New opportunities to enhance this integration exist with the emerging technologies of artificial intelligence (AI), blockchain, and robotics. Complex agricultural data can be analyzed through chalcogenide sensors using AI to model and aid decision-making. The blockchain provides an opportunity to trace inputs and outputs in a safe way and inspire trust and adherence to regulations. Applications and monitoring of nanomaterials are also possible through robotics, and this minimizes human error and risks. The combination of these technologies can be used to speed up innovation, improve resource utilization, and increase safety. Multi-disciplinary projects, pilot studies, and scalable implementation may focus on future research to bring chalcogenide materials and digital devices simultaneously to achieve the full potential of smart and durable agricultural systems.

FINDINGS

The chapter brings to the fore the modulatory capacity of sulfur- and selenium-based chalcogenide nanomaterials in the agricultural sector that is eco-friendly. Their nanoscale properties of high surface area, redox potential, antimicrobial properties and bioavailability enable the precision of nutrient delivery, recreational pathogen control, and classification of soil fertility (Ates et al., 2021; Maity et al., 2022). Green production of plants and agro-based waste offers cheap and recyclable production pathways and may be integrated into the agricultural models that are resource constrained (Sagadevan, 2024; Khouri et al., 2024). The chapter has found application in crop nutrition, livestock health, aquaculture, forestry, and smart sensor technologies to demonstrate how they can be applied across a range of agricultural systems (Pandikumar and Rameshkumar, 2020; Nath et al., 2018). It is important to note that preparations based on selenium improve biofortification and animal immunity, whereas sulfur nanoparticles are effective to enrich soils and make them more resilient to stress (Liu et al., 2020; Shao et al., 2021). The IoT-based precision farming also makes it possible, using chalcogenide sensors, to make decisions based on real-time data (Enamala et al., 2021; Beheshti-Marnani et al., 2025). Taken together, the results validate the fact that they can be used to enhance productivity with minimal effects on the environment.

CONCEPTUAL FRAMEWORK

This chapter is organized based on a multi-pillar conceptual framework of connecting chalcogenide nanomaterials with sustainable agriculture. Material innovation is the first pillar, in which sulfur, selenium, and tellurium chalcogenides offer exclusive physicochemical attributes, including tunable bandgaps, antimicrobial effects, and bioavailability of nutrients that have direct benefits to plant, animal and soil systems (Masikane et al., 2020; Maity et al., 2022). The second pillar is green synthesis, which is concerned with the possibility of its production using plant extracts, microbial agents, and agro-waste without considering its production to be harmful to the environment to make it accessible and environmentally friendly (Kaur et al., 2022; Jancik-Prochazkova and Pumera, 2023). Agricultural uses are the third pillar, and they involve biofortification, livestock supplementation, livestock aquaculture feed, forestry rehabilitation, and controlled-release fertilizers that have less waste and are more resilient (Pandikumar and Rameshkumar, 2020; Abzal et al., 2025). Fourth is the digital integration, where the chalcogenide-based sensors will be utilized in digitalizing the precision agriculture and, as a result, will be effective in irrigation, fertilization and pest control (Enamala et al., 2021; Nath et al., 2018). Lastly, the framework contextualizes these developments in the global sustainability agenda since it aligns with SDGs on hunger, responsible production, and climate action (Khatavkar et al., 2023). There are feedback loops, i.e., a higher level of efficiency can decrease an ecological burden, whereas the potential hazards, e.g., the toxicity of nanoparticles, will require regulation and education of farmers (C. Wang et al., 2024).

DISCUSSION

As per the discussion, the steering force of climate-smart and resource-efficient agriculture can be played by the chalcogenide nanomaterials. Their ability to produce controlled nutrient release will solve fertilizer inefficiencies, decreasing the runoff and environmental harm (Tang et al., 2021; C. Wang et al., 2024). Fertilizers containing selenium also overcome the deficiency of micronutrients by biofortification in the context of hidden hunger (Xu et al., 2023; Zhang et al., 2021). Selenium and sulfur nanoparticles in aquaculture and animal nutrition have been observed to enhance immunity, growth, and food efficiency and improve food security (Jia et al., 2021; Zheng et al., 2024). Real-time tracking of the soil and crops is a feature of chalcogenide-enabled sensors that enhance precision farming (Beheshti-Marnani et al., 2025). However, major problems still exist. Bioaccumulation, potential toxicity, and absence of knowledge among farmers are the issues that can limit its uptake (Ratha & Samantara, 2020; C. Wang et al., 2024). Moreover, to be able to scale green synthesis to an industrial scale, economic validation has to be done. Therefore, although it has potential, a successful execution will have to strike a balance between innovation and the environment, control, and development.

RECOMMENDATIONS

The following chapter must be carried out to ensure that chalcogenide nanomaterials of various types are introduced safely and effectively into agriculture, and to begin with, the scale-up of green synthesis methods should be given top priority to guarantee its cost-effective nature and hence availability to small-scale farmers (Kaur et al., 2022; Sagadevan, 2024). Secondly, governments and regulatory bodies should establish clear safety criteria concerning the preparation, use, and disposal of these materials, including toxicity risk (C. Wang et al., 2024). Thirdly, there might be educational programs to make farmers aware of the benefits of nanomaterials, their safe application, and understanding of data through sensing (Jariwala & Ravindra, 2014). Fourthly, the tests must be based on field trials across diverse agroecological domains for testing productivity, safety, and scalability (Abzal et al., 2025). Fifthly, there must be digital integration through defining and enabling IoT-based chalcogenide sensors to facilitate precision farming and minimize the wastage of labor, capital, and natural resources (Nath et al., 2018). And lastly, policymakers should put in a conscious effort to link these innovations in explicit terms with SDGs toward food security, sustainable production, and climate action (Khatavkar et al., 2023).

CONCLUSION

The chapter provides a general roadmap on future development of agriculture due to the innovative use of chalcogenide-based materials. The materials, which are produced using cost-efficient and eco-friendly processes, can transform the agricultural sector since they can facilitate more efficient, adaptable, and eco-friendly production processes. The main sectors of its use are crop farming, livestock and poultry feed, fisheries and aquaculture, and forestry and plantation farming. An inclusion of the chalcogenide nanomaterials in these various farming systems can enhance productivity and stability in the countryside and the city. Chalcogenide-based technologies resolve issues like pest and disease control, environmental degradation, and the effects of climate change. Precision agriculture, optimal use of resources, and enhanced animal health are facilitated by the use of smart sensors, nanofertilizers, and advanced feed additives, resulting in higher yields and higher-quality production at a reduced ecological footprint. Also, the implementation of chalcogenide materials intensifies green aquaculture and forestry, which enhances biodiversity protection and carbon sequestration. This chapter highlights that cross-sector partnership, educating the farmers, and regulating its adoption as a requirement are vital to the crucial aspects of safe, effective, and fair adoption. With the help of promoting interdisciplinary research and adapting to the use of new digital technologies like IoT, AI, and blockchain, the agricultural sector will be able to reach the full potential of chalcogenide-based breakthroughs. After all, nanotechnology-based solutions will enable world agriculture to respond to the increasing food demand without compromising the integrity of the environment.

FUTURE STUDIES

Addressing knowledge gaps concerning the long-term ecological and toxicological effects of chalcogenide nanoparticles on soils, aquatic systems, and food chains is, therefore, necessary (Ratha & Samantara, 2020; C. Wang et al., 2024). Another approach could be to define safe dose-response thresholds to regulate standards while considering biodegradable composites that minimize risk as alternatives, such as biochar- or chitosan-based carriers (Yang, 2024). Further inquiry into urban agriculture, hydroponics, and rooftop farming could play a crucial role in enhancing city food security (Nath et al., 2018). In aquaculture, there is a need for systematic trials to investigate impacts on fish immunity, water quality, and microbial ecology under practical farming conditions (Sajjad et al., 2021; Shankar & Prabhu, 2023). Besides, in the case of livestock, it is imperative to assess the economic and nutritional impacts of nano-selenium and sulfur feed additives (Zheng et al., 2024). Finally, it is vital to combine AI, IoT, and blockchain with chalcogenide-based sensors for data-driven predictive farming (Enamala et al., 2021; Beheshti-Marnani et al., 2025).

REFERENCES

Abzal, S. M., Kalyan, K., Janga, S. L., Arshad Ahamed, A., & Dash, J. K. (2025). Fabrication and Functionalization of Bismuth Chalcogenide Nanostructures for Biomedical and Toxic Gas Sensing Applications. In Uddin, I. (Ed.), *Smart Nanosensors* (pp. 399–422). Springer Nature., DOI: 10.1007/978-981-96-3878-9_16

Alsaggaf, A. (2015). *Investigation of Non-Vacuum Deposition Techniques in Fabrication of Chalcogenide-Based Solar Cell Absorbers*. http://hdl.handle.net/10754/565108

Andrea, S. (2022). *Sviluppo di celle fotovoltaiche bifacciali o a film sottile basate su calcogenuri complessi con deposizioni fisiche da vuoto e da fase liquida*. https://tesidottorato.depositolegale.it/handle/20.500.14242/196702

Ates, M., Yılmaz, E., & Tanaydın, M. K. (2021). 14—Challenges, novel applications, and future prospects of chalcogenides and chalcogenide-based nanomaterials for photocatalysis. In Khan, M. M. (Ed.), *Chalcogenide-Based Nanomaterials as Photocatalysts* (pp. 307–337). Elsevier., DOI: 10.1016/B978-0-12-820498-6.00014-7

Beheshti-Marnani, A., Rohani, T., Kermani, M. A., & Mohammadi, S. zia. (2025). A sensitive chalcogenide-based electrochemical sensor for ultra-level detection of Mospilan residues in real samples. *Scientific Reports*, *15*(1), 5966. DOI: 10.1038/s41598-025-89256-x PMID: 39966467

Buffiere, M., Dhawale, D. S., & El-Mellouhi, F. (2019). Chalcogenide Materials and Derivatives for Photovoltaic Applications. *Energy Technology (Weinheim)*, *7*(11), 1900819. DOI: 10.1002/ente.201900819

Chua, H. J. (2022). *A chalcogenide-based colorimetric sensor on paper for heavy metal ions detection* [Nanyang Technological University]. DOI: 10.32657/10356/163946

Enamala, M. K., Chavali, M., Tangellapally, A., Pasumarthy, D., Murthy, M. K., Kuppam, C., Chaudhary, V., Mishra, R., & Naradasu, D. (2021). 10—Use of chalcogenide-based nanomaterials for wastewater treatment, including bacterial disinfection and organic contaminant degradation. In Khan, M. M. (Ed.), *Chalcogenide-Based Nanomaterials as Photocatalysts* (pp. 243–259). Elsevier., DOI: 10.1016/B978-0-12-820498-6.00010-X

Gupta, V., Chandra, N., Tari, V., & Kannan, K. (2024). A sustainable approach for the green synthesis of silver nanoparticles using *Brassica oleracea* subsp. *botrytis* (L.) leaves and its antibacterial activity. *Physics and Chemistry of Solids*, *25*(2), 303–310. DOI: 10.15330/pcss.25.2.303-310

Haque, F., Daeneke, T., Kalantar-zadeh, K., & Ou, J. Z. (2017). Two-Dimensional Transition Metal Oxide and Chalcogenide-Based Photocatalysts. *Nano-Micro Letters, 10*(2), 23. DOI: 10.1007/s40820-017-0176-y PMID: 30393672

Jancik-Prochazkova, A., & Pumera, M. (2023). Light-powered swarming phoretic antimony chalcogenide-based microrobots with "on-the-fly" photodegradation abilities. *Nanoscale, 15*(12), 5726–5734. DOI: 10.1039/D3NR00098B PMID: 36866684

Jariwala, B., & Ravindra, N. M. (2014). Process, property and performance of chalcogenide-based thermoelectric materials. *Nanomaterials and Energy, 3*(3), 68–81. DOI: 10.1680/nme.14.00006

Jia, H., Peng, L., Yu, C., Dong, L., Cheng, S., & Xie, J. (2021). Chalcogenide-based inorganic sodium solid electrolytes. *Journal of Materials Chemistry. A, Materials for Energy and Sustainability, 9*(9), 5134–5148. DOI: 10.1039/D0TA10994K

Kaur, K. Nisika, & Kumar, M. (2022). Flexible Solar Cells Based on Chalcogenides. In *Smart and Flexible Energy Devices*. CRC Press.

Khan, M. M. (2021). 1—Introduction and fundamentals of chalcogenides and chalcogenides-based nanomaterials. In Khan, M. M. (Ed.), *Chalcogenide-Based Nanomaterials as Photocatalysts* (pp. 1–6). Elsevier., DOI: 10.1016/B978-0-12-820498-6.00001-9

Khatavkar, S. N., Khatavkar, S. N., & Sartale, S. D. (2023). Chemically Deposited Iron Chalcogenide-Based Carbon Composites for Supercapacitor Applications. In Ezema, F. I., Lokhande, C. D., & Lokhande, A. C. (Eds.), *Chemically Deposited Metal Chalcogenide-based Carbon Composites for Versatile Applications* (pp. 83–121). Springer International Publishing., DOI: 10.1007/978-3-031-23401-9_3

Khouri, M., Escalier, R., Heran, D., Myara, M., Kribich, R., Vigreux, C., & Bendoula, R. (2024). Simulating and Fabricating Chalcogenide-based Waveguides for Agro-Environmental Applications. *2024 24th International Conference on Transparent Optical Networks (ICTON)*, 1–4. DOI: 10.1109/ICTON62926.2024.10647576

Kujur, V. S., Wadhwa, G., Singh, K., Aslam, S., & Kumar, R. (2023). Advanced of Chalcogenides Based as Hazardous Gas Sensing. In Sonker, R. K., Singh, K., & Sonkawade, R. (Eds.), *Advanced Functional Materials for Optical and Hazardous Sensing: Synthesis and Applications* (pp. 47–67). Springer Nature., DOI: 10.1007/978-981-99-6014-9_3

Kumar, B., & Debut, A. (2022). *Green Chemistry: New Perspectives*. BoD – Books on Demand.

Li, Y., Zhang, Y., Ma, H., Wan, Y., & Zhao, T. (2025). Low-dimensional metal chalcogenides for wearable gas sensing. *Nano Convergence, 12*(1), 34. DOI: 10.1186/s40580-025-00500-6 PMID: 40637935

Liu, K., Liu, K., Liu, J., Ren, Q., Zhao, Z., Wu, X., Li, D., Yuan, F., Ye, K., & Li, B. (2020). Copper chalcogenide materials as photothermal agents for cancer treatment. *Nanoscale, 12*(5), 2902–2913. DOI: 10.1039/C9NR08737K PMID: 31967164

Maity, S., Parvin, U. K. S., Das, S., & Chatterjee, K. (2022). Polymer chalcogenides—New smart materials for thermoelectric applications. *Smart Materials and Structures, 31*(7), 073001. DOI: 10.1088/1361-665X/ac7595

Masikane, S. C., Revaprasadu, N., Masikane, S. C., & Revaprasadu, N. (2020). Indium Chalcogenide Nanomaterials in the Forefront of Recent Technological Advancements. In *Post-Transition Metals*. IntechOpen., DOI: 10.5772/intechopen.94558

Muslih, E. Y., Munir, B., & Khan, M. M. (2021). 2—Advances in chalcogenides and chalcogenides-based nanomaterials such as sulfides, selenides, and tellurides. In Khan, M. M. (Ed.), *Chalcogenide-Based Nanomaterials as Photocatalysts* (pp. 7–31). Elsevier., DOI: 10.1016/B978-0-12-820498-6.00002-0

Nath, M., Masud, J., Swesi, A. T., Nath, M., Silva, U. D., Liyanage, W. P. R., Umapathi, S., & Amin, B. G. (2018). Designing Smart Materials for Efficient Electrosynthesis of Fuels and Environmental Remediation: The Story of Transition Metal Chalcogenides. *ECS Meeting Abstracts, MA2018-01*(28), 1635. DOI: 10.1149/MA2018-01/28/1635

Özşirvan, G., Yalçın, M., Turgut, N., Tari, V., & Turgut, C. (2024). Insights into the uptake, translocation, and accumulation dynamics of cyantraniliprole and thiamethoxam seed coating pesticides in maize plants. *Environmental Science and Pollution Research International*, (Special Issue). Advance online publication. DOI: 10.1007/s11356-024-34135-7 PMID: 38954337

Pandikumar, A., & Rameshkumar, P. (2020). *Graphene-Based Electrochemical Sensors for Toxic Chemicals*. Materials Research Forum LLC.

Pimpang, P., Hongsith, K., Choopun, S., & Wongrerkdee, S. (2025). A Review on Chalcogenide-Based Materials for Counter Electrode Applications in Dye-Sensitized Solar Cells: Sulfides, Selenides, and Tellurides. *ACS Omega, 10*(25), 26293–26310. DOI: 10.1021/acsomega.5c03194 PMID: 40620954

Priya, A., Sagadevan, S., Priya, A., & Sagadevan, S. (2024). Chalcogenides-Based Nanomaterials for Contaminant Removal in Wastewater Treatment. In *Structural and Chemical Features of Chalcogenides*. IntechOpen., DOI: 10.5772/intechopen.1005357

Rahman, A., & Khan, M. M. (2021). Chalcogenides as photocatalysts. *New Journal of Chemistry, 45*(42), 19622–19635. DOI: 10.1039/D1NJ04346C

Ratha, S., & Samantara, A. K. (2020). Metal Chalcogenide-Based Electrochemical Capacitors. In *Electrochemical Energy Conversion and Storage Systems for Future Sustainability*. Apple Academic Press.

Sagadevan, S. (2024). *Structural and Chemical Features of Chalcogenides*. BoD – Books on Demand.

Sajitha, D. R., Stephen, B., Nakamura, A., Selvaraj, M., Salammal, S. T., & Hussain, S. (2024). The emergence of chalcogenides: A new era for thin film solar absorbers. *Progress in Solid State Chemistry, 76*, 100490. DOI: 10.1016/j.progsolidstchem.2024.100490

Sajjad, M., Cheng, F., & Lu, W. (2021). Research progress in transition metal chalcogenide based anodes for K-ion hybrid capacitor applications: A mini-review. *RSC Advances, 11*(41), 25450–25460. DOI: 10.1039/D1RA02445K PMID: 35478910

Shankar, M. R., & Prabhu, A. N. (2023). A review on structural characteristics and thermoelectric properties of mid-temperature range Chalcogenide-based thermoelectric materials. *Journal of Materials Science, 58*(43), 16591–16633. DOI: 10.1007/s10853-023-09028-8

Shao, X., Zhang, X., Liu, Y., Qiao, J., Zhou, X.-D., Xu, N., Malcombe, J. L., Yi, J., & Zhang, J. (2021). Metal chalcogenide-associated catalysts enabling CO2 electroreduction to produce low-carbon fuels for energy storage and emission reduction: Catalyst structure, morphology, performance, and mechanism. *Journal of Materials Chemistry. A, Materials for Energy and Sustainability, 9*(5), 2526–2559. DOI: 10.1039/D0TA09232K

Sheldrick, W. S. (2000). Network self-assembly patterns in Main Group metal chalcogenide-based materials. *Journal of the Chemical Society, Dalton Transactions: Inorganic Chemistry, 18*, 3041–3052. DOI: 10.1039/B003762L

Sivasamy, R., Kaliyan, G., Kaliyamoorthy, S., Mosquera, E., Sivasamy, R., Kaliyan, G., Kaliyamoorthy, S., & Mosquera, E. (2024). Chalcogenide Materials for Sustainable Energy and Environmental Applications. In *Structural and Chemical Features of Chalcogenides*. IntechOpen., DOI: 10.5772/intechopen.1003048

Tang, Y., Qin, Z., Yin, S., & Sun, H. (2021). Transition metal oxide and chalcogenide-based nanomaterials for antibacterial activities: An overview. *Nanoscale, 13*(13), 6373–6388. DOI: 10.1039/D1NR00664A PMID: 33885521

Tari, V. Kannan, K., & Vishwakarma, V. (2022). Chapter 8: Emerging antiviral technologies. In P. Nunthavarawong, S. M. Rangappa, S. Siengchin, & M. Thoppil-Mathew (Eds.), *Antimicrobial and antiviral materials: Polymers, metals, ceramics, and applications* (1st ed., pp. 137–150). CRC Press. DOI: 10.1201/9781003143093-8

Tari, V. Kannan, K., Koliar, J., Gurushankar, K., & Wahyuni, D. K. (2025). Influence of zinc oxide nanomaterials on crop plants: A prospective analysis. In D. Jabbrova, D. Sarkar, Rakesh, S. Singh, & R. Datta (Eds.), *Smart nanomaterials in agroforestry systems* (Chapter 8). Springer. DOI: 10.1007/978-981-96-1337-3_8

Tari, V., Siddiqui, N., Rathi, D., Siddiqui, N. N., & Wahyuni, D. K. (2024). Need for an evolved groundwater justice in rural area Uttar Pradesh, India. *Nature. Environment and Pollution Technology, 23*(1), 125–137. DOI: 10.46488/NEPT.2024.v23i01.009

Tari, V. S., & Patil, P. Y. (2017). Challenge of contamination of pesticide for Alphonso in Ratnagiri district, Maharashtra, India. *Research Journal of Chemistry and Environment, 21*(12), 57–67.

Tari, V. S., & Patil, P. Y. (2017a). Transfer factor of heavy metal in soil to plant from pesticide-contaminated area (mango orchards), Ratnagiri district, Maharashtra. *Research Journal of Chemistry and Environment, 21*(1), 26–32.

Wang, C., Nie, R., Dai, Y., Tai, H., Zhu, B., Zhao, L., Wu, Y., Guo, W., & Il Seok, S. (2024). Enhancing the inherent stability of perovskite solar cells through chalcogenide-halide combinations. *Energy & Environmental Science, 17*(4), 1368–1386. DOI: 10.1039/D3EE03612J

Wang, Y., Lin, P., Lou, Q., Zhang, Z., Huang, S., Lu, Y., & He, J. (2021). Design guidelines for chalcogenide-based flexible thermoelectric materials. *Materials Advances, 2*(8), 2584–2593. DOI: 10.1039/D0MA01018A

Xu, Y., Li, R., Bai, S., Li, Y., Jia, Z., Yang, Y., Cui, E., Yao, F., Wang, D., Lei, C., & Lin, Q. (2023). Chalcogenide-Based Narrowband Photodetectors for Imaging and Light Communication. *Advanced Functional Materials, 33*(11), 2212523. DOI: 10.1002/adfm.202212523

Yalçın, M., Turgut, N., Gökbulut, C., Mermer, S., Sofuoğlu, S. C., Tari, V., & Turgut, C. (2022). Removal of pesticide residues from apple and tomato cuticle. *Environmental Science and Pollution Research International*. Advance online publication. DOI: 10.1007/s11356-022-23269-1 PMID: 36171324

Yang, G. (2024). *Composite Strategy to Enhance the Thermoelectric Performance of Chalcogenide-based Materials* [Thesis, University of Wollongong]. https://ro.uow.edu.au/articles/thesis/Composite_Strategy_to_Enhance_the_Thermoelectric_Performance_of_Chalcogenide-based_Materials/27669117/1

Zhang, D., Li, L., & Zhang, Y. (2021). Metal chalcogenides-based materials for high-performance metal ion capacitors. *Journal of Alloys and Compounds*, *869*, 159352. DOI: 10.1016/j.jallcom.2021.159352

Zheng, J., Ma, M., Yang, G., Wu, Y., & Mei, D. (2024). Progress in the study of binary chalcogenide-based thermoelectric compounds. *Journal of Solid State Chemistry*, *334*, 124617. DOI: 10.1016/j.jssc.2024.124617

Chapter 17
Enhancing Pharmaceutics With AI Through Predictive Modeling of Crystal Structures and Atom Properties for Improved Solubility and Bioavailability

Sekar Kidambi Raju
https://orcid.org/0000-0003-4300-2041
School of Computing, SASTRA University, Thanjavur, India

Subhash Kannan
K. Ramakrishnan College of Engineering, Samayapuram, India

Ganesh Karthikeyan Varadarajan
School of Computing, SASTRA University, Thanjavur, India

Raj Anand Sundaramoorthy
School of Computing, SASTRA University, Thanjavur, India

Marwa M. Eid
Jadara University Research Center, Jadara University, Jordan

El-Sayed M. El-Kenawy
Applied Science Research Center, Applied Science Private University, Amman, Jordan

ABSTRACT

The research study explores how sophisticated machine learning algorithms, i.e., our suggested model SENET, coupled with SVM and ADAM, could enhance drug solubility and bioavailability through the prediction of crystallographic structures

DOI: 10.4018/979-8-3373-3962-7.ch017

and atom property classification. The current research aims to revolutionize pharmaceutical drug development processes by embracing AI-based preclinical Testing. We investigate the sophisticated field of crystallographic structures and dissect atomistic data in depth to reveal determining factors influencing crystal forms. The aim of the research is to enhance drug solubility and bioavailability by conducting this study, which will significantly transform pharmaceutical practice. The study shows that sophisticated machine learning methods can potentially forecast drug behavior in complicated situations. This study discloses the potential for groundbreaking drug discovery by speeding up the progress in pharmaceutical science, reducing the development cost, and optimizing patient benefit through enhanced drug delivery and action.

INTRODUCTION

One of the most significant predictive determinants of properties and behaviors of pharmaceutical drugs with a significant impact on their dimensions of solubility and bioavailability is crystallographic structures. The stability and dissolution rate of a drug and its overall efficacy will be reliant on the coordination of atoms in their crystalline state and, to a large extent, deterred by the complexities surrounding them. Understanding the complexities of crystallographic structures has been a daunting challenge in pharmaceutical science, largely due to the emphasis placed on these parameters. We address the complexity of crystallography using advanced machine learning algorithms, including not only our induced SENET model but also SVM and ADAM algorithms, to decode the specific details and order in these atomic structures while ensuring that crystal structures have careful atomistic data and that the factors influencing crystal geometries are at the level required. We are hoping that through deeper levels of scrutiny, we will overcome these traditional boundaries and reach new horizons that will predict drug solubility and bioavailability. We are not only researching to enhance our knowledge of pharmaceutical practice but also, hopefully, preparing the ground for new drug discovery that will, in the end, contribute to progress in pharmaceutical science and direct benefits such as reduced costs in drug development, as well as improved health outcomes for patients through improved delivery of the drug and its efficacy.

In an efficient manner, researchers propose a new methodology called Quick-CSP to predict crystal structures of similarly structured molecules in an effective manner. The methodology relies on valid and transferable custom-designed force fields and is also computationally less expensive Mattei, A., et al., 2022). . Crystal formation and crystal structure criteria are important to consider when developing a drug delivery system. Understanding the parameters in crystal formation and controlling them are important for the stability, release characteristics, and general activity of a drug. Taylor, L. S., Braun, D. E., & Steed, J. W. (2021). There has been a shift in the field with recent developments in the ability to predict the structure of molecular crystals. In this article, we will discuss how scientists have and are using these methods to develop new drugs and functional organic materials with specific properties Beran, G. J., 2023). . Crystal engineering provides us with a solution to one of the most common problems in drug development: low drug solubility and bioavailability. Crystal engineering improves drug absorption into the body by specifically changing the crystal form of the drug, which improves its effectiveness Varshosaz, J., Ghassami, E., & Ahmadipour, S. (2018).

The contributions of the research are

- We present a new machine learning algorithm (SENET-SVM) for predicting the crystal structure and atom properties of various drug crystal structures, which could help develop drugs with better solubility and bioavailability.
- It could change practices across the pharmaceutical industry with its applications in using algorithms to drastically lower the cost of drug discovery and to support those navigating the lengthy, time-consuming drug delivery process before a drug gets to the patient.

The article discusses pharmaceutical co-crystals as a green way to increase solubility and stability of drugs, and thereby therapeutic efficacy, over more traditional alteration of drugs. This is because this strategy does not interfere with the drug's inherent property Chettri, A., et al. (2024). Crystal engineering is one of those techniques in the pharmaceutical industry that has come to the rescue to address the solubility problem of drugs. This method increases the solubility of the drugs and their effectiveness by regulating their crystal form Sugiyama, H., & Uekusa, H. (2025). The crystal habit, or the familiar form of appearance of faces on a crystal, could gain control of the solubility of ticagrelor. It has been found that ticagrelor crystals containing predominantly (001) faces are the most soluble ones, as such faces expose hydrophilic propyl groups Ren, Y., et al. (2019). Crystallization Research was performed to improve the solubility of a natural product, vanillin, in medicine and food. Researchers obtained cocrystals by combining vanillin with some amide drugs that were more soluble and dissolved more quickly than the raw vanillin. DFT computations described the origin of the formation and stability of the cocrystals Ouyang, J., et al. (2022). This paper discusses the concept of co-crystallization as a new strategy for improving drugs. The combination of a drug with another molecule into a new structure to enhance its properties, such as enhanced solubility and possibly even synergy between drugs, is known as co-crystallization Ngilirabanga, J. B., & Samsodien, H. (2021). Researchers are studying a potentially useful approach of enhancing the absorption of insoluble drugs. It incorporates the co-crystallization process that alters the structure of the drug to enhance its solubility and nanocrystallization, reducing the particle size to increase the rate of dissolution. Huang, Z., Staufenbiel, S., & Bodmeier, R. (2022). To increase the solubility and activity of two antithrombotic drugs, apixaban and quercetin, which have low solubility, researchers made a cocrystal of the two in order to improve their solubility and activity within the body. The approach will offer a means to maximize the efficacy of such medications Zhang, L., et al. (2021). X-ray crystallography is an effective instrument used to determine the atom-scale composition of the pharmaceutical cocrystals, drugs that are acquired by the type of interlinkage between a drug molecule and another molecule. The prediction and design of the physicochemical characteristics of these cocrystals, such as solubility, stability, and rate of dissolution, are of most significance to this structural information Aitipamula, S., & Vangala, V. R. (2017). In most cases, co-crystals have the potential to improve the solubility and bioavailability of poorly soluble drugs and, in certain circumstances, improve efficacy. This is done by creating a new form of crystalline that has superior qualities by introducing a drug molecule and a second molecule Al-Dulaimi, A. F., Al-kotaji, M., & Abachi, F. T. (2022).

Research Key Activities In The Work

The programs of cheminformatics play a major role in forecasting and evaluating drug candidate solubility. A standard format is used to feed the drug molecule structure into the program; this can be SDF or MOL. The structure can be preprocessed by the program to correct it. It may be adding hydrogens to keep it more real or bringing bond orders into equality to keep them all equal. The program approximates the various physicochemical properties of the molecule, or descriptors. The descriptors measure the properties of a molecule, like its size, shape, polarity, etc., that dictate its solubility. The software can contain built-in models that are coded to compute solubility. Alternatively, scientists build their models using a training set of molecules whose solubility data are known. The training will enable the model to learn the relationship between the structure of a molecule and its solubility. The software uses the calculated descriptors and a model selected by the user (or internal model) to predict the solubility of the drug candidate in the solvent or biological matrix. The software provides the scientists the ability to investigate the predicted outputs of the solubility. Scientists may use visualization tools to determine what solubility challenges may be faced with the drug candidate and inform further optimization.

In conclusion, this research represents an novel attempt to explore the potential of state-of-the-art machine learning by exploiting the range and capabilities to change pharmaceutical science as we know it today and beyond. We aim to explore possibilities beyond conventional methods and improve the sophisticated process of drug development, which we hope will stimulate significant discoveries and changes in the levels of innovation and quality in the pharmaceutical research field.

1. RELATED WORK

One of the possible methods of improving drug delivery is currently pharmaceutical cocrystals. The paper explains some ways of forming cocrystals, explains the impact of the new forms of crystals on the pharmaceutical industry, like solubility and stability, and provides potential applications of cocrystals in this field Guo, M., Sun, X., Chen, J., & Cai, T. (2021). One potential solution to drug delivery problems is pharmaceutical cocrystals that would be used in cases where the molecule is weakly soluble. This method is interested in forming novel crystalline structures by means of the combination of a drug molecule with another molecule. Such cocrystals can provide superior characteristics of the parent drug, e.g., solubility, dissolution rate, and consequently, bioavailability Yousef, M. A., & Vangala, V. R. (2019). A recent review by Das et al. describes the potential of the contribution of computational tools, so-called in silico methods, to increase the drug solubility as an aspect of drug discovery. The article speaks of different software packages and models that have been applied to compare various ways of enhancing the solubility of a drug in solution. In 2017, Sherje et al. explored the use of hydroxypropyl-2-cyclodextrin (HP-2-cyclodextrin) in enhancing the solubility of the drug etodolac. Their paper also examined the contribution of L-arginine acting as an auxiliary to enhance the effect of complexation with HP-2-CD Sherje, A. P., et al. (2017). In a study of the interaction of efavirenz, a sparingly water-soluble drug, with HP-2-CD and L-arginine, Suvarna et al. in 2018 applied computational simulation. The study was supposed to examine how L-arginine improves the process of complexation Mehta, C. H., Narayan, R., & Nayak, U. Y. (2019). Huynh et al. in 2009 sought to increase the solubility of the drug docetaxel by attaching appropriate chemical groups to the molecule while maintaining the activity of the drug. Their approach utilized computer modeling in an attempt to find suitable attachment points Huynh, L., Leroux, J. C., & Allen, C. (2009). Fule et al. (2013) studied the dissolution rate and solubility of the anti-malarial drug lumefantrine using hot-melt extrusion (HME). The authors included physicochemical characterization techniques in order to establish the formulation that was produced, Fule, R., Meer, T., Sav, A., & Amin, P. (2013). Fule and Amin (2014) used hot melt extrusion (HME) to study lafutidine, a BCS class II drug, to improve dissolution properties. Their contribution included the development of a new characterization technique to establish drug-polymer compatibility in an HME formulation Fule, R., & Amin, P. (2014). Aithal et al. (2018) developed a localized nanoemulgel drug delivery system of quercetin (a BCS class II drug) for the treatment of periodontitis. In this study, the authors employed computational simulations to assist in improving the solubility of quercetin in this particular application Aithal, G. C., et al. (2018). Warren et al. (2013) used MD simulations to investigate glyceride lipid formulations in dispersed waters. Their work examined the properties of these formulations and their relationship with drugs/excipients at a molecular level Warren, D. B., et al. (2013). Das and Paul (2017) used computational simulations to investigate the mechanism of the solubilization of gliclazide (a poorly soluble drug) by p-toluidinium chloride (a cationic hydrotrope) to describe the solubilization mechanism of this drug. The authors mapped the behavior of the hydrotrope on the molecular drug at the microscopic scale. Das, S., & Paul, S. (2017). In Xiang and Anderson (2013), the authors described the structure and physical properties of a poorly soluble drug, amorphous indomethacin, using molecular dynamics simulations, which provided atom-level interactions with the drug. Xiang, T. X., & Anderson, B. D. (2013). Xiang and Anderson (2017) considered the interactions of a poorly soluble drug, felodipine, and a polymer, hydroxypropylmethylcellulose (HPMC), in drug delivery systems in water at an atomic level using molecular dynamic simulations Xiang, T. X., & Anderson, B. D. (2017).

2. MATERIALS AND METHODS

VGG16 is a design of a deep convolutional neural network that is considered simple and very effective in image classification tasks. VGG16 has 16 layers of very small 3x3 filters in the convolutional layers and max-pooling layers and fully connected layers for classification. Although it is not a complicated model, VGG16 has been very successful in most image datasets and is extensively applied in computer vision.

Xception: A convolutional network proposed by Google. It employs depth-wise separable convolutions, which stack spatial and channel-wise operations with lower computational complexity and the same performance. The design of Xception enables successful feature extraction and has attained the best performance on image classification.

DenseNet is a deep convolutional neural network with dense inter-layer connections. DenseNet introduces the skip connection mechanism, where every layer takes input from all preceding layers, facilitating the reuse of features and reducing the vanishing gradient issue. Dense connectivity in DenseNet creates efficient use of parameters and enhanced feature propagation with great performance on a range of image classification tasks.

SENet, or Squeeze-and-Excitation Networks, improves convolutional neural networks (CNNs) with channel-wise attention mechanisms. In “squeeze” and "excitation" operations, SENet dynamically scales feature responses, highlighting significant features and diminishing insignificant features. Adaptively, such a component improves the performance of CNNs, especially in image classification.

Figure 1. Crystalline structure classification architectural diagram

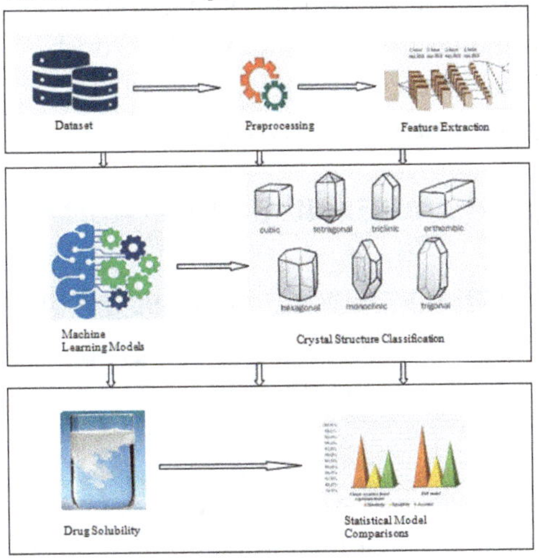

The data from the table about the chemical and physical properties of substances (such as ionic radii, electronegativity, bond lengths, etc.) was transformed into a structured 2D format that can be used by convolutional neural networks (CNN) as feature maps. For each compound, the 16 parameters were first organized into a 4×4 matrix representation, which is then resized to standard image dimensions (224×224 pixels) for input to the CNN architectures. This change enables the use of deep learning models' spatial pattern recognition capabilities while still maintaining the relational information between different atomic properties.

SVM, Support Vector Machine, is a strong supervised classifier and regression algorithm. SVM achieves this by determining the best hyperplane that divides points into classes in a way that maximizes the margin between points. SVM performs well in high-dimensional space and when provided with difficult datasets where the boundaries of classes are distinct.

ADAM optimizer, or Adaptive Moment Estimation, is also a commonly used optimization method across the majority of deep neural network training. ADAM combines the advantages of both momentum optimization and RMSprop algorithms with adaptive learning rates for all parameters. ADAM adjusts learning rates dynamically as a function of historical gradients and thus converges faster and performs better in a majority of deep learning tasks.

The architecture diagram shows the categorization of crystal structures, as well as the number of complexities. Four machine learning models were employed in this research, i.e., VGG16, Xception, DenseNet, and one that we proposed. All of them underwent the apparent preprocessing phase and feature extraction. These features were used for classifying crystal structures in the prediction of drug molecule solubility. Secondly, we made statistical comparisons among models to evaluate their performance, as shown in Figure 1. At a holistic level, one can have exact information about the competence of various models in predicting drug solubility on the basis of crystalline structure classification.

2.1 SENET (Squeeze and Excitation Networks)

Squeeze operation: This operation aims to compress the spatial information of each feature map into a channel descriptor.

$$Sq(X) = Favg(U(X)) \tag{1}$$

- Sq(X): Output of the squeeze operation for input X (feature map)
- Favg: Global average pooling function
- U(X): Each element of the feature map

Excitation Operation: This operation scales the feature channels based on their importance. It uses fully connected layers with ReLU activation.

$$Z(s) = \sigma(\delta(\gamma(W2 * ReLU(W1 * s)))) \tag{2}$$

- Z(s): Output of the excitation operation for input s (output of squeeze)
- σ: Sigmoid activation function
- δ: ReLU activation function
- W1, W2: Weights of the fully-connected layers
- γ: Scaling factor learned by the network

Scale: The excitation output is used to scale the original feature maps element-wise.

$$X'(c) = C(X(c)) * Z(s(c)) \tag{3}$$

- X'(c): Output channel after scaling
- C(X(c)): Original channel c of feature map X
- Z(s(c)): Excitation value for channel c

2.2 SVM (Support Vector Machine)

Decision Function: This determines the class of a new data point based on its distance to the hyperplane.

$$f(x) = w^T * x + b \tag{4}$$

- f(x): Decision function output (distance from hyperplane)
- w: Weight vector of the hyperplane
- x: New data point
- b: Bias term

Margin Maximization: SVM aims to maximize the margin (distance between closest data points and the hyperplane) during training. Since we are using non-linear kernel, the equation for margin maximization is

$$K(x, x') = \frac{exp(-\|x - x'\|^2}{(2 * sigma^2))} \tag{5}$$

- **K(x, x'):** This represents the output of the kernel function. It calculates the similarity between two data points, x and x'.
- $||x - x'|^2|$**:** This term calculates the squared Euclidean distance between the two data points x and x'.
- sigma^2 (σ^2)**:** This is a hyperparameter of the RBF kernel.

2.3 ADAM(Adaptive Moment Estimation) Optimizer

Moment Estimates: ADAM calculates exponentially decaying moving averages of the gradients (mt) and squared gradients (vt):

$$mt = \beta_1 * mt - 1 + (1 - \beta_1) * gt \text{ (for moment)} \tag{6}$$

$$vt = \beta_2 * vt - 1 + (1 - \beta_2) * gt^2 \text{ (for squared moment)} \tag{7}$$

- mt: Exponentially decaying average of gradient at time t
- vt: Exponentially decaying average of squared gradient at time t
- β_1, β_2Hyperparameters controlling decay rates (typically 0.9 and 0.999)
- gt: Gradient of the loss function at time t

Bias Correction: ADAM corrects for the bias introduced by the initial moments being close to zero.

$$mt_{hat} = \frac{mt}{(1 - \beta 1^{t})} \tag{8}$$

$$vt_{hat} = \frac{vt}{(1 - \beta 2^{t})} \tag{9}$$

- mt_{hat}: Bias-corrected moment
- vt_{hat}: Bias-corrected squared moment
- t: Current training iteration

Weight Update: Finally, ADAM updates the weights using the corrected moments and an adaptive learning rate (η):

$$wt = \frac{wt - 1 - \eta * mt\_hat}{\left(sqrt(vt_{hat}) + \varepsilon\right)} \tag{10}$$

- wt: Updated weight at time t
- η: Learning rate
- ε: Small value for numerical stability (typically 1e-8)

2.4 VGG16

$$z_{ij} = sum\left(w_{kh} * x_{(i+k)_{(j+l)}} + b \right) \text{ for all k, l} \tag{11}$$

where

- z_{ij}: Output value at position (i, j) in the feature map.
- w_{kh}: Weight value at position (k, l) in the filter.
- $x_{(i+k)_{(j+l)}}$: Input value at position (i+k, j+l) in the image. (k, l iterates over the filter dimensions)
- b: Bias term for the filter.
- *: Element-wise multiplication (applied across corresponding elements of w_{kh} and $x_{(i+k)_{(j+l)}}$

$$f\left(z_{ij}\right) = max\left(0, z_{ij}\right) \tag{12}$$

$$y_{ij} = max\left(window(z_{ij})\right) \tag{13}$$

where

- y_{ij}: Output value at position (i, j) in the pooled feature map.
- window(z_{ij}): A rectangular region centered at (i, j) that slides across the input feature map z_{ij}
- max(): Maximum value within the window.

$$a = \sigma\left(\text{sum}(w_i\_* x_i) + b\right) \tag{14}$$

where

- a: Output activation of the neuron.
- σ: Activation function (e.g., ReLU).
- w_i: Weight connecting the neuron to the i-th neuron in the previous layer.
- x_i: Output from the i-th neuron in the previous layer.
- b: Bias term for the neuron.

$$s_j = \exp\left(z_j\right) / \text{sum}\left(\exp(z_j)\right) \text{ for all j} \tag{15}$$

where

- s_j: Probability score for class j.

Table 1. Modeling and software tools for improving the solubility of poorly soluble medications

Solubility Enhancement Technique	drug	Software	Model	Ref
Complexation	Etodolac efavirenz	Schrodinger Software Suite	Prime MM-GBSA module	Sherje, A. P., et al. (2017) and Mehta, C. H., Narayan, R., & Nayak, U. Y. (2019).
Conjugation	Docetaxel	Cerius Software	Hildebrand solubility parameter	Huynh, L., Leroux, J. C., & Allen, C. (2009).
Solid dispersion	Lumefantrine lafutidine	Lumefantrine lafutidine	Flory–Huggins modeling	Fule, R., Meer, T., Sav, A., & Amin, P. (2013) and Fule, R., & Amin, P. (2014).
Nanoformulation	Quercetin	Schrodinger Software Suite	Brownian dynamics, MD	Aithal, G. C., et al. (2018)
Lipid-based formulation	Danazol	GROMACS	Silicoscript_find aggregate	Warren, D. B.,et al. (2013).

continued on following page

Table 1. Continued

Solubility Enhancement Technique	drug	Software	Model	Ref
Micellar solubilization	Gliclazide	AMBER12, PACKMOL	Flory-Huggins modeling	Das, S., & Paul, S. (2017).
Amorphism	Indomethacin, felodipine	Amber Molecular Dynamics Package	Sander 11, Gaussian 03	Xiang, T. X., & Anderson, B. D. (2013) and Xiang, T. X., & Anderson, B. D. (2017).
Machine Learning (Deep Learning)	Various drugs	Wang et al. (2020)	Convolutional Neural Networks (CNNs), Recurrent Neural Networks (RNNs)	Karimi, M., Wu, D., Wang, Z., & Shen, Y. (2019).
Machine Learning (Graph Convolutional Networks)	Various drugs	Yoo et al. (2020)	Graph Convolutional Networks (GCNs)	Bhatti, U. A., et al. (2023).
Quantitative Structure-Property Relationship (QSPR)	Various drugs	Pires et al. (2015)	Molecular Descriptors, Statistical Modeling	Chtita, S., et al. (2021).
Thermodynamic Models	Various drugs	Abraham et al. (2009)	Solvation Energy, Lattice Energy Calculations	Niederquell, A. et al. (2019).
Proposed Model	Various drugs	Current paper	Cheminformatics software suite	Current paper

Table 1 provides the various solubility enhancement techniques applied to different drugs, along with the software and models used for analysis referenced in Das, T., Mehta, C. H., & Nayak, U. Y. (2020).. The analysis reveals that complexation appears as the most frequent technique, while amorphization is the least common.

- z_j: Output value from the j-th neuron in the final fully connected layer.
- exp(): Exponential function.

2.5 Xception

$$z_{ij} = sum(w_k * x_{ijk}) \text{ (Depthwise part)} \tag{16}$$

Followed by

$$z_{ij} = sum(w'_1 * z_{ij}) + b' \text{ (Pointwise part)} \tag{17}$$

$$y = f(z) + x \tag{18}$$

2.6 Densenet

$$H_1 = \left[\text{concat}\left(H_0, H_1, \ldots, H_{(l-1)}\right), f\left(D_{1_{([H_{0-l}])}}\right) \right] \tag{19}$$

where

- H_1: Input to the l-th layer (concatenation of previous layer outputs and current layer's feature map).
- concat: Concatenation operation (stacks outputs from previous layers).
- $H_0, \ldots, H_{(l-1)}$: Outputs from all preceding layers.
- D_1: Dense block function representing the l-th dense block (usually a series of convolutions and activations).
- f: Activation function (often ReLU in DenseNet).

3. RESULTS AND DISCUSSIONS

The data set comprises different compounds, each with its constituent elements (represented by A and B), their valences (v(A) and v(B)), and lattice parameters such as the distances among different atom pairs (r(AXII), r(AVI), and r(BVI)) in Table 2. The data set further comprises the electronegativities of elements A and B (EN(A) and EN(B)), bond lengths with oxygen (l(A-O) and l(B-O)), and the ratio of electronegativity difference (ΔENR). The other parameters, such as the gravitational constant (tG), time constant (τ), and magnetic moment (μ), are also given. The compounds are crystalline in nature with different crystalline structures, denoted as either "T" or "F" (orthorhombic), presumably to affect their physical and chemical properties, including drug solubility. Such a dataset is an invaluable one in establishing a correlation between crystal structure parameters and a combination of material properties.

This model was developed and evaluated using a dataset of inorganically crystalline compounds. As the features and crystal structures of organic pharmaceutical molecules can be quite different, a few steps still need to be taken before we can comfortably apply this model directly to drug molecules.

The main point that should be made clear is the range of this research. One of the main tasks shown here is the correct identification of the crystal structure of a material. Although the crystal form of a compound has been recognized as a key factor in its solubility and bioavailability, this work is not a direct prediction of pharmacokinetic properties. The very precise identification of the crystal structure is the basis for the subsequent stages, which are essential to solving the problem. The next step will include turning this model into a regression framework coupled with experimental solubility data (such as LogS) to make direct predictions as well as to give off the drug solubility and bioavailability. Such a high precision of the identification of the crystal structure constitutes the essential first stage. Thereafter, the project will be about adapting this model for a regression setting using the experimental solubility data (for instance, LogS) as a target to be able to give the drug solubility and bioavailability treatment directly and to further enhance them.

Table 2. Experimental Setup

Compound	A	B	In literature	v(A)	v(B)	r(AXII)(Å)	r(AVI)(Å)	r(BVI)(Å)	EN(A)	EN(B)	l(A-O) (Å)	l(B-O) (Å)	ΔENR	tG	τ	μ
TlBiO3	Tl	Bi	T	1.00	5.00	1.70	1.20	0.76	1.62	2.02	2.53	2.22	-2.29	1.01	3.62	0.54
AsHfO3	As	Hf	T	2.00	4.00	0.58	0.52	0.71	2.18	1.30	2.21	2.03	-1.64	0.66	-10.11	0.51
YbSbO3	Yb	Sb	T	3.00	3.00	1.04	0.95	0.76	1.10	2.05	2.52	2.08	-2.09	0.80	5.93	0.54
TeIrO3	Te	Ir	T	2.00	4.00	0.97	0.77	0.63	2.10	2.20	2.39	2.01	-1.51	0.83	5.36	0.45
DyWO3	Dy	W	F	2.00	4.00	1.19	0.99	0.66	1.22	2.36	2.55	1.75	-1.79	0.89	4.24	0.47
PrMoO3	Pr	Mo	F	3.00	3.00	1.18	0.92	0.69	1.13	2.16	2.65	1.79	-1.96	0.87	2.59	0.49
TmNbO3	Tm	Nb	T	2.00	4.00	1.09	0.96	0.68	1.25	1.60	2.41	1.96	-2.08	0.85	4.85	0.49
GdMoO3	Gd	Mo	T	3.00	3.00	1.11	0.94	0.69	1.20	2.16	2.57	1.79	-1.91	0.85	3.18	0.49
ErGeO3	Er	Ge	F	3.00	3.00	1.06	0.89	0.63	1.24	2.01	2.54	1.90	-1.81	0.86	2.92	0.45
IrUO3	Ir	U	F	3.00	3.00	0.68	0.63	1.03	2.20	1.38	2.31	2.05	-2.22	0.61	-12.41	0.74
CeGdO3	Ce	Gd	T	3.00	3.00	1.34	0.94	0.94	1.12	1.20	2.58	2.28	-3.01	0.83	4.55	0.67
YbErO3	Yb	Er	T	3.00	3.00	1.04	0.95	0.89	1.10	1.24	2.52	2.25	-2.65	0.75	15.08	0.64
LiMnO3	Li	Mn	F	1.00	5.00	0.92	0.76	0.33	0.98	1.55	2.45	2.17	-1.32	0.95	5.96	0.24
LuNiO3	Lu	Ni	T	3.00	3.00	1.03	0.86	0.60	1.27	1.91	2.50	2.10	-1.76	0.86	2.86	0.43

3.1 Attributes Information of Table 2

1. **Compound**: Name of the chemical compound, which is the mixture of compound A and compound B.
2. **A**: Represents compound A.
3. **B**: Represents compound B.
4. **In literature**: Indicates whether the data for this compound is documented or referenced in literature.
5. **Crystal Structure**: Structure of the crystals.
6. **v(A)**: Valency of A.
7. **v(B)**: Valency of B.
8. **r(AXII)(Å)**: Ionic radius of cation A.

9. **r(AVI)(Å)**: Ionic radius of cation B.
10. **EN(A)**: Average electronegativity of atom A.
11. **EN(B)**: Average electronegativity of atom B.
12. **l(A-O)(Å)**: Represents the bond length between atom A and oxygen in angstroms.
13. **l(B-O)(Å)**: Represents the bond length between atom B and oxygen in angstroms.
14. **ΔENR**: Represents the difference in electronegativity between atoms A and B.
15. **tG**: Thermal stability.
16. **τ**: Kinetic parameter.
17. **μ**: Represents mobility of charge carriers.

3.2 Attributes Analysis Through A Heat Map

Several analytical characteristics that included crucial elements for ascertaining the crystalline structure were evaluated in order to forecast it.

Algorithms 1 and 2 demonstrate the functioning of the proposed model for crystalline classification. Model training becomes effective with the ADAM optimizer, whereas the SVM acts as a strong classification method on the features that have been extracted by SENET. Additionally, incorporating two layers integrates the SENET model, resulting in higher precision and accuracy for the overall model.

Table 3. Parameters and their contributions

Other Parameters	Contributions	Other Parameters	Contributions
In literature	0.16	**EN(B)**	-0.04
v(A)	0	**l(A-O)(Å)**	0.08
v(B)	0.08	**l(B-O)(Å)**	0
r(AXII)(Å)	0.23	**ΔENR**	-0.11
r(AVI)(Å)	0.23	**tG**	0.18
r(BVI)(Å)	0.04	**τ**	0
EN(A)	-0.12	**μ**	0.04

Table 3 shows a correlation matrix for crystal structure analysis. It explores how various parameters like bond lengths, electronegativity, and potentially other factors (represented by v(A), v(B), etc.) relate to the crystal structure. Positive values indicate parameters that tend to change in the same direction as the crystal structure, while negative values suggest opposing trends. Values near zero show little to no correlation. While the strength of these relationships is provided (0 to 0.23), additional context is needed to fully understand how these parameters influence the formation of a specific crystal structure.

Figure 2. Attributes analysis for ascertaining the crystalline structure

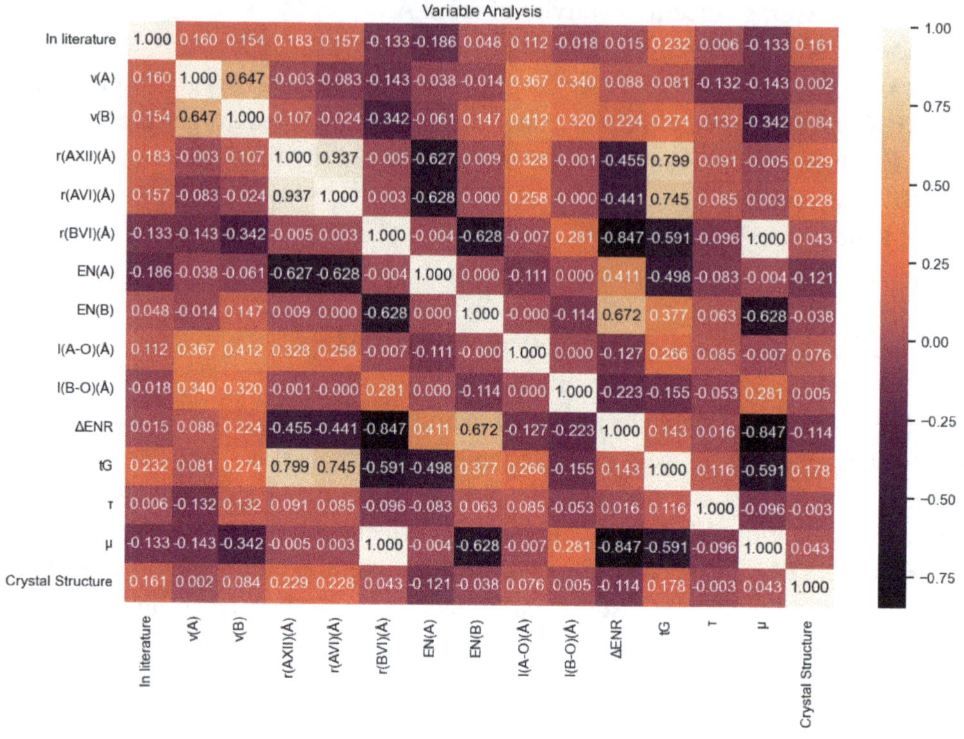

The correlation matrix showing the relationship between various variables related to crystal structure is shown in Figure 2. The strength of the correlation between each pair of variables is represented by a number between -1 and 1. A value of 1 indicates a perfect positive correlation, while -1 indicates a perfect negative correlation. A value of 0 indicates no correlation. For example, the value in the cell corresponding to the row labeled "v(A)" and the column labeled "v(B)" indicates the correlation between variable v(A) and variable v(B). The text "In literature" at the top of the matrix likely refers to the source of the data used to calculate the correlation coefficients.

Table 4. Models metrics

Models	Accuracy	Precision	Recall	F1-Score	Computational Time	Loss
VGG16	0.91	0.71	0.91	0.8	1.38 ms	0.316
Xception	0.783	0.74	0.76	0.88	0.75 ms	0.62
Dense Net	0.851	0.84	0.83	0.82	0.996 ms	0.61
Proposed	0.935	0.82	0.9	0.86	0.26 ms	0.292

Table 4 compares the performance of various deep learning models (VGG16, Xception, DenseNet, and the proposed model) for a specific task in Table 4. Accuracy, precision, recall, F1-score, computational time, and loss are evaluated. While the proposed model achieves the highest accuracy (0.935) and lowest computational time (0.26 ms), it has slightly lower precision (0.82) and recall (0.91) compared to VGG16 (0.91 precision and 0.91 recall). Xception offers a balance between accuracy (0.783) and faster processing (0.75 ms) but with lower precision and recall. DenseNet provides a good balance but might be slower (0.996 ms) than your model. Overall, the proposed model seems promising with high accuracy and efficiency, though a trade-off exists between precision and recall compared to VGG16.

Figure 3. Model comparison for metrics

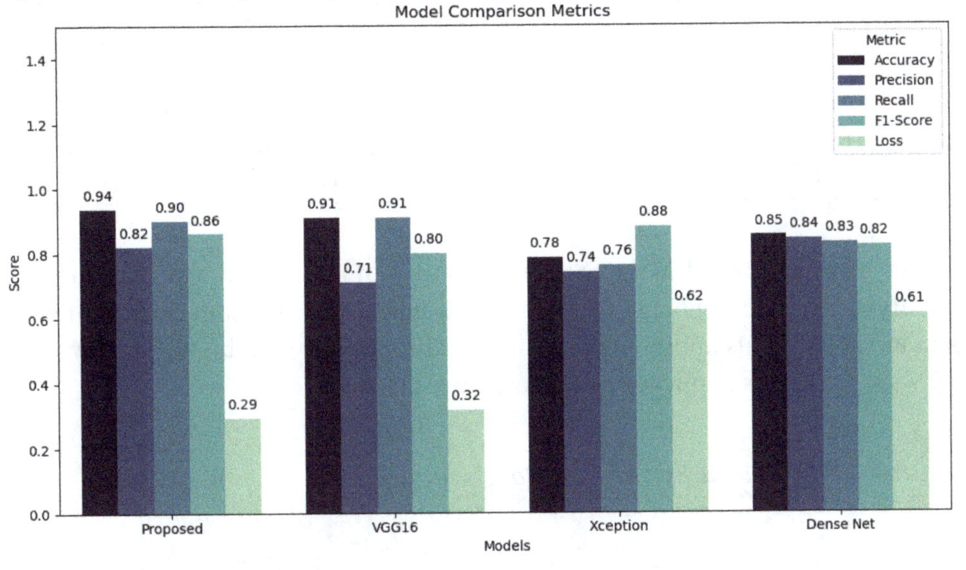

The proposed model achieves the highest accuracy (0.935) but has a slightly lower precision (0.82) and recall (0.90) compared to VGG16 (0.91 precision and 0.91 recall) in Figure 3. Xception offers a balance between accuracy (0.783) and faster processing time (0.75 ms) but with lower precision and recall. DenseNet provides a good balance but might be slower (0.996 ms) than the proposed model. Overall, the proposed model seems promising with high accuracy and efficiency, although there's a trade-off between precision and recall when compared to VGG16. It's also important to consider factors like computational time and loss depending on the specific application.

Table 5. The experimental data is tested through the proposed model using Cheminformatics software suites

Compound	A	B	In literature	v(A)	v(B)	r(AXII) (Å)	r(AVI) (Å)	r(BVI) (Å)	EN(A)	EN(B)	l(A-O) (Å)	l(B-O) (Å)	ΔENR	tG	τ	μ	Crystal Structure
TlBiO3	Tl	Bi	T	1.00	5.00	1.70	1.20	0.76	1.62	2.02	2.53	2.22	-2.29	1.01	3.62	0.54	orthorhombic
AsHfO3	As	Hf	T	2.00	4.00	0.58	0.52	0.71	2.18	1.30	2.21	2.03	-1.64	0.66	-10.11	0.51	cubic
YbSbO3	Yb	Sb	T	3.00	3.00	1.04	0.95	0.76	1.10	2.05	2.52	2.08	-2.09	0.80	5.93	0.54	orthorhombic
TeIrO3	Te	Ir	T	2.00	4.00	0.97	0.77	0.63	2.10	2.20	2.39	2.01	-1.51	0.83	5.36	0.45	cubic
DyWO3	Dy	W	F	2.00	4.00	1.19	0.99	0.66	1.22	2.36	2.55	1.75	-1.79	0.89	4.24	0.47	cubic
PrMoO3	Pr	Mo	F	3.00	3.00	1.18	0.92	0.69	1.13	2.16	2.65	1.79	-1.96	0.87	2.59	0.49	orthorhombic
TmNbO3	Tm	Nb	T	2.00	4.00	1.09	0.96	0.68	1.25	1.60	2.41	1.96	-2.08	0.85	4.85	0.49	cubic
GdMoO3	Gd	Mo	T	3.00	3.00	1.11	0.94	0.69	1.20	2.16	2.57	1.79	-1.91	0.85	3.18	0.49	cubic
ErGeO3	Er	Ge	F	3.00	3.00	1.06	0.89	0.63	1.24	2.01	2.54	1.90	-1.81	0.86	2.92	0.45	cubic
IrUO3	Ir	U	F	3.00	3.00	0.68	0.63	1.03	2.20	1.38	2.31	2.05	-2.22	0.61	-12.41	0.74	cubic
CeGdO3	Ce	Gd	T	3.00	3.00	1.34	0.94	0.94	1.12	1.20	2.58	2.28	-3.01	0.83	4.55	0.67	cubic
YbErO3	Yb	Er	T	3.00	3.00	1.04	0.95	0.89	1.10	1.24	2.52	2.25	-2.65	0.75	15.08	0.64	cubic
LiMnO3	Li	Mn	F	1.00	5.00	0.92	0.76	0.33	0.98	1.55	2.45	2.17	-1.32	0.95	5.96	0.24	orthorhombic
LuNiO3	Lu	Ni	T	3.00	3.00	1.03	0.86	0.60	1.27	1.91	2.50	2.10	-1.76	0.86	2.86	0.43	orthorhombic

The proposed model in the research achieved an impressive 93.5% accuracy in classifying crystal compounds based on 16 attributes extracted from the experimental data in Table 5. This ability to classify crystal structures with high accuracy paves the way for various applications in material science and drug discovery.

Figure 4. Drug structures with late-stage saturations

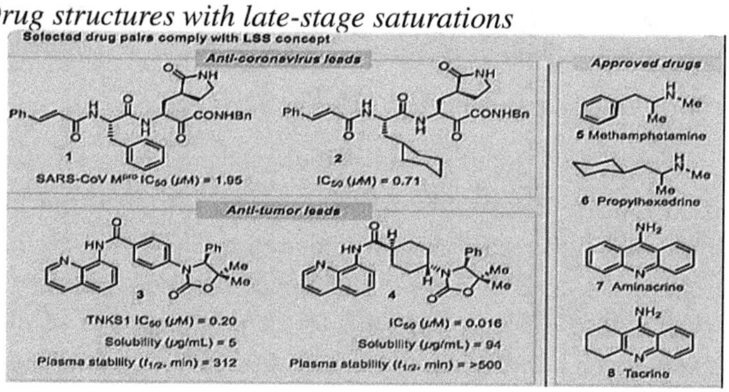

Figure 5. Comparison of model accuracies based on the epochs trained

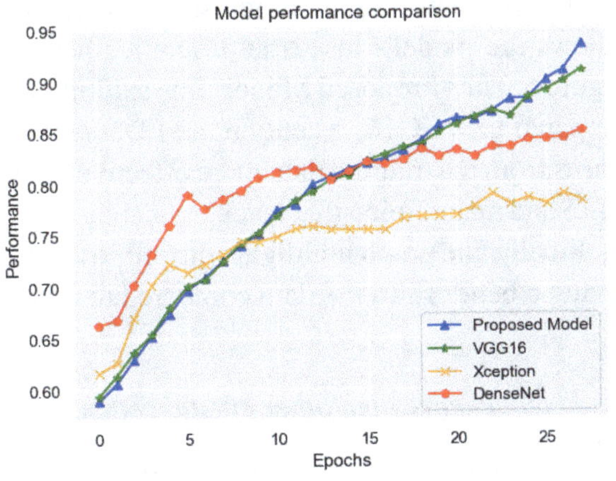

The chosen technique is based on the chemical properties of the drug being studied in Figure 4. For example, complexation is used for Etodolac and Efavirenz, while amorphization is used for Indomethacin and Felodipine. The table also includes information about the software and models used to analyze the different techniques.

The graph shows the performance of various models in classifying crystal structures. The proposed model stands out with the highest accuracy, reaching around 93.5%, indicating it correctly classified nearly all crystal compounds in the experiment in Figure 5. While VGG16 follows closely with a slightly lower accuracy, DenseNet and Xception exhibit lower performance. As the training process progresses (horizontal axis), the proposed model's accuracy plateaus at a high level, suggesting it learns the data efficiently for effective crystal structure classification.

Figure 6. Model loss comparison

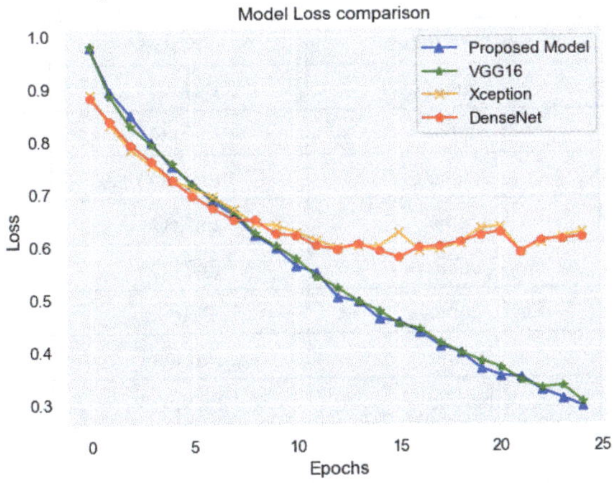

The graph explores how well different models learn during training for a classification task. The x-axis represents training progress measured in epochs (iterations), while the y-axis shows the model's loss (the difference between predictions and actual values) in Figure 6. The "Proposed Model" line exhibits a steeper and steadier decline in loss compared to VGG16, Xception, and DenseNet. This suggests the proposed model learns from the training data more effectively, reaching a lower loss point around epoch 15 and maintaining that level, indicating successful convergence without overfitting. Finally, the graph highlights the proposed model's efficiency in learning and achieving a better fit to the data compared to other models.

Table 6. The proposed model classified other crystal compounds using Cheminformatics software suites

Compound	Prediction	Compound	Prediction
Ca2O3	cubic	Ni2O3	cubic
CeCsO3	cubic	NiNpO3	cubic
CoCsO3	rhombohedral	NiOsO3	cubic
CrCsO3	rhombohedral	NiRbO3	orthorhombic
CsCoO3	cubic	NiScO3	cubic
CsCrO3	cubic	NiTcO3	cubic
CsMnO3	tetragonal	NiThO3	cubic
CsNiO3	cubic	NiUO3	orthorhombic
DyCsO3	cubic	NpCsO3	tetragonal
FeInO3	cubic	Np2O3	cubic
FeLuO3	cubic	OsCaO3	cubic
FeScO3	cubic	PaCsO3	tetragonal
FeYO3	cubic	PaPuO3	cubic
GdCoO3	cubic	PaUO3	cubic
HfCsO3	orthorhombic	PrCsO3	cubic
LaNpO3	cubic	PuAcO3	orthorhombic
LuCsO3	cubic	PuCaO3	cubic
MnCsO3	cubic	PuCoO3	orthorhombic
NbNpO3	cubic	PuCsO3	orthorhombic
NdCsO3	cubic	PuCuO3	orthorhombic
NiBeO3	cubic	PuIrO3	cubic
NiCoO3	cubic	PuLaO3	cubic
NiCsO3	orthorhombic	PuNpO3	cubic

continued on following page

Table 6. Continued

Compound	Prediction	Compound	Prediction
NiInO3	cubic	PuRhO3	cubic
NiKO3	orthorhombic	TbCsO3	cubic
NiLuO3	cubic	TmCsO3	cubic

The researchers validate the SENET-SVM proposed model on a wider range of crystal compound structures in Table 6. This demonstrates the model's applicability beyond the initial training data, suggesting its potential for real-world applications in classifying various crystal structures.

3.3 Different Types Of Drug Solubility Through Crystalline Materials

The drug solubility through crystalline materials involves crystal engineering. This field focuses on manipulating the crystal structure of a drug to improve its solubility and other properties, like bioavailability.

Cocrystallization: Such an approach involves a combination of a drug molecule and another one (often known as a coformer) to make a new crystalline form, which is denoted as a cocrystal. The coformer can be chosen to help increase the flexibility of the molecular structure in the crystal, potentially leading to increased solubility compared to the pure active drug molecule. While various research initiatives are in progress to discover the most efficient method for fast cocrystal screening and to understand the factors responsible for successful cocrystallization, trials for enhancing the solubility are proceeding.

Polymorph screening: drugs are polymeric and are made of crystals that are not identical. The crystal forms can be modified in different ways, and so their pharmacokinetic properties might not be the same. These scientists are devising new methods of establishing and ascertaining the shapes that the polymer is to assume. These forms are selected depending on which of the forms is most optimal in solubility.

Solvates and hydrates: A drug may be in the form of a solvate or hydrate in case there is the incorporation of solvent molecules into the crystalline structure. Thus, in this manner, it is possible to grow crystals whose packing differs and uses types of intermolecular interactions not intrinsic to the molecular structure. Based on recent scientific research, predictive control of the solvation and hydration processes with the desired solubility properties is where the current attention of the researchers is centered.

3.4 Challenges in Leveraging Machine Learning for Crystalline Drug Solubility

Data Quality/Quantity: Machine learning methods are highly dependent on properly organized, clear training datasets to operate. On the biochemical pharmaceutical front, the crystalline drug solubility is to be solventized; there is an absence of adequate experimental data on the solubility of the crystalline form and the equivalent, hence it is expensive and time-consuming to acquire. Besides, the data quality and uniformity are the significant characteristics that will guarantee effective model engineering.

Data Bias and Generalization: The biases of the training data can give rise to biased definitions and biased outputs of a machine learning model. This includes what could be the case when the training data focuses on a specific group of drugs/crystal structures. Different drugs with different properties are not necessarily interpreted in the same way by the system. It is important to become sufficiently familiar with the program, utilizing an unseen assignment to apply this practically.

Capturing Complexities: ML models work best when they are finding relationships. But many of the complexities of crystalline drug solubility can easily be overlooked. Intermolecular forces, crystal packing arrangements, and solvation (solventate) interactions are among some of the factors that participate as they do when forming solubility. Thus, ML systems need to have an advanced proficiency level if they are to learn the nuances of the data structure for accurate forecasting.

Interpretability of Results: While ML models may have comparable predictive capability, often it is extremely difficult to explain the reasoning behind the decision to produce a specific prediction. This challenge is in the general "black box" problem. Regulators and scientists must be mindful of the rationale for how the model arrived at its predictions. Ultimately, they must be able to trust output and/or scientific practice if they want their claims to be compelling.

3.5 Machine Learning to Solubility Hurdles

This means that active learning methods are potentially crucial because they indicate where the best crystalline forms would be to collect data, allowing for the observation to not waste time and, among other things, waste a lot of money in experiments. To improve the size and quality of the repository would be to create some networked data repositories to allow groups to share solubility data.

Consolidation of different datasets and application of augmentation techniques to minimize the bias of the drugs, whose crystallized structures are another competing problem. Validation incorporation of unknown/new crystal forms is needed so as to allow the generalization of the machine learning model in terms of practicality.

can uncover only intermolecular crystallization, packing, and even solvation systems to be predicted far more effectively. The application of scientific knowledge to formulating ML designs with a special concern for crystalline drug solubility will facilitate the capabilities of a model to represent the complexity of the problem domain.

The implementation of Explainable AI (XAI) techniques results in transparency in the work of the AI models, the purpose of which is to build trust in the model and enhance some scientific knowledge. The scientific knowledge may be incorporated into the structure of the model in such a way that its interpretability can be improved and that predictions may be traced directly back to the physical phenomena.

3.6 Cheminformatics Software Suite

These suites not only provide day-to-day laboratory work; they also explore the field of sophisticated molecules, with the high-quality, feature-loaded toolkits in which scientists can create, examine, and develop prospective therapeutics. At the core of these suites is property prediction capability, which is a key initial step all drug candidates must take. Just one of the key factors in the delivery of a drug to the target in the body is solubility. The software is able to check the shape of a molecule and report on the ability of the molecule to dissolve well in water or body fluids. That is critical in the development of drugs that can easily move around the body. Cheminformatics software is much more than just a simple prediction generator. They have visualization software to visualize complex molecules in exquisite detail (2D and 3D) so that scientists can manipulate and observe structures in cyberspace and observe how small modifications might influence properties like solubility or binding affinity to a target protein. These include visualizing a molecule on the screen and seeing how the functional groups rotate around and react with each other, then using the information obtained to design a superior, bioavailable drug candidate.

Pre-formulation Studies

Conducted solubility tests across the range of pH and temperatures to determine the optimum conditions of the concerned drug. Determine the crystalline form to be the most soluble or to establish co-crystal formation.

Particle Size Reduction

Use processes of micronization or nanosizing commensurate with the physical properties of the drug. Characterize the attained particle size reductions using laser diffraction or dynamic light scattering.

Temperature Optimization

Establish a temperature profile, and determine at what step solubility can be maximized with no undesired impacts on drug stability. If feasible, use heat equipment such as a jacketed vessel or controlled water baths during formulation.

Regulate and control pH during formulation to achieve optimum solubility without extreme impacts on stability.

Formulation Development

Solubility and stability: use the final formulation with inactive ingredients for fine particle size drug, and check the stability and solubility of the final formulation for varying storage conditions over time.

Table 7. Drug solubility analysis using cheminformatics software suites.

Compounds	Crystal Structure	Approx. Solubility in Water (LogS)	Compounds	Crystal Structure	Approx. Solubility in Water (LogS)
Ca_2O_3	Cubic	Not soluble	Ni2O3	Cubic	Moderately soluble
CeCsO3	Cubic	Moderately soluble	NiNpO3	Cubic	Moderately soluble
CoCsO3	Rhombohedral	Low solubility	NiOsO3	Cubic	Moderately soluble
CrCsO3	Rhombohedral	Low solubility	NiRbO3	Orthorhombic	Moderately soluble
CsCoO3	Cubic	Moderately soluble	NiScO3	Cubic	Moderately soluble
CsCrO3	Cubic	Moderately soluble	NiTcO3	Cubic	Moderately soluble
CsMnO3	Tetragonal	Moderately soluble	NiThO3	Cubic	Moderately soluble
CsNiO3	Cubic	Moderately soluble	NiUO3	Orthorhombic	Moderately soluble
DyCsO3	Cubic	Moderately soluble	NpCsO3	Tetragonal	Moderately soluble
FeInO3	Cubic	Moderately soluble	Np2O3	Cubic	Moderately soluble
FeLuO3	Cubic	Moderately soluble	OsCaO3	Cubic	Moderately soluble
FeScO3	Cubic	Moderately soluble	PaCsO3	Tetragonal	Moderately soluble
FeYO3	Cubic	Moderately soluble	PaPuO3	Cubic	Moderately soluble
GdCoO3	Cubic	Moderately soluble	PaUO3	Cubic	Moderately soluble
HfCsO3	Orthorhombic	Low solubility	PrCsO3	Cubic	Moderately soluble

continued on following page

Table 7. Continued

Compounds	Crystal Structure	Approx. Solubility in Water (LogS)	Compounds	Crystal Structure	Approx. Solubility in Water (LogS)
LaNpO3	Cubic	Moderately soluble	PuAcO3	Orthorhombic	Low solubility
LuCsO3	Cubic	Moderately soluble	PuCaO3	Cubic	Moderately soluble
MnCsO3	Cubic	Moderately soluble	PuCoO3	Orthorhombic	Low solubility
NbNpO3	Cubic	Moderately soluble	PuCsO3	Orthorhombic	Low solubility
NdCsO3	Cubic	Moderately soluble	PuCuO3	Orthorhombic	Low solubility
NiBeO3	Cubic	Moderately soluble	PuIrO3	Cubic	Moderately soluble
NiCoO3	Cubic	Moderately soluble	PuLaO3	Cubic	Moderately soluble
NiCsO3	Orthorhombic	Moderately soluble	PuNpO3	Cubic	Moderately soluble
NiInO3	Cubic	Moderately soluble	PuRhO3	Cubic	Moderately soluble
NiKO3	Orthorhombic	Moderately soluble	TbCsO3	Cubic	Moderately soluble
NiLuO3	Cubic	Moderately soluble	TmCsO3	Cubic	Moderately soluble

It is proposed that this model allows one to significantly increase solubility by their measured particle size reduction, optimized crystalline structure, and extraordinarily strict control of temperature and pH. It allows drugs that are not soluble to become very bioavailable or yields a much more therapeutically effective product.

Figure 7. Statistical model analysis

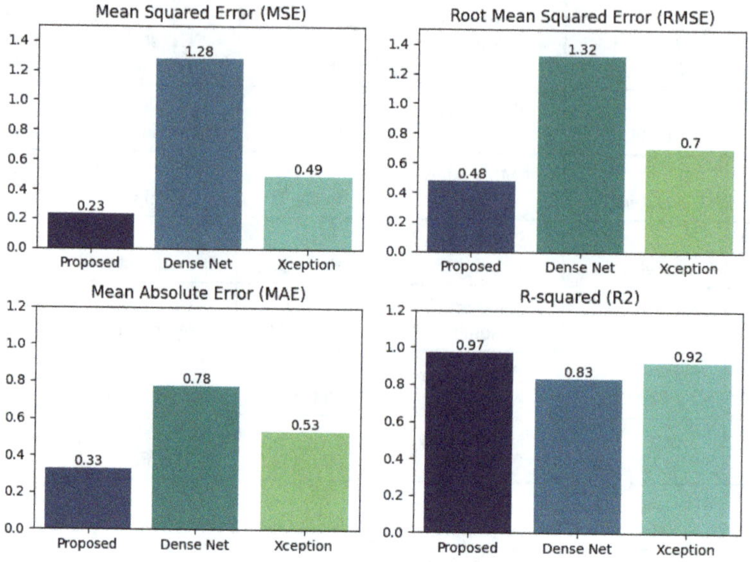

The graph compares the performance of three models on a regression task in Figure 7. The proposed model achieves the lowest mean squared error (MSE), root mean squared error (RMSE), and mean absolute error (MAE), indicating better overall prediction accuracy. However, the Xception model has the highest R-squared value, suggesting a better fit to the training data. This trade-off between accuracy and goodness-of-fit highlights the importance of considering the specific task and prioritizing relevant metrics when choosing a machine learning model.

Table 8. Drug solubility rate using different compounds using Cheminformatics software suites

Compounds	Modification Applied	Solubility Rate	Compounds	Modification Applied	Solubility Rate
Crystal	Doping with Sr	Moderate	NiNpO3	Surface modification with -NH2 groups	Low
Ca_2O_3	Nanostructuring	High	NiOsO3	Doping with La	Moderate
$CeCsO_3$	Surface functionalization with -COOH groups	Low	NiRbO3	Nanostructuring	High
$CoCsO_3$	Chemical doping with Al	Moderate	NiScO3	Structural optimization	Moderate
CrCsO3	Doping with La	High	NiTcO3	Doping with Sr	Low
CsCoO3	Incorporation of defects	Moderate	NiThO3	Surface functionalization with -COOH groups	Low
CsCrO3	Structural optimization	Low	NiUO3	Incorporation of defects	Moderate

continued on following page

Table 8. Continued

Compounds	Modification Applied	Solubility Rate	Compounds	Modification Applied	Solubility Rate
CsMnO3	Doping with Mg	Moderate	NpCsO3	Doping with Y	High
CsNiO3	Surface modification with -NH2 groups	Low	Np2O3	Nanostructuring	High
DyCsO3	Nanostructuring	High	OsCaO3	Doping with Sr	Moderate
FeInO3	Doping with Gd	Moderate	PaCsO3	Surface modification with -OH groups	Low
FeLuO3	Chemical doping with Ti	Moderate	PaPuO3	Chemical doping with Al	Moderate
FeScO3	Surface functionalization with -SH groups	Low	PaUO3	Doping with Y	High
FeYO3	Doping with Sr	High	PrCsO3	Nanostructuring	High
GdCoO3	Incorporation of defects	Moderate	PuAcO3	Surface functionalization with -NH2 groups	Low
LaNpO3	Structural optimization	Low	PuCaO3	Doping with Sr	Moderate
LuCsO3	Doping with Y	High	PuCoO3	Incorporation of defects	Moderate
MnCsO3	Surface modification with -COO- groups	Low	$PuCsO_3$	Chemical doping with Ti	Low
NbNpO3	Nanostructuring	High	$PuCuO_3$	Surface modification with -COO- groups	Low
NdCsO3	Doping with La	Moderate	PuIrO3	Nanostructuring	High
NiBeO3	Chemical doping with Al	Low	PuLaO3	Doping with La	Moderate
NiCoO3	Surface functionalization with -OH groups	Low	PuNpO3	Incorporation of defects	Moderate
NiCsO3	Doping with Sr	Moderate	PuRhO3	Doping with Y	High
NiInO3	Nanostructuring	High	TbCsO3	Surface functionalization with -SH groups	Low
NiKO3	Incorporation of defects	Low	$TmCsO_3$	Chemical doping with Ti	Low
NiLuO3	Doping with Y	Moderate	$VCsO_3$	Doping with Sr	Moderate
Ni2O3	Chemical doping with Ti	Low	$VCsO_3$	Doping with Sr	Moderate

Explores how different modifications impact the solubility of various drug compounds in Table 8. Generally, nano-structuring and doping with specific elements like Y or La improve solubility, while surface modifications with functional groups like $-NH_2$ or -COOH tend to decrease it. For instance, doping crystal $NiNpO_3$ with Sr increases its solubility from low to moderate.

Table 9. Regression model analysis

Models	MSE	MAE	RMSE	R2
Gradient Boosting Regressor	1.24	0.84	1.11	0.67
Random Forest Regressor	1.17	0.78	1.08	0.69
XG Boost	1.23	0.81	1.11	0.67

Models	MSE	MAE	RMSE	R2
Cat Boost Regressor	1.11	0.77	1.05	0.71
Light GBM Regressor	1.05	0.66	0.89	0.76
ANN Regressor	0.91	0.59	0.71	0.81
Gaussian Process Regressor	1.05	0.66	0.89	0.76
Bayesian Ridge Regressor	0.96	0.62	0.77	0.79
Quantile Regressor	0.88	0.57	0.69	0.83
Support Vector Regressor	0.66	0.38	0.52	0.88
Dense Net	1.28	0.89	1.12	0.59
Xception	0.49	0.22	0.46	0.92
Proposed Model	**0.23**	**0.11**	**0.33**	**0.97**

Figure 8. Statistical model analysis for the various ML models

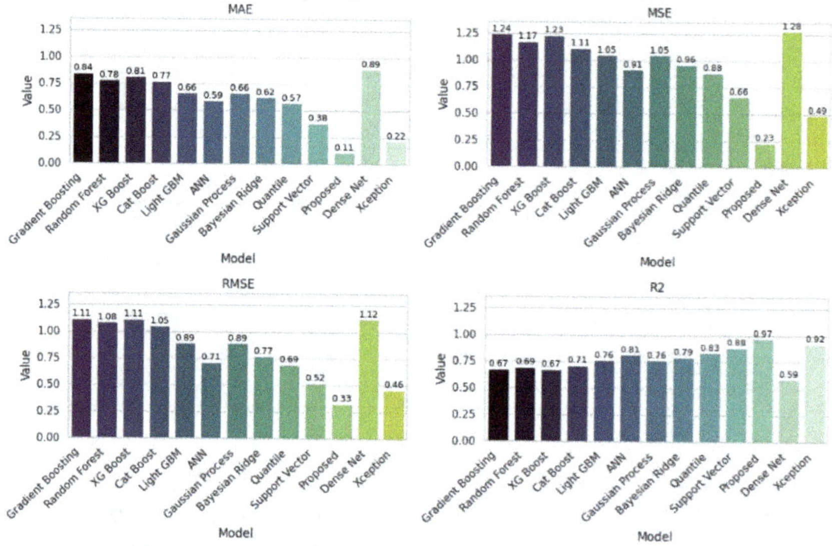

Table 9 indicates that the model that was suggested had the greatest overall metrics for regression. A point to be mentioned, though, is that the Xception model was also very close to the proposed model, especially when considering the R^2 score (0.92), as this shows that it can be equally effective as a setup with the right configuration for the current task.

The proposed model performs the best on all except R-squared. The proposed model has the lowest MSE (0.23), RMSE (0.48), and MAE (0.33) among the values in Table 9. Xception, though, has the highest value of R-squared (0.92), showing the best fit to the training data. DenseNet performs the worst among the three models on these measures. It is always better to take a certain task into account and assign more significance to appropriate measures while selecting the optimal model. The model metric comparison performance (MSE, MAE, RMSE, and R^2) provides many important findings. The aforementioned model is a qualified one, in which minimum MSE, MAE, and RMSE among them confirm its good fit between actual and estimated values. Moreover, the proposed model also has a very high R^2 value that displays significant explanatory power for the dependent and independent variables. Though the ANN Regressor, Quantile Regressor, Support Vector Regressor, and Xception models are competitive on all the measures as well, further studies are needed to validate these results. Specifically, testing the models on a larger data set and comparing their performance on unseen new data would add solidity to the whole analysis. Overall, the suggested model has tremendous potential for inorganic crystal solubility prediction, but more targeted analyses need to be performed in order to solidify these preliminary results.

Although the proposed model matches the lowest error statistics (MSE, MAE, RMSE) and the highest R^2 value, i.e., it fits the regression problem better (e.g., inorganic crystal solubility prediction), other models, including gradient boosting regressors, also perform well in Figure 8, and support vector regressors and deep neural networks (DenseNet, Xception) have lower R 2, i.e., a worse fit, and are subject to further research or tuning for this specific task. In general, the given model has prospects, and a more profound analysis is of paramount importance to make a final decision.

4. CONCLUSION

This research suggests a novel machine learning algorithm as an estimated framework (SENET-SVM) to predict drug crystal frameworks and atom qualities from the perspective of enhancing drug solubility and bioavailability. The investigation of the complexity of crystallographic structures and fine details of a form sets the course of the research by giving a fresh approach to the development of drugs in the pharmaceutical sector. The study reveals that more sophisticated algorithms can accelerate the process of drug discovery, reduce expenses, and eventually improve patient outcomes by having an improved system of drug delivery. The perspective of a more efficient and brighter future of drug discovery brings the highlights of the potential of sophisticated algorithms in the field of pharmaceutical science to the bright and crisp focus in this article.

REFERENCES

Aithal, G. C., Nayak, U. Y., Mehta, C., Narayan, R., Gopalkrishna, P., Pandiyan, S., & Garg, S. (2018). Localized in situ nanoemulgel drug delivery system of quercetin for periodontitis: Development and computational simulations. *Molecules (Basel, Switzerland), 23*(6), 1363. DOI: 10.3390/molecules23061363 PMID: 29882751

Aitipamula, S., & Vangala, V. R. (2017). X-ray crystallography and its role in understanding the physicochemical properties of pharmaceutical cocrystals. *Journal of the Indian Institute of Science, 97*(2), 227–243. DOI: 10.1007/s41745-017-0026-4

Beran, G. J. (2023). Frontiers of molecular crystal structure prediction for pharmaceuticals and functional organic materials. *Chemical Science (Cambridge), 14*(46), 13290–13312. DOI: 10.1039/d3sc03903j PMID: 38033897

Bhatti, U. A., Tang, H., Wu, G., Marjan, S., & Hussain, A. (2023). Deep learning with graph convolutional networks: An overview and latest applications in computational intelligence. *International Journal of Intelligent Systems, 2023*(1), 8342104. DOI: 10.1155/2023/8342104

Chettri, A., Subba, A., Singh, G. P., & Bag, P. P. (2024). Pharmaceutical co-crystals: A green way to enhance drug stability and solubility for improved therapeutic efficacy. *The Journal of Pharmacy and Pharmacology, 76*(1), 1–12. DOI: 10.1093/jpp/rgad097 PMID: 37934904

Chtita, S., Belhassan, A., Bakhouch, M., Taourati, A. I., Aouidate, A., Belaidi, S., & Lakhlifi, T. (2021). QSAR study of unsymmetrical aromatic disulfides as potent avian SARS-CoV main protease inhibitors using quantum chemical descriptors and statistical methods. *Chemometrics and Intelligent Laboratory Systems, 210*, 104266. DOI: 10.1016/j.chemolab.2021.104266 PMID: 33558778

Das, S., & Paul, S. (2017). Hydrotropic action of cationic hydrotrope p-toluidinium chloride on the solubility of sparingly soluble gliclazide drug molecule: A computational study. *Journal of Chemical Information and Modeling, 57*(6), 1461–1473. DOI: 10.1021/acs.jcim.7b00182 PMID: 28530396

Das, T., Mehta, C. H., & Nayak, U. Y. (2020). Multiple approaches for achieving drug solubility: An in silico perspective. *Drug Discovery Today, 25*(7), 1206–1212. DOI: 10.1016/j.drudis.2020.04.016 PMID: 32353425

Fule, R., & Amin, P. (2014). Development and evaluation of lafutidine solid dispersion via hot melt extrusion: Investigating drug-polymer miscibility with advanced characterisation. asian journal of pharmaceutical sciences, 9(2), 92-106. https://doi.org/DOI: 10.1016/j.ajps.2013.12.004

Fule, R., Meer, T., Sav, A., & Amin, P. (2013). Solubility and dissolution rate enhancement of lumefantrine using hot melt extrusion technology with physicochemical characterisation. *Journal of Pharmaceutical Investigation, 43*(4), 305–321. DOI: 10.1007/s40005-013-0078-z

Guo, M., Sun, X., Chen, J., & Cai, T. (2021). Pharmaceutical cocrystals: A review of preparations, physicochemical properties and applications. *Acta Pharmaceutica Sinica. B, 11*(8), 2537–2564. DOI: 10.1016/j.apsb.2021.03.030 PMID: 34522597

Huang, Z., Staufenbiel, S., & Bodmeier, R. (2022). Combination of co-crystal and nanocrystal techniques to improve the solubility and dissolution rate of poorly soluble drugs. *Pharmaceutical Research, 39*(5), 949. DOI: 10.1007/s11095-022-03243-9 PMID: 35552985

Huynh, L., Leroux, J. C., & Allen, C. (2009). Enhancement of docetaxel solubility via conjugation of formulation-compatible moieties. *Organic & Biomolecular Chemistry, 7*(17), 3437–3446. DOI: 10.1039/b906862g PMID: 19675898

Karimi, M., Wu, D., Wang, Z., & Shen, Y. (2019). DeepAffinity: Interpretable deep learning of compound–protein affinity through unified recurrent and convolutional neural networks. *Bioinformatics (Oxford, England), 35*(18), 3329–3338. DOI: 10.1093/bioinformatics/btz111 PMID: 30768156

Mattei, A., Hong, R. S., Dietrich, H., Firaha, D., Helfferich, J., Liu, Y. M., & Sheikh, A. Y. (2022). Efficient crystal structure prediction for structurally related molecules with accurate and transferable tailor-made force fields. *Journal of Chemical Theory and Computation, 18*(9), 5725–5738. DOI: 10.1021/acs.jctc.2c00451 PMID: 35930763

Mehta, C. H., Narayan, R., & Nayak, U. Y. (2019). Computational modeling for formulation design. *Drug Discovery Today, 24*(3), 781–788. DOI: 10.1016/j.drudis.2018.11.018 PMID: 30502513

Ngilirabanga, J. B., & Samsodien, H. (2021). Pharmaceutical co-crystal: An alternative strategy for enhanced physicochemical properties and drug synergy. *Nano Select, 2*(3), 512–526. DOI: 10.1002/nano.202000201

Niederquell, A., Wyttenbach, N., Kuentz, M., & Panayiotou, C. (2019). Partial solvation parameters of drugs as a new thermodynamic tool for pharmaceutics. *Pharmaceutics, 11*(1), 17. DOI: 10.3390/pharmaceutics11010017 PMID: 30621192

Ouyang, J., Xing, X., Zhou, L., Zhang, C., & Heng, J. Y. (2022). Cocrystal design of vanillin with amide drugs: Crystal structure determination, solubility enhancement, DFT calculation. *Chemical Engineering Research & Design*, *183*, 170–180. DOI: 10.1016/j.cherd.2022.05.009

Ren, Y., Shen, J., Yu, K., Phan, C. U., Chen, G., Liu, J., & Feng, J. (2019). Impact of crystal habit on solubility of ticagrelor. *Crystals*, *9*(11), 556. DOI: 10.3390/cryst9110556

Sherje, A. P., Kulkarni, V., Murahari, M., Nayak, U. Y., Bhat, P., Suvarna, V., & Dravyakar, B. (2017). Inclusion complexation of etodolac with hydroxypropyl-beta-cyclodextrin and auxiliary agents: Formulation characterization and molecular modeling studies. *Molecular Pharmaceutics*, *14*(4), 1231–1242. DOI: 10.1021/acs.molpharmaceut.6b01115 PMID: 28248111

Sugiyama, H., & Uekusa, H. (2025). Crystal Engineering of the Photochromic Organic Compounds. In Advances in Organic Crystal Chemistry: Comprehensive Reviews 2025 on Crystal Structures (pp. 273-296). Singapore: Springer Nature Singapore. DOI: 10.1007/978-981-96-5984-5_12

Taylor, L. S., Braun, D. E., & Steed, J. W. (2021). Crystals and crystallization in drug delivery design. *Molecular Pharmaceutics*, *18*(3), 751–753. DOI: 10.1021/acs.molpharmaceut.0c01157 PMID: 33594892

Varshosaz, J., Ghassami, E., & Ahmadipour, S. (2018). Crystal engineering for enhanced solubility and bioavailability of poorly soluble drugs. *Current Pharmaceutical Design*, *24*(21), 2473–2496. DOI: 10.2174/1381612824666180712104447 PMID: 29998799

Warren, D. B., King, D., Benameur, H., Pouton, C. W., & Chalmers, D. K. (2013). Glyceride lipid formulations: Molecular dynamics modeling of phase behavior during dispersion and molecular interactions between drugs and excipients. *Pharmaceutical Research*, *30*(12). Advance online publication. DOI: 10.1007/s11095-013-1206-1 PMID: 24092054

Xiang, T. X., & Anderson, B. D. (2013). Molecular dynamics simulation of amorphous indomethacin. *Molecular Pharmaceutics*, *10*(1), 102–114. DOI: 10.1021/mp3000698 PMID: 23116319

Xiang, T. X., & Anderson, B. D. (2017). Molecular dynamics simulation of amorphous hydroxypropylmethylcellulose and its mixtures with felodipine and water. *Journal of Pharmaceutical Sciences*, *106*(3), 803–816. DOI: 10.1016/j.xphs.2016.10.026 PMID: 27986292

Yousef, M. A., & Vangala, V. R. (2019). Pharmaceutical cocrystals: Molecules, crystals, formulations, medicines. *Crystal Growth & Design*, *19*(12), 7420–7438. DOI: 10.1021/acs.cgd.8b01898

Zhang, L., Kong, D., Wang, H., Jiao, L., Zhao, X., Song, J., & Lu, Y. (2021). Cocrystal of apixaban–quercetin: Improving solubility and bioavailability of drug combination of two poorly soluble drugs. *Molecules (Basel, Switzerland)*, *26*(9), 2677. DOI: 10.3390/molecules26092677 PMID: 34063645

Compilation of References

Abbas, M., & Bang, J. H. (2015). Rising again: Opportunities and challenges for platinum-free electrocatalysts. *Chemistry of Materials*. Advance online publication. DOI: 10.1021/acs.chemmater.5b03331

Abd El-Ghany, W. A. (2025). Review on the optical and electrical properties of chalcogenide thin films: Challenges and applications. *Physical Chemistry Chemical Physics*, *27*(9), 4567–4586. DOI: 10.1039/D4CP04473H PMID: 39957264

Abdullah, R., Jalil, A. A., Asmadi, M., Hassan, N. S., Bahari, M. B., Alhassan, M., & Karimi-Maleh, H. (2024). Recent advances in zinc oxide-based photoanodes for photoelectrochemical water splitting. *International Journal of Hydrogen Energy*. Advance online publication. DOI: 10.1016/j.ijhydene.2024.05.461

Abrutis, A., Plausinaitiene, V., Skapas, M., Wiemer, C., Salicio, O., Longo, M., Pirovano, A., Siegel, J., Gawelda, W., Rushworth, S., & Giesen, C. (2008). Chemical vapor deposition of chalcogenide materials for phase-change memories. *Microelectronic Engineering*, *85*(12), 2338–2341. DOI: 10.1016/J.MEE.2008.09.014

Abu Elella, M. H., Hassan, S. S., Abdallah, H. M., Mostafa, M. S., & Abu-Thabit, N. Y. (2024). Metal chalcogenides and their nanocomposites in water purification systems. *Metal-Chalcogenide Nanocomposites: Fundamentals, Properties and Industrial Applications*, 59–82. https://doi.org/DOI: 10.1016/B978-0-443-18809-1.00004-3

Abu Elella, M. H., Saad, L., & Samy, M. (2024). Two-Dimensional Nanomaterials-Based Polymer Nanocomposites for Solar Energy Applications. *Two-Dimensional Nanomaterials-Based Polymer Nanocomposites: Processing, Properties and Applications*, 419-464.

Abzal, S. M., Kalyan, K., Janga, S. L., Arshad Ahamed, A., & Dash, J. K. (2025). Fabrication and Functionalization of Bismuth Chalcogenide Nanostructures for Biomedical and Toxic Gas Sensing Applications. In Uddin, I. (Ed.), *Smart Nanosensors* (pp. 399–422). Springer Nature., DOI: 10.1007/978-981-96-3878-9_16

Adabala, S., & Dutta, D. P. (2022). A review on recent advances in metal chalcogenide-based photocatalysts for CO_2 reduction. *Journal of Environmental Chemical Engineering, 10*(3), 107763.

Adam, D. B., Tsai, M.-C., Awoke, Y. A., Huang, W.-H., Lin, C.-H., Alamirew, T., Ayele, A. A., Yang, Y.-W., Pao, C.-W., Su, W.-N., & Hwang, B. J. (2022). Engineering self-supported ruthenium-titanium alloy oxide on 3D web-like titania as iodide oxidation reaction electrocatalyst to boost hydrogen production. *Applied Catalysis B: Environmental, 316*, 121608. DOI: 10.1016/j.apcatb.2022.121608

Adhikari, G. (2024). Nexus between Financial Literacy and Financial Behavior: Theoretical and Empirical Insights. *IIMRJBC, 2*(1), 1–14.

Adhikari, S., Das, S., & Johari, P. (2025). Post-transition metal Sn-based chalcogenide perovskites: A promising lead-free and transition metal alternative for stable, high-performance photovoltaics. *Journal of Materials Chemistry. C, Materials for Optical and Electronic Devices, 13*(15), 7792–7805. DOI: 10.1039/D4TC04701J

Agarwal, A., Tolani, R., & Sankapal, B. R. (2024). Conducting Polymers–Metal Chalcogenides Hybrid Composite: Current Trends and Future Prospects toward Supercapacitor Applications. *Energy Technology (Weinheim), 12*(8), 2400133.

Agrawal, A., Siddiqui, S. A., Soni, A., & Sharma, G. D. (2022). Advancements, frontiers and analysis of metal oxide semiconductor, dye, electrolyte and counter electrode of dye sensitized solar cell. *Solar Energy, 233*, 378–407. DOI: 10.1016/j.solener.2022.01.027

Agu, E. E., Abhulimen, A. O., Obiki-Osafiele, A. N., Osundare, O. S., Adeniran, I. A., & Efunniyi, C. P. (2024). Proposing strategic models for integrating financial literacy into national public education systems. *International Journal of Finance and Management ResearchS, 3*(2), 1–17.

Aguiar, I., Oreggioni, D., Pérez Barthaburu, M.E. (2021). Tin-Based Compounds for Water Remediation. Water Pollution and Remediation: Heavy Metals, 281–312.

Ahluwalia, S., Prakash, N. T., Prakash, R., & Pal, B. (2016). Improved degradation of methyl orange dye using bio-co-catalyst Se nanoparticles impregnated ZnS photocatalyst under UV irradiation. *Chemical Engineering Journal, 306*, 1041–1048. DOI: 10.1016/j.cej.2016.07.060

Ahmad Ramli, S., Mohd Zin, R., Ahmad, M. K., Azyan, N. I., Hamed, N. K. A., Saputri, D. G., & Ahmad, M. Y. (2025). MoS_2-decorated etched-TiO_2 rods film for superior photocatalytic dye degradation. *Journal of Sol-Gel Science and Technology*, •••, 1–19. DOI: 10.1007/s10971-025-06702-y

Ahmad, K. S., Jaffri, S. B., Makawana, B., Gupta, R. K., Nadeem, A., & Attia, S. M. (2024). Developing sustainable chalcogenide for energy storage: Investigating the semiconducting MnS2: CuS: CoS2: Ni2S3 quaternary complexes as supercapacitor electrode material. *Microchemical Journal*, *207*, 112218. DOI: 10.1016/J.MICROC.2024.112218

Aina, S., Scott, N., Lobera, M. P., & Navarro, M. B. (2024). Sustainable solution-processed solar cells based on environmentally friendly nanocrystals. In *Photovoltaics Beyond Silicon* (pp. 437–477). Elsevier.

Aithal, G. C., Nayak, U. Y., Mehta, C., Narayan, R., Gopalkrishna, P., Pandiyan, S., & Garg, S. (2018). Localized in situ nanoemulgel drug delivery system of quercetin for periodontitis: Development and computational simulations. *Molecules (Basel, Switzerland)*, *23*(6), 1363. DOI: 10.3390/molecules23061363 PMID: 29882751

Aitipamula, S., & Vangala, V. R. (2017). X-ray crystallography and its role in understanding the physicochemical properties of pharmaceutical cocrystals. *Journal of the Indian Institute of Science*, *97*(2), 227–243. DOI: 10.1007/s41745-017-0026-4

Akman, E., & Karapinar, H. S. (2022). electrochemically stable, costeffective and facile produced selenium@ activated carbon composite counter electrodes for dye-sensitized solar cells. *Solar Energy*, *234*, 368–376. DOI: 10.1016/j.solener.2022.02.011

Al Abri, R., Al Marzouqi, F., Kuvarega, A. T., Meetani, M. A., Al Kindy, S. M., Karthikeyan, S., Kim, Y., & Selvaraj, R. (2019). Nanostructured cerium-doped ZnO for photocatalytic degradation of pharmaceuticals in aqueous solution. *Journal of Photochemistry and Photobiology A Chemistry*, *384*, 112065.

Alagesan, V., & Venugopal, S. (2019). Green synthesis of selenium nanoparticle using leaves extract of *Withania somnifera* and its biological applications and photocatalytic activities. *BioNanoScience*, *9*(1), 105–116. DOI: 10.1007/s12668-018-0566-8

Alam, H., & Ramakrishna, S. (2013). A review on the enhancement of figure of merit from bulk to nano-thermoelectric materials. *Nano Energy*, *2*(2), 190–212. DOI: 10.1016/j.nanoen.2012.10.005

Alasiri, , A. & Shanmugan, S. (2025). Experimental performance assessment of M-shaped solar distillation for efficient thermal energy storage using directional aluminum can configurations. *Separation and Purification Technology*, *376*(Part 3), 134172. DOI: 10.1016/j.seppur.2025.134172

Alasiri, Abdulaziz, & Shanmugan. S. (2025). The role of Mn_2O_3-Cu NanoPCM in aluminum cans for enhancing thermal performance and freshwater yield of M-shaped basin solar desalination system: An experimental analysis. *Separation and Purification Technology*, *377*(Part 1), 134267. DOI: 10.1016/j.seppur.2025.134267

Al-Fartoos, M. M. R., Roy, A., Mallick, T. K., & Tahir, A. A. (2023). Advancing Thermoelectric Materials: A Comprehensive Review Exploring the Significance of One-Dimensional Nano Structuring. *Nanomaterials (Basel, Switzerland)*, *13*(13), 2011. DOI: 10.3390/nano13132011 PMID: 37446526

Alhaddad, M., & Shawky, A. (2020). Superior photooxidative desulfurization of thiophene by reduced graphene oxide-supported MoS2 nanoflakes under visible light. *Fuel Processing Technology*, *205*, 106453.

Ali, A., & Andriyana, A. (2020). Properties of multifunctional composite materials based on nanomaterials: A review. *RSC Advances*, *10*(28), 16390–16403. DOI: 10.1039/C9RA10594H PMID: 35498826

Ali, A., Shaukat, H., Bibi, S., Altabey, W. A., Noori, M., & Kouritem, S. A. (2023). Recent progress in energy harvesting systems for wearable technology. *Energy Strategy Reviews*, *49*, 101124. DOI: 10.1016/j.esr.2023.101124

Ali, I., Islam, M. R., Yin, J., Eichhorn, S. J., Chen, J., Karim, N., & Afroj, S. (2024). Advances in smart photovoltaic textiles. *ACS Nano*, *18*(5), 3871–3915.

Alijani, H. Q., Pourseyedi, S., Torkzadeh Mahani, M., & Khatami, M. (2019). Green synthesis of zinc sulfide (ZnS) nanoparticles using *Stevia rebaudiana* Bertoni and evaluation of its cytotoxic properties. *Journal of Molecular Structure*, *1175*, 214–218. DOI: 10.1016/j.molstruc.2018.08.081

Alizadeh, A., Roudgar-Amoli, M., Shariatinia, Z., Abedini, E., Asghar, S., & Imani, S. (2023). Recent developments of perovskites oxides and spinel materials as platinum-free counter electrodes for dye-sensitized solar cells: A comprehensive review. *Renewable & Sustainable Energy Reviews*, *187*, 113770. DOI: 10.1016/j.rser.2023.113770

Alizadeh, S. R., Seyedabadi, M., Montazeri, M., Khan, B. A., & Ebrahimzadeh, M. A. (2023). *Allium paradoxum* extract mediated green synthesis of SeNPs: Assessment of their anticancer, antioxidant, iron chelating activities, and antimicrobial activities against fungi, ATCC bacterial strains, *Leishmania* parasite, and catalytic reduction of methylene blue. *Materials Chemistry and Physics*, *296*, 127240. DOI: 10.1016/j.matchemphys.2022.127240

Al-Jawad, S.M.H., Aboud, K.H., & Imran, N.J, Sally Yakoob Taher. (2025). Copper Doping of CdS Nanoflakes and Nanoflowers for Efficient Photocatalytic Degradation of MB and MV Dyes. *Plasmonics*, *20*, 983–1002.

Alrasheedi, N. H., Saravanavalli, K. J., Panchal, H., & Shanmugan, S. (2025). Using Aluminum Cans and Conch Shells to Increase Thermic Efficiency of a solar distiller: A Comparison Study with Decision Tree and Random Forest Predictions. *Case Studies in Thermal Engineering*, •••, 106984.

Alsaggaf, A. (2015). *Investigation of Non-Vacuum Deposition Techniques in Fabrication of Chalcogenide-Based Solar Cell Absorbers*. http://hdl.handle.net/10754/565108

Alsaggaf, M. S., Elbaz, A. F., El Badawy, S., & Moussa, S. H. (2020). Anticancer and antibacterial activity of cadmium sulfide nanoparticles by *Aspergillus niger*. *Advances in Polymer Technology*, *4909054*. Advance online publication. DOI: 10.1155/2020/4909054

Al-Shetwi, A. Q. (2022). Sustainable development of renewable energy integrated power sector: Trends, environmental impacts, and recent challenges. *The Science of the Total Environment*, *822*, 153645. DOI: 10.1016/j.scitotenv.2022.153645 PMID: 35124039

Althomali, R. H., Al-Hawary, S. I. S., Gehlot, A., Qasim, M. T., Abdullaeva, B., Sapaev, I. B., & Alsalamy, A. (2023). A novel Pt-free counter electrode based on $MoSe_2$ for cost effective dyesensitize solar cells (DSSCs): Effect of Ni doping. *Journal of Physics and Chemistry of Solids*, *182*, 111597. DOI: 10.1016/j.jpcs.2023.111597

Al-Tohamy, R., Ali, S. S., Li, F., Okasha, K. M., Mahmoud, Y. A. G., Elsamahy, T., & Sun, J. (2022). A critical review on the treatment of dye-containing wastewater: Ecotoxicological and health concerns of textile dyes and possible remediation approaches for environmental safety. *Ecotoxicology and Environmental Safety*, *231*, 113160.

Ameer, M. F., Hussain, M., Abbas, K., Sufyan, M., Khan, S. U., & Kolsi, L. (2024). Graphene and post-transition metal sulfide-based photodetectors for enhanced visible light photodetection: A review. *Physica Scripta*. Advance online publication. DOI: 10.1088/1402-4896/ada3f7

Ameri, A., Shakibaie, M., Ameri, A., Faramarzi, M. A., Amir Heidari, B., & Forootanfar, H. (2016). Photocatalytic decolorization of bromothymol blue using biogenic selenium nanoparticles synthesized by terrestrial actinomycete *Streptomyces griseobrunneus* strain FSHH12. *Desalination and Water Treatment, 57*(48), 21552–21563. DOI: 10.1080/19443994.2015.1112291

Aminorroaya Yamini, S., Santos, R., Fortulan, R., Gazder, A. A., Malhotra, A., Vashaee, D., Serhiienko, I., & Mori, T. (2023). Room-Temperature Thermoelectric Performance of n-Type Multiphase Pseudobinary Bi_2Te_3–Bi_2S_3 Compounds: Synergic Effects of Phonon Scattering and Energy Filtering. *ACS Applied Materials & Interfaces, 15*(15), 19220–19229. DOI: 10.1021/acsami.3c01956 PMID: 37014987

Amon, A., Jagri, T., & Oplotnik, Ž. (2024). The neobanks and sustainability. *Journal of Information and Organizational Sciences, 48*(9 Andolfi, L., Marxen, H., & Frank, M. (2024, June). Are you Flexible Enough? The Impact of Energy Literacy and Environmental Values on Flexibility Provision. In *2024 11th International Conference on the European Energy Market (EEM)* (pp. 1-5). IEEE. (Andolfi, Marxen, & Frank, 2024, June).

Amrillah, T. (2023). Enhancing the value of environment-friendly CZTS compound for next generation photovoltaic device: A review. *Solar Energy, 263*, 111982. DOI: 10.1016/j.solener.2023.111982

Amrillah, T., Prasetio, A., Supandi, A. R., Sidiq, D. H., Putra, F. S., Nugroho, M. A., & Azmi, R. (2023). Environment-friendly copper-based chalcogenide thin film solar cells: Status and perspectives. *Materials Horizons, 10*(2), 313–339.

Andrea, S. (2022). *Sviluppo di celle fotovoltaiche bifacciali o a film sottile basate su calcogenuri complessi con deposizioni fisiche da vuoto e da fase liquida*. https://tesidottorato.depositolegale.it/handle/20.500.14242/196702

Andrei, E. Y., & MacDonald, A. H. (2020). Graphene bilayers with a twist. *Nature Materials, 19*(12), 1265–1275.

Annveer, , & Gautam, , R., Kumar, A., Kumar, A., Gautam, Y. K., Saroj, A. L., & Singh, R. P. (2021). Study of optoelectronic and thermoelectric spectra of Tl(Nd/Gd)S2. *Journal of Materials Science Materials in Electronics, 32*(1), 727–744. DOI: 10.1007/s10854-020-04852-z

Annveer, , & Gautam, , R., Kumar, A., Kumar, A., Singh, P. K., & Singh, R. P. (2020). Magneto-optical effects in half metallic ferromagnets: Rare earth thallium tellurides (TlXTe2; X = Tb-Er). *Optik (Stuttgart), 223*, 165317. DOI: 10.1016/j.ijleo.2020.165317

Anthony, J. E. (2014). Addressing challenges. *Nature Materials, 13*(8), 773–775. DOI: 10.1038/nmat4034 PMID: 24997739

Anucha, C. B., Altin, I., Bacaksiz, E., & Stathopoulos, V. N. (2022). Titanium dioxide (TiO_2)-based photocatalyst materials activity enhancement for contaminants of emerging concern (CECs) degradation: In the light of modification strategies. *Chemical Engineering Journal Advances, 10*, 100262.

Anu, K., Singaravelu, G., Murugan, K., & Benelli, G. (2017). Green synthesis of selenium nanoparticles using garlic cloves (*Allium sativum*): Biophysical characterization and cytotoxicity on Vero cells. *Journal of Cluster Science, 28*(1), 551–563. DOI: 10.1007/s10876-016-1124-2

Arachchige, I. U., & Brock, S. L. (2007). Sol–Gel Methods for the Assembly of Metal Chalcogenide Quantum Dots. *Accounts of Chemical Research, 40*(9), 801–809. DOI: 10.1021/AR600028S PMID: 17441681

Arif Dar, M., Satgunam, M., Batoo, K. M., Kalpnan, S., Arularasan, P., Ali Dar, U., ... & Majid, S. R. (2025). Future Photovoltaic Applications of Nanostructured Metal Chalcogenides. *Comments on Inorganic Chemistry*, 1-54.

Arrondel, L. (2021). Financial Literacy and French Behaviour on the Stock Market. *Journal of Financial Transformation, 14*(1), 88–100.

Arslan, Z., Ates, M., McDuffy, W., Agachan, M. S., Farah, I. O., Yu, W., & Bednar, A. J. (2011). Probing metabolic stability of CdSe nanoparticles: Alkaline extraction of free cadmium from liver and kidney samples of rats exposed to CdSe nanoparticles. *Journal of Hazardous Materials, 192*, 192–199.

Asghar, H., Riaz, T., Mannan, H. A., Khan, S. M., & Butt, O. M. (2024). Rheology and modeling insights into dye-sensitized solar cells (DSSCs) material: Bridging the gap to solar energy advancements. *Renewable & Sustainable Energy Reviews, 193*, 114298.

Asha, S., Shanmugan, S., Venkateswarlu, M., Meenachi, M., Sangeetha, A., & Rao, M. C. (2022). Thermal potential porous materials and challenges of improving solar still using TiO_2/Jackfruit peel-enhanced energy storage material. *Materials Today: Proceedings, 66*, 3616–3625. DOI: 10.1016/j.matpr.2022.07.142

Asiri, A. M., Adeosun, W. A., Khan, S. B., Alamry, K. A., Marwani, H. M., & Zakeeruddin, S. M.. (2021). NiCuCoS3 chalcogenide as an efficient electrocatalyst for hydrogen and oxygen evolution. *Journal of Materials Research and Technology*, *15*, 4826–4837. DOI: 10.1016/j.jmrt.2021.09.122

Aslam, S., Awais, M., Servat, N., & Safdar, M. (2023). Synthesis and characterization of Ag2ZnTiS4 nanostructures prepared by a hot-injection method towards low-cost electrocatalytic oxygen evolution. *New Journal of Chemistry*, *47*, 20537–20545. DOI: 10.1039/D3NJ03483

Ates, M., Tanaydın, M. K., & Yılmaz, E. (2021a). Challenges, novel applications, and future prospects of chalcogenides and chalcogenide-based nanomaterials for photocatalysis. *Chalcogenide-Based Nanomaterials as Photocatalysts*, 307–337. https://doi.org/DOI: 10.1016/B978-0-12-820498-6.00014-7

Ates, M., Demir, V. Z., Arslan, Z., Daniels, J., Farah, I. O., & Bogatu, C. (2015). Evaluation of alpha and gamma aluminum oxide nanoparticle accumulation,toxicity and depuration in artemia salina larvae. *Environmental Toxicology*, *30*, 109–118.

Ates, M., Demir, V., Arslan, Z., Kaya, H., Yılmaz, S., & Camas, M. (2016). Chronic exposure of tilapia (*Oreochromisniloticus*) to iron oxide nanoparticles: Effects of particle morphology on accumulation, elimination, hematology and immune responses. *Aquatic Toxicology (Amsterdam, Netherlands)*, *177*, 22–32.

Ates, M., Yılmaz, E., & Tanaydın, M. K. (2021). Challenges, novel applications, and future prospects of chalcogenides and chalcogenide-based nanomaterials for photocatalysis. In *Chalcogenide-based nanomaterials as photocatalysts* (pp. 307–337). Elsevier.

Attipoe, V., Chukwuma-Eke, E. C., Lawal, C. I., Friday, S. C., Isibor, N. J., & Akintobi, A. O. (2025). The Role of Financial Institutions in Promoting Sustainable Energy Practices for SMEs: A Strategic Approach. *Contemporary Scientific and Engineering Investigations*, *2*(1), 28–49.

Augustynski, J., Alexander, B., & Solarska, R. (2011). Metal oxide photoanodes for water splitting. *Topics in Current Chemistry*, *303*, 1–38. DOI: 10.1007/128_2011_135 PMID: 21506001

Awfa, D., Ateia, M., Fujii, M., & Yoshimura, C. (2019). Novel Magnetic Carbon Nanotube-TiO_2 Composites for Solar Light Photocatalytic Degradation of Pharmaceuticals in the Presence of Natural Organic Matter. *Journal of Water Process Engineering*, *31*, 100836.

Ayaz, M., Hijji, M., Alatawi, A. S., Namazi, M. A., & Ershath, M. M. (2024). Enhancing photovoltaic performance in dye-sensitized solar cells using nanostructured NiS/MoS2 composite counter electrodes. *Materials Science in Semiconductor Processing*, *173*, 108172. DOI: 10.1016/j.mssp.2024.108172

Bacaksiz, C., Šabani, D., Menezes, R. M., & Milošević, M. V. (2021). Distinctive magnetic properties of CrI_3 and $CrBr_3$ monolayers caused by spin-orbit coupling. *Physical Review. B*, *103*(12), 125418.

Badran, G., & Lazarov, V. K. (2025, February). From Waste to Resource: Exploring the Current Challenges and Future Directions of Photovoltic Solar Cell Recycling. In *Solar* (Vol. 5, No. 1, p. 4). MDPI.

Bajpai, P. K., Yadav, S., Tiwari, A., & Virk, H. S. (2015). Recent advances in the synthesis and characterization of chalcogenide nanoparticles. *Diffusion and Defect Data, Solid State Data. Part B, Solid State Phenomena*, *222*, 187–233.

Bal, S., & Rani, N. R. A. (2025). Next generation building materials for energy efficiency and climate responsive design. *Discover Applied Sciences*, *7*(8), 796. DOI: 10.1007/s42452-025-07360-z

Banik, D., Lionel, C., Das, S., & Koley, S. (2024). Comparative study of the electronic and optical properties of Rhenium based Chalcogenides. *Physica Scripta*, *99*(8), 085975.

Baraneedharan, P., Sekar, S., Murugesan, S., Ahamada, D., Mohamed, S. A. B., Lee, Y., & Lee, S. (2024). Recent Advances and Remaining Challenges in Perovskite Solar Cell Components for Innovative Photovoltaics. *Nanomaterials (Basel, Switzerland)*, *14*(23), 1867.

Bard, A. J., Faulkner, L. R., & White, H. S. (2022). *Electrochemical methods: fundamentals and applications*. Wiley.

Barker, A. S., Ditzenberger, J. A.Jr, & DiSalvo, F. J. (1975). Infrared study of the electronic instabilities in tantalum disulfide and tantalum diselenide. *Physical Review. B*, *12*, 2049–2054.

Bartlett, P. N., De Groot, C. H. K., Greenacre, V. K., Huang, R., Noori, Y. J., Reid, G., & Thomas, S. (2025). Molecular precursors for the electrodeposition of 2D-layered metal chalcogenides. *Nature Reviews. Chemistry*, *9*(2), 88–101. DOI: 10.1038/s41570-024-00671-6 PMID: 39775527

Bastami, T. R., Ahmadpour, A., & Hekmatikar, F. A. (2017). Synthesis of Fe_3O_4/Bi_2WO_6 nanohybrid for the photocatalytic degradation of pharmaceutical ibuprofen under solar light. *Journal of Industrial and Engineering Chemistry*, *51*, 244–254.

Beheshti-Marnani, A., Rohani, T., Kermani, M. A., & Mohammadi, S. zia. (2025). A sensitive chalcogenide-based electrochemical sensor for ultra-level detection of Mospilan residues in real samples. *Scientific Reports*, *15*(1), 5966. DOI: 10.1038/s41598-025-89256-x PMID: 39966467

Beran, G. J. (2023). Frontiers of molecular crystal structure prediction for pharmaceuticals and functional organic materials. *Chemical Science (Cambridge)*, *14*(46), 13290–13312. DOI: 10.1039/d3sc03903j PMID: 38033897

Bhagwan, J., & Han, J. I. (2022). Promotive effect of MWCNTs on alpha-NiS microstructure and their application in aqueous asymmetric supercapacitor. *Energy & Fuels*, *36*, 15210–15220. DOI: 10.1021/acs.energyfuels.2c03080

Bhanja, P., Kim, Y., & Paul, B.. (2020). Facile synthesis of nanoporous transition metal-based phosphates for oxygen evolution reaction. *ChemCatChem*, *12*, 2091–2096. DOI: 10.1002/cctc.201901803

Bhatti, U. A., Tang, H., Wu, G., Marjan, S., & Hussain, A. (2023). Deep learning with graph convolutional networks: An overview and latest applications in computational intelligence. *International Journal of Intelligent Systems*, *2023*(1), 8342104. DOI: 10.1155/2023/8342104

Bhushan, M., JHA, R., Bhardwaj, R., & Sharma, R. (2021). Visible light emission and enhanced electrocatalytic activity of pure ZnS nanoparticles synthesized via thermal decomposition route. *Bulletin of Materials Science*, *44*(4), 261. DOI: 10.1007/s12034-021-02546-8

Bhushan, M., Jha, R., & Bhardwaj, R. (2019). Reduced band gap and diffusion controlled spherical n-type ZnS nanoparticles for absorption of UV-Vis region of solar spectrum. *Journal of Physics and Chemistry of Solids*, *135*, 109021. DOI: 10.1016/j.jpcs.2019.05.018

Bhushan, M., Jha, R., Bhardwaj, R., & Sharma, R. (2021). Graphene-doped ZnS nanoparticles synthesized via hydrothermal route for enhanced electrocatalytic performance. *International Journal of Applied Ceramic Technology*, *18*(5), 1510–1526. DOI: 10.1111/ijac.13801

Bidikoudi, M., & Stathatos, E. (2024a). Chalcogenides in Perovskite Solar Cells with a Carbon Electrode: State of the Art and Future Prospects. *Nanomaterials 2024, Vol. 14, Page 1783, 14*(22), 1783. https://doi.org/DOI: 10.3390/NANO14221783

Bidikoudi, M., & Stathatos, E. (2024). Chalcogenides in Perovskite Solar Cells with a Carbon Electrode: State of the Art and Future Prospects. *Nanomaterials (Basel, Switzerland)*, *14*(22), 1783.

Biney, J., Madou, M., Jabbour, G., & Park, J. (2025). 2D Composite Materials for Electrodes in Dye-Sensitized Solar Cells— An Overview. *ACS Applied Materials & Interfaces*. Advance online publication. DOI: 10.1021/acsami.4c13963

Bishnoi, S., Datt, R., Tsoi, W. C., & Gupta, V. (2025).Development of Solar Cells. In *Solar Cells Development and Fabrication* (pp. 1–21). CRC Press.

Blasch, J., Boogen, N., Daminato, C., & Filippini, M. (2021). Empower the Consumer! Energy-related Financial Literacy and its Implications for Economic Decision Making. *Energy Journal*, *42*(2).

Boogen, N., Cattaneo, C., Filippini, M., & Obrist, A. (2021). Energy efficiency and the role of energy-related financial literacy: Evidence from the European residential sector. *Review of Environmental Economics and Policy*, *15*(2), 335–356.

Borisenko, S. V., Kordyuk, A. A., Yaresko, A. N., Zabolotnyy, V. B., Inosov, D. S., Schuster, R., Büchner, B., Weber, R., & Follath, R. (2008). Pseudogap and charge density waves in two dimensions. *Physical Review Letters*, *100*(19), 196402.

Borovaya, M., Naumenko, A., Horiunova, I., Plokhovska, S., Blume, Y., & Yemets, A. (2020). "Green" synthesis of Ag_2S nanoparticles, study of their properties and bioimaging applications. *Applied Nanoscience*, *10*, 4931–4940. DOI: 10.1007/s13204-020-01465-3

Boudouma, A., Layachi, O. A., Hrir, H., Nejmi, Y., Taoufyk, A., Moujib, A., Elbrouzi, A., & Khoumri, E. (2025). Eliminating the sulfurization step: Optimizing the growth of CFTS thin films in the stannite phase by electrodeposition. *Ionics*, *31*, 3847–3859. DOI: 10.1007/s11581-025-06144-0

Bouroushian, M. (2010). *Electrochemistry of Metal Chalcogenides*. https://doi.org/DOI: 10.1007/978-3-642-03967-6

Brauer, H. E., Starnberg, H. I., Holleboom, L. J., Hughes, H. P., & Strocov, V. N. (2001). Na and Cs intercalation of 2H-$TaSe_2$ studied by photoemission. *Journal of Physics Condensed Matter*, *13*(44), 9879.

Britto, J., Barani, P., Vanaja, M., Pushpalaksmi, E., Samraj, J. J., & Annadurai, G. (2021). Adsorption of dyes by chitosan–selenium nanoparticles: Recent developments and adsorption mechanisms. *Nature Environment and Pollution Technology*, *20*(2), 467–479.

Buckingham, M. A., Qu, J., Ward-O'Brien, B., & Lewis, D. J. (2022). High entropy metal chalcogenides: Synthesis, properties, applications and future directions. *Chemical Communications*, *58*(61), 8025–8036. DOI: 10.1039/D2CC01796B PMID: 35770747

Buffiere, M., Dhawale, D. S., & El-Mellouhi, F. (2019). Chalcogenide Materials and Derivatives for Photovoltaic Applications. *Energy Technology (Weinheim)*, *7*(11), 1900819. DOI: 10.1002/ente.201900819

Byeon, H., Devi, V. P., Murugan, A., Tonk, A., Haribabu, K., Patil, H., & Vincy, V. G. A. G. (2024). Evaluating the potential capability of selenium nanoparticles on degradation of hazardous textile dye using environmental friendly approach. *Global NEST Journal*, *26*, 05633. DOI: 10.30955/gnj.005633

Cadavid, D., Wei, K., Liu, Y., Zhang, Y., Li, M., Genç, A., Berestok, T., Ibáñez, M., Shavel, A., Nolas, G. S., & Cabot, A. (2021). Synthesis, Bottom up Assembly and Thermoelectric Properties of Sb-Doped PbS Nanocrystal Building Blocks. *Materials (Basel)*, *14*(4), 853. DOI: 10.3390/ma14040853 PMID: 33578981

Cai, M., Yang, J., Lu, X., & Lu, X. (2024). Layer-by-layer self-assembly strategies of atomically thin two-dimensional nanomaterials: Principles, methods, and functional applications. *ACS Applied Nano Materials*, *7*(24), 27940–27959.

Cao, S., Li, H., Zhou, X., Guo, H., & Chen, Y. (2022). NiS/activated carbon composite derived from sodium lignosulfonate for long cycle-life asymmetric supercapacitors. *Journal of Alloys and Compounds*, *900*, 163546. DOI: 10.1016/j.jallcom.2021.163546

Cao, Y.. (2018). Unconventional superconductivity in magic-angle graphene superlattices. *Nature*, *556*, 43–50.

Carrasco, V., Amarelle, V., Lagos-Moraga, S., Quezada, C. P., Espinoza-González, R., Faccio, R., Fabiano, E., & Pérez-Donoso, J. M. (2021). Production of cadmium sulfide quantum dots by the lithobiontic Antarctic strain *Pedobacter* sp. UYP1 and their application as photosensitizer in solar cells. *Microbial Cell Factories*, *20*, 41. DOI: 10.1186/s12934-021-01548-3 PMID: 33568151

Casaluci, S., Gemmi, M., Pellegrini, V., Di Carlo, A., & Bonaccorso, F. (2016). Graphene-based large area dye-sensitized solar cell modules. *Nanoscale*, *8*(9), 5368–5378. DOI: 10.1039/C5NR07971C PMID: 26883743

Castro Neto, A. H.. (2009). The electronic properties of graphene. *Reviews of Modern Physics*, *81*, 109–162.

Cerovic, L., Malnar, A., Sinčić, D. (2024). Energy Literacy of Economics Students in Rijeka: Knowledge, Attitudes, and Behavioral Approach. *Energies* 17 DO - DOI: 10.3390/en17081840

Chanda, D., Kannan, K., Gautam, J., Meshesha, M. M., Jang, S. G., Dinh, V. A., & Yang, B. L. (2023). Effect of the interfacial electronic coupling of nickel-iron sulfide nanosheets with layer Ti3C2 MXenes as efficient bifunctional electrocatalysts for anion-exchange membrane water electrolysis. *Applied Catalysis B: Environmental*, *321*, 122039. DOI: 10.1016/j.apcatb.2022.122039

Chao, C., Jun, Z., Xiaoshan, X., Shuai, Y., & Jiacen, L. (2022). High Photoelectrochemical and Hydrogen Evolution Reaction Activities from 0D/2D ZnSe/(001)CuSe Heterojunction Catalysts Constructed by ZnSe Nanoparticles and {001} Facets-Exposed CuSe Nanosheets. *ACS Applied Energy Materials*, *5*(7), 8952–8962. DOI: 10.1021/acsaem.2c01443

Chatterjee, A., Kumar, G. K., Roymahapatra, G., Das, H. S., Jaishree, G., & Rao, T. S. (2024). Zinc chalcogenide nanostructures: Synthesis methodologies and applications—a review. *Frontiers in Nanotechnology*, *6*, 1433591.

Chawla, A., Sudhaik, A., Kumar, R., Raizada, P., Khan, A. A. P., Ahamad, T., & Singh, P. (2024). a Recent advances in synthesis methods and surface structure manipulating strategies of copper selenide (CuSe) nanoparticles for photocatalytic environmental and energy applications. *Journal of Environmental Chemical Engineering*, •••, 113125.

Che, L., Dong, Y., Wu, M., Zhao, Y., Liu, L., & Zhou, H. (2017). Characterization of selenite reduction by *Lysinibacillus* sp. ZYM-1 and photocatalytic performance of biogenic selenium nanospheres. *ACS Sustainable Chemistry & Engineering*, *5*(3), 2535–2543. DOI: 10.1021/acssuschemeng.6b02525

Chen, D. H., & Wu, S. H. (2000). Synthesis of nickel nanoparticles in water-in-oil microemulsions. *Chemistry of Materials*, *12*(5), 1354–1360. DOI: 10.1021/cm991167y

Chen, M., Wang, J., Yin, F., Du, Z., Belfiore, L. A., & Tang, J. (2021). Strategically integrating quantum dots into organic and perovskite solar cells. *Journal of Materials Chemistry. A, Materials for Energy and Sustainability*, *9*(8), 4505–4527.

Chen, S., Duan, J., Jaroniec, M., & Qiao, S. Z. (2013). Three-dimensional N-doped graphene hydrogel/NiCo double hydroxide electrocatalysts for highly efficient oxygen evolution. *Angewandte Chemie International Edition*, *52*, 13567–13570. DOI: 10.1002/anie.201306166

Chen, S., Gong, X. G., Walsh, A., & Wei, S.-H. (2009). Crystal and Electronic Band Structure of Cu2ZnSnX4 (X = S and Se) Photovoltaic Absorbers: First-Principles Insights. *Applied Physics Letters*, *94*, 041903. DOI: 10.1063/1.3074499

Chen, W.-C., Hung, C.-W., Chang, C.-H., Liang, F.-C., Benas, J.-S., Yan, Z.-L., Lin, B.-H., Lin, J.-H., & Kuo, C.-C. (2022). Crystal orientation and insulating ligand of quasi-two dimensional perovskite optimized through silver ion doping for realizing efficient light emitting diodes. *Chemical Engineering Journal, 443*, 136496. DOI: 10.1016/j.cej.2022.136496

Chen, Z., Gao, Y., & Zhang, Q.. (2019). TiO2/NiO/reduced graphene oxide nanocomposites as anode materials for high-performance lithium-ion batteries. *Journal of Alloys and Compounds, 774*, 873–878. DOI: 10.1016/j.jallcom.2018.10.010

Chettri, A., Subba, A., Singh, G. P., & Bag, P. P. (2024). Pharmaceutical co-crystals: A green way to enhance drug stability and solubility for improved therapeutic efficacy. *The Journal of Pharmacy and Pharmacology, 76*(1), 1–12. DOI: 10.1093/jpp/rgad097 PMID: 37934904

Chhowalla, M.. (2013). The chemistry of two-dimensional layered transition metal dichalcogenide nanosheets. *Nature Chemistry, 5*, 263–275.

Chib, S., Mehta, K., Selvakumar, P., Saravanan, A., Mishra, B. R., & Manjunath, T. C. (2025). Mental Helth Challenges and Solutions in High-Pressure Work Environments. In Harriott, S. (Ed.), *Innovative Approaches to Managing Conflict and Change in Diverse Work Environments* (pp. 247–272). IGI Global Scientific Publishing., DOI: 10.4018/979-8-3693-9556-1.ch011

Chinnaiah, K., Kannan, K., Chen, Y. S., & Gurushankar, K. (2025). Exploring Ag/Mn3O4 composite nanorods as an attractive battery-type electrode material for supercapacitors. *Journal of Physics and Chemistry of Solids, 196*, 112310.

Chinnaiah, K., Kannan, K., Krishnamoorthi, R., Palko, N., Grishina, M., Chen, Y. S., & Gurushankar, K. (2024). Enhancing supercapacitor and antimicrobial performance of bioengineered Ag/Mn3O4 composite nanorods. *Journal of Industrial and Engineering Chemistry*.

Chinnaiah, K., Kannan, K., Palko, N., Grishina, M., Chen, Y.-S., & Gurushankar, K. (2024). Exploring the potential of Withania somnifera-mediated Ag/Mn_3O_4 nanocomposites as electrode material for high-performance supercapattery device. *Journal of the Taiwan Institute of Chemical Engineers, 157*, 105441. DOI: 10.1016/j.jtice.2024.105441

Chinnaiah, K., Karthik, K., Krishnamoorthi, R., Nadezhda, P., Maria, G., Yong-Song, C., & Gurushankar, K. (2025). Enhancing supercapacitor and Antimicrobial Performance of Bioengineered Ag/Mn_3O_4 Composite Nanorods. *Journal of Industrial and Engineering Chemistry, 147*, 329–340. DOI: 10.1016/j.jiec.2024.12.023

Choi, J., Song, K., Kim, J.-I., Im, W. B., & Ahn, C. (2022). Enhanced thermal and electrical properties of hybrid polymer composites containing Al2O3 microspheres and nanowires. *Ceramics International*, *48*(21), 32081–32088. DOI: 10.1016/j.ceramint.2022.07.147

Choubey, A., & Yadav, A. (2024). MXene and transition metal chalcogenides-based 2D nanomaterials for next-generation supercapacitors. *Journal of Energy Storage*, *79*, 110131. DOI: 10.1016/j.est.2023.110131

Chtita, S., Belhassan, A., Bakhouch, M., Taourati, A. I., Aouidate, A., Belaidi, S., & Lakhlifi, T. (2021). QSAR study of unsymmetrical aromatic disulfides as potent avian SARS-CoV main protease inhibitors using quantum chemical descriptors and statistical methods. *Chemometrics and Intelligent Laboratory Systems*, *210*, 104266. DOI: 10.1016/j.chemolab.2021.104266 PMID: 33558778

Chua, H. J. (2022). *A chalcogenide-based colorimetric sensor on paper for heavy metal ions detection* [Nanyang Technological University]. DOI: 10.32657/10356/163946

Chusnutdinow, S., Kazakov, A., Jakieła, R., Szot, M., Schreyeck, S., Brunner, K., & Karczewski, G. (2023). Effect of Manganese Alloying on Infrared Detectors Made of $Pb_{1-x}Mn_xTe$/CdTe Multilayer Composite. *Materials (Basel)*, *16*(12), 4211. DOI: 10.3390/ma16124211 PMID: 37374395

Cittrarasu, V., Kaliannan, D., Dharman, K., Maluventhen, V., Easwaran, M., Liu, W. C., Balasubramanian, B., & Arumugam, M. (2021). Green synthesis of selenium nanoparticles mediated from *Ceropegia bulbosa* Roxb extract and its cytotoxicity, antimicrobial, mosquitocidal and photocatalytic activities. *Scientific Reports*, *11*, 1032. DOI: 10.1038/s41598-020-80366-0 PMID: 33441811

Cui, D., Liang, T., & Sun, L.. (2018). Green synthesis of selenium nanoparticles with extract of hawthorn fruit induced HepG2 cells apoptosis. *Pharmaceutical Biology*, *56*(1), 528–534. DOI: 10.1080/13880209.2018.1500525 PMID: 30387372

Cui, X., Kong, Z., Gao, E., Huang, D., Hao, Y., Shen, H., Di, C., Xu, Z., Zheng, J., & Zhu, D. (2018). Rolling up transition metal dichalcogenide nanoscrolls via one drop of ethanol. *Nature Communications*, *9*(1), 1301. DOI: 10.1038/s41467-018-03752-5 PMID: 29615627

da Silveira Salla, J., Dotto, G. L., Hotza, D., Landers, R., da Boit Martinello, K., & Foletto, E. L. (2020). Enhanced catalytic performance of CuFeS2 chalcogenide prepared by microwaveassisted route for photo-Fenton oxidation of emerging pollutant in water. *Journal of Environmental Chemical Engineering*, *8*, 104077.

Dahiya, R., & Selvakumar, P. Vidyadhara, & Manjunath, T. C. (2025). Design and Aerodynamics: Principles and Advances. In M. Mellal (Ed.), Innovations and Developments in Unmanned Aerial Vehicles (pp. 145-170). IGI Global Scientific Publishing. DOI: 10.4018/979-8-3693-8462-6.ch007

Dahiya, Y., Hariram, M., Kumar, M., Jain, A., & Sarkar, D. (2022). Modified transition metal chalcogenides for high performance supercapacitors: Current trends and emerging opportunities. *Coordination Chemistry Reviews*, *451*, 214265. DOI: 10.1016/j.ccr.2021.214265

Dakua, P. K., Trabelsi, Y., Dudekula, U., Tripathi, R., Laidouci, A., Pal, A., Panda, D. K., Misra, R. K., & Bhattarai, S. (2024). Optimization of efficiency of CZTS-based solar cell through exclusive BSF layer engineering method. *Journal of Physics and Chemistry of Solids*, *193*, 112156. DOI: 10.1016/j.jpcs.2024.112156

Dardel, B., Grioni, M., Malterre, D., Weibel, P., Baer, Y., & Levy, F. (1993). Spectroscopic observation of charge-density-wave-induced changes in the electronic structure of 2H-$TaSe_2$. *Journal of Physics Condensed Matter*, *5*(33), 6111.

Das, S., & Paul, S. (2017). Hydrotropic action of cationic hydrotrope p-toluidinium chloride on the solubility of sparingly soluble gliclazide drug molecule: A computational study. *Journal of Chemical Information and Modeling*, *57*(6), 1461–1473. DOI: 10.1021/acs.jcim.7b00182 PMID: 28530396

Das, T., Mehta, C. H., & Nayak, U. Y. (2020). Multiple approaches for achieving drug solubility: An in silico perspective. *Drug Discovery Today*, *25*(7), 1206–1212. DOI: 10.1016/j.drudis.2020.04.016 PMID: 32353425

Deogratias, G., Al-Qurashi, O. S., & Wazzan, N. (2023). Optical and electronic properties enhancement via chalcogenides: Promising materials for DSSC applications. *Journal of Molecular Modeling*, *29*(4), 86. DOI: 10.1007/s00894-023-05472-0

Dhanasekar, N., Vanitha, M., Shankar, N., & Bindhu, M. (2024). Design and fabrication of NiSe2/g-C3N4 hybrid composite counter electrode for Pt-free dye-sensitized solar cells. *Journal of Materials Science Materials in Electronics*, *35*(6), 443.

Di Salvo, F. J., & Graebner, J. E. (1977). The Low Temperature Electrical Properties of 1T-TaS_2. *Solid State Communications*, *23*, 825–828.

Dias, G. D. S., Costa, J. M., & Almeida Neto, A. F. D. (2023). Transition metal chalcogenides carbon-based as bifunctional cathode electrocatalysts for rechargeable zinc-air battery: An updated review. *Advances in Colloid and Interface Science*, *315*, 102891. DOI: 10.1016/j.cis.2023.102891 PMID: 37058836

Digraskar, R. V., Mali, S. M., Tayade, S. B., Ghule, A. V., & Sathe, B. R. (2019a). Overall noble metal free Ni and Fe doped Cu 2 ZnSnS 4 (CZTS) bifunctional electrocatalytic systems for enhanced water splitting reactions. *International Journal of Hydrogen Energy*, *44*, 8144–8155. DOI: 10.1016/j.ijhydene.2019.02.054

Ding, L., Li, D., Shen, H., Qiao, X., Shen, H., & Shi, W. (2021). 2D β-NiS as electron harvester anchors on 2D ZnIn2S4 for boosting photocatalytic hydrogen production. *Journal of Alloys and Compounds*, *853*, 157328. DOI: 10.1016/j.jallcom.2020.157328

Dutkova, E., Baláž, M., Kováč, J., Daneu, N., Kashimbetova, A., Briančin, J., Kováč, J., Kováčová, S., & Čelko, L. (2024a). Optical and Optoelectrical Properties of Ternary Chalcogenide CuInS2/TiO2 Nanocomposite Prepared by Mechanochemical Synthesis. *Crystals 2024, Vol. 14, Page 324*, *14*(4), 324. https://doi.org/DOI: 10.3390/CRYST14040324

Ebrahimzadeh, M. A., Moradsomarein, M., Lalerdi, F. S., & Alizadeh, S. R. (2023). Biogenic synthesis of selenium nanoparticles using *Hibiscus esculentus* L. extract: Catalytic degradation of organic dye and its anticancer, antibacterial, and antifungal activities. *European Journal of Chemistry*, *14*(2), 144–154. DOI: 10.5155/eurjchem.14.2.144-154.2416

Edison, T. N. J. I., Atchudan, R., Karthik, N., Ganesh, K., Xiong, D., & Lee, Y. R. (2020). A novel binder-free electro-synthesis of hierarchical nickel sulfide nanostructures on nickel foam as a battery-type electrode for hybrid-capacitors. *Fuel*, *276*, 118077. DOI: 10.1016/j.fuel.2020.118077

Ehsan, M. F., Qudoos, S., Ahmad, Z., Hamid, S., Arfan, M., Zia, A., & Tyagi, D. (2019). ZnTe/ZnSe heterostructures: In-situ synthesis, characterization and photocatalytic activity for Congo Red degradation. *SN Applied Sciences*, *1*(3), 197.

Eidsvåg, H., Bentouba, S., & Vajeeston, P.. (2021). TiO2 as a photocatalyst for water splitting—An experimental and theoretical review. *Molecules (Basel, Switzerland)*, *26*, 1687. DOI: 10.3390/molecules26061687 PMID: 33802911

Eisa, T., Abdelkareem, M. A., Jadhav, D. A., Mohamed, H. O., Sayed, E. T., Olabi, A. G., Castaño, P., & Chae, K. J. (2023). Critical review on the synthesis, characterization, and application of highly efficient metal chalcogenide catalysts for fuel cells. *Progress in Energy and Combustion Science*, *94*, 101044. DOI: 10.1016/J.PECS.2022.101044

El Hammoumi, A., Chtita, S., Motahhir, S., & El Ghzizal, A. (2022). Solar PV energy: From material to use, and the most commonly used techniques to maximize the power output of PV systems: A focus on solar trackers and floating solar panels. *Energy Reports*, *8*, 11992–12010.

Ellappan, B., Madhavan, V. R., Uzorka, A., & Shanmugan, S. (2025). Leveraging Grewia optiva leaf extract as a doping phase change material for enhanced solar desalination performance: Simulation prediction and diverse technologies. *Case Studies in Thermal Engineering*, •••, 106853.

El-Sayed, E.-S. R., Abdelhakim, H. K., & Ahmed, A. S. (2020). Solid state fermentation for enhanced production of selenium nanoparticles by gamma-irradiated *Monascus purpureus* and their biological evaluation and photocatalytic activities. *Bioprocess and Biosystems Engineering*, *43*(6), 797–809. DOI: 10.1007/s00449-020-02279-1 PMID: 31898764

Enamala, M. K., Chavali, M., Tangellapally, A., Pasumarthy, D., Murthy, M. K., Kuppam, C., Chaudhary, V., Mishra, R., & Naradasu, D. (2021). 10—Use of chalcogenide-based nanomaterials for wastewater treatment, including bacterial disinfection and organic contaminant degradation. In Khan, M. M. (Ed.), *Chalcogenide-Based Nanomaterials as Photocatalysts* (pp. 243–259). Elsevier., DOI: 10.1016/B978-0-12-820498-6.00010-X

Fang, C., Wang, H., Shen, Z., Shen, H., Wang, S., Ma, J., Wang, J., Luo, H., & Li, D. (2019). High-Performance Photodetectors Based on Lead-Free 2D Ruddlesden–Popper Perovskite/MoS_2 Heterostructures. *ACS Applied Materials & Interfaces*, *11*(8), 8419–8427. DOI: 10.1021/acsami.8b20538 PMID: 30702273

Fang, C., Xu, B., Li, M., Han, J., Yang, Y., & Liu, X. (2024). Advanced Design of Fibrous Flexible Actuators for Smart Wearable Applications. *Advanced Fiber Materials*, *6*(3), 622–657. DOI: 10.1007/s42765-024-00386-9

Fang, X., Wei, P., Wang, L., Wang, X., Chen, B., He, Q., Yue, Q., Zhang, J., Zhao, W., Wang, J., Lu, G., Zhang, H., Huang, W., Huang, X., & Li, H. (2018). Transforming Monolayer Transition-Metal Dichalcogenide Nanosheets into One-Dimensional Nanoscrolls with High Photosensitivity. *ACS Applied Materials & Interfaces*, *10*(15), 13011–13018. DOI: 10.1021/acsami.8b01856 PMID: 29600705

Fan, X., Shi, Y., Cui, M., Ren, S., Wang, H., & Pu, J. (2021). MoS_2/WS_2 Nanosheet-based composite films irradiated by atomic oxygen: Implications for lubrication in space. *ACS Applied Nano Materials*, *4*, 10307–10320.

Farokhi, A., Shahroosvand, H., Delle Monache, G., Pilkington, M., & Nazeeruddin, M. K. (2022). The evolution of triphenylamine hole transport materials for efficient perovskite solar cells. *Chemical Society Reviews*, *51*(14), 5974–6064.

Fazekas, P., & Tosatti, E. (1979). Electrical, structural and magnetic properties of pure and doped 1T-TaS_2. *Philosophical Magazine Part B*, *39*, 229–244.

Feng, J., Dong, H., Pang, B., Chen, Y., Yu, L., & Dong, L. (2019). Tuning the electronic and optical properties of graphene quantum dots by selective boronization. *Journal of Materials Chemistry. C, Materials for Optical and Electronic Devices*, *7*(2), 237–246. DOI: 10.1039/C8TC03863E

Feng, J., Xiong, S., Qian, Y., & Yin, L. (2014). Synthesis of nanosized cadmium oxide (CdO) as a novel high capacity anode material for Lithium-ion batteries: Influence of carbon nanotubes decoration and binder choice. *Electrochimica Acta*, *129*, 107–112. DOI: 10.1016/j.electacta.2014.02.085

Feng, L., Ding, R., Chen, Y., Wang, J., & Xu, L. (2020). Zeolitic imidazolate framework-67 derived ultra-small CoP particles incorporated into N-doped carbon nanofiber as efficient bifunctional catalysts for oxygen reaction. *Journal of Power Sources*, *452*, 227837. DOI: 10.1016/j.jpowsour.2020.227837

Feng, L., Vrubel, H., Bensimon, M., & Hu, X. (2014). Easily-prepared dinickel phosphide (Ni2P) nanoparticles as an efficient and robust electrocatalyst for hydrogen evolution. *Physical Chemistry Chemical Physics*, *16*, 5917–5921. DOI: 10.1039/C4CP00482E

Feng, Q., Li, M., Wang, T., Chen, Y., Wang, X., Zhang, X., Li, X., Yang, Z., Feng, L., Zheng, J., Xu, H., Zhai, T., & Jiang, Y. (2020). Low-temperature growth of Three dimensional ReS2/ReO2 metal-semiconductor heterojunctions on Graphene/polyimide film for enhanced hydrogen evolution reaction. *Applied Catalysis B: Environmental*, *271*, 118924. DOI: 10.1016/j.apcatb.2020.118924

Flasck, R., & Rockstad, H. K. (1973). The thermal conductivity of some chalcogenide glasses. *Journal of Non-Crystalline Solids*, *12*(3), 353–356. DOI: 10.1016/0022-3093(73)90007-0

Fo, Y., Wang, M., & Ma, Y.. (2020). Origin of highly efficient photocatalyst NiO/SrTiO3 for overall water splitting: Insights from density functional theory calculations. *Journal of Solid State Chemistry*, *292*, 121683. DOI: 10.1016/j.jssc.2020.121683

Frumar, M., Jedelský, J., Frumarová, B., Wágner, T., & Hrdlička, M. (2003). Optically and thermally induced changes of structure, linear and non-linear optical properties of chalcogenides thin films. *Journal of Non-Crystalline Solids*, *326–327*, 399–404. DOI: 10.1016/S0022-3093(03)00446-0

Fu, H. (2018). Environmentally friendly and earth-abundant colloidal chalcogenide nanocrystals for photovoltaic applications. *Journal of Materials Chemistry. C, Materials for Optical and Electronic Devices*, *6*(3), 414–445.

Fule, R., & Amin, P. (2014). Development and evaluation of lafutidine solid dispersion via hot melt extrusion: Investigating drug-polymer miscibility with advanced characterisation. asian journal of pharmaceutical sciences, 9(2), 92-106. https://doi.org/DOI: 10.1016/j.ajps.2013.12.004

Fule, R., Meer, T., Sav, A., & Amin, P. (2013). Solubility and dissolution rate enhancement of lumefantrine using hot melt extrusion technology with physicochemical characterisation. *Journal of Pharmaceutical Investigation*, *43*(4), 305–321. DOI: 10.1007/s40005-013-0078-z

Fu, Q., Han, J., Wang, X., Xu, P., Yao, T., Zhong, J., Zhong, W., Liu, S., Gao, T., Zhang, Z., Xu, L., & Song, B. (2021). 2D Transition Metal Dichalcogenides: Design, Modulation, and Challenges in Electrocatalysis. *Advanced Materials*, *33*(6), 1907818. DOI: 10.1002/adma.201907818 PMID: 32578254

Fu, S., Yuan, W., Yan, Y., Liu, H., Shi, X., Zhao, F., & Zhou, J. (2019). Highly efficient visible-light photoactivity of Z-scheme MoS2/Ag2CO3 photocatalysts for organic pollutants degradation and bacterial inactivation. *Journal of Environmental Management*, *252*, 109654.

Gallegos, A., Valencia-Arias, A., Rodrguez Zavala, L., Rodrguez-Correa, P., Martnez Rojas, E., & Castaeda Rodrguez, L. D. R. (2025). Adoption of financial literacy: Systematic literature review and proposal of a theoretical model. *Journal of Innovation and Entrepreneurship*, *14*(1), 1–22.

Gaman, V. I., Kalygina, V. M., & Panin, A. V. (1999). Effect of chalcogenide elements on the electrical characteristics of GaAs MIS structures. *Solid-State Electronics*, *43*(3), 583–588. DOI: 10.1016/S0038-1101(98)00320-7

Ganguly, P., Mathew, S., Clarizia, L., Kumar, R. S., Akande, A., Hinder, S. J., Breen, A., & Pillai, S. C. (2020). Ternary Metal Chalcogenide Heterostructure (AgInS2-TiO2) Nanocomposites for Visible Light Photocatalytic Applications. *ACS Omega*, *5*(1), 406–421. DOI: 10.1021/ACSOMEGA.9B02907/ASSET/IMAGES/LARGE/AO9B02907_0007.JPEG PMID: 31956788

Ganguly, P., Mathew, S., Clarizia, L., Kumar, R. S., Akande, A., Hinder, S. J., Breen, A., Suresh, C., & Pillai, S. C. (2020). Ternary Metal Chalcogenide Heterostructure ($AgInS_2-TiO_2$) Nanocomposites for Visible Light Photocatalytic Applications. *ACS Omega*, *5*, 406–421.

Gao, M. R., Jiang, J., & Yu, S. H. (2012). Solution-based synthesis and design of late transition metal chalcogenide materials for oxygen reduction reaction. *Small*, *8*, 13–27.

Gao, M., Jiang, J., & Yu, S. (2012). Solution-based synthesis and design of late transition metal chalcogenide materials for oxygen reduction reaction (ORR). *Small*, *8*(13), 13–27. PMID: 21972127

Gao, M.-R., Xu, Y.-F., Jiang, J., & Yu, S.-H. (2013). Nanostructured metal chalcogenides: Synthesis, modification, and applications in energy conversion and storage devices. *Chemical Society Reviews*, *42*(7), 2986. DOI: 10.1039/c2cs35310e

Gao, M., Yu, L., Lv, Q., Kang, F., Huang, Z.-H., & Lv, R. (2023). Photoluminescence manipulation in two-dimensional transition metal dichalcogenides. *Journal of Materiomics*, *9*(4), 768–786. DOI: 10.1016/j.jmat.2023.02.005

Gas, K., Kret, S., Zaleszczyk, W., Kamińska, E., Sawicki, M., Wojtowicz, T., & Szuszkiewicz, W. (2021). Oxidation of MBE-Grown ZnTe and ZnTe/Zn Nanowires and Their Structural Properties. *Materials (Basel)*, *14*(18), 5252. DOI: 10.3390/ma14185252 PMID: 34576476

Gautam, J., Kannan, K., Meshesha, M. M., Dahal, B., Subedi, S., Ni, L., Wei, Y., & Yang, B. L. (2022). Heterostructure of polyoxometalate/zinc-iron-oxide nanoplates as an outstanding bifunctional electrocatalyst for the hydrogen and oxygen evolution reaction. *Journal of Colloid and Interface Science*, *618*, 419–430. DOI: 10.1016/j.jcis.2022.03.103 PMID: 35364543

Gautam, R., Kumar, A., & Singh, R. P. (2017). First Principle Investigations on Electronic, Magnetic, Thermodynamic, and Transport Properties of TlGdX 2 (X = S, Se, Te). *Acta Physica Polonica A*, *132*(4), 1371–1378. DOI: 10.12693/APhysPolA.132.1371

Geim, A. K., & Grigorieva, I. V. (2013). Van der Waals heterostructures. *Nature*, *499*, 419–425.

Ge, J., Chu, J., Yan, Y., Jiang, J., & Yang, P. (2015). Co-Electroplated kesterite bifacial thin-film solar cells: A study of sulfurization temperature. *ACS Applied Materials & Interfaces*, *7*, 10414–10428. DOI: 10.1021/acsami.5b01641

Gericke, M., & Pinches, A. (2006). Biological synthesis of metal nanoparticles. *Hydrometallurgy*, *83*(1–4), 132–140. DOI: 10.1016/j.hydromet.2006.03.019

Ge, X., Sumboja, A., Wuu, D., An, T., Li, B., Goh, F. W. T., Hor, T. S. A., Zong, Y., & Liu, Z. (2015a). Oxygen Reduction in Alkaline Media: From Mechanisms to Recent Advances of Catalysts. *ACS Catalysis*, *5*(8), 4643–4667. DOI: 10.1021/acscatal.5b00524

Ghandourah, E. I., Sangeetha, A., Shanmugan, S., Zayed, M. E., Moustafa, E. B., Tounsi, A., & Elsheikh, A. H. (2022). Performance assessment of a novel solar distiller with a double slope basin covered by coated wick with lanthanum cobalt oxide nanoparticles. *Case Studies in Thermal Engineering*, *32*, 101859.

Ghasemipour, P., Fattahi, M., Rasekh, B., & Yazdian, F. (2020). Developing the ternary ZnO doped MoS2 nanostructures grafted on CNT and reduced graphene oxide (RGO) for photocatalytic degradation of aniline. *Scientific Reports*, *10*(1), 4414.

Ghediya, P. R., Joshi, K. K., Kasela, P. A., Chaudhuri, T. K., & Kandoliya, M. (2019). Effect of solvents on physical properties of direct-coated Cu2CoSnS4 films. *Materials Research Express*, *6*, 106419. DOI: 10.1088/2053-1591/ab3ae5

Ghorui, U. K., Mondal, P., Adhikary, B., Mondal, A., & Sarkar, A. (2022). Newly Designed One-Pot In-Situ Synthesis of VS_2 /rGO Nanocomposite to Explore Its Electrochemical Behavior towards Oxygen Electrode Reactions**. *ChemElectroChem*, *9*(15), e202200526. DOI: 10.1002/celc.202200526

Ghosh, A. B., Saha, N., Sarkar, A., Srivastava, D. N., Paul, P., & Adhikary, B. (2015). Solvent assisted and solvent free orientation of growth of nanoscaled lanthanide sulfides: Tuning of morphology and manifestation of photocatalytic behaviour. *RSC Advances*. Advance online publication. DOI: 10.1039/C5RA19959J

Giri, A., Park, G., & Jeong, U. (2023). Layer-Structured Anisotropic Metal Chalcogenides: Recent Advances in Synthesis, Modulation, and Applications. *Chemical Reviews*, *123*(7), 3329–3442. DOI: 10.1021/acs.chemrev.2c00455 PMID: 36719999

Giroux, M. S., Zahra, Z., Salawu, O. A., Burgess, R. M., Ho, K. T., & Adeleye, A. S. (2022). Assessing the environmental effects related to quantum dot structure, function, synthesis and exposure. *Environmental Science. Nano*, *9*(3), 867–910. DOI: 10.1039/D1EN00712B

Giuffredi, G., Asset, T., Liu, Y., Atanassov, P., & Di Fonzo, F. (2021). Transition Metal Chalcogenides as a Versatile and Tunable Platform for Catalytic CO_2 and N_2 Electroreduction. *ACS Materials Au*, *1*(1), 6–36. DOI: 10.1021/acsmaterialsau.1c00006 PMID: 36855615

Gong, F., Xu, X., Li, Z., Zhou, G., & Wang, Z. S. (2013). $NiSe_2$ as an efficient electrocatalyst for a Pt-free counter electrode of dye-sensitized solar cells. *Chemical Communications (Cambridge)*, *49*(14), 1437–1439. DOI: 10.1039/C2CC38621F PMID: 23321537

Goud, K. G., Veldurthi, N. K., Vithal, M., & Reddy, G. (2016). Characterization and evaluation of biological and photocatalytic activities of selenium nanoparticles synthesized using yeast fermented broth. *Journal of Materials NanoScience*, *3*(2), 33–40.

Guan, J., Chen, W., Fang, Y., Wang, L., Fu, Y., Guo, B., & Zhang, M. (2022). Electrospinning derivative fabrication of sandwich-structured CNF/Co3S4/MoS2 as self-supported electrodes to accelerate electron transport in HER. *International Journal of Hydrogen Energy*, *47*(33), 14930–14941. DOI: 10.1016/j.ijhydene.2022.03.013

Guannan He, G., Yimin Zhang, Y., & Qinyu He, Q. (2019). MoS_2/CdS Heterostructure for Enhanced Photoelectrochemical Performance under Visible Light. *Catalysts*, *9*, 379.

Gul, M. M., Khuram Shahzad Ahmad,K.S.,(2022). Nanocomposite Zr2S3-BaS-Cr2S3 ternary-metal chalcogenide: An impressive supercapacitor electrode and environmental remediant of toxic pollutants. *International Journal of Energy Research*. Advance online publication. DOI: 10.1002/er.8489

Guo, M., Sun, X., Chen, J., & Cai, T. (2021). Pharmaceutical cocrystals: A review of preparations, physicochemical properties and applications. *Acta Pharmaceutica Sinica. B*, *11*(8), 2537–2564. DOI: 10.1016/j.apsb.2021.03.030 PMID: 34522597

Guo, S., Luo, H., Li, Y., Chen, J., Mou, B., Shi, X., & Sun, G. (2021). Structure-controlled threedimensional BiOI/MoS_2 microspheres for boosting visible-light photocatalytic degradation of tetracycline. *Journal of Alloys and Compounds*, *852*, 157026.

Guo, T., Li, L., & Wang, Z. (2022). Recent Development and Future Perspectives of Amorphous Transition Metal-Based Electrocatalysts for Oxygen Evolution Reaction. *Advanced Energy Materials*, *12*(24), 2200827. DOI: 10.1002/aenm.202200827

Guo, Y., Zhou, X., & Tang, J.. (2020). Multiscale structural optimization: Highly efficient hollow iron-doped metal sulfide heterostructures as bifunctional electrocatalysts for water splitting. *Nano Energy*, *75*, 104913. DOI: 10.1016/j.nanoen.2020.104913

Gupta, B., Hossain, M. A., Riaz, A., Sharma, A., Zhang, D., Tan, H. H., & Karuturi, S. (2022). Recent advances in materials design using atomic layer deposition for energy applications. *Advanced Functional Materials*, *32*(3), 2109105.

Gupta, V., Chandra, N., Tari, V., & Kannan, K. (2024). A sustainable approach for the green synthesis of silver nanoparticles using *Brassica oleracea* subsp. *botrytis* (L.) leaves and its antibacterial activity. *Physics and Chemistry of Solid State*, *25*(2), 303–310. DOI: 10.15330/pcss.25.2.303-310

Gurushankar, K., Karuppaiya, C., Tari, V., & Kannan, K. (2024). Exploring MXenes and MBenes in metal sulfur and selenium battery technologies. In Kannan, K., & Tari, V. (Eds.), *Novel energy storage and conversion technologies for two-dimensional MXenes and MBenes*. IGI Global., DOI: 10.4018/979-8-3693-9546-2.ch005

Gurushankar, K., Karuppaiya, C., Tari, V., & Kannan, K. (2025). Exploring MXenes and MBenes in Metal Sulfur and Selenium Battery Technologies. In *Novel Energy Storage and Conversion Technologies for Two-Dimensional MXenes and MBenes* (pp. 105–120). IGI Global Scientific Publishing.

Gutierrez, Z.-B. Zayas-Bazán, P., De Melo, O., De Moure-Flores, F., Andraca-Adame, J. A., Moreno-Ruiz, L. A., Martínez-Gutiérrez, H., Gallardo, S., Sastré-Hernández, J., & Contreras-Puente, G. (2018). CdS/CdTe Heterostructures for Applications in Ultra-Thin Solar Cells. *Materials (Basel)*, *11*(10). Advance online publication. DOI: 10.3390/ma11101788

Hagfeldt, A., & Grätzel, M. (1995). Light-induced redox reactions in nanocrystalline systems. *Chemical Reviews*, *95*(1), 49–68. DOI: 10.1021/cr00033a003

Hairan, R., & Azizi, A. T. (2023). Synthesis and X-ray Diffraction Study of Se79 Te20 Pb1 Chalcogen Glass. *Journal for Research in Applied Sciences and Biotechnology*, *2*(5), 89–92. DOI: 10.55544/JRASB.2.5.14

Hamam, K. A., Mohery, M., & Aly, K. A. (2025). New quaternary Ge3S6Ag1-xSnx ($0.0 \leq x \leq 1$) chalcogenide glasses for optical devices. *Ceramics International*, *51*(3), 2880–2890. DOI: 10.1016/J.CERAMINT.2024.11.265

Han, H., Zhao, L., Wu, X., Zuo, B., Bian, S., Li, T., Liu, X., Jiang, Y., Chen, C., Bi, J., Xu, J., & Yu, L. (2024). Advancements in thermoelectric materials: Optimization strategies for enhancing energy conversion. *Journal of Materials Chemistry. A, Materials for Energy and Sustainability*, *12*(36), 24041–24083. DOI: 10.1039/D4TA03666B

Han, J. H., Kwak, M., Kim, Y., & Cheon, J. (2018a). Recent Advances in the Solution-Based Preparation of Two-Dimensional Layered Transition Metal Chalcogenide Nanostructures. *Chemical Reviews*, *118*(13), 6151–6188. DOI: 10.1021/acs.chemrev.8b00264 PMID: 29926729

Hannappel, T., Shekarabi, S., Jaegermann, W., Runge, E., Hofmann, J. P., van de Krol, R., & Zhang, H. (2024). Integration of Multijunction Absorbers and Catalysts for Efficient Solar-Driven Artificial Leaf Structures: A Physical and Materials Science Perspective. *Solar RRL*, *8*(11), 2301047.

Han, T.-H., Jang, K. Y., Dong, Y., Friend, R. H., Sargent, E. H., & Lee, T.-W. (2022). A roadmap for the commercialization of perovskite light emitters. *Nature Reviews. Materials*, *7*(10), 757–777. DOI: 10.1038/s41578-022-00459-4

Hao, Y., Wang, L., & Huang, L.-F. (2023). Lanthanide-doped MoS2 with enhanced oxygen reduction activity and biperiodic chemical trends. *Nature Communications*, *14*(1), 3256. DOI: 10.1038/s41467-023-39100-5 PMID: 37277362

Haque, F., Daeneke, T., Kalantar-zadeh, K., & Ou, J. Z. (2017). Two-Dimensional Transition Metal Oxide and Chalcogenide-Based Photocatalysts. *Nano-Micro Letters*, *10*(2), 23. DOI: 10.1007/s40820-017-0176-y PMID: 30393672

Hashem, A. H., & Salem, S. S. (2021). Green and ecofriendly biosynthesis of selenium nanoparticles using *Urtica dioica* (stinging nettle) leaf extract: Antimicrobial and anticancer activity. *Biotechnology Journal*, *17*(2), 2100432. Advance online publication. DOI: 10.1002/biot.202100432 PMID: 34747563

Hassanien, R., Abed-Elmageed, A. A., & Husein, D. Z. (2019). Eco-friendly approach to synthesize selenium nanoparticles: Photocatalytic degradation of sunset Yellow azo dye and anticancer activity. *ChemistrySelect*, *4*(29), 9018–9026. DOI: 10.1002/slct.201902193

Hassan, T., Salam, A., Khan, A., Khan, S. U., Khanzada, H., Wasim, M., Khan, M. Q., & Kim, I. S. (2021). Functional nanocomposites and their potential applications: A review. *Journal of Polymer Research*, *28*(2), 36. DOI: 10.1007/s10965-021-02408-1

Ha, T. D. C., Lee, H., Vamvasakis, I., Armatas, G. S., Oh, Y., & Kim, M. G. (2023). Recent developments in porous metal chalcogenides for environmental remediation and sustainable energy. *EcoMat*, *5*(12), e12419. DOI: 10.1002/EOM2.12419;SUBPAGE:STRING:FULL

He, G., Zhang, Y., & He, Q. (2019). MoS_2/CdS heterostructure for enhanced photoelectrochemical performance under visible light. *Catalysts*, *9*, 379.

Hobson, A., & Maxwell, B. Kinga Káplár-Kodácsy, & Hotham, E. H. (2020, December). *The Nature and Impact of Effective Mentoring Training, Education and Development (MTED)*. The University of Brighton; Education and Training Foundation. https://research.brighton.ac.uk/en/publications/the-nature-and-impact-of-effective-mentoring-training-education-a

Holkar, C.R., Jadhav, A.J., Pinjari, D.V., & Naresh, M. Mahamuni, N.M., & B. Pandit, A.B. (2016). A critical review on textile wastewater treatments: Possible approaches. *Journal of Environmental Management*, *182*, 351–366.

Hong, J., Chen, X., Li, P., Koshino, M., Li, S., Xu, H., Hu, Z., Ding, F., & Suenaga, K. (2022). Multiple 2D Phase Transformations in Monolayer Transition Metal Chalcogenides. *Advanced Materials, 34*(19), 2200643. DOI: 10.1002/adma.202200643 PMID: 35307877

Hong, W. T., Risch, M., Stoerzinger, K. A., Grimaud, A., Suntivich, J., & Shao-Horn, Y. (2015). Toward the rational design of non-precious transition metal oxides for oxygen electrocatalysis. *Energy & Environmental Science, 8*, 1404–1427. DOI: 10.1039/c4ee03869j

Hossain, N., Mobarak, M. H., Mimona, M. A., Islam, M. A., Hossain, A., Zohura, F. T., & Chowdhury, M. A. (2023). Advances and significances of nanoparticles in semiconductor applications – A review. *Results in Engineering, 19*, 101347. DOI: 10.1016/J.RINENG.2023.101347

Hsu, Y. K., Mondal, A., Su, Y.-Z., Sofer, Z., Anuratha, K. S., & Lin, J. Y. (2022). Highly hydrophilic electrodeposited NiS/Ni_3S_2 interlaced nanosheets with surface-enriched Ni^{3+} sites as binder-free flexible cathodes for high-rate hybrid supercapacitors. *Applied Surface Science, 579*, 151923. DOI: 10.1016/j.apsusc.2021.151923

Huang, H. H., Fan, X., Singh, D. J., & Zheng, W. T. (2020). Recent progress of TMD nanomaterials: Phase transitions and applications. *Nanoscale, 12*(3), 1247–1268.

Huang, S., Yue, C., Uvdal, K., & Hu, Z. (2025). Recent advances in irradiation-mediated synthesis and tailoring of inorganic nanomaterials for photo-/electrocatalysis. *Nanoscale Advances.*

Huang, W., Luo, X., Gan, C. K., Quek, S. Y., & Liang, G. (2014). Theoretical study of thermoelectric properties of few-layer MoS2 and WSe2. *Physical Chemistry Chemical Physics, 16*(22), 10866. DOI: 10.1039/c4cp00487f PMID: 24760342

Huang, X. H., Tu, J. P., Xia, X. H., Wang, X. L., Xiang, J. Y., Zhang, L., & Zhou, Y. (2009). Morphology effect on the electrochemical performance of NiO films as anodes for lithium ion batteries. *Journal of Power Sources, 188*(2), 588–591. DOI: 10.1016/j.jpowsour.2008.11.111

Huang, Y. L., Chen, W., & Wee, A. T. S. (2021). Two-dimensional magnetic transition metal chalcogenides. *SmartMat, 2*(2), 139–153. DOI: 10.1002/smm2.1031

Huang, Y., Zhu, M., Huang, Y., Pei, Z., Li, H., Wang, Z., Xue, Q., & Zhi, C. (2016). Multifunctional Energy Storage and Conversion Devices. *Advanced Materials, 28*(38), 8344–8364. DOI: 10.1002/adma.201601928

Huang, Z., Staufenbiel, S., & Bodmeier, R. (2022). Combination of co-crystal and nanocrystal techniques to improve the solubility and dissolution rate of poorly soluble drugs. *Pharmaceutical Research*, *39*(5), 949. DOI: 10.1007/s11095-022-03243-9 PMID: 35552985

Hua, Q., & Shen, G. (2024). Low-dimensional nanostructures for monolithic 3D-integrated flexible and stretchable electronics. *Chemical Society Reviews*, *53*(3), 1316–1353. DOI: 10.1039/D3CS00918A PMID: 38196334

Hu, Q., Zhang, S., Zou, X., Hao, J., Bai, Y., Yan, L., & Li, W. (2021). Coordination agent-dominated phase control of nickel sulfide for high-performance hybrid supercapacitor. *Journal of Colloid and Interface Science*, *607*, 45–52. DOI: 10.1016/j.jcis.2021.08.185 PMID: 34492352

Hu, S., Ren, Z., Djurišić, A. B., & Rogach, A. L. (2021). Metal Halide Perovskites as Emerging Thermoelectric Materials. *ACS Energy Letters*, *6*(11), 3882–3905. DOI: 10.1021/acsenergylett.1c02015

Hu, S., Wang, S., & Feng, C.. (2020). Novel MOF-derived nickel nitride as high-performance bifunctional electrocatalysts for hydrogen evolution and urea oxidation. *ACS Sustainable Chemistry & Engineering*, *8*, 7414–7422. DOI: 10.1021/acssuschemeng.0c01450

Hu, W.-H., Shang, X., Han, G.-Q., Dong, B., Liu, Y.-R., & Li, X.. (2016). MoSx supported graphene oxides with different degree of oxidation as efficient electrocatalysts for hydrogen evolution. *Carbon*, *100*, 236–242. DOI: 10.1016/j.carbon.2016.01.019

Hu, W., Quang, N. D., Majumder, S., Jeong, M. J., Park, J. H., Cho, Y. J., & Chang, H. S. (2021). Three-dimensional nanoporous SnO2/CdS heterojunction for high-performance photoelectrochemical water splitting. *Applied Surface Science*, *560*, 149904.

Huynh, L., Leroux, J. C., & Allen, C. (2009). Enhancement of docetaxel solubility via conjugation of formulation-compatible moieties. *Organic & Biomolecular Chemistry*, *7*(17), 3437–3446. DOI: 10.1039/b906862g PMID: 19675898

Hwang, Y., & Shin, N. (2021). Colloidal Synthesis of $MoSe_2$/WSe_2 Heterostructure Nanoflowers via Two-Step Growth. *Materials (Basel)*, *14*(23), 7294. DOI: 10.3390/ma14237294 PMID: 34885449

Imran, M., Abdullah, A. Z., Khan, M. E., & Mohammad, A. (2025). A focused review on photocatalytic potential of graphitic carbon nitride (g-C3N4) based metal oxide-nanostructures for effective remediation of most overused antibiotics. *Journal of Environmental Management*, *373*, 123759. DOI: 10.1016/j.jenvman.2024.123759 PMID: 39708681

Iorember, P. T., Hounkanrin, D. O., Diyoke, K., & Tang, C. F. (2025). Tracing financial inclusion pathway to sustainable development: The criticality of population growth and energy intensity using a machine learning approach. *African Journal of Economic and Management Studies*, *16*(2), 241–259.

Isaev, A. B., Shabanov, N. S., & Sobola, D.. (2022). ZnO/Chalcogenides semiconductor heterostructures for photoelectrochemical water splitting. In Kasinathan, K., Elshikh, M. S., & Al Farraj, D. A.-A. (Eds.), *Nanomaterials for energy conversion, biomedical and environmental applications* (pp. 3–35).

Islam, S., Weerasinghe, H., Prado, D. M., Gonzaga, A. N., & Burda, C. (2025). Diversifying the Materials and Technologies for the Future of Energy Storage. *Energy & Fuels*.

Island, J. O., Biele, R., Barawi, M., Clamagirand, J. M., Ares, J. R., Sánchez, C., Herre, S., van der Zant, J., Ferrer, I. J., D'Agosta, R., & Castellanos-Gomez, A. (2016). Titanium trisulfide (TiS_3): A 2D semiconductor with quasi-1D optical and electronic properties. *Scientific Reports*, *6*(1), 22214.

Jabeen, N., Aslam, S., Awais, M., Mirza, M., & Safdar, M. (2023). Construction of quaternary chalcogenide Ag2BaSnS4 nanograins for HER/OER performances and supercapacitor applications in alkaline media. *ACS Applied Electronic Materials*, *5*, 6531–6541. DOI: 10.1021/acsaelm.3c01030

Jacob, J. M., Lens, P. N. L., & Balakrishnan, R. M. (2016). Microbial synthesis of chalcogenide semiconductor nanoparticles: A review. *Microbial Biotechnology*, *9*(1), 11–21. DOI: 10.1111/1751-7915.12297 PMID: 26110980

Jadoun, S., Arif, R., Jangid, N. K., & Meena, R. K. (2020). Green synthesis of nanoparticles using plant extracts: A review. *Environmental Chemistry Letters*, *19*(1), 355–374. DOI: 10.1007/s10311-020-01074-x

Jana, S., Karingula, S., & Sajeevan, A., V. V. N., P. K., & Kotagiri, Y. G. (2025). *In situ* synthesis of bimetallic chalcogenides with highly conductive carbon nanotubes for efficient symmetric hybrid supercapacitors. *Nanoscale*, *17*(21), 13283–13297. DOI: 10.1039/D5NR00340G PMID: 40293371

Jancik-Prochazkova, A., & Pumera, M. (2023). Light-powered swarming phoretic antimony chalcogenide-based microrobots with "on-the-fly" photodegradation abilities. *Nanoscale*, *15*(12), 5726–5734. DOI: 10.1039/D3NR00098B PMID: 36866684

Jang, S.-M., & Kim, I.-H. (2025). Solid-State Synthesis and Thermoelectric Properties of $CuFeSe_2$–$CuFeS_2$ Solid Solutions. *Materials (Basel)*, *18*(6), 1366. DOI: 10.3390/ma18061366 PMID: 40141648

Jang, S., Moon, K., Park, Y., Park, S., & Park, K. H. (2020). Recent studies on multifunctional electrocatalysts for fuel cell by various nanomaterials. *Catalysts*, *10*, 621. DOI: 10.3390/catal10060621

Jariwala, B., & Ravindra, N. M. (2014). Process, property and performance of chalcogenide-based thermoelectric materials. *Nanomaterials and Energy*, *3*(3), 68–81. DOI: 10.1680/nme.14.00006

Jayawardane, J. T. S. T., Akmal, U. K. D. M., Hu, D., Abeygunawardhana, P. K. W., & Sewvandi, G. A. (2025). Eco-friendly bismuth halide chalcogenide perovskites for solar energy harvesting. *Sustainable Energy & Fuels*. Advance online publication. DOI: 10.1039/D4SE01523A

Jhonsi, M. A., & Kathiravan, A. (2017). Photoinduced interaction of arylamine dye with carbon quantum dots ensued from *Centella asiatica*. *Journal of Luminescence*, *192*, 321–327. DOI: 10.1016/j.jlumin.2017.06.056

Jiabin, L., Shuai, Y., Hui, Z., Chao, W., David, B., François, V., Shuhui, S., Zhiming, M. W., Jiming, B., Haiguang, Z., & Gurpreet, S. S. (2023). Federico Rosei, Efficient Photoelectrochemical Hydrogen Generation Using Eco-Friendly "Giant" InP/ZnSe Core/Shell Quantum Dots. *ACS Applied Materials & Interfaces*, *15*(29), 34797–34808. DOI: 10.1021/acsami.3c04900 PMID: 37433096

Jia, H., Peng, L., Yu, C., Dong, L., Cheng, S., & Xie, J. (2021). Chalcogenide-based inorganic sodium solid electrolytes. *Journal of Materials Chemistry. A, Materials for Energy and Sustainability*, *9*(9), 5134–5148. DOI: 10.1039/D0TA10994K

Jia, J., Wu, J., Tu, Y., Huo, J., Zheng, M., & Lin, J. (2015). Transparent nickel selenide used as counter electrode in high efficient dye-sensitized solar cells. *Journal of Alloys and Compounds*, *640*, 29–33. DOI: 10.1016/j.jallcom.2015.03.233

Jiang, J., Wang, G., Shao, Y., Wang, J., Zhou, S., & Su, Y. (2022). Step-scheme ZnO@ZnS hollow microspheres for improved photocatalytic H2 production performance. *Chinese Journal of Catalysis*, *43*(2), 329–338. DOI: 10.1016/S1872-2067(21)63889-5

Jian, J., Kumar, R., & Sun, J. (2020). Cu2O/ZnO p–n junction decorated with NiOx as a protective layer and cocatalyst for enhanced photoelectrochemical water splitting. *ACS Applied Energy Materials*, *3*, 10408–10414. DOI: 10.1021/acsaem.0c01198

Jin, C., Zhou, L., Fu, L., Zhu, J., Li, D., & Yang, W. (2017). The acceleration intermediate phase (NiS and Ni 3 S 2) evolution by nanocrystallization in Li/NiS 2 thermal batteries with high specific capacity. *Journal of Power Sources*, *352*, 83–89. DOI: 10.1016/j.jpowsour.2017.03.119

Jingyun, C., Zhulin, N., Zhipeng, Z., Xiangdong, P., Shuo, Z., Hongqiang, W., Dan, L., & Zaiping, G. (2023). Enhanced Ion/Electron Migration and Sodium Storage Driven by Different MoS_2-$ZnIn_2S_4$ Heterointerfaces. *Advanced Energy Materials, 13*,5, 2023, 2203248. DOI: 10.1002/aenm.202203248

Joo, J., Kim, T., Lee, J., Choi, S., & Lee, K. (2019). Morphology-Controlled Metal Sulfides and Phosphides for Electrochemical Water Splitting. *Advanced Materials*, *31*(14), 1806682. DOI: 10.1002/adma.201806682 PMID: 30706578

Joshi, K. K., Pataniya, P. M., Bhadu, G., & Sumesh, C. K. (2023). Monometallic, bimetallic, and trimetallic chalcogenide-based electrodes for electrocatalytic hydrogen evolution reaction. *International Journal of Hydrogen Energy*, *48*(20), 7260–7272. DOI: 10.1016/j.ijhydene.2022.11.088

Junjie, H., Haochen, L., Jun, M., Rui, L., Ziming, Z., Bingxin, Z., Bin, X., Jiaji, C., Kai, W., & Marie-Helene, D. (2019). A Facile Route to Synthesize CdSe/ZnS Thick-shell Quantum Dots with Precisely Controlled Green Emission Properties: Towards QDs Based LED Applications. *Scientific Reports*, *9*, 12048. DOI: 10.1038/s41598-019-48469-7 PMID: 31427624

Ju, Z., Lv, R., Ansari, A. A., & Lin, J. (2025). Recent advances in additive manufacturing for solar cell based on organic/inorganic hybrid materials. *InfoMat*, *7*(7), e70017. DOI: 10.1002/inf2.70017

Kalaiselvi, C., Kannan, K., Pandian, M. S., Devi, N., Arpornwichanop, A., Chandar, N. K., & Chen, Y.-S. (2025). Ti3C2Tx MXene decorated with CoMnO3 / CoMn2O4 nanoparticles as electrocatalysts for anion exchange membrane water electrolysis. *International Journal of Hydrogen Energy*, *137*, 202–213. DOI: 10.1016/j.ijhydene.2025.04.402

Kalmi, P., Trotta, G., & Kaukauskas, A. (2021). Energy-related financial literacy and electricity consumption: Survey-based evidence from Finland. *The Journal of Consumer Affairs*, *55*(6), 1698–1729.

Kamaraki, C., Klug, M. T., Green, T., Miranda Perez, L., & Case, C. (2021). Perovskite/silicon tandem photovoltaics: Technological disruption without business disruption. *Applied Physics Letters*, *119*(7), 070501. DOI: 10.1063/5.0054086

Kamran, M. A., Raza, M., Ullah, S., Alarifi, S. S., & Aziz, M. H. (2024). Facile synthesis of chromium doped cadmium sulfide/zinc telluride nanocomposites for enhanced electrochemical energy storage and photocatalytic applications. *Materials Chemistry and Physics*, *322*, 129531.

Kannan, K., & Tari, V. (Eds.). (2025). Novel energy storage and conversion technologies for two-dimensional MXenes and MBenes. https://doi.org/ ISBN: 9798369395462, eISBN: 9798369395486DOI: 10.4018/979-8-3693-9546-2

Kannan, K., & Tari, V. (Eds.). (2025). *Novel energy storage and conversion technologies for two-dimensional MXenes and MBenes*. https://www.amazon.com/dp/9798369395462

Kannan, K., & Tari, V. (Eds.). (2025). Novel energy storage and conversion technologies for two-dimensional MXenes and MBenes. https://www.amazon.com/dp/9798369395462 ISBN: 9798369395462, eISBN: 9798369395486

Kannan, K., Chanda, D., Gautam, J., Behera, A., Meshesha, M. M., Jang, S. G., & Yang, B. (2023). Hydrothermally synthesized mixed metal oxide nanocomposites for electrochemical water splitting and photocatalytic hydrogen production. *International Journal of Hydrogen Energy*, *48*(93), 36412–36426.

Kannan, K., Chanda, D., Gautam, J., Meshesha, M. M., Jang, S. G., & Yang, B. L. (2023). Facial synthesis of pp heterojunction composites: Evaluation of their electrochemical properties with photovoltaics-electrolyzer water splitting using two-electrode system. *International Journal of Hydrogen Energy*, *48*(37), 13814–13826.

Kannan, K., Chanda, D., Gautam, J., Meshesha, M. M., Jang, S. G., & Yang, B. L. (2023). Facial synthesis of p-p heterojunction composites: Evaluation of their electrochemical properties with photovoltaics-electrolyzer water splitting using two-electrode system. *International Journal of Hydrogen Energy*, *48*(37), 13814–13826. DOI: 10.1016/j.ijhydene.2022.12.235

Kannan, K., Chanda, D., Meshesha, M. M., & Yang, B. (2024). Impressive efficiency of zinc oxide-manganese oxide/MAX composite in two-electrode system for photovoltaic-electrolyzer water splitting. *Colloids and Surfaces. A, Physicochemical and Engineering Aspects*, *689*, 133599. DOI: 10.1016/j.colsurfa.2024.133599

Kannan, K., Gautam, J., Chanda, D., Meshesha, M. M., Jang, S. G., & Yang, B. L. (2023). Two-dimensional MAX supported copper oxide/nickel Oxide/MAX as an efficient and novel photocatalyst for hydrogen evolution. *International Journal of Hydrogen Energy*, *48*(20), 7273–7283.

Kannan, K., Radhika, D., Gnanasangeetha, D., Lakkaboyana, S. K., Sadasivuni, K. K., Gurushankar, K., & Hanafiah, M. M. (2021). Photocatalytic and antimicrobial properties of microwave synthesized mixed metal oxide nanocomposite. *Inorganic Chemistry Communications*, *125*, 108429.

Kannan, K., Sliem, M. H., Abdullah, A. M., Sadasivuni, K. K., & Kumar, B. (2020). Fabrication of ZnO-Fe-MXene based nanocomposites for efficient CO2 reduction. *Catalysts*, *10*(5), 549.

Karade, V. C., He, M., Abudulimu, A., Song, Z., Park, Y., Song, D., & Suryawanshi, M. P. (2025). Opportunities and Challenges for Emerging Inorganic Chalcogenide-Silicon Tandem Solar Cells. *Energy & Environmental Science*.

Karimi, M., Wu, D., Wang, Z., & Shen, Y. (2019). DeepAffinity: Interpretable deep learning of compound–protein affinity through unified recurrent and convolutional neural networks. *Bioinformatics (Oxford, England)*, *35*(18), 3329–3338. DOI: 10.1093/bioinformatics/btz111 PMID: 30768156

Karol, V., Sharma, P., Singh, A., Goel, D., & Kaur, S. (2024a). Review of Energy Storage Devices: Fuel Cells, Hydrogen Storage Fuel Cells, Rechargeable Batteries, PV Solar Cells. *ACS Symposium Series, 1488*, 1–15. https://doi.org/DOI: 10.1021/BK-2024-1488.CH001/ASSET/IMAGES/LARGE/BK-2023-00657M_G005.JPEG

Karumanchi, L. (n.d.). The Role of Green Financing in Accelerating Sustainable Economic Development in Emerging Economies. *International Journal of Technology and Business Management, 14*(3).

Katsnelson, M. I. (2007). Graphene: Carbon in two dimensions. *Materials Today*, *10*(1-2), 20–27.

Kaur, G., Vij, A., & Kumar, A. (2022). Recent Advances in Earth Abundant and Environmentally Green Semiconducting Chalcogenide Nanomaterials for Photovoltaics Applications. *Advanced Nanomaterials*, 21-50.

Kaur, K. Nisika, & Kumar, M. (2022). Flexible Solar Cells Based on Chalcogenides. In *Smart and Flexible Energy Devices*. CRC Press.

Kaur, M., Selvakumar, P., & Dinesh, N. Mohit, Chandel, P. S., & Manjunath, T. C. (2025). Reconstruction and Preservation Strategies for Cultural Heritage. In S. Oflazoğlu & M. Dora (Eds.), Preserving Cultural Heritage in Post-Disaster Urban Renewal (pp. 385-412). IGI Global Scientific Publishing. DOI: 10.4018/979-8-3693-9097-9.ch016

Kaur, A., Bajaj, B., Kaushik, A., Saini, A., & Sud, D. (2022). A review on template assisted synthesis of multi-functional metal oxide nanostructures: Status and prospects. *Materials Science and Engineering B*, *286*. Advance online publication. DOI: 10.1016/J.MSEB.2022.116005

Kaur, M., & Dhiman, B. A. (2024). Framework for Exploring the Integration of AI and Financial Literacy toward a Sustainable Financial Ecosystem. *Educational Administration: Theory and Practice*, *30*(4), 8750–8759.

Kaysar Hamid, K., & Mukhlish, M,Z.B., & Uddin, Md.T. (2024). Sunlight-activated heterostructure MoS2/CdS nanocomposite photocatalyst with enhanced photocatalytic activity: Band alignment and mechanism study. *RSC Advances*, *14*, 38908–38923.

Kebede, W. L., Hau, K. D., & Xiaoyun, C. (2024). Organic pollutant degradation over highly efficient Mn–Mo (O, S)$_2$ oxysulfide photocatalyst under visible light illumination. *Optical Materials*, *155*, 115828.

Kebede, W. L., Kuo, D. H., Ahmed, K. E., & Abdullah, H. (2020). Dye degradation over the multivalent charge-and solid solution-type n-MoS2/p-WO3 based diode catalyst under dark condition with a self-supporting charge carrier transfer mechanism. *Advanced Powder Technology*, *31*(7), 2629–2640.

Khan, A., Danish, M., Alam, U., Zafar, S., & Muneer, M. (2020). Facile synthesis of a Z-scheme ZnIn2S4/MoO3 heterojunction with enhanced photocatalytic activity under visible light irradiation. *ACS Omega*, *5*, 8188–8199.

Khan, B. A., Rafaqat Hussain, A., Shah, A. M., Shah, M. Z. U., Ismail, J., Rahman, S., Sajjad, M., Assiri, M. A., Imran, M., & Javed, M. S. (2022). NiSe$_2$ nanocrystals intercalated rGO sheets as a high-performance asymmetric supercapacitor electrode. *Ceramics International*, *48*(4), 5509–5517. DOI: 10.1016/j.ceramint.2021.11.095

Khan, F. S. A., Mubarak, N. M., Tan, Y. H., Karri, R. R., Khalid, M., Walvekar, R., Abdullah, E. C., Mazari, S. A., & Nizamuddin, S. (2020). Magnetic nanoparticles incorporation into different substrates for dyes and heavy metals removal—A Review. *Environmental Science and Pollution Research International*, *27*(35), 43526–43541.

Khan, M. M. (2021). *Introduction and fundamentals of chalcogenides and chalcogenides-based nanomaterials*. Elsevier., DOI: 10.1016/B978-0-12-820498-6.00001-9

Khan, S., Mahmood, A., Shah, A. U. H. A., Rahman, G., Khan, A., & Ullah, N. (2022). Challenges and innovative strategies related to synthesis and electrocatalytic/energy storage applications of metal sulfides and its derivatives. *Journal of Electroanalytical Chemistry (Lausanne, Switzerland)*, *923*, 116760.

Kharboot, L. H., Fadil, N. A., Bakar, T. A. A., Najib, A. S. M., Nordin, N. H., & Ghazali, H. (2023). A review of transition metal sulfides as counter electrodes for dye-sensitized and quantum dot-sensitized solar cells. *Materials (Basel)*, *16*(7), 2881.

Khatavkar, S. N., Khatavkar, S. N., & Sartale, S. D. (2023). Chemically Deposited Iron Chalcogenide-Based Carbon Composites for Supercapacitor Applications. In Ezema, F. I., Lokhande, C. D., & Lokhande, A. C. (Eds.), *Chemically Deposited Metal Chalcogenide-based Carbon Composites for Versatile Applications* (pp. 83–121). Springer International Publishing., DOI: 10.1007/978-3-031-23401-9_3

Khodamorady, M., & Bahrami, K. (2023). A Novel ZnS-CdS Nanocomposite as a Visible Active Photocatalyst for Degradation of Synthetic and Real Wastewaters. *Scientific Reports*, *13*, 2177. DOI: 10.1038/s41598-023-28725-7 PMID: 36750747

Khouri, M., Escalier, R., Heran, D., Myara, M., Kribich, R., Vigreux, C., & Bendoula, R. (2024). Simulating and Fabricating Chalcogenide-based Waveguides for Agro-Environmental Applications. *2024 24th International Conference on Transparent Optical Networks (ICTON)*, 1–4. DOI: 10.1109/ICTON62926.2024.10647576

Kim, Y., Woo, W. J., Kim, D., Lee, S., Chung, S., Park, J., & Kim, H. (2021). Atomic-Layer-Deposition-Based 2D Transition Metal Chalcogenides: Synthesis, Modulation, and Applications. *Advanced Materials*, *33*(47), 2005907. DOI: 10.1002/adma.202005907 PMID: 33749055

Kinoshita, S. (2025). An empirical analysis of the impact of energy literacy on individual purchase behavior of energy-efficient homes. *International Journal of Energy Sector Management*, *19*(3), 661–677.

Kiran, R., Pashov, D., van Schilfgaarde, M., Katsnelson, M. I., Taraphder, A., & Acharya, S. (2025). Possible Realization of Hyperbolic Plasmons in Few-Layered Rhenium Disulfide. *Condensed Matter*, *10*, 40.

Koley, S. (2023). Intercalation in 2H-$TaSe_2$ for modulation of electronic properties and electrochemical energy storage. *Physica B, Condensed Matter*, *669*, 415312.

Kommineni, M., Chundru, S., Maroju, P. K., & Selvakumar, P. (2025). Ethical Implications of AI in Sustainable Development Pedagogy. In Chinedu, C., Mafarja, N., Hussein, M., & Saleem, A. (Eds.), *Rethinking the Pedagogy of Sustainable Development in the AI Era* (pp. 17–36). IGI Global Scientific Publishing., DOI: 10.4018/979-8-3693-9062-7.ch002

Korir, B. K., Kibet, J. K., & Ngari, S. M. (2024). A review on the current status of dye-sensitized solar cells: Toward sustainable energy. *Energy Science & Engineering*, *12*(8), 3188–3226. DOI: 10.1002/ese3.1815

Kotla, D. P., Anna, V. R., Praveenkumar, S., Saleh, S. M., & Shanmugan, S. (2025). Optimizing solar still performance: A study of TiO2 nanofluid derived from Saccharum officinarum L. *Separation and Purification Technology*, *359*, 130584.

Kottayi, R., Maurya, D. K., Sittaramane, R., & Angaiah, S. (2022). Recent developments in metal chalcogenides-based quantum dot sensitized solar cells. *ES Energy & Environment.*, *18*, 1–40.

Kräuter, G., & Rees, W. S.Jr. (1995). Cadmium bis(alkylthiolate) complexes as precursors for cadmium sulfide: A mild route to hawleyite. *Journal of Materials Chemistry*, *5*(8), 1265–1267. DOI: 10.1039/JM9950501265

Kujur, V. S., Wadhwa, G., Singh, K., Aslam, S., & Kumar, R. (2023). Advanced of Chalcogenides Based as Hazardous Gas Sensing. In Sonker, R. K., Singh, K., & Sonkawade, R. (Eds.), *Advanced Functional Materials for Optical and Hazardous Sensing: Synthesis and Applications* (pp. 47–67). Springer Nature., DOI: 10.1007/978-981-99-6014-9_3

Kuldeep, K., Munish, S., & Ravindra, P. (2025). Tunability in Electronic and Optical Properties of GaS/PbS vdW Heterostructure. *Journal of Physics and Chemistry of Solids*, *196*, 112348. DOI: 10.1016/j.jpcs.2024.112348

Kumar, B., Malali, P., Sadasivuni, K. K., & Kannan, K. (2022). Two-dimensional Nanomaterials Design and Reactor Engineering of Different Methods for CO2 Electrochemical Conversion Process.

Kumar, S., Muthu, S., Sekar, S., Bathula, C., Kaliamurthy, A. K., & Lee, S. (2022). Metal chalcogenide-based counter electrodes for dye-sensitized solar cells. *Oxide Free Nanomaterials for Energy Storage and Conversion Applications*, 259-286. https://doi.org/DOI: 10.1016/B978-0-12-823936-0.00015-2

Kumar, A., Chowdhuri, A. R., Laha, D., Mahto, T. K., Karmakar, P., & Sahu, S. K. (2017). Green synthesis of carbon dots from *Ocimum sanctum* for effective fluorescent sensing of Pb^{2+} ions and live cell imaging. *Sensors and Actuators. B, Chemical, 242*, 679–686. DOI: 10.1016/j.snb.2016.11.109

Kumar, A., Guatam, R., Chand, S., Kumar, A., & Singh, R. P. (2019). First principle electronic, magnetic and thermodynamic characterization of heavy fermion ternary rare earth metal alloys. *Materials Physics and Mechanics, 42*(1), 112–130. https://doi.org/DOI: 10.18720/MPM.4212019_10

Kumar, A., Kumar, A., Jain, P., Pundir, S. K., & Singh, N. (2024). TB-mBJ for doping concentration effects on magneto-optical properties in spintronics materials. *Optik (Stuttgart), 315*, 172039. DOI: 10.1016/j.ijleo.2024.172039

Kumar, A., Kumar, A., Kumar, K., Singh, R. P., Singh, R., & Kumar, R. (2023). The Electronic and Thermodynamic Properties of Ternary Rare Earth Metal Alloys. *East European Journal of Physics, 1*, 109–117. DOI: 10.26565/2312-4334-2023-1-13

Kumar, A., Kumar, R., Saxena, D., Nautiyal, V. K., Kumar, A., & Iram, N. (2024). Structural, electronic, magnetic and optical properties of GdCuX2 (X = S, Se and Te) compounds. *Optical and Quantum Electronics, 56*(10), 1742. DOI: 10.1007/s11082-024-07298-3

Kumaravelan, S., Seshadri, S., Suresh, R., Ravichandran, K., Sathishkumar, P., Shanthaseelan, K., & Suganthi, N. (2021). Effect of Zn dopant on SnO_2 nano-pyramids for photocatalytic degradation. *Chemical Physics Letters, 769*, 138352.

Kumar, B., & Debut, A. (2022). *Green Chemistry: New Perspectives*. BoD – Books on Demand.

Kumar, K., Jamnuch, S., Majidi, L., Misal, S., Ahmadiparidari, A., Dato, M. A., Sterbinsky, G. E., Wu, T., Salehi-Khojin, A., Pascal, T. A., & Cabana, J. (2023). Active states during the reduction of CO_2 by a MoS_2 electrocatalyst. *The Journal of Physical Chemistry Letters, 14*(13), 3222–3229. DOI: 10.1021/acs.jpclett.2c03942 PMID: 36972067

Kumar, P., Gupta, M., & Singh, K. (2020). Performance Evaluation of Transition Metal Dichalcogenides Based Steep Subthreshold Slope Tunnel Field Effect Transistor. *Silicon, 12*(8), 1857–1864. DOI: 10.1007/s12633-019-00285-4

Kumar, Ravinder, Chanda, Jayanta, Elsheikh, Ammar H., Ongar, Bulbul, Khidolda, Yerkin, Kumar, Seepana Praveen, & Panchal, Hitesh, & Shanmugan, S. (2023). Performance improvement of single and double effect solar stills with silver balls/nanofluids for bioactivation: An experimental analysis. *Solar Energy*, *259*, 452–463. DOI: 10.1016/j.solener.2023.05.012

Kumar, V. P., Raut, P., Panda, D. K., & Rashed, A. N. Z. (2025). Exploring Next-Generation TMDC Materials: A Comprehensive Review of Their Classifications, Properties, and Applications. *Silicon*, *17*(6), 1199–1221. DOI: 10.1007/s12633-025-03274-y

Kumeda, T., Laverdure, L., Honkala, K., Melander, M. M., & Sakaushi, K. (2023). Cations Determine the Mechanism and Selectivity of Alkaline Oxygen Reduction Reaction on Pt(111)**. *Angewandte Chemie*, *135*(51), e202312841. DOI: 10.1002/ange.202312841 PMID: 37983729

Kwon, I. S., Debela, T. T., Kwak, I. H., Park, Y. C., Seo, J., Shim, J. Y., Yoo, S. J., Kim, J., Park, J., & Kang, H. S. (2020a). Ruthenium Nanoparticles on Cobalt-Doped 1T′ Phase MoS_2 Nanosheets for Overall Water Splitting. *Small*, *16*(13), 2000081. DOI: 10.1002/smll.202000081 PMID: 32147958

Lahariya, V., Pandey, P. P., & Ramrakhiani, M. (2024). II–VI semiconductor metal chalcogenide nanomaterials and polymer composites: fundamentals, properties, and applications. In *Metal-Chalcogenide Nanocomposites* (pp. 187–220). Woodhead Publishing., DOI: 10.1016/B978-0-443-18809-1.00011-0

Lai, X., Li, W., Gu, X., Chen, H., Zhang, Y., Li, G., Zhang, R., Fan, D., He, F., Zheng, N., Yu, J., Chen, R., Kyaw, A. K. K., & Sun, X. W. (2022). High-performance quasi-2D perovskite solar cells with power conversion efficiency over 20% fabricated in humidity-controlled ambient air. *Chemical Engineering Journal*, *427*, 130949. DOI: 10.1016/j.cej.2021.130949

Lamiel, C., Hussain, I., Rabiee, H., Ogunsakin, O. R., & Zhang, K. (2023). Metal-organic framework-derived transition metal chalcogenides (S, Se, and Te): Challenges, recent progress, and future directions in electrochemical energy storage and conversion systems. *Coordination Chemistry Reviews*, *480*, 215030. DOI: 10.1016/j.ccr.2023.215030

Lana, G. M., Bello, I. T., Adedokun, O. M., Adenigba, V. O., Jubu, P. R., Adedokun, O., & Awodugba, A. O. (2024). One-Dimensional TiO2 Nanocomposite-based Photoanode for Dye-Sensitized solar Cells: A review. *Solar Energy*, *279*, 112850. DOI: 10.1016/j.solener.2024.112850

Larquet, C., Nguyen, A. M., Glais, E., Paulatto, L., Sassoye, C., Selmane, M., Lecante, P., Maheu, C., Geantet, C., Cardenas, L., Chanéac, C., Gauzzi, A., Sanchez, C., & Carenco, S. (2019). Bandgap Engineering from Cation Balance: The Case of Lanthanide Oxysulfide Nanoparticles. *Chemistry of Materials*. Advance online publication. DOI: 10.1021/acs.chemmater.9b00450

Le Bahers, T., & Takanabe, K. (2019). Combined theoretical and experimental characterizations of semiconductors for photo electrocatalytic applications. *Journal of Photochemistry and Photobiology A Chemistry*, *40*, 212–233. DOI: 10.1016/j.jphotochemrev.2019.01.001

Leal, J. F., Cruz, S. M. A., & Almeida, B. T. A.. (2020). TiO2–rGO nanocomposite as an efficient catalyst to photodegrade formalin in aquaculture waters, under solar light. *Environmental Sciences (Ruse)*, *6*, 1018–1027. DOI: 10.1039/C9EW00950G

Liang, T., Liu, Y., Cheng, Y., Ma, F., & Dai, Z. (2020). Scalable Synthesis of a MoS_2 /Black Phosphorus Heterostructure for pH-Universal Hydrogen Evolution Catalysis. *ChemCatChem*, *12*(10), 2840–2848. DOI: 10.1002/cctc.202000139

Li, B., Liu, X. J., Zhu, H. W., Guan, H. P., & Guo, R. T. (2025). Recent Progress of Round-the-Clock Photocatalytic System in Environmental and Energy Applications: A Review. *ChemPhysChem*, *202401144*. Advance online publication. DOI: 10.1002/cphc.202401144

Li, C., Cao, Q., Wang, F., Xiao, Y., Li, Y., Delaunay, J. J., & Zhu, H. (2018). Engineering graphene and TMDs based van der Waals heterostructures for photovoltaic and photoelectrochemical solar energy conversion. *Chemical Society Reviews*, *47*(13), 4981–5037.

Li, C.-H. A., Zhou, Z., Vashishtha, P., & Halpert, J. E. (2019). The Future Is Blue (LEDs): Why Chemistry Is the Key to Perovskite Displays. *Chemistry of Materials*, *31*(16), 6003–6032. DOI: 10.1021/acs.chemmater.9b01650

Li, H., Chen, F., Wu, X., Wang, D., Ren, Y., & Li, Y. (2024). Improved HER/OER Performance of NiS_2/MoS_2 Composite Modified by CeO_2 and LDH. *Materials (Basel)*, *17*(19), 4876. DOI: 10.3390/ma17194876 PMID: 39410447

Li, H., Cheng, C., & Yang, Z. (2022). Encapsulated CdSe/CdS Nanorods in Double-shelled Porous Nanocomposites for Efficient Photocatalytic CO_2 Reduction. *Nature Communications*, *13*, 6466. DOI: 10.1038/s41467-022-34263-z PMID: 36309504

Li, J., Xia, Z., Ma, D., Liu, G., Song, N., Xiang, D., Xin, Y., Zhang, G., & Chen, Q. (2021). Improving photocatalytic activity by construction of immobilized Z-scheme CdS/Au/TiO2 nanobelt photocatalyst for eliminating norfloxacin from water. *Journal of Colloid and Interface Science*, *586*, 243–256.

Li, M., Cai, B., Tian, R., Yu, X., Breese, M. B. H., Chu, X., Han, Z., Li, S., Joshi, R., Vinu, A., Wan, T., Ao, Z., Yi, J., & Chu, D. (2021). Vanadium doped 1T MoS2 nanosheets for highly efficient electrocatalytic hydrogen evolution in both acidic and alkaline solutions. *Chemical Engineering Journal*, *409*, 128158. DOI: 10.1016/j.cej.2020.128158

Li, S., Tuo, P., Xie, J., Zhang, X., Xu, J., Bao, J., Pan, B., & Xie, Y. (2018). Ultra-thin MXene nanosheets with rich fluorine termination groups realizing efficient electrocatalytic hydrogen evolution. *Nano Energy*, *47*, 512–518. DOI: 10.1016/j.nanoen.2018.03.022

Li, T., & Galli, G. (2007). Electronic properties of MoS_2 nanoparticles. *The Journal of Physical Chemistry. C, Nanomaterials and Interfaces*, *111*(44), 16192–16196.

Liu, B., Abbas, A., & Zhou, C. (2017). Two-Dimensional Semiconductors: From Materials Preparation to Electronic Applications. *Advanced Electronic Materials*, *3*(7), 1700045. DOI: 10.1002/aelm.201700045

Liu, C., Li, F., Ma, L., & Cheng, H. (2010). Advanced Materials for Energy Storage. *Advanced Materials*, *22*(8). Advance online publication. DOI: 10.1002/adma.200903328

Liu, F., Wu, R., Wei, J., Nie, W., Mohite, A. D., Brovelli, S., Manna, L., & Li, H. (2022). Recent Progress in Halide Perovskite Radiation Detectors for Gamma-Ray Spectroscopy. *ACS Energy Letters*, *7*(3), 1066–1085. DOI: 10.1021/acsenergylett.2c00031

Liu, J., Lin, H., Dong, Y., He, Y., & Liu, C. (2022). MoS_2 nanosheets loaded on collapsed structure zeolite as a hydrophilic and efficient photocatalyst for tetracycline degradation and synergistic mechanism. *Chemosphere*, *287*, 132211.

Liu, K., Liu, K., Liu, J., Ren, Q., Zhao, Z., Wu, X., Li, D., Yuan, F., Ye, K., & Li, B. (2020). Copper chalcogenide materials as photothermal agents for cancer treatment. *Nanoscale*, *12*(5), 2902–2913. DOI: 10.1039/C9NR08737K PMID: 31967164

Liu, L., An, W., Gu, F., Cui, L., He, X., & Fan, M. (2023). 2D layered materials: Structures, synthesis, and electrocatalytic applications. *Green Chemistry*, *25*(16), 6149–6169. DOI: 10.1039/D3GC01822A

Liu, Q., Ding, X., Pang, Y., Cao, Y., Lei, J., Wu, J., & Zhang, T. (2022). New insights into the safety assessment of quantum dots: Potential release pathways, environmental transformations, and health risks. *Environmental Science. Nano*, *9*(9), 3277–3311.

Liu, R.-J., Chang, L.-Y., Lin, F.-S., Lee, Y.-H., Yeh, M.-H., & Ho, K.-C. (2024). Multifunctional Structure-Modified Quaternary Compounds Co_9Se_8–$CuSe_2$–WSe_2 Mixed with MWCNT as a Counter Electrode Material for Dye-Sensitized Solar Cells. *ACS Applied Materials & Interfaces*, *16*(3), 3476–3488. DOI: 10.1021/acsami.3c16527 PMID: 38207165

Liu, R., Olson, C. G., Tonjes, W. C., & Frindt, R. F. (1998). Momentum dependent spectral changes induced by the charge density wave in 2H-$TaSe_2$ and the implication on the CDW mechanism. *Physical Review Letters*, *80*, 5762–5765.

Liu, X.-K., Xu, W., Bai, S., Jin, Y., Wang, J., Friend, R. H., & Gao, F. (2021). Metal halide perovskites for light-emitting diodes. *Nature Materials*, *20*(1), 10–21. DOI: 10.1038/s41563-020-0784-7 PMID: 32929252

Liu, X., Xu, F., Li, Z., Liu, Z., Yang, W., & Zhang, Y.. (2022). Design strategy for MXene and metal chalcogenides/oxides hybrids for supercapacitors, secondary batteries and electro/photocatalysis. *Coordination Chemistry Reviews*, *464*, 214544. DOI: 10.1016/j.ccr.2022.214544

Liu, Y., He, K., Chen, G., Leow, W. R., & Chen, X. (2017). Nature-Inspired Structural Materials for Flexible Electronic Devices. *Chemical Reviews*, *117*(20), 12893–12941. DOI: 10.1021/acs.chemrev.7b00291 PMID: 28991450

Liu, Y., & Huang, C. (2025). Recent advances in metal chalcogenide catalysts for electrochemical carbon dioxide reduction. *Advanced Functional Materials*. Advance online publication. DOI: 10.1002/adfm.202505245

Li, X., Chen, C., Yang, Y., Lei, Z., & Xu, H. (2020). 2D Re-based transition metal chalcogenides: Progress, challenges, and opportunities. *Advancement of Science*, *7*(23), 2002320.

Li, X., Su, M., Zhu, G., Zhang, K., Zhang, X., & Fan, J. (2018). Fabrication of a novel few-layer WS2/Bi2MoO6 plate-on-plate heterojunction structure with enhanced visible-light photocatalytic activity. *Dalton Transactions (Cambridge, England)*, *47*(30), 10046–10056.

Li, Y., Liu, Y., Gao, G., Zhu, Y., Wang, D., Ding, M., Yao, T., Liu, M., & You, W. (2021). L-cysteine and urea synergistically-mediated one-pot one-step self-transformed hydrothermal synthesis of p-Ag2S/n-AgInS2 core-shell hetero nano-flowers for photocatalytic MO degradation. *Applied Surface Science*, *548*, 149279.

Li, Y., Wang, H., Shu, T., Yuan, J., Lu, G., Lin, B., Gao, Z., Wei, F., Ma, C., Qi, J., & Sui, Y. (2022). Two-dimensional hierarchical MoS2 lamella inserted in CoS2 flake as an advanced supercapacitor electrode. *Journal of Energy Storage*, *51*, 104299. DOI: 10.1016/j.est.2022.104299

Li, Y., Zhang, Y., Ma, H., Wan, Y., & Zhao, T. (2025). Low-dimensional metal chalcogenides for wearable gas sensing. *Nano Convergence*, *12*(1), 34. DOI: 10.1186/s40580-025-00500-6 PMID: 40637935

Li, Y., Zhao, Y., Wang, X., Liu, W., He, J., Luo, X., Liu, J., & Liu, Y. (2023). Precise Construction and Growth of Submillimeter Two-Dimensional WSe_2 and $MoSe_2$ Monolayers. *Materials (Basel)*, *16*(13), 4795. DOI: 10.3390/ma16134795 PMID: 37445110

Li, Z., Wu, R., Bao, E., Du, S., Xu, C., Zhu, J., Mao, H., & Chen, H. (2023). Assembly of high-performance hybrid supercapacitors using $FeCo_2O_4$ microflowers as battery-type cathode materials. *Ceramics International*, *49*, 10411–10419. DOI: 10.1016/j.ceramint.2022.11.223

Lopez-Delgado, R., Zhou, Y., & Zazueta-Raynaud, A. (2017). Enhanced Conversion Efficiency in Si Solar Cells Employing Photoluminescent Down-shifting CdSe/CdS Core/shell Quantum Dots. *Scientific Reports*, *7*, 14104. DOI: 10.1038/s41598-017-14269-0 PMID: 29074855

Lozano, G. (2018). The Role of Metal Halide Perovskites in Next-Generation Lighting Devices. *The Journal of Physical Chemistry Letters*, *9*(14), 3987–3997. DOI: 10.1021/acs.jpclett.8b01417 PMID: 29952563

Lu, M., Zhang, Y., Wang, S., Guo, J., Yu, W. W., & Rogach, A. L. (2019). Metal Halide Perovskite Light-Emitting Devices: Promising Technology for Next-Generation Displays. *Advanced Functional Materials*, *29*(30), 1902008. DOI: 10.1002/adfm.201902008

Luo, X., Zhong, J., & Zhou, Q.. (2018). Cationic reduced graphene oxide as self-aligned nanofiller in the epoxy nanocomposite coating with excellent anticorrosive performance and its high antibacterial activity. *ACS Applied Materials & Interfaces*, *10*, 18400–18415. DOI: 10.1021/acsami.8b01982 PMID: 29727162

Lu, R. C., Zhang, K., Liu, X. H., Chen, Q. Y., & Wang, Y. H. (2022). Investigation of the charge transport kinetics and photoelectrochemical properties of CuInS2/R-TiO2 NTs heterostructures photoanode. *Applied Surface Science*, *585*, 152742.

Lu, S., Yang, W., Zhou, M., Qiu, L., Tao, B., Zhao, Q., Wang, X., Zhang, L., Xie, Q., & Ruan, Y. (2022). Nitrogen- and oxygen-doped carbon with abundant micropores derived from biomass waste for all-solid-state flexible supercapacitors. *Journal of Colloid and Interface Science*, *610*, 1088–1099. DOI: 10.1016/j.jcis.2021.11.164 PMID: 34876262

Lusardi, A., & Mitchell, O. S. (2014). The economic importance of financial literacy: Theory and evidence. *Journal of Economic Literature*, *52*(1), 5–44.

Lv, J., Wang, L., Li, R., Zhang, K., Zhao, D., Li, Y., Li, X., Huang, X., & Wang, G. (2021). Constructing a Hetero-interface Composed of Oxygen Vacancy-Enriched Co_3 O_4 and Crystalline–Amorphous NiFe-LDH for Oxygen Evolution Reaction. *ACS Catalysis*, *11*(23), 14338–14351. DOI: 10.1021/acscatal.1c03960

Maalouf, A., Okoroafor, T., Jehl, Z., Babu, V., & Resalati, S. (2023). A comprehensive review on life cycle assessment of commercial and emerging thin-film solar cell systems. *Renewable & Sustainable Energy Reviews*, *186*, 113652.

Machín, A., & Márquez, F. (2024). Advancements in Photovoltaic Cell Materials: Silicon, Organic, and Perovskite Solar Cells. *Materials (Basel)*, *17*(5), 1165. DOI: 10.3390/ma17051165 PMID: 38473635

Ma, D., Hu, B., Wu, W., Liu, X., Zai, J., Shu, C., Tadesse Tsega, T., Chen, L., Qian, X., & Liu, T. L. (2019). Highly active nanostructured CoS2/CoS heterojunction electrocatalysts for aqueous polysulfide/iodide redox flow batteries. *Nature Communications*, *10*(1), 3367. DOI: 10.1038/s41467-019-11176-y PMID: 31358738

Maeda, K., & Domen, K. (2007). New Non-Oxide Photocatalysts Designed for Overall Water Splitting under Visible Light. *The Journal of Physical Chemistry. C, Nanomaterials and Interfaces*, *111*(22), 7851–7861. DOI: 10.1021/jp070911w

Ma, G., Ye, J., Qin, M., Sun, T., Tan, W., Fan, Z., Huang, L., & Xin, X. (2023). Mn-doped NiCoP nanopin arrays as high-performance bifunctional electrocatalysts for sustainable hydrogen production via overall water splitting. *Nano Energy*, *115*, 108679. DOI: 10.1016/j.nanoen.2023.108679

Mahajan, M., Kallatt, S., Dandu, M., Sharma, N., Gupta, S., & Majumdar, K. (2019). Light emission from the layered metal 2H-$TaSe_2$ and its potential applications. *Communications on Physics*, *2*(1), 88.

Maheskumar, V., Gnanaprakasam, P., Selvaraju, T., & Vidhya, B. (2018a). Comparative studies on the electrocatalytic hydrogen evolution property of Cu2SnS3 and Cu4SnS4 ternary alloys prepared by solvothermal method. *International Journal of Hydrogen Energy*, *43*, 3967–3975. DOI: 10.1016/j.ijhydene.2017.07.194

Mahmood, N., Zhang, C., & Hou, Y. (2013). Nickel sulfide/nitrogen-doped graphene composites: Phase-controlled synthesis and high performance anode materials for lithium ion batteries. *Small, 9*(8), 1321–1328. DOI: 10.1002/smll.201203032 PMID: 23494938

Maity, S., Parvin, U. K. S., Das, S., & Chatterjee, K. (2022). Polymer chalcogenides—New smart materials for thermoelectric applications. *Smart Materials and Structures, 31*(7), 073001. DOI: 10.1088/1361-665X/ac7595

Maji, S. K., & Chakraborty, P. (2024). Energy-related financial literacy and energy consumption: A study of residential households in West Bengal, India. *International Journal of Energy Sector Management, 18*(6), 1489–1510.

Mak, K. F.. (2012). Control of valley polarization in monolayer MoS_2 by optical helicity. *Nature Nanotechnology, 7*, 494–498.

Mak, K. F., Lee, C., Hone, J., Shan, J., & Heinz, T. F. (2010). Atomically Thin MoS 2: A New Direct-Gap Semiconductor. *Physical Review Letters, 105*(13), 136805. DOI: 10.1103/PhysRevLett.105.136805 PMID: 21230799

Malarkodi, C., & Annadurai, G. (2013). A novel biological approach on extracellular synthesis and characterization of semiconductor zinc sulfide nanoparticles. *Applied Nanoscience, 3*(5), 389–395. DOI: 10.1007/s13204-012-0119-9

Malarkodi, C., Rajeshkumar, S., Paulkumar, K., Vanaja, M., Gnanajobitha, G., & Annadurai, G. (2014). Biosynthesis and antimicrobial activity of semiconductor nanoparticles against oral pathogens. *Bioinorganic Chemistry and Applications, 347167*. Advance online publication. DOI: 10.1155/2014/347167 PMID: 24860280

Mal, J., Nancharaiah, Y. V., Van Hullebusch, E. D., & Lens, P. N. L. (2016). Metal chalcogenide quantum dots: Biotechnological synthesis and applications. *RSC Advances, 6*(47), 41477–41495. DOI: 10.1039/C6RA08447H

Mandal, S., Mendhe, A. B., Rakhade, H. M., Barse, N. S., Roy, M., Rosaiah, P., Park, T., Lee, H. S., Mendhe, A. C., & Kim, D. (2025). Recent advancement and design in supercapacitor hybrid electrode materials: Bridging the gap between energy and power density. *Chemical Engineering Journal Advances, 21*, 100690. DOI: 10.1016/J.CEJA.2024.100690

Mane, R. S., & Lokhande, C. D. (2000). Chemical deposition method for metal chalcogenide thin films. *Materials Chemistry and Physics, 65*, 1–31.

Manikandan, R., Justin Raj, C., Nagaraju, G., Velayutham, R., Moulton, S. E., & Puigdollers, J.. (2021). Selenium enriched hybrid metal chalcogenides with enhanced redox kinetics for high-energy density supercapacitors. *Chemical Engineering Journal, 414*, 128924. DOI: 10.1016/j.cej.2021.128924

Manjunatha, C., Rastogi, C. K., Manmadha Rao, B., Girish Kumar, S., Varun, S., Raitani, K., & Khosla, A. (2024). Advances in hierarchical inorganic nanostructures for efficient solar energy harvesting systems. *ChemSusChem, 17*(11), e202301755.

Manzeli, S.. (2017). 2D transition metal dichalcogenides. *Nature Reviews. Materials, 2*, 17033.

Maogang, G., Qingfeng, L., Ryan, G., Dan, E., Matthew, C., Alex, S., Alan, E., & Judy, Z. W. (2017). Printable Nanocomposite FeS_2–PbS Nanocrystals/Graphene Heterojunction Photodetectors for Broadband Photodetection. *ACS Applied Materials & Interfaces, 9*(33), 27801–27808. DOI: 10.1021/acsami.7b08226 PMID: 28758390

Marchini, E., Caramori, S., Bignozzi, C. A., & Carli, S. (2021). On the use of PEDOT as a catalytic counter electrode material in dye-sensitized solar cells. *Applied Sciences (Basel, Switzerland), 11*(9), 3795.

Martin, J., & Nolas, G. S. (2014). SYNTHESIS AND CHARACTERIZATION OF CHALCOGENIDE NANOCOMPOSITES. *Progress in Nanotechnology Processing*, 367–372. https://doi.org/DOI: 10.1002/9780470588246.CH50;PAGE:STRING:ARTICLE/CHAPTER

Martins, A., Madaleno, M., & Dias, M. F. (2019). Energy literacy: What is out there to know? *Energy for Sustainable Development, 53*, 1–10.

Martins, A., Madaleno, M., & Dias, M. F. (2020). Energy Literacy: does age matters? *AHFE International Conference. AHFE Open Access.*

Martins, A., Madaleno, M., & Dias, M. F. (2020). Financial Knowledge Role in Portuguese Energy Literacy. *Energies, 13*(13), 3412.

Ma, S., Yuan, G., Zhang, Y., Yang, N., Li, Y., & Chen, Q. (2022). Development of encapsulation strategies towards the commercialization of perovskite solar cells. *Energy & Environmental Science, 15*(1), 13–55.

Masikane, S. C., Revaprasadu, N., Masikane, S. C., & Revaprasadu, N. (2020). Indium Chalcogenide Nanomaterials in the Forefront of Recent Technological Advancements. In *Post-Transition Metals*. IntechOpen., DOI: 10.5772/intechopen.94558

Masud, & Kim, H. K. (2023). Redox shuttle-based electrolytes for dye-sensitized solar cells: comprehensive guidance, recent progress, and future perspective. *ACS omega, 8*(7), 6139-6163.

Mattei, A., Hong, R. S., Dietrich, H., Firaha, D., Helfferich, J., Liu, Y. M., & Sheikh, A. Y. (2022). Efficient crystal structure prediction for structurally related molecules with accurate and transferable tailor-made force fields. *Journal of Chemical Theory and Computation, 18*(9), 5725–5738. DOI: 10.1021/acs.jctc.2c00451 PMID: 35930763

Matyszczak, G., Plocinski, T., Dluzewski, P., Fidler, A., Jastrzebski, C., Lawniczak-Jablonska, K., & Krawczyk, K. (2024). Sonochemical synthesis of SnS and SnS2 quantum dots from aqueous solutions, and their photo-and sonocatalytic activity. *Ultrasonics Sonochemistry, 105*, 106834. DOI: 10.1016/j.ultsonch.2024.106834

Maurya, A., Shanmugan, S., Kabeel, A. E., Akanovna, Z. A., Sarsenbayev, Y., & Panchal, H. (2024). Use of absorber plate built of ZnO/PVC/Bioactivation modified epoxy nanocomposites to improvement of double-effect Solar Distiller productivity analyzing the Energy, Exergo-environment and Enviro-economical. *Journal of Cleaner Production, 434*, 139601.

Ma, X., & Cheng, H. (2023). ReS2 with unique trion behavior as a co-catalyst for enhanced sunlight hydrogen production. *Journal of Colloid and Interface Science, 634*, 32–43. DOI: 10.1016/j.jcis.2022.12.014 PMID: 36528969

McDowell, L. L., Rastkar Mirzaei, M., & Shi, Z. (2023). Epitaxial CdSe/PbSe Heterojunction Growth and MWIR Photovoltaic Detector. *Materials (Basel), 16*(5), 1866. DOI: 10.3390/ma16051866 PMID: 36902982

McGovern, C. (2010). Commoditization of nanomaterials. *Nanotechnology Perceptions, 6*, 155–178.

McWhan, D. B., Fleming, R. M., Moncton, D. E., & DiSalvo, F. J. (1980). Reentrant lock-in transition of the charge-density wave in 2H-$TaSe_2$ at high pressure. *Physical Review Letters, 45*, 269–272.

Medina-Cruz, D., Tien-Street, W., & Zhang, B.. (2019). Citric juice mediated synthesis of tellurium nanoparticles with antimicrobial and anticancer properties. *Green Chemistry, 21*(8), 1982–1998. DOI: 10.1039/C8GC03897K PMID: 31156349

Medina-Cruz, D., Vernet-Crua, A., & Mostafavi, E. (2021). *Aloe vera*-mediated Te nanostructures: Highly potent antibacterial agents and moderated anticancer effects. *Nanomaterials (Basel, Switzerland), 11*(2), 514. DOI: 10.3390/nano11020514 PMID: 33670538

Medina-Ramírez, I., Hernández-Ramírez, A., & Maya-Trevino, M. L. (2015). Synthesis methods for photocatalytic materials. In Photocatalytic semiconductors: Synthesis, characterization, and environmental applications (pp. 69–102).

Mehta, C. H., Narayan, R., & Nayak, U. Y. (2019). Computational modeling for formulation design. *Drug Discovery Today*, *24*(3), 781–788. DOI: 10.1016/j.drudis.2018.11.018 PMID: 30502513

Mendhe, A. C. (2023). Spin coating: Easy technique for thin films. In B. R. Sankapal, A. Ennaoui, G. R. B., & L. C. D. (Eds.), *Simple chemical methods for thin film deposition: Synthesis and applications* (pp. 387–424). Springer Nature Singapore. https://doi.org/DOI: 10.1007/978-981-99-0961-2_9

Mendhe, A. C., Kim, M., Lekshmi, S., Jadhav, S. B., Barse, N. S., & Lee, H. (2025). Recent progress in metal chalcogenide-MXene and MOF-derived composites for supercapacitors: Synthesis, challenges, and future solutions. *Progress in Materials Science*, *156*, 101558. DOI: 10.1016/j.pmatsci.2025.101558

Mensah-Darkwa, K., Nframah Ampong, D., Agyekum, E., de Souza, F. M., & Gupta, R. K. (2022). Recent advancements in chalcogenides for electrochemical energy storage applications. *Energies*, *15*(11), 4052.

Meshesha, M. M., Gautam, J., Chanda, D., Kannan, K., Jang, S. G., & Yang, B. L. (2022). Effect of reaction temperature on morphology and photoelectrochemical performance of titanium dioxide. *Materials Letters*, *329*, 133176. DOI: 10.1016/j.matlet.2022.133176

Meshesha, M. M., Kannan, K., Chanda, D., Gautam, J., Jang, S. G., & Yang, B. L. (2022). Remarkable photoelectrochemical activity of titanium dioxide nanorod arrays sensitized with transition metal sulfide nanoparticles for solar hydrogen production. *Materials Today. Chemistry*, *26*, 101216. DOI: 10.1016/j.mtchem.2022.101216

Metcalf, H. C., Williams, J. E., & Caskta, J. F. (1982). *Modern chemistry*. Holt, Rinehart and Winston.

Miah, Md. H., Khandaker, M. U., Hossen, Md. J., Noor-E-Ashrafi, Jahan, I., Shahinuzzaman, Md., Nur-E-Alam, M., Hanfi, M. Y., Ullah, Md. H., & Islam, M. A. (2025). Lead-free alternatives and toxicity mitigation strategies for sustainable perovskite solar cells: A critical review. *Materials Advances*, *6*(9), 2718–2752. DOI: 10.1039/D5MA00010F

Milošević, I., Nikolić, B., Dobardžić, E., Damnjanović, M., Popov, I., & Seifert, G. (2007). Electronic properties and optical spectra of MoS_2 and WS_2 nanotubes. *Physical Review B: Condensed Matter and Materials Physics*, *76*(23), 233414.

Ming, H., Luo, Z.-Z., Zou, Z., & Kanatzidis, M. G. (2025). Strategies and Prospects for High-Performance Te-Free Thermoelectric Materials. *Chemical Reviews*, *125*(7), 3932–3975. DOI: 10.1021/acs.chemrev.4c00786 PMID: 40105866

Mirershadi, S., Sattari, F., Hadi, S., Rasoulzadehfard, S., & Mortazavi, S. Z. (2023). Efficiency enhancement of silicon-based solar cells by solar energy harvesting in WS2 nano-flake/perovskite layer. *Journal of Materials Science Materials in Electronics*, *34*(26), 1794. DOI: 10.1007/s10854-023-11205-z

Mishra, T. T., Chakraborty, M., Behera, J. N., & Roy, D. (2024). Comprehensive review on bimetallic (Ni and Co)-chalcogenides and phosphides with three-dimensional architecture for hybrid supercapacitor application. *Energy & Fuels*, *38*(11), 9186–9217.

Mitchell, K., & Ibers, J. A. (2002). Rare-Earth Transition-Metal Chalcogenides. *Chemical Reviews*, *102*(6), 1929–1952. DOI: 10.1021/cr010319h

Mohammadi, S., Kaur, N., & Radu, D. R. (2024). Nanoscale Cu2ZnSnSxSe (4− x) (CZTS/Se) for Sustainable Solutions in Renewable Energy, Sensing, and Nanomedicine. *Crystals*, *14*(5), 479.

Mohammed, R., Ali, M. E. M., Gomaa, E., & Mohsen, M. (2022). Copper sulfide and zinc oxide hybrid nanocomposite for wastewater decontamination of pharmaceuticals and pesticides. *Scientific Reports*, *12*(1), 18153.

Mohd. S., Kamlesh, V., Chandekar., Aslam K., Ahmed Mohamed E.T., AlFaify, S. (2020). A Facile Synthesis of Bi@PbS Nanosheets and their Key Physical Properties Analysis for Optoelectronic Technology, *Materials Science in Semiconductor Processing*. 107, 104807, DOI: 10.1016/j.mssp.2019.104807

Moncton, D. E., Axe, J. D., & Disalvo, F. J. (1975). Study of Superlattice Formation in 2H-$NbSe_2$ and 2H-$TaSe_2$ by Neutron Scattering. *Physical Review Letters*, *34*, 734–737.

Moon, J. W., Ivanov, I. N., Joshi, P. C., Armstrong, B. L., Wang, W., Jung, H., Rondinone, A. J., Jellison, G. E.Jr, Meyer, H. M.III, & Jang, G. G.. (2014). Scalable production of microbially mediated zinc sulfide nanoparticles and application to functional thin films. *Acta Biomaterialia*, *10*(10), 4474–4483. DOI: 10.1016/j.actbio.2014.06.003 PMID: 24932768

Moradialvand, Z., Parseghian, L., & Rajabi, H. R. (2025). Green Synthesis of Quantum Dots: Synthetic Methods, Applications, and Toxicity. *Journal of Hazardous Materials Advances*, *100697*. Advance online publication. DOI: 10.1016/j.hazadv.2025.100697

Moreira, T. F. M., Santana, I. L., Moura, M. N., Ferreira, S. A. D., Lelis, M. F. F., & Freitas, M. B. J. G. (2017). Recycling of negative electrodes from spent Ni-Cd batteries as CdO with nanoparticle sizes and its application in remediation of azo dye. *Materials Chemistry and Physics*, *195*, 19–27. DOI: 10.1016/j.matchemphys.2017.04.009

More, P. B., Jagtap, C. V., Kadam, V. S., Naushad, M., Naik, N., Hiremath, P., & Pathan, M., H. (2024). Synthesis and performance evaluation of ZnO/CdS photoanodes with copper sulfide (Cu2S) and carbon counter electrodes. *Scientific Reports*, *14*(1), 31551.

Mortazavi, B. (2025). Recent Advances in Machine Learning-Assisted Multiscale Design of Energy Materials. *Advanced Energy Materials*, *15*(9), 2403876.

Mousa, G., Basem, A., Shanmugan, S., Karmiris-Obratański, P., Faqeha, H., Alshareef, R. S., & Elsheikh, A. H. (2024). Harnessing fluorescence resonance energy transfer for improved solar still performance with zinc oxide nanoparticles and activated carbon. *Applied Materials Today*, *38*, 102196.

Mu, F., Liu, Z., Zi, W., Cao, Y., Lu, X., Li, Y., Zhao, Z., Xiao, Z., & Cheng, N. (2021). CZTS nanoparticles as an effective hole-transport layer for Sb2Se3 thin-film solar cells. *Solar Energy*, *226*, 154–160. DOI: 10.1016/j.solener.2021.08.044

Mu, G., Tan, Y., & Bi, C. (2024). Visible to Mid-wave Infrared PbS/HgTe Colloidal Quantum Dot Imagers. *Nature Photonics*, *18*, 1147–1154. DOI: 10.1038/s41566-024-01492-1

Munonde, T. S., Selahle, S. K., & Nomngongo, P. N. (2024). Chalcogenides and their nanocomposites in heavy metal decontamination. *Metal-Chalcogenide Nanocomposites: Fundamentals, Properties and Industrial Applications*, 117–133. https://doi.org/DOI: 10.1016/B978-0-443-18809-1.00007-9

Munonde, T. S., & Nomngongo, P. N. (2023). Review on Metal Chalcogenides and Metal Chalcogenide-Based Nanocomposites in Photocatalytic Applications. *Chemistry Africa*, *6*(3), 1127–1143. DOI: 10.1007/s42250-022-00577-0

Muslih, E. Y., Munir, B., & Khan, M. M. (2021). Advances in chalcogenides and chalcogenides-based nanomaterials such as sulfides, selenides, and tellurides. In *chalcogenide-based nanomaterials as photocatalysts* (pp. 7-31). Elsevier.

Muthu, P., Rajagopal, S., Saju, D., Kesavan, V., Dellus, A., Sadhasivam, L., & Chandrasekaran, N. (2024). Review of transition metal chalcogenides and halides as electrode materials for thermal batteries and secondary energy storage systems. *ACS Omega*, *9*(7), 7357–7374.

Nabera, A., Istrate, I. R., Martín, A. J., Pérez-Ramírez, J., & Guillén-Gosálbez, G. (2023). Energy crisis in Europe enhances the sustainability of green chemicals. *Green Chemistry, 25*(17), 6603–6611. DOI: 10.1039/D3GC01053H PMID: 38013722

Nady, A., Ahmed, M. H. M., Numan, A., Ramesh, S., Latiff, A. A., Ooi, C. H. R., Arof, H., & Harun, S. W. (2017). Passively Q-switched erbium-doped fibre laser using cobalt oxide nanocubes as a saturable absorber. *Journal of Modern Optics, 64*(13), 1315–1320. DOI: 10.1080/09500340.2017.1286398

Naguib, M.. (2011). Two-dimensional nanocrystals produced by exfoliation of Ti_3AlC_2. *Advanced Materials, 23*, 4248–4253.

Nakanishi, K., & Shiba, H. (1977). Domain-like incommensurate charge-density-wave states and the first-order incommensurate-commensurate transitions in layered tantalum dichalcogenides. I. 1T-polytype. *Journal of the Physical Society of Japan, 43*(6), 1839–1847.

Nallusamy, S., Vasanthi, V., Kavinkumar, T., Bhaviripudi, V. R., Mangalaraja, R. V., Srinivasan, R., Chidhambaram, N., Dhanabalan, S. S., Asaithambi, P., & Thirumurugan, A. (2025). Advances in optoelectronics for environmental and energy sustainability. *Next Energy, 9*, 100387. DOI: 10.1016/j.nxener.2025.100387

Nashmi, H. (2025). Alrasheedi, Padamata Varshini, Afam Uzorka, & Shanmugan, S., (2025). Impact of microwave irradiation on the thermophysical and energy storage properties of MXene-Ag-Syzygiumcumini dye nanofluids for solar distillation. *Energy Nexus, 19*, 100525. DOI: 10.1016/j.nexus.2025.100525

Nath, M., Masud, J., Swesi, A. T., Nath, M., Silva, U. D., Liyanage, W. P. R., Umapathi, S., & Amin, B. G. (2018). Designing Smart Materials for Efficient Electrosynthesis of Fuels and Environmental Remediation: The Story of Transition Metal Chalcogenides. *ECS Meeting Abstracts, MA2018-01*(28), 1635. DOI: 10.1149/MA2018-01/28/1635

Nautiyal, H., Lohani, K., Mukherjee, B., Isotta, E., Malagutti, M. A., Ataollahi, N., & Scardi, P. (2023). Mechanochemical synthesis of sustainable ternary and quaternary nanostructured Cu2SnS3, Cu2ZnSnS4, and Cu2ZnSnSe4 chalcogenides for thermoelectric applications. *Nanomaterials (Basel, Switzerland), 13*(2), 366.

Negi, P., & Jaiswal, A. (2024). Impact of financial literacy on consumer financial behavior: A systematic review and research agenda using TCCM framework. *International Journal of Consumer Studies.*

Ngilirabanga, J. B., & Samsodien, H. (2021). Pharmaceutical co-crystal: An alternative strategy for enhanced physicochemical properties and drug synergy. *Nano Select*, *2*(3), 512–526. DOI: 10.1002/nano.202000201

Niederquell, A., Wyttenbach, N., Kuentz, M., & Panayiotou, C. (2019). Partial solvation parameters of drugs as a new thermodynamic tool for pharmaceutics. *Pharmaceutics*, *11*(1), 17. DOI: 10.3390/pharmaceutics11010017 PMID: 30621192

Nie, L., & Zhang, Q. (2017). Recent progress in crystalline metal chalcogenides as efficient photocatalysts for organic pollutant degradation. *Inorganic Chemistry Frontiers*, *4*, 1953–1962.

Nie, T., Fang, Z., Ren, X., Duan, Y., & Liu, S. (2023). Recent advances in wide-bandgap organic-inorganic halide perovskite solar cells and tandem application. *Nano-Micro Letters*, *15*(1), 70. DOI: 10.1007/s40820-023-01040-6 PMID: 36943501

Nimai, M., Noah, J. O., Feng, W., Zhongjian, H., Sachidananda, K., Anton, V. M., Joanna, L. C., Han, H., Milan, S., & Jennifer, A. H. (2017). Using Shape to Turn off Blinking for Two-colour Multiexciton Emission in CdSe/CdS Tetrapods. *Nature Communications*, *8*, 15083. DOI: 10.1038/ncomms15083 PMID: 28497776

Niyitanga, T., & Jeong, H. K. (2017). Graphite oxide and molybdenum disulfide composite for hydrogen evolution reaction. *Chemical Physics Letters*, *685*, 451–456. DOI: 10.1016/j.cplett.2017.08.015

Njema, G. G., & Kibet, J. K. (2025). A review of chalcogenide-based perovskites as the next novel materials: Solar cell and optoelectronic applications, catalysis and future perspectives. *Next Nanotechnology*, *7*, 100102. DOI: 10.1016/J.NXNANO.2024.100102

Novak, T. G., Kim, K., & Jeon, S. (2019). 2D and 3D nanostructuring strategies for thermoelectric materials. *Nanoscale*, *11*(42), 19684–19699. DOI: 10.1039/C9NR07406F PMID: 31617541

Novoselov, K. S.. (2004). Electric field effect in atomically thin carbon films. *Science*, *306*, 666–669.

Novoselov, K. S.. (2007). Room-temperature quantum Hall effect in graphene. *Science*, *315*(5817), 1379–1379.

O'Regan, B., & Grätzel, M. (1991). A low-cost, high-efficiency solar cell based on dye-sensitized colloidal TiO_2 films. *Nature*, *353*, 737–740. DOI: 10.1038/353737a0

Obodo, K. O., Ouma, C. N. M., Obodo, J. T., & Braun, M. (2017). Influence of transition metal doping on the electronic and optical properties of ReS_2 and $ReSe_2$ monolayers. *Physical Chemistry Chemical Physics*, *19*(29), 19050–19057.

Olaniyan, O. T., Adetunji, C. O., Adeniyi, M. J., Dare, A., & Adewuyi, O. A. (2024). Morphology, rheology, properties, and applications of nanostructured metal chalcogenide-filled polymer blends. In *Nanofillers for Binary Polymer Blends* (pp. 401–422). Elsevier.

Olawale, F., Ariatti, M., & Singh, M. (2021). Biogenic synthesis of silver-core selenium-shell nanoparticles using *Ocimum tenuiflorum* L.: Response surface methodology-based optimization and biological activity. *Nanomaterials (Basel, Switzerland)*, *11*, 2516. DOI: 10.3390/nano11102516 PMID: 34684956

Olawale, F., Oladimeji, O., Ariatti, M., & Singh, M.. (2022). Emerging roles of green-synthesized chalcogen and chalcogenide nanoparticles in cancer theranostics. *Journal of Nanotechnology*, *2022*. Advance online publication. DOI: 10.1155/2022/6176610

Oni, A. M., Mohsin, A. S., Rahman, M. M., & Bhuian, M. B. H. (2024). A comprehensive evaluation of solar cell technologies, associated loss mechanisms, and efficiency enhancement strategies for photovoltaic cells. *Energy Reports*, *11*, 3345–3366. DOI: 10.1016/j.egyr.2024.03.007

Órdenes-Aenishanslins, N., Anziani-Ostuni, G., Monrás, J. P., Tello, A., Bravo, D., Toro-Ascuy, D., Soto-Rifo, R., Prasad, P. N., & Pérez-Donoso, J. M. (2020). Bacterial synthesis of ternary CdSAg quantum dots through cation exchange: Tuning the composition and properties of biological nanoparticles for bioimaging and photovoltaic applications. *Microorganisms*, *8*(5), 631. DOI: 10.3390/microorganisms8050631 PMID: 32349316

Órdenes-Aenishanslins, N., Anziani-Ostuni, G., Quezada, C. P., Espinoza-González, R., Bravo, D., & Pérez-Donoso, J. M. (2019). Biological synthesis of CdS/CdSe core/shell nanoparticles and its application in quantum dot sensitized solar cells. *Frontiers in Microbiology*, *10*, 1587. DOI: 10.3389/fmicb.2019.01587 PMID: 31354676

Ouyang, J., Xing, X., Zhou, L., Zhang, C., & Heng, J. Y. (2022). Cocrystal design of vanillin with amide drugs: Crystal structure determination, solubility enhancement, DFT calculation. *Chemical Engineering Research & Design*, *183*, 170–180. DOI: 10.1016/j.cherd.2022.05.009

Özşirvan, G., Yalçın, M., Turgut, N., Tari, V., & Turgut, C. (2024). Insights into the uptake, translocation, and accumulation dynamics of cyantraniliprole and thiamethoxam seed coating pesticides in maize plants. *Environmental Science and Pollution Research International*, (Special Issue). Advance online publication. DOI: 10.1007/s11356-024-34135-7

Pacchioni, G. (2022). Sustainable flexible supercapacitors. *Nature Reviews. Materials*, *7*, 844. DOI: 10.1038/s41578-022-00508-y

Paitandi, R. P., Wan, Y., Aftab, W., Zhong, R., & Zou, R. (2023). Pristine Metal–Organic Frameworks and their Composites for Renewable Hydrogen Energy Applications. *Advanced Functional Materials*, *33*(8), 2203224. DOI: 10.1002/adfm.202203224

Pakštas, V., Grincienė, G., Selskis, A., Balakauskas, S., Talaikis, M., Bruc, L., Curmei, N., Niaura, G., & Franckevičius, M. (2022). Improvement of CZTSSe film quality and superstrate solar cell performance through optimized post-deposition annealing. *Scientific Reports*, *12*(1). Advance online publication. DOI: 10.1038/s41598-022-20670-1 PMID: 36171263

Palaniappan, M., El-Shafay, A. S., & Shanmugan, S. (2024). Improving heat retention properties of steeped M-shape basin solar distillers utilizing paraffin RT50-enhanced silver nanoparticles and Manihot esculenta extracts. *Desalination*, *586*, 117836.

Palchoudhury, S., Ramasamy, K., & Gupta, A. (2020). Multinary copper-based chalcogenide nanocrystal systems from the perspective of device applications. *Nanoscale Advances*, *2*(8), 3069–3082.

Palchoudhury, S., Ramasamy, K., Han, J., Chen, P., & Gupta, A. (2023). Transition metal chalcogenides for next-generation energy storage. *Nanoscale Advances*, *5*(10), 2724–2742. DOI: 10.1039/D2NA00944G PMID: 37205287

Pandikumar, A., & Rameshkumar, P. (2020). *Graphene-Based Electrochemical Sensors for Toxic Chemicals*. Materials Research Forum LLC.

Panigrahy, S. K., Nandha, A., Chaturvedi, M., & Mishra, P. K. (2024). Novel nanocomposites with advanced materials and their role in waste water treatment. *Next Sustainability*, *4*, 100042. DOI: 10.1016/j.nxsust.2024.100042

Pan, Y., Liu, Y., & Liu, C. (2015). Phase- and morphology-controlled synthesis of cobalt sulfide nanocrystals and comparison of their catalytic activities for hydrogen evolution. *Applied Surface Science*, *357*, 1133–1140. DOI: 10.1016/j.apsusc.2015.09.125

Pareek, D., Balasubramaniam, K. R., & Sharma, P. (2015). Synthesis and characterization of bulk Cu2ZnSnX4 (X: S, Se) via thermodynamically supported mechano-chemical process. *Materials Characterization*, *103*, 42–49. DOI: 10.1016/j.matchar.2015.03.014

Park, J., Yoo, G., & Heo, J. (2019). CdSe/ZnS Quantum Dot Encapsulated MoS_2 Phototransistor for Enhanced Radiation Hardness. *Scientific Reports*, *9*, 1411. DOI: 10.1038/s41598-018-37902-y PMID: 30723221

Patel, P., & Pathak, N. (2025). Sustainable Lifestyle Choices in Chhattisgarh, India: Financial Implications and Consumer Behavior. *International Journal of Financial Management & Research*, *7*(2), 119–126.

Patil, I. G., Thakur, K., Nath, S. S., & Sundriyal, P. (2024). 3D Printed Energy Harvesting Devices for Flexible and Wearable Electronics. *Sustainable Energy & Fuels*.

Patil, S. J., Chodankar, N. R., Hwang, S.-K., Shinde, P. A., Rama Raju, G. S., Ranjith, K. S., Karekar, S. V., Huh, Y.-S., & Han, Y.-K. (2023). Two-dimensional nanosheets of bimetallic chalcogenide-tagged nitrogen-doped carbon as a cathode for high-performance and durable zinc-ion capacitors. *Journal of Materials Chemistry. A, Materials for Energy and Sustainability*, *11*(10), 5112–5126. DOI: 10.1039/D2TA07524E

Patil, S., Raut, S., Gore, R., & Sankapal, B. (2015). One-dimensional cadmium hydroxide nanowires towards electrochemical supercapacitor. *New Journal of Chemistry*, *39*(12), 9124–9131. DOI: 10.1039/C5NJ02022K

Paul, A.K., Kuiri, M., Saha, D. (2017). Photo-Tunable Transfer Characteristics in $MoTe_2$–MoS_2 Vertical Heterostructure. *2D Mater Appl*, *1*, 17. DOI: 10.1038/s41699-017-0017-3

Peng, D., Wang, Y., Shi, H., Jiang, W., Jin, T., Jin, Z., & Chen, Z. (2022). Fabrication of novel Cu2WS4/NiTiO3 heterostructures for efficient visible-light photocatalytic hydrogen evolution and pollutant degradation. *Journal of Colloid and Interface Science*, *613*, 194–206.

Peng, Q., Iram, N., Sharma, R., Kumar, A., Saad Alsubaie, A., & Rehman, J. (2024). Ab-initio investigation of structural, electronic, thermoelectric and optical properties of Full-Heusler X2MnB (X = Ti, Zr) for energy harvesting applications. *Inorganic Chemistry Communications*, *170*, 113240. DOI: 10.1016/j.inoche.2024.113240

Peng, Q., Yang, W., Li, X., Yang, H., Cao, J., Wang, J., Zheng, Y., Pan, J., & Li, C. (2023). The NiS/CdS hollow octahedral composites towards enhanced photocatalytic hydrogen evolution and photodegradation via active sites synergism. *International Journal of Hydrogen Energy*, *48*, 36427–36438. DOI: 10.1016/j.ijhydene.2023.06.045

Pérez, L. M., Aghoutane, N., Laroze, D., Díaz, P., El-Yadri, M., & Feddi, E. M. (2023). Unveiling the Role of Donor Impurity Position on the Electronic Properties in Strained Type I and Type II Core/Shell Quantum Dots under Magnetic Field. *Materials (Basel)*, *16*(19), 6535. DOI: 10.3390/ma16196535 PMID: 37834672

Perryman, J. T., & Velazquez, J. M. (2021). Design principles for multinary metal chalcogenides: Toward programmable reactivity in energy conversion. *Chemistry of Materials*, *33*(18), 7133–7147.

Phogat, P. Shreya, Jha, R., & Singh, S. (2023a). Diffusion Controlled Features of Microwave Assisted ZnS/ZnO Nanocomposite with Reduced Band Gap. *ECS Journal of Solid State Science and Technology, 12*(3), 034004. https://doi.org/DOI: 10.1149/2162-8777/acc426

Phogat, P., Shreya, Jha, R., & Singh, S. (2023b). Electrochemical analysis of thermally treated two dimensional zinc sulphide hexagonal nano-sheets with reduced band gap. *Physica Scripta*, *98*(12), 125962. DOI: 10.1088/1402-4896/ad0d93

Phogat, P., Jha, R., & Singh, S. (2024). Emerging advances and future prospects of two dimensional nanomaterials based solar cells. *Journal of Alloys and Compounds*, •••, 175063.

Phogat, P., Shreya, S., Jha, R., & Singh, S. (2024). Chalcogenide Nanocomposites for Energy Materials. *Engineering and Technology Journal*, *09*(07). Advance online publication. DOI: 10.47191/ETJ/V9I07.29

Pickett, N. L., Foster, D. F., & Cole-Hamilton, D. J. (1996). Synthesis of metal chalcogenide clusters from metal carbonyls and elemental chalcogen. *Journal of Materials Chemistry*, *6*(3), 507–511. DOI: 10.1039/JM9960600507

Pimpang, P., Hongsith, K., Choopun, S., & Wongrerkdee, S. (2025). A Review on Chalcogenide-Based Materials for Counter Electrode Applications in Dye-Sensitized Solar Cells: Sulfides, Selenides, and Tellurides. *ACS Omega*, *10*(25), 26293–26310. DOI: 10.1021/acsomega.5c03194 PMID: 40620954

Pirarath, R., Shivashanmugam, P., Syed, A., Elgorban, A. M., Anandan, S., & Ashokkumar, M. (2021). Mercury removal from aqueous solution using petal-like MoS_2 nanosheets. *Frontiers of Environmental Science & Engineering*, *15*, 1–10.

Pishgar, S., Gulati, S., Strain, J. M., Liang, Y., Mulvehill, M. C., & Spurgeon, J. M. (2021). In situ analytical techniques for the investigation of material stability and interface dynamics in electrocatalytic and photoelectrochemical applications. *Small Methods*, *5*(7), 2100322.

Premkumar, N., RadhaMadhavi, M., Kitmo, K, & Shanmugan. S. (2024). Utilizing the lignocellulosicfbers from Pineapple Crown Leaves extract for enhancing TiO_2 interfacial bonding in dye-sensitized solar cell photoanodes. *Materials for Renewable and Sustainable Energy*, *13*, 13–25. DOI: 10.1007/s40243-023-00245-4

Priya, A., Sagadevan, S., Priya, A., & Sagadevan, S. (2024). Chalcogenides-Based Nanomaterials for Contaminant Removal in Wastewater Treatment. In *Structural and Chemical Features of Chalcogenides*. IntechOpen., DOI: 10.5772/intechopen.1005357

Priyadarshini, P., Das, S., & Naik, R. (2022). A review on metal-doped chalcogenide films and their effect on various optoelectronic properties for different applications. *RSC Advances*, *12*(16), 9599–9620. DOI: 10.1039/D2RA00771A PMID: 35424915

Priyanka, U., Akshay Gowda, K. M., Elisha, M. G., Surya Teja, B., Nitish, N., & Raj Mohan, B. (2017). Biologically synthesized PbS nanoparticles for the detection of arsenic in water. *International Biodeterioration & Biodegradation*, *119*, 78–86. DOI: 10.1016/j.ibiod.2017.01.026

Qamar, S., & Ela, S. E. (2024). Dye-sensitized solar cells (DSSC): Principles, materials and working mechanism. *Current Opinion in Colloid & Interface Science*, ●●●, 101871.

Qazi, U. Y. (2022). Future of hydrogen as an alternative fuel for next-generation industrial applications. *Chall Expect Oppor Energ*, *15*, 4741. DOI: 10.3390/en15134741

Qian, X., Liu, H., Huang, Y., Ren, Z., Yu, Y., Xu, C., & Hou, L. (2019). Co-Ni-MoSx yolk-shell nanospheres as superior Pt-free electrode catalysts for highly efficient dye-sensitized solar cells. *Journal of Power Sources*, *412*, 568–574. DOI: 10.1016/j.jpowsour.2018.11.067

Qian, Y., Yu, J., Zhang, F., Fei, Z., Shi, H., Kang, D. J., & Pang, H. (2023). Hierarchical binary metal sulfides nanoflakes decorated on graphene with precious-metal-like activity for water electrolysis. *Chemical Engineering Journal*, *470*, 144372. DOI: 10.1016/j.cej.2023.144372

Qiao, F., Liu, W., Yang, J., Liu, Y., & Yuan, J. (2024). Fabrication of ZnO/CuInS2 heterojunction for boosting photocatalytic hydrogen production. *International Journal of Hydrogen Energy*, *53*, 840–847.

Qi, P., Zhang, D., Zeng, Y., & Wan, Y. (2016). Biosynthesis of CdS nanoparticles: A fluorescent sensor for sulfate-reducing bacteria detection. *Talanta*, *147*, 142–146. DOI: 10.1016/j.talanta.2015.09.054 PMID: 26592588

Queffurus, M., & Barnes, S.-J. (2015). A review of sulfur to selenium ratios in magmatic nickel–copper and platinum-group element deposits. *Ore Geology Reviews*, *69*, 301–324.

Qureshi, S. S., Shah, V., Nizamuddin, S., Mubarak, N. M., Karri, R. R., Dehghani, M. H., Ramesh, S., Khalid, M., & Rahman, M. E. (2022). Microwave-assisted synthesis of carbon nanotubes for the removal of toxic cationic dyes from textile wastewater. *Journal of Molecular Liquids*, *356*, 119045.

R. Fareza, A., Ardy Nugroho, F.A, F. Abdi, F., & Fauzia V. (2022). Nanoscale metal oxides–2D materials heterostructures for photoelectrochemical water splitting—A review. *Journal of Materials Chemistry. A, Materials for Energy and Sustainability*, *10*, 8656–8686. DOI: 10.1039/D1TA10203F

Rabell, G. O., Cruz, M. R. A., & Juárez-Ramírez, I. (2021). Hydrogen production of ZnO and ZnO/Ag films by photocatalysis and photoelectrocatalysis. *Materials Science in Semiconductor Processing*, *134*, 105985. DOI: 10.1016/j.mssp.2021.105985

Radhakrishnan, J., Kareem, A., Senthilkumar, S., & Biswas, K. (2022b). Facile synthesis of S-vacancy induced electrochemical HER activity in multilayered Sn doped MoS2. *Journal of Alloys and Compounds*, *917*, 165444. DOI: 10.1016/j.jallcom.2022.165444

Rahman, A., Khan, F., & Jennings, J. R. (2024). Microwave-assisted Synthesis of ZnS@$CuIn_xS_y$ for Photocatalytic Degradation of Coloured and Non-coloured Pollutants. *Scientific Reports*, *14*, 16155. DOI: 10.1038/s41598-024-66100-2 PMID: 38997296

Rahman, A., & Khan, M. M. (2021). Chalcogenides as photocatalysts. *New Journal of Chemistry*, *45*(42), 19622–19635. DOI: 10.1039/D1NJ04346C

Rahman, M. M., Karim, M. R., & Alam, M. M. (2020). Facile and Efficient 3-chlorophenol Sensor Development Based on Photolumenescent Core-shell CdSe/ZnS Quantum Dots. *Scientific Reports*, *10*, 557. DOI: 10.1038/s41598-019-57091-6 PMID: 31953448

Rahman, M., Davey, K., & Qiao, S. Z. (2017). Advent of 2D rhenium disulfide (ReS_2): Fundamentals to applications. *Advanced Functional Materials*, *27*(10), 1606129.

Rahman, Md. F., Monnaf, Md. A., Amami, M., Ben Farhat, L., & Rahman, Md. A. (2024). Improving the efficiency above 35% of MoS2-based solar cells by through optimization of various wide-bandgap S-chalcogenides ETL. *Journal of Physics and Chemistry of Solids*, *194*, 112216. DOI: 10.1016/j.jpcs.2024.112216

Rahman, S., Haleem, A., Siddiq, M., Hussain, M. K., Qamar, S., Hameed, S., & Waris, M. (2023). Research on dye sensitized solar cells: Recent advancement toward the various constituents of dye sensitized solar cells for efficiency enhancement and future prospects. *RSC Advances*, *13*(28), 19508–19529.

Rajamathi, M., & Seshadri, R. (2002). Oxide and chalcogenide nanoparticles from hydrothermal/solvothermal reactions. *Current Opinion in Solid State and Materials Science*, *6*(4), 337–345. DOI: 10.1016/S1359-0286(02)00029-3

Rajasekar, R., SenthilKumar, M., Shanmugan, S., & Nagarajan, M. (2022). The influence of Cu2ZnSnS4 thin films with characteristics of treatment conditions on spray pyrolysis technique for solar cells applications. *Indian Journal of Physics*, *96*(3), 707–716.

Rajeshkumar, S., Ponnanikajamideen, M., Malarkodi, C., Malini, M., & Annadurai, G. (2014). Microbe-mediated synthesis of antimicrobial semiconductor nanoparticles by marine bacteria. *Journal of Nanostructure in Chemistry*, *4*, 96. DOI: 10.1007/s40097-014-0096-7

Rajkumar, S., Venkatraman, M. R., Arunkumar, A., & Jayaprakash, K. (2025). Anthocyanin and Betalain Pigments Assisted Green Synthesis of TiO_2: A Sustainable and Cost-Effective Approach for Dye-Sensitized Solar Cells Photoanodes. *Optical Materials*, •••, 116710.

Ramachandran, K., Gaidi, M., Columbus, S., Daoudi, K., & Hammouche, J. (2023). Fabrication of noble metal–based antimicrobial nanosystems. *Antimicrobial Nanosystems: Fabrication and Development*, 353–375. https://doi.org/DOI: 10.1016/B978-0-323-91156-6.00023-3

Ramamurthy, C., Sampath, K. S., & Arunkumar, P.. (2013). Green synthesis and characterization of selenium nanoparticles and its augmented cytotoxicity with doxorubicin on cancer cells. *Bioprocess and Biosystems Engineering*, *36*(8), 1131–1139. DOI: 10.1007/s00449-013-0898-7 PMID: 23446776

Ramzan, M. A., Favre, R., Steinmann, S. N., Le Bahers, T., & Raybaud, P. (2024). Electrocatalytic reduction mechanisms of CO_2 on MoS_2 edges using grand-canonical DFT: From CO_2 adsorption to HCOOH or CO. *The Journal of Physical Chemistry. C, Nanomaterials and Interfaces*, *128*(24), 10025–10034. DOI: 10.1021/acs.jpcc.4c03266

Rana, D., Soni, A., Sharma, A., Katoch, A., & Jamwal, D. (2018a). Nanocomposites of Chalcogenide and their Applications. *Nano Hybrids and Composites*, *20*, 46–64. DOI: 10.4028/WWW.SCIENTIFIC.NET/NHC.20.46

Rani, S., Rajan, S. T., Shanthi, J., Ayeshamariam, A., & Jayachandran, M. (2015). Review on the materials properties and photoelctrochemical (PEC) solar cells of CdSe, Cd1-x ZnxSe, Cd1-x InxSe, thin films. *Materials Science Forum*, *832*, 1–27. DOI: 10.4028/WWW.SCIENTIFIC.NET/MSF.832.1

Raphae, S., & Stefan, M. K., lesya, Y., Alexander, D., Maksym, Y., Annina, M., Shadi, N., Yuriy, F., & Vanessa, W. L. (2023). Highly Responsive Mid-Infrared Metamaterial Enhanced Heterostructure Photodetector Formed out of Sintered PbSe/PbS Colloidal Quantum Dots. *ACS Applied Materials & Interfaces*, *15*(8), 10847–10857. DOI: 10.1021/acsami.2c23050 PMID: 36795914

Rasal, A. S., Yadav, S., Kashale, A. A., Altaee, A., & Chang, J. Y. (2022). Stability of quantum dot-sensitized solar cells: A review and prospects. *Nano Energy*, *94*, 106854. DOI: 10.1016/j.nanoen.2021.106854

Rashid, U., Ma, X., Zhu, Y., Cao, C., & Zou, M. (2025). The Comprehensive Review of Cobalt Based Electrocatalysts Synthesized via new Microwave Assisted Methodology. *Materials Chemistry Frontiers*. Advance online publication. DOI: 10.1039/D4QM00929K

Ratha, S., & Samantara, A. K. (2020). Metal Chalcogenide-Based Electrochemical Capacitors. In *Electrochemical Energy Conversion and Storage Systems for Future Sustainability*. Apple Academic Press.

Ren, Y., Shen, J., Yu, K., Phan, C. U., Chen, G., Liu, J., & Feng, J. (2019). Impact of crystal habit on solubility of ticagrelor. *Crystals*, *9*(11), 556. DOI: 10.3390/cryst9110556

Resalati, S., Okoroafor, T., Maalouf, A., Saucedo, E., & Placidi, M. (2022). Life cycle assessment of different chalcogenide thin-film solar cells. *Applied Energy*, *313*, 118888.

Ridhova, A., Puspasari, V., & Amal, M. I. (2021). Synthesis methods for chalcogenides and chalcogenides-based nanomaterials for photocatalysis. *Chalcogenide-Based Nanomaterials as Photocatalysts*, 105–134. https://doi.org/DOI: 10.1016/B978-0-12-820498-6.00005-6

Ridley, P., Gallano, C., Andris, R., Shuck, C. E., Gogotsi, Y., & Pomerantseva, E. (2020). MXene-derived bilayered vanadium oxides with enhanced stability in Li-ion batteries. *ACS Applied Energy Materials*, *3*(11), 10892–10901.

Roebben, G., Rauscher, H., Amenta, V., Ashberger, K., Sanfeliu, B., & Calzolai, L. (2024). Towards a review of the EC Recommendation for a definition of the term “nanomaterial” Part 2: Assessment of collected information concerning the experience with the definition 2014. In Roebben, G., & Rauscher, H. (Eds.), *JRC Scientific and Policy Report, EUR 26744 EN* (pp. 1–83). Publications Office of the European Union.

Rong, Y., Hu, Y., Mei, A., Tan, H., Saidaminov, M. I., Seok, S. I., McGehee, M. D., Sargent, E. H., & Han, H. (2018). Challenges for commercializing perovskite solar cells. *Science, 361*(6408), eaat8235. Advance online publication. DOI: 10.1126/science.aat8235 PMID: 30237326

Rossnagel, K. (2011). On the origin of charge-density waves in select layered transition-metal dichalcogenides. *Journal of Physics Condensed Matter*, *23*(21), 213001.

Roy, D., Panigrahi, K., Das, B. K., Ghorui, U. K., Bhattacharjee, S., Samanta, M., Sarkar, S., & Chattopadhyay, K. K. (2021). Boron vacancy: A strategy to boost the oxygen reduction reaction of hexagonal boron nitride nanosheet in hBN–MoS_2 heterostructure. *Nanoscale Advances*, *3*(16), 4739–4749. DOI: 10.1039/D1NA00304F PMID: 36134305

Rudnik, E., & Nikiel, M. (2007). Hydrometallurgical recovery of cadmium and nickel from spent Ni–Cd batteries. *Hydrometallurgy*, *89*(1–2), 61–71. DOI: 10.1016/j.hydromet.2007.05.006

Saadat, M., & Amiri, O. (2024). Chalcogenide perovskites for solar energy applications: The role of Sn substitution in BaZrS3-based photovoltaic devices. *Inorganic Chemistry Communications*, *168*, 112977. DOI: 10.1016/J.INOCHE.2024.112977

Sadasivuni, K. K., Kannan, K., Abdullah, A. M., & Kumar, B. (Eds.). (2022). *2D nanomaterials for CO2 conversion into chemicals and fuels*. Royal Society of Chemistry.

Sadiq, I., Ali, S. A., & Ahmad, T. (2025). Graphene Derivative-Based Particulate Photocatalysts for Energy Conversion Applications: Current Achievements, Bottlenecks, and Future Outlook. *ACS Applied Energy Materials*.

Said, Z., Pandey, A. K., Tiwari, A. K., Kalidasan, B., Jamil, F., Thakur, A. K., Tyagi, V. V., Sarı, A., & Ali, H. M. (2024). Nano-enhanced phase change materials: Fundamentals and applications. *Progress in Energy and Combustion Science*, *104*, 101162. DOI: 10.1016/J.PECS.2024.101162

Saied, E., Mekky, A. E., Al-Askar, A. A., Hagag, A. F., Elbana, A. A., Ashraf, M., Walid, A., Nour, T., Fawzi, M. M., & Arishi, A. A. (2023). *Aspergillus terreus*-mediated selenium nanoparticles and their antimicrobial and photocatalytic activities. *Crystals, 13*(4), 450. DOI: 10.3390/cryst13040450

Sai, G., Anna, V. R., Swapna, K., Prasad, M. S., Ch, R. P., Krishna, B. C., & Shanmugan, S. (2025). Evaluating the effects of sugarcane juice-mediated ZnO nanofluids on solar light activation for enhancing double-slope solar still performance. *Applied Materials Today*, *42*, 102542.

Sajitha, D. R., Stephen, B., Nakamura, A., Selvaraj, M., Salammal, S. T., & Hussain, S. (2024). The Emergence of Chalcogenides: A New Era for Thin Film Solar Absorbers. *Progress in Solid State Chemistry*, •••, 100490.

Sajitha, D. R., Stephen, B., Nakamura, A., Selvaraj, M., Salammal, S. T., & Hussain, S. (2024). The emergence of chalcogenides: A new era for thin film solar absorbers. *Progress in Solid State Chemistry*, *76*, 100490. DOI: 10.1016/j.progsolidstchem.2024.100490

Sajjad, M., Cheng, F., & Lu, W. (2021). Research progress in transition metal chalcogenide based anodes for K-ion hybrid capacitor applications: A mini-review. *RSC Advances*, *11*(41), 25450–25460. DOI: 10.1039/D1RA02445K PMID: 35478910

Salahodjaev, R., & Sadikov, A. (2025). Financial literacy and environmental sustainability: A cross-country test. *Frontiers in Sustainability*, *8*, 1514393.

Salam, Z., Vijayakumar, E., Subramania, A., Sivasankar, N., & Mallick, S. (2015). Graphene quantum dots decorated electrospun TiO_2 nanofibers as an effective photoanode for dye sensitized solar cells. *Solar Energy Materials and Solar Cells*, *143*, 250–259. DOI: 10.1016/j.solmat.2015.07.001

Samal, A., & Misra, L. I., Mohit, Pougajendy, S., Chandel, P. S., Manjunath, T. C., & Selvakumar, P. (2025). Musealization and Tourism Practices. In S. Oflazoğlu & M. Dora (Eds.), Preserving Cultural Heritage in Post-Disaster Urban Renewal (pp. 115-142). IGI Global Scientific Publishing. DOI: 10.4018/979-8-3693-9097-9.ch006

Sampaio, M. J., Oliveira, J. W. L., Sombrio, C. I. L., Baptista, D. L., Teixeira, S. R., Carabineiro, S. A. C., Silva, C. G., & Faria, J. L. (2016). Photocatalytic performance of Au/ZnO nanocatalysts for hydrogen production from ethanol. *Applied Catalysis A, General*, *518*, 198–205. DOI: 10.1016/j.apcata.2015.10.013

Sangeetha, A., Shanmugan, S., & Gorjian, S. (2022). Experimental evaluation and thermodynamic Gibbs free energy analysis of a double-slope U-shaped stepped basin solar still using activated carbon with ZnO nanoparticlesJournal of Cleaner Production. 380, Part 2, 135118. https://doi.org/DOI: 10.1016/j.jclepro.2022.135118

Sangeetha, A., & Shanmugan, S., & Alasiri, A. (2024). ZnO/nZVI Nanoparticle-Enhanced Double-Slope U-Shaped Solar Distillation: A Thermodynamic Investigation of Cephalexin Adsorption. *Materials Today Sustainability.*, *100983*. Advance online publication. DOI: 10.1016/j.mtsust.2024.100983

Santhi, K., Harish, S., Navaneethan, M., & Ponnusamy, S. (2023). Enhanced Photocatalytic activities of TiO_2 through Metal chalcogenides-based Nanocomposites (ZnS/TiO_2) for Methylene Blue degradation. *Surfaces and Interfaces*, *41*, 103205. DOI: 10.1016/j.surfin.2023.103205

Santhosh, C., Balasubramanian, B., Vino, P., Viji, M., Rejeeth, C., Kannan, S., Ullah, H., Rengasamy, K. R., Daglia, M., & Maruthupandian, A. (2022). Biofabricated selenium nanoparticles mediated from *Goniothalamus wightii* gains biomedical applications and photocatalytic degrading ability. *Journal of King Saud University. Science*, *34*, 102331. DOI: 10.1016/j.jksus.2021.102331

Santilln, O. S., & Cedano, K. G. (2023). Energy Literacy: A Systematic Review of the Scientific Literature. *Energies*, *16*(21), 7235.

Santos, T. F., Carvalho, B. R., Santos, E. V., Pereira, R. F., Souza, D. F., & Nascimento, J. H. (2025). Trends and developments in flexible solar cells: Insights from a bibliometric analysis. *Sustainable Energy & Fuels*.

Sarkar, A., Chatterjee, P., & Chakraborty, A. K. (2022). Sulfides and selenides as electrodes for dye-sensitized solar cells. In *Sulfide and Selenide Based Materials for Emerging Applications* (pp. 243–267). Elsevier.

Sarkar, B. K., Verma, A. S., Sharma, S., & Kundu, S. K. (2014). First-principles calculations of the structural, phonon and thermal properties of ZnX (X = S, Se,Te) chalcogenides. *Physica Scripta*, *89*(7), 075704. DOI: 10.1088/0031-8949/89/7/075704

Sarkar, P., Nisha, Kumar, P., & Katiyar, R. S. (2023). Enhanced Opto-Electrical Properties of Chalcogenide-Rich Tin Selenide Thin Film after Incorporating Sulfur Yielding Tin Sulfoselenide. *ChemistrySelect*, *8*(32), e202301545. DOI: 10.1002/SLCT.202301545;WGROUP:STRING:PUBLICATION

Sarma, N., & Saikia, P. (2024). Thin Film Approach For Scalable Production of Hydrogen through Photocatalysis. In *Towards Sustainable and Green Hydrogen Production by Photocatalysis: Scalability Opportunities and Challenges (Volume 1)* (pp. 165-183). American Chemical Society.

Sarma, L., & Thirumal, M. (2021). Coupling of $MoSe_2$ and graphitic carbon nitride (g-C_3N_4) by facile hydrothermal route for enhanced visible-light photocatalytic properties. *Materials Today: Proceedings, 36*, 679–688. DOI: 10.1016/j.matpr.2020.04.649

Sasikumar, R., Thirumalaisamy, S., Kim, B., & Hwang, B. (2024). Dye-sensitized solar cells: Insights and research divergence towards alternatives. *Renewable & Sustainable Energy Reviews, 199*, 114549.

Satpathy, S., Panigrahi, L. L., Samal, P., Sahoo, K. K., & Arakha, M. (2024). Biogenic synthesis of selenium nanoparticles from *Nyctanthes arbor-tristis* L. and evaluation of their antimicrobial, antioxidant and photocatalytic efficacy. *Heliyon, 10*, e32499. DOI: 10.1016/j.heliyon.2024.e32499 PMID: 39183842

Sawal, M. H., Jalil, A. A., Khusnun, N. F., Hassan, N. S., & Bahari, M. B. (2023). A review of recent modification strategies of TiO2-based photoanodes for efficient photoelectrochemical water splitting performance. *Electrochimica Acta, 467*, 143142. DOI: 10.1016/j.electacta.2023.143142

Sekar, R., Sivasamy, R., Ricardo, B., & Manidurai, P. (2021). Ultrasonically synthesized TiO_2/ZnS nanocomposites to improve the efficiency of dye sensitized solar cells. *Materials Science in Semiconductor Processing, 132*, 105917.

Sekiyama, H., Kosugi, N., Kuroda, H., & Ohta, T. (1986). Sulfur K-Edge Absorption Spectra of Na2SO4, Na2SO3, Na2S2O3, and Na2S2O *x* (*x* =5–8). *Bulletin of the Chemical Society of Japan, 59*(2), 575–579. DOI: 10.1246/bcsj.59.575

Sengupta, A., Afroz, M. A., Sharma, B., Choudhary, S., Pant, N., Gulia, Y., & Satapathi, S. (2025). Commercialization of perovskite solar cells: Opportunities and Challenges. *Sustainable Energy & Fuels.*

Shahid, M. U., Muhsan, A. S., Mohamed, N. M., Zaine, S. N. A., Baig, M. M. A., Ahmad, W., & Ayub, H. M. U. (2025). Graphene-Modified Photoelectrode for Efficient and Cost-Effective Dye-Sensitized Solar Cells. *International Journal of Energy Research, 2025*(1), 6672843. DOI: 10.1155/er/6672843

Shahshahanipour, M., Rezaei, B., Ensafi, A. A., & Etemadifar, Z. (2019). An ancient plant for the synthesis of a novel carbon dot and its applications as an antibacterial agent and probe for sensing of an anti-cancer drug. *Materials Science and Engineering C, 98*, 826–833. DOI: 10.1016/j.msec.2019.01.041 PMID: 30813088

Shahzad, N., Perveen, T., Pugliese, D., Haq, S., Fatima, N., Salman, S. M., & Shahzad, M. I. (2022). Counter electrode materials based on carbon nanotubes for dye-sensitized solar cells. *Renewable & Sustainable Energy Reviews*, *159*, 112196. DOI: 10.1016/j.rser.2022.112196

Shahzad, U., Saeed, M., Marwani, H. M., Al-Humaidi, J. Y., & Rehman, S. ur, Al-thomali, R. H., & Rahman, M. M. (2024). Transition metal-based chalcogenides as electrocatalysts for overall water splitting in hydrogen energy production. *International Journal of Hydrogen Energy*, *65*, 215–224. DOI: 10.1016/J.IJHYDENE.2024.03.275

Shaikh, S. A. L., Mendhe, A. C., Nadimetla, D. N., Biradar, M. R., Vijayanand, P., & Puyad, A. L.. (2024). Benzothiazole functionalized diketopyrrolopyrrole photosensitizer for CdS nanowire based DSSC applications. *Journal of Photochemistry and Photobiology A Chemistry*, *447*, 115220. DOI: 10.1016/j.jphotochem.2023.115220

Shankar, M. R., & Prabhu, A. N. (2023). A review on structural characteristics and thermoelectric properties of mid-temperature range Chalcogenide-based thermoelectric materials. *Journal of Materials Science*, *58*(43), 16591–16633. DOI: 10.1007/s10853-023-09028-8

Shanmugan, S., Djuansjah, J., Ahmadein, M., Alsaleh, N.A., Parsa, S.M., & Elsheikh, A.H. (2023). Chemical potential of different phases inside the pyramid stepped basin solar still through Gibbs free energy. *Case Studies in Thermal Engineering*, *49*, 103277. DOI: 10.1016/j.csite.2023.103277

Shanmugan, S., Madupu, R.K., Selvaraju, P., Ravichandran, S. (2022). Systematic growth on antibacterial activities use of silver nanoparticles with citrus aurantifolia. *Materials Today: Proceedings*, *51*(Part 1), 998–1005. DOI: 10.1016/j.matpr.2021.07.055

Shanmugan, S., Selvaraju, P., Nagaraj, J., Sivakumar, S., & Ravichandran, S. (2022). Biogenic silver nanoparticles of antibacterial activities for poly-herbal extracts in novel medicine. *Materials Today: Proceedings*, *51*(Part 1), 1107–1114. DOI: 10.1016/j.matpr.2021.07.107

Shanmugan, S., Selvaraju, P., Sivakumar, S., Nagaraj, J., Srinivas Prasad, M. V. V. K., & Abhiram, N. (2022). Solar cells absorption viewpoint of Mie theory: Experimental analysis of TiO_2 doping V/Ce. *Materials Today: Proceedings*, *51*(Part 1), 1124–1128. DOI: 10.1016/j.matpr.2021.07.110

Shao, X., Zhang, X., Liu, Y., Qiao, J., Zhou, X.-D., Xu, N., Malcombe, J. L., Yi, J., & Zhang, J. (2021). Metal chalcogenide-associated catalysts enabling CO2 electroreduction to produce low-carbon fuels for energy storage and emission reduction: Catalyst structure, morphology, performance, and mechanism. *Journal of Materials Chemistry. A, Materials for Energy and Sustainability*, *9*(5), 2526–2559. DOI: 10.1039/D0TA09232K

Sharma, S., Phogat, P., Jha, R., & Singh, S. (2023). Electrochemical and Optical Properties of Microwave Assisted MoS2 Nanospheres for Solar Cell Application. *International Journal of Smart Grid and Clean Energy*, 66–72. https://doi.org/DOI: 10.12720/sgce.12.3.66-72

Sheebha, I., David, C. I., Vidhya, B., & Nandhakumar, R. (2023). *Earth-abundant quaternary chalcogenide copper zinc tin sulfide-based nanocomposites CZTS-gCN and CZTS-GO (graphene oxide) as photocatalysts for degradation of methylene blue*. Inorganic and Nano-Metal Chemistry., DOI: 10.1080/24701556.2024.2355511

Sheela, S. E., Sekar, R., Maurya, D. K., Paulraj, M., & Angaiah, S. (2023). Progress in transition metal chalcogenides-based counter electrode materials for dye-sensitized solar cells. *Materials Science in Semiconductor Processing*, *156*, 107273. DOI: 10.1016/j.mssp.2022.107273

Sheldrick, W. S. (2000). Network self-assembly patterns in Main Group metal chalcogenide-based materials. *Journal of the Chemical Society, Dalton Transactions: Inorganic Chemistry*, *18*, 3041–3052. DOI: 10.1039/B003762L

Shen, Y., Zhang, Y., Huo, J., Li, X., Yan, Z., Pan, Y., Sun, W., Deng, N., & Kang, W. (2023). Two-dimensional SnSe material for solar cells and rechargeable batteries. *Journal of Energy Storage*, *69*, 107958.

Sherje, A. P., Kulkarni, V., Murahari, M., Nayak, U. Y., Bhat, P., Suvarna, V., & Dravyakar, B. (2017). Inclusion complexation of etodolac with hydroxypropyl-beta-cyclodextrin and auxiliary agents: Formulation characterization and molecular modeling studies. *Molecular Pharmaceutics*, *14*(4), 1231–1242. DOI: 10.1021/acs.molpharmaceut.6b01115 PMID: 28248111

Shirvani, S., Ghashghaee, M., & Smith, K. J. (2023). Two-dimensional Nanomaterials in Thermocatalytic Reactions: Transition Metal Dichalcogenides, Metal Phosphorus Trichalcogenides and MXenes. *Catalysis Reviews*, *65*(1), 1–51. DOI: 10.1080/01614940.2021.1899605

Shivaji, K., Mani, S., & Ponmurugan, P. (2018). Green-synthesis derived CdS quantum dots using tea leaf extract: Antimicrobial, bioimaging, and therapeutic applications in lung cancer cells. *ACS Applied Nano Materials*, *1*(4), 1683–1693. DOI: 10.1021/acsanm.8b00220

Shivashankarappa, A., & Sanjay, K. R. (2020). *Escherichia coli*-based synthesis of cadmium sulfide nanoparticles, characterization, antimicrobial and cytotoxicity studies. *Brazilian Journal of Microbiology : Publication of the Brazilian Society for Microbiology*, *51*(3), 939–948. DOI: 10.1007/s42770-020-00237-6 PMID: 32067210

Shi, Y., Sturm, C., & Kleinke, H. (2019). Chalcogenides as thermoelectric materials. *Journal of Solid State Chemistry*, *270*, 273–279. DOI: 10.1016/J.JSSC.2018.10.049

Shoukat, K., Khan, M. M., Bukhari, S., Shah, S. I. A., Bashir, I., Alothman, A. A., & Allakhverdiev, S. I. (2024). Study on oxygen evolution reaction efficiency demonstrated by Ce-E (E= S, Se and te) electrocatalyst. *International Journal of Hydrogen Energy*, *69*, 11–20.

Shreya, P. P., Jha, R., & Singh, S. (2023). *Elevated Refractive Index of MoS2 Amorphous Nanoparticles with a Reduced Band Gap Applicable for Optoelectronics* (pp. 431–439). https://doi.org/DOI: 10.1007/978-981-99-2349-6_39

Siddiqui, N., Faiyaz, T., & Tari, V. (2022). Ganga and Yamuna: Through the lens of National Green Tribunal. *Nature. Environment and Pollution Technology*, *21*(4), 1657–1664. DOI: 10.46488/NEPT.2022.v21i04.018

Siegmund, D., Blanc, N., Smialkowski, M., Tschulik, K., & Apfel, U. (2020). Metal-rich chalcogenides for electrocatalytic hydrogen evolution: Activity of electrodes and bulk materials. *ChemElectroChem*, *7*, 1514–1527.

Singh, A. K., & Jen, T.-C. (2021). *Chalcogenide : Carbon Nanotubes and Graphene Composites*. Chalcogenide., DOI: 10.1201/9780429260827

Singh, A. K., Kim, J., Park, J. T., & Sangunni, K. S. (2015). Properties of the chalcogenide–carbon nano tubes and graphene composite materials. *Journal of Alloys and Compounds*, *627*, 468–475. DOI: 10.1016/J.JALLCOM.2014.11.210

Singh, J., Thareja, R., & Malik, P. (2025a). Exploring the Potential of Quantum Dot-Sensitized Solar Cells: Innovation and Insights. *ChemPhysChem*, *26*(10), e202400800. DOI: 10.1002/CPHC.202400800;REQUESTEDJOURNAL:JOURNAL:14397641;PAGEGROUP:STRING:PUBLICATION PMID: 39964946

Singh, N. K., Kumar, A., Kumar, S., Nautiyal, V. K., Kumar, R., Kumar, A., & Iram, N. (2025). A comprehensive theoretical analysis of Cs2KGaX6 (X = Cl, Br, I): For green energy solutions. *Solid State Communications*, *401*, 115929. DOI: 10.1016/j.ssc.2025.115929

Sipos, B., Kusmartseva, A. F., Akrap, A., Berger, H., Forró, L., & Tutiš, E. (2008). From Mott state to superconductivity in 1T-TaS_2. *Nature Materials*, *7*(12), 960–965.

Sivasamy, R., Amirthaganesan, S., Gonzalez, R. E., Quero, F., & Batoo, K. M. (2021). First-principles investigation of the electronic structure, optical and thermodynamic properties on monolayer $Sn_{0.5}Ge_{0.5}Se$ nanosheet. *Physica E, Low-Dimensional Systems and Nanostructures*, *126*, 114454.

Sivasamy, R., Kaliyan, G., & Kaliyamoorthy, S. (2024). *Edgar Mosquera*. Chalcogenide Materials for Sustainable Energy and Environmental Applications. InteChOpen., DOI: 10.5772/intechopen.1003048

Sivasamy, R., Kaliyan, G., Kaliyamoorthy, S., Mosquera, E., Sivasamy, R., Kaliyan, G., Kaliyamoorthy, S., & Mosquera, E. (2024a). *Chalcogenide Materials for Sustainable Energy and Environmental Applications*. Structural and Chemical Features of Chalcogenides., DOI: 10.5772/INTECHOPEN.1003048

Sliem, M. H., Kannan, K., Maurya, M. R., Jlassi, K., Sadasivuni, K. K., Kumar, B., & Abdullah, A. M. (2025). Rational synthesis of mixed metal oxide clusters supported on a partially etched MAX phase for efficient electrocatalytic CO2 conversion. *Topics in Catalysis*, *68*(13), 1382–1397.

Smart, L., & Moore, E. (1996). *Solid state chemistry* (2nd ed.). Chapman and Hall.

Sokolikova, M. S., & Mattevi, C. (2020). Direct synthesis of metastable phases of 2D transition metal dichalcogenides. *Chemical Society Reviews*, *49*(12), 3952–3980. DOI: 10.1039/D0CS00143K PMID: 32452481

Song, R., Cho, S., Khan, S., Park, I., & Gao, W. (2025). Lighting the Path to Precision Healthcare: Advances and Applications of Wearable Photonic Sensors. *Advanced Materials, 2419161*. Advance online publication. DOI: 10.1002/adma.202419161 PMID: 39865847

Sonia, T. S., Anjali, P., Roshny, S., Lakshmi, V., Ranjusha, R., Subramanian, K. R. V., Nair, S. V., & Balakrishnan, A. (2014). Nano/micro-hybrid NiS cathodes for lithium-ion batteries. *Ceramics International*, *40*(6), 8351–8356. DOI: 10.1016/j.ceramint.2014.01.041

Sowbakkiyavathi, E. S., Arunachala Kumar, S. P., Maurya, D. K., Balakrishnan, B., Guo, J. Z., & Subramania, A. (2024). Research progress in the development of transition metal chalcogenides and their composite-based electrode materials for supercapacitors. *Advanced Composites and Hybrid Materials*, *7*(4), 130. DOI: 10.1007/s42114-024-00913-7

Spangler, L. C., Lu, L., Kiely, C. J., Berger, B. W., & McIntosh, S. (2016). Biomineralization of PbS and PbS–CdS core–shell nanocrystals and their application in quantum dot sensitized solar cells. *Journal of Materials Chemistry. A, Materials for Energy and Sustainability*, *4*(16), 6107–6115. DOI: 10.1039/C5TA10650A

Splendiani, A.. (2010). Emerging photoluminescence in monolayer MoS_2. *Nano Letters*, *10*, 1271–1275.

Sreekanthaswamy, N., Anitha, S., Singh, A., Jayadeva, S. M., Gupta, S., Manjunath, T. C., & Selvakumar, P. (2025). Digital Tools and Methods. In Çela, E., Paliktzoglou, V., Sengar, S., Eappen, P., & Vajjhala, N. (Eds.), *Enhancing School Counseling With Technology and Case Studies* (pp. 25–48). IGI Global Scientific Publishing., DOI: 10.4018/979-8-3693-8392-6.ch002

Srivastava, K. V., Srivastava, P., Srivastava, A., Maurya, R. K., Singh, Y. P., & Srivastava, A. (2025). 1D TiO 2 photoanodes: A game-changer for high-efficiency dye-sensitized solar cells. *RSC Advances*, *15*(7), 4789–4819.

Stamford, L., & Azapagic, A. (2019). Environmental impacts of copper-indium-gallium-selenide (CIGS) photovoltaics and the elimination of cadmium through atomic layer deposition. *The Science of the Total Environment*, *688*, 1092–1101. DOI: 10.1016/J.SCITOTENV.2019.06.343 PMID: 31726540

Subudhi, P., & Punetha, D. (2023). Progress, challenges, and perspectives on polymer substrates for emerging flexible solar cells: A holistic panoramic review. *Progress in Photovoltaics: Research and Applications*, *31*(8), 753–789.

Suganthi, N., Karthik, K., & Thangavel, S. (2022). *Nanostructured Metal Oxide and Hybrid Photocatalyst for Water Treatment and Hydrogen Generation*. Jenny Stanford Publishing.

Suganthi, N., Karthik, K., Thangavel, S., & Gurushankar, K. (2022). Electrochemical and magnetic properties of 3D porous NiS/CuS nanocomposites. *Applied Surface Science Advances*, *7*, 100209.

Suganthi, N., & Pushpanathan, K. (2018a). Photocatalytic degradation and antimicrobial activity of transition metal doped mesoporous ZnS nanoparticles. *International Journal of Environmental Science and Technology*. Advance online publication. DOI: 10.1007/s13762-018-1811-y

Suganthi, N., & Pushpanathan, K. (2018b). Spherical and dumbbell shape biphasic paramagnetic ZnS: Fe nanoparticles on ferromagnetic ZnS host background. *Journal of Electronic Materials*, *47*, 7343–7357.

Suganthi, N., & Pushpanathan, K. (2018c). Photocatalytic degradation and ferromagnetism in mesoporous La doped ZnS nanoparticles. *Journal of Materials Science Materials in Electronics*, *29*(16), 13970–13983.

Suganthi, N., Thangavel, S., & Karthik, K. (2020). Hibiscus subdariffa leaf extract mediated 2-D fern-like ZnO/TiO_2 hierarchical nanoleaf for photocatalytic degradation. *FlatChem*, *24*, 100197.

Sugiyama, H., & Uekusa, H. (2025). Crystal Engineering of the Photochromic Organic Compounds. In Advances in Organic Crystal Chemistry: Comprehensive Reviews 2025 on Crystal Structures (pp. 273-296). Singapore: Springer Nature Singapore. DOI: 10.1007/978-981-96-5984-5_12

Sungwoo, K., Taehoon, K., Meejae, K., Seong, K. K., Tae, W. Y., Lee Soon, P., Ilseung, Y., Sunjin, H., Jung Eun, L., Seong Keun, K., & Sang-Wook, K. (2012). Highly Luminescent InP/GaP/ZnS Nanocrystals and Their Application to White Light-Emitting Diodes. *Journal of the American Chemical Society*, *134*(8), 3804–3809. DOI: 10.1021/ja210211z PMID: 22303916

Sun, Q., Du, C., & Chen, G. (2025). Thermoelectric materials and devices: Applications in enhancing building energy conversion and efficiency. *Advanced Nanocomposites*, *2*, 15–31. DOI: 10.1016/j.adna.2024.11.001

Sun, Y. L., Xie, D., & Sun, M. X. (2018). Hybrid Graphene/cadmium-Free ZnSe/ZnS Quantum Dots Phototransistors for UV detection. *Scientific Reports*, *8*, 5107. DOI: 10.1038/s41598-018-23507-y PMID: 29572513

Syrrokostas, G., Bhorkar, K., Sygellou, L., Dracopoulos, V., Leftheriotis, G., & Yannopoulos, S. N. (2023). PtSe2 outperforms Pt as a counter electrode in dye sensitized solar cells. *Materials Chemistry and Physics*, *305*, 127994.

Tai, Y., Yang, B., Li, J., Meng, L., Xing, P., & Wang, S. (2024). Design and Preparation of Heterostructured Cu2O/TiO2 Materials for Photocatalytic Applications. *Molecules (Basel, Switzerland)*, *29*(21), 5028.

Tamilselvan, S. N., & Shanmugan, S. (2024). Towards sustainable solar cells: Unveiling the latest developments in bio-nano materials for enhanced DSSC efficiency. *Clean Energy*, *8*(3), 238–257.

Tamilselvan, S. N., Shanmugan, S., Basem, A., Faqeha, H., Alsaati, A. A., Rajkumar, V., & Elsheikh, A. H. (2024). Revolutionizing dye-sensitized solar cells: Chassalia curviflora fruit extract as photosensitizer and charge recombination inhibitor. *Inorganic Chemistry Communications*, *170*, 113292.

Tamrakar, E., Singla, R., Patel, R. N., Kumar, A., Tamrakar, R. K., & Butolia, S. (2025). Review on energy storage applications using new developments in solar photovoltaic materials and systems. *Chemical Papers*, *79*(1), 21–42.

Tang, Y., Qin, Z., Yin, S., & Sun, H. (2021). Transition metal oxide and chalcogenide-based nanomaterials for antibacterial activities: An overview. *Nanoscale*, *13*(13), 6373–6388. DOI: 10.1039/D1NR00664A PMID: 33885521

Tang, Y., Yang, C., Xu, X., Kang, Y., Henzie, J., Que, W., & Yamauchi, Y. (2022). MXene Nanoarchitectonics: Defect-Engineered 2D MXenes towards Enhanced Electrochemical Water Splitting. *Advanced Energy Materials*, *12*(12), 2103867. DOI: 10.1002/aenm.202103867

Tang, Y., Zheng, Z., Sun, X., Li, X., & Li, L. (2019). Ternary CdS-MoS_2 coated ZnO nanobrush photoelectrode for one-dimensional acceleration of charge separation upon visible light illumination. *Chemical Engineering Journal*, *368*, 448–458.

Tao, H., Zhou, M., Wang, K., Cheng, S., & Jiang, K. (2017). Nickel sulfide nanospheres anchored on reduced graphene oxide in situ doped with sulfur as a high performance anode for sodium-ion batteries. *Journal of Materials Chemistry. A, Materials for Energy and Sustainability*, *5*(19), 9322–9328. DOI: 10.1039/c7ta02194

Taraphder, A., Koley, S., Vidhyadhiraja, N. S., & Laad, M. S. (2011). Preformed excitonic liquid route to a charge density wave in 2H-TaSe2. *Physical Review Letters*, *106*(23), 236405.

Tari, V. Gupta, V., Joshi, A., Kurve, P., & Chandra, N. (2024). Chapter 19: Biodiversity loss due to forest diversion. In A. Rathore (Ed.), *Impact of societal development and infrastructure on biodiversity decline* (pp. 137–165). IGI Global. https://doi.org/ DOI: 10.4018/979-8-3693-6950-0.ch010

Tari, V. Kannan, K., & Vishwakarma, V. (2022). Chapter 8: Emerging antiviral technologies. In P. Nunthavarawong, S. M. Rangappa, S. Siengchin, & M. Thoppil-Mathew (Eds.), *Antimicrobial and antiviral materials: Polymers, metals, ceramics, and applications* (1st ed., pp. 137–150). CRC Press. DOI: 10.1201/9781003143093-8

Tari, V. Kannan, K., Koliar, J., Gurushankar, K., & Wahyuni, D. K. (2025). Influence of zinc oxide nanomaterials on crop plants: A prospective analysis. In D. Jabbrova, D. Sarkar, Rakesh, S. Singh, & R. Datta (Eds.), *Smart nanomaterials in agroforestry systems* (Chapter 8). Springer. https://doi.org/DOI: 10.1007/978-981-96-1337-3_8

Tariq, S., Saleem, F., & Shahzad, F. (2022). Does digital finance and financial inclusion strengthen environmental sustainability: Evidence from Asia. *Pakistan Social Sciences Review*, *4*(4), 847–861.

Tari, V. (2015a). Earth, man, and environment. In *Definition principles and scope of environment* (pp. 1–20). DPS Publishing House.

Tari, V. S., & Patil, P. Y. (2017). Challenge of contamination of pesticides for Alphonso in Ratnagiri district, Maharashtra, India. *Research Journal of Chemistry and Environment*, *21*(12), 57–67.

Tari, V. S., & Patil, P. Y. (2017a). Transfer factor of heavy metal in soil to plant from pesticide-contaminated area (mango orchards), Ratnagiri district, Maharashtra. *Research Journal of Chemistry and Environment*, *21*(1), 26–32.

Tari, V. S., & Patil, P. Y. (2017b). Challenge of contamination of pesticide for Alphonso in Ratnagiri district, Maharashtra, India. *Research Journal of Chemistry and Environment*, *21*(12), 57–67.

Tari, V., Gupta, R., & Siddiqui, N. (2021). Impact of climate change on Upper Ganga Ramsar Site, UP, India. In *Monitoring and evaluating the ecological health of wetlands* (1st ed., pp. 92–105). IGI Global., DOI: 10.4018/978-1-7998-9498-8.ch006

Tari, V., & Kannan, K. (2022). Microbial bioremediation of polythenes and plastics – A green sustainable approach. In *Microbial biotechnology in green remediation* (1st ed.). Elsevier., DOI: 10.1016/B978-0-323-90452-0.00003-7

Tari, V., Siddiqui, N., Rathi, D., Siddiqui, N. N., & Wahyuni, D. K. (2024). Need for an evolved groundwater justice in rural area Uttar Pradesh, India. *Nature. Environment and Pollution Technology*, *23*(1), 125–137. DOI: 10.46488/NEPT.2024.v23i01.009

Taylor, L. S., Braun, D. E., & Steed, J. W. (2021). Crystals and crystallization in drug delivery design. *Molecular Pharmaceutics*, *18*(3), 751–753. DOI: 10.1021/acs.molpharmaceut.0c01157 PMID: 33594892

Thakur, A., & Devi, P. (2022). based flexible devices for energy harvesting, conversion and storage applications: A review. *Nano Energy*, *94*, 106927.

Thangamathi, R., Kumaresan, N., Praveen Kumar, M., Mangalaraja, R. V., Herrera Diaz, F. V., Farhang Sahlevani, S., Ferreira De Oliveira, T., Pabba, D. P., & Sivakumar, P. (2024). Heterostructure-induced interfacial charge transfer interaction in CoS/CoO@NAC nanosheets as a bi-functional electrocatalyst for water-splitting application. *International Journal of Hydrogen Energy*, *64*, 69–79. DOI: 10.1016/j.ijhydene.2024.03.240

Tian, P., Min, L., Xin, L., Yaqin, Z., Tian, Z., Pu, S., & Fuzhi, L. (2024). Microflower-like CuS/Co_3S_4 Composite as High-capacity and Excellent-stability Material for Hybrid Supercapacitors. *Journal of Alloys and Compounds*, *1008*, 176563. DOI: 10.1016/j.jallcom.2024.176563

Tian, Z., Tsui, G. C.-P., Tang, Y.-M., Wong, C.-H., Tang, C.-Y., & Ko, C.-C. (2026). Additive Manufacturing for Nanogenerators: Fundamental Mechanisms, Recent Advancements, and Future Prospects. *Nano-Micro Letters*, *18*(1), 30. DOI: 10.1007/s40820-025-01874-2 PMID: 40788321

Tidman, J. P., Singh, O., Curzon, A. E., & Frindt, R. F. (1974). The phase transition in 2H-TaS_2 at 75 K. *Philosophical Magazine*, *30*, 1191–1194.

Tiwari, A. P., Azam, A., Novak, T. G., Prakash, O., & Jeon, S. (2018). Chemical strain formation through anion substitution in Cu2WS4 for efficient electrocatalysis of water dissociation. *Journal of Materials Chemistry. A, Materials for Energy and Sustainability*, *6*, 7786–7793. DOI: 10.1039/c8ta01061g

Todorov, L., Aleksandrova, A., & Ismailov, T. (2023). Relation between Financial Literacy and Carbon Footprint: Review on Implications for Sustainable Development. *Economics*, *2*(2), 2–11.

Toe, C. Y., Zhou, S., Gunawan, M., Lu, X., Ng, Y. H., & Amal, R. (2021). Recent advances and the design criteria of metal sulfide photocathodes and photoanodes for photoelectrocatalysis. *Journal of Materials Chemistry. A, Materials for Energy and Sustainability*, *9*(36), 20277–20319.

Tongay, S., Sahin, H., Ko, C., Luce, A., Fan, W., Liu, K., & Wu, J. (2014). Monolayer behaviour in bulk ReS2 due to electronic and vibrational decoupling. *Nature Communications*, *5*(1), 3252.

Toshniwal, A., & Kheraj, V. (2017). Development of organic-inorganic tin halide perovskites: A review. *Solar Energy*, *149*, 54–59. DOI: 10.1016/j.solener.2017.03.077

Tripathi, R. M., Hameed, P., Rao, R. P., Shrivastava, N., Mittal, J., & Mohapatra, S. (2020). Biosynthesis of highly stable fluorescent selenium nanoparticles and the evaluation of their photocatalytic degradation of dye. *BioNanoScience, 10*(2), 389–396. DOI: 10.1007/s12668-020-00718-0

Tumram, P. V., Nafdey, R., Kautkar, P. R., Agnihotri, S. V., Khaparde, R. A., Wankhede, S. P., & Moharil, S. V. (2024). Solar cell performance enhancement using nanostructures. *Materials Science and Engineering B, 307*, 117504.

Turna, T., Solmaz, A., & Baran, A. (2024). Green synthesis study: Adsorption of Congo red dye with selenium nanoparticles obtained from *Prunus armeniaca* L. leaf waste. *ChemistrySelect, 9*(27), e202403106. DOI: 10.1002/slct.202403106

Tu, Y., Shi, Y., Yan, X., Jiang, Y., Xiao, W., Zhu, S., Lv, P., & Li, C. (2024). Synergistic enhancement of bifunctional electrocatalysis for ORR and OER: Unveiling the role of Cr-doped CoS2 and N-doped carbon. *Colloids and Surfaces. A, Physicochemical and Engineering Aspects, 697*, 134457. DOI: 10.1016/j.colsurfa.2024.134457

Ursi, F., Virga, S., Garcìa-Espejo, G., Masciocchi, N., Martorana, A., & Giannici, F. (2022). Long-Term Stability of TiS_2–Alkylamine Hybrid Materials. *Materials (Basel), 15*(23), 8297. DOI: 10.3390/ma15238297 PMID: 36499793

Vaizo˘ gullar, A, I. (2020). Ternary CdS/MoS_2/ZnO photocatalyst: synthesis, characterization and degradation of ofloxacin under visible light irradiation. J. Inorg. Organomet. Polym Mater. 30, 4129- 41413.

Valla, T., Fedorov, A. V., Johnson, P. D., Xue, J., Smith, K. E., & DiSalvo, F. J. (2000). Charge-density-wave-induced modifications to the quasiparticle self-energy in 2H-$TaSe_2$. *Physical Review Letters, 85*(22), 4759.

Van Maaren, M. H., & Harland, H. B. (1969). An energy band model of Nb-and Ta-dichalcogenide superconductors. *Physics Letters. [Part A], 29*(9), 571–573.

Vanalakar, S. A., Agawane, G. L., Shin, S. W., Suryawanshi, M. P., Gurav, K. V., & Jeon, K. S.. (2015). A review on pulsed laser deposited CZTS thin films for solar cell applications. *Journal of Alloys and Compounds, 619*, 109–121. DOI: 10.1016/j.jallcom.2014.09.018

Varadarajan, S., Kavitha, A., Selvaraju, P., Muthu, S. E., Gurushankar, K., Shanmugan, S., & Kannan, K. (2024). Enhanced Photocatalytic Hydrogen Evolution by TiO2: A Synergistic Approach with Defect-Rich SnS2 and Ti3C2 MXene Cocatalysts. *Hydrogen, 5*(4), 940–957.

Varshosaz, J., Ghassami, E., & Ahmadipour, S. (2018). Crystal engineering for enhanced solubility and bioavailability of poorly soluble drugs. *Current Pharmaceutical Design*, *24*(21), 2473–2496. DOI: 10.2174/13816128246661807121044 47 PMID: 29998799

Vaskivskyi, I.. (2015). Controlling the metal-to-insulator relaxation of the metastable hidden quantum state in 1T-TaS_2. *Science Advances*, *1*(6), e1500168.

Velayutham, L., Parvathiraja, C., Anitha, D. C., Mahalakshmi, K., Jenila, M., Gupta, J. K., Wabaidur, S. M., Siddiqui, M. R., Aftab, S., & Lai, W.-C. (2023). Antibacterial and photocatalytic dye degradation activities of green synthesized NiSe nanoparticles from *Hibiscus rosa-sinensis* leaf extract. *Water (Basel)*, *15*(7), 1380. DOI: 10.3390/w15071380

Venkatesan, J., Rekha, P. D., Anil, S., Bhatnagar, I., Sudha, P. N., Dechsakulwatana, C., Kim, S.-K., & Shim, M. S. (2018). Hydroxyapatite from Cuttlefish Bone: Isolation, Characterizations, and Applications. *Biotechnology and Bioprocess Engineering; BBE*, *23*(4), 383–393. DOI: 10.1007/s12257-018-0169-9

Vescoli, V., Degiorgi, L., Berger, H., & Forró, L. (1998). Dynamics of correlated two-dimensional materials: The 2H-TaSe2 case. *Physical Review Letters*, *81*, 453–456.

Voigt, D., Primavera, G., Uphoff, H., Rethmeier, J. A., Schepp, L., & Bredol, M. (2022). Ternary Chalcogenide-Based Quantum Dots and Carbon Nanotubes: Establishing a Toolbox for Controlled Formation of Nanocomposites. *The Journal of Physical Chemistry. C, Nanomaterials and Interfaces*, *126*(21), 9076–9090. DOI: 10.1021/ACS.JPCC.2C01142/ASSET/IMAGES/LARGE/JP2C01142_0014.JPEG PMID: 35686224

Wang, C., He, M., Liu, X., & Malen, J. A. (2022). Modeling the tunable thermal conductivity of intercalated layered materials with three-directional anisotropic phonon dispersion and relaxation times. *Journal of Materials Chemistry. C, Materials for Optical and Electronic Devices*, *10*(32), 11686–11696. DOI: 10.1039/D1TC05369H

Wang, C., Nie, R., Dai, Y., Tai, H., Zhu, B., Zhao, L., & Seok, S. I. (2024). Enhancing the inherent stability of perovskite solar cells through chalcogenide-halide combinations. *Energy & Environmental Science*, *17*(4), 1368–1386.

Wang, G., Zhang, L., & Zhang, J. (2012). A review of electrode materials for electrochemical supercapacitors. *Chemical Society Reviews*, *41*, 797–828. DOI: 10.1039/c1cs15060j PMID: 21779609

Wang, H.. (2017). High-quality monolayer superconductor $NbSe_2$ grown by chemical vapour deposition. *Nature Communications*, *8*(1), 394.

Wang, H., Li, X., Zhao, X., Li, C., Song, X., Zhang, P., & Huo, P. (2022). A review on heterogeneous photocatalysis for environmental remediation: From semiconductors to modification strategies. *Chinese Journal of Catalysis*, *43*(2), 178–214. DOI: 10.1016/S1872-2067(21)63910-4

Wang, H., Tsai, C., Kong, D., Chan, K., Abild-Pedersen, F., & Nørskov, J. K.. (2015). Transition-metal doped edge sites in vertically aligned MoS2 catalysts for enhanced hydrogen evolution. *Nano Research*, *8*, 566–575. DOI: 10.1007/s12274-014-0677-7

Wang, H., Ye, W., Yin, B., Wang, K., Riaz, M. S., Xie, B., Zhong, Y., & Hu, Y. (2023). Modulating Cation Migration and Deposition with Xylitol Additive and Oriented Reconstruction of Hydrogen Bonds for Stable Zinc Anodes. *Angewandte Chemie International Edition*, *62*(10), e202218872. DOI: 10.1002/anie.202218872 PMID: 36647214

Wang, J., Xie, F., Cao, X.-H., An, S.-C., Zhou, W.-X., Tang, L.-M., & Chen, K.-Q. (2017). Excellent Thermoelectric Properties in monolayer WSe2 Nanoribbons due to Ultralow Phonon Thermal Conductivity. *Scientific Reports*, *7*(1), 41418. DOI: 10.1038/srep41418 PMID: 28120912

Wang, J., Yue, X., Yang, Y., Sirisomboonchai, S., Wang, P., & Ma, X.. (2020). Earth-abundant transition-metal-based bifunctional catalysts for overall electrochemical water splitting: A review. *Journal of Alloys and Compounds*, *819*, 153346. DOI: 10.1016/j.jallcom.2019.153346

Wang, J., Zheng, X., Wang, G., Cao, Y., Ding, W., Zhang, J., Wu, H., Ding, J., Hu, H., Han, X., Ma, T., Deng, Y., & Hu, W. (2022). Defective Bimetallic Selenides for Selective CO_2 Electroreduction to CO. *Advanced Materials*, *34*(3), 2106354. DOI: 10.1002/adma.202106354 PMID: 34699632

Wang, M., Dong, C.-L., Huang, Y.-C., & Shen, S. (2020). Operando Spectral and Electrochemical Investigation into the Heterophase Stimulated Active Species Transformation in Transition-Metal Sulfides for Efficient Electrocatalytic Oxygen Evolution. *ACS Catalysis*, *10*(3), 1855–1864. DOI: 10.1021/acscatal.9b05170

Wang, S., He, P., Jia, L., He, M., Zhang, T., Dong, F., Liu, M., Liu, H., Zhang, Y., Li, C., & Gao, J. (2019). Nanocoral-like composite of nickel selenide nanoparticles anchored on two-dimensional multi-layered graphitic carbon nitride: A highly efficient electrocatalyst for oxygen evolution reaction. *Applied Catalysis B: Environmental*, *243*, 463–469. DOI: 10.1016/j.apcatb.2018.10.071

Wang, X., Chen, A., Wu, X., Zhang, J., Dong, J., & Zhang, L. (2024). Synthesis and Modulation of Low-Dimensional Transition Metal Chalcogenide Materials via Atomic Substitution. *Nano-Micro Letters*, *16*(1), 163. DOI: 10.1007/s40820-024-01378-5 PMID: 38546814

Wang, X., Feng, J., Bai, Y., Zhang, Q., & Yin, Y. (2016). Synthesis, Properties, and Applications of Hollow Micro-/Nanostructures. *Chemical Reviews*, *116*(18), 10983–11060. DOI: 10.1021/acs.chemrev.5b00731 PMID: 27156483

Wang, X., Huang, H., Liang, B., Liu, Z., Chen, D., & Shen, G. (2013). ZnS Nanostructures: Synthesis, Properties, and Applications. *Critical Reviews in Solid State and Material Sciences*, *38*(1), 57–90. DOI: 10.1080/10408436.2012.736887

Wang, X., Xie, Y., Pan, K., Wu, J., Xiao, Y., Yu, P., Zhou, W., & Fu, H. (2018). Morphology effect of NiSe hierarchical microspheres on the performance of dye-sensitized solar cells. *ACS Applied Nano Materials*, *1*(9), 4900–4909. DOI: 10.1021/acsanm.8b01069

Wang, X., Yu, J., & Chen, R. (2018). Optical Characteristics of ZnS Passivated CdSe/CdS Quantum Dots for High Photostability and Lasing. *Scientific Reports*, *8*, 17323. DOI: 10.1038/s41598-018-35768-8 PMID: 30470827

Wang, Y. F., Feng, H., Deng, Y. R., Xin, F. F., Li, D. J., Hu, Z. F., & Liu, R. P. (2022). Solvothermal growth of Zn 2 SnO 4 for efficient dye-sensitized solar cells. *Rare Metals*, •••, 1–9.

Wang, Y., Bourgès, C., Rajamathi, R., Nethravathi, C., Rajamathi, M., & Mori, T. (2022). The Effect of Reactive Electric Field-Assisted Sintering of MoS_2/Bi_2Te_3 Heterostructure on the Phase Integrity of Bi_2Te_3 Matrix and the Thermoelectric Properties. *Materials (Basel)*, *15*(1), 53. DOI: 10.3390/ma15010053 PMID: 35009201

Wang, Y., Lin, P., Lou, Q., Zhang, Z., Huang, S., Lu, Y., & He, J. (2021). Design guidelines for chalcogenide-based flexible thermoelectric materials. *Materials Advances*, *2*(8), 2584–2593. DOI: 10.1039/D0MA01018A

Wang, Y., Zhong, Y., Zi, J., & Lian, Z. (2023). Type-I CdSe@CdS@ZnS Heterostructured Nanocrystals with Long Fluorescence Lifetime. *Materials (Basel)*, *16*(21), 7007. DOI: 10.3390/ma16217007 PMID: 37959604

Wang, Y., Zhu, Q., Tao, L., & Su, X. (2011). Controlled-synthesis of NiS hierarchical hollow microspheres with different building blocks and their application in lithium batteries. *Journal of Materials Chemistry*, *21*(25), 9248. DOI: 10.1039/c1jm10271k

Wang, Z., Cao, S.-W., Loo, S. C. J., & Xue, C. (2013). Nanoparticle heterojunctions in ZnS–ZnO hybrid nanowires for visible-light-driven photocatalytic hydrogen generation. *CrystEngComm*, *15*(28), 5688. DOI: 10.1039/c3ce40523k

Wang, Z., Tu, Q., Sim, A., Yu, J., Duan, Y., Poon, S., Liu, B., Han, Q., Urban, J. J., Sedlak, D., & Mi, B. (2020). Superselective removal of lead from water by two-dimensional MoS_2 nanosheets and layer-stacked membranes. *Environmental Science & Technology*, *54*, 12602–12611.

Wani, S. A., Tari, V., & Mansoor, S. I. U. (2025). A Comprehensive Analysis of Wastewater Management Challenges in India: Infrastructure and Policy Perspectives. In Dhiman, S., & Mukherjee, G. (Eds.), *Resource Resurgence. Springer Water*. Springer., DOI: 10.1007/978-3-032-02750-4_11

Warren, D. B., King, D., Benameur, H., Pouton, C. W., & Chalmers, D. K. (2013). Glyceride lipid formulations: Molecular dynamics modeling of phase behavior during dispersion and molecular interactions between drugs and excipients. *Pharmaceutical Research*, *30*(12). Advance online publication. DOI: 10.1007/s11095-013-1206-1 PMID: 24092054

Wenda, Z., Shihao, D., Weidong, Z., Dan, W., Pai, L., Xiangwei, Q., Haochen, L., Hongcheng, Y., Zhenghui, W., Kai, W., & Xiao, W.S. (2020). InP/ZnS/ZnS Core/Shell Blue Quantum Dots for Efficient Light-Emitting Diodes, *Adv. Funct. Mater.* *30*,49, 2005303. DOI: 10.1002/adfm.202005303

Werlinger, F., Segura, C., Martínez, J., Osorio-Roman, I., Jara, D., Yoon, S. J., & Gualdrón-Reyes, A. F. (2023). Current progress of efficient active layers for organic, chalcogenide and perovskite-based solar cells: A perspective. *Energies*, *16*(16), 5868.

Weshahy, A. R., Sakr, A. K., Gouda, A. A., Atia, B. M., Somaily, H. H., Hanfi, M. Y., Sayyed, M. I., El Sheikh, R., El-Sheikh, E. M., Radwan, H. A., Cheira, M. F., & Gado, M. A. (2022). Selective Recovery of Cadmium, Cobalt, and Nickel from Spent Ni–Cd Batteries Using Adogen® 464 and Mesoporous Silica Derivatives. *International Journal of Molecular Sciences*, *23*(15), 8677. DOI: 10.3390/ijms23158677

Williams, P., Keshavarz-Moore, E., & Dunnill, P. (2002). Efficient production of microbially synthesized cadmium sulphide quantum semiconductor crystallite. *Enzyme and Microbial Technology*, *30*(3), 354–362. DOI: 10.1016/S0141-0229(01)00500-1

Wilson, J. A., Di Salvo, F. J., & Mahajan, S. (1975). Charge-density waves and superlattices in the metallic layered transition metal dichalcogenides. *Advances in Physics*, *24*(2), 117–201.

Wilson, J. A., & Yoffe, A. D. (1969). The transition metal dichalcogenides discussion and interpretation of the observed optical, electrical and structural properties. *Advances in Physics*, *18*(73), 193–335.

Won, Y. H., Cho, O., & Kim, T. (2019). Highly Efficient and Stable InP/ZnSe/ZnS Quantum Dot Light-emitting Diodes. *Nature*, *575*, 634–638. DOI: 10.1038/s41586-019-1771-5 PMID: 31776489

Wu, H., Wang, Z., Li, Z., Ma, Y., Ding, F., Li, F., Bian, H., Zhai, Q., Ren, Y., Shi, Y., Yang, Y., Deng, Y., Tang, S., & Meng, X. (2023). Medium-Entropy Metal Selenides Nanoparticles with Optimized Electronic Structure as High-Performance Bifunctional Electrocatalysts for Overall Water Splitting. *Advanced Energy Materials*, *13*(28), 2300837. DOI: 10.1002/aenm.202300837

Wu, L., Li, Y., Liu, G. Q., & Yu, S. H. (2024). Polytypic metal chalcogenide nanocrystals. *Chemical Society Reviews*. Advance online publication. DOI: 10.1039/d3cs01095c PMID: 39212091

Wu, L., Shen, X., Ji, Z., Yuan, J., Yang, S., Zhu, G., Chen, L., Kong, L., & Zhou, H. (2023). Facile Synthesis of Medium-Entropy Metal Sulfides as High-Efficiency Electrocatalysts toward Oxygen Evolution Reaction. *Advanced Functional Materials*, *33*(3), 2208170. DOI: 10.1002/adfm.202208170

Wu, Y., Xing, H., Lian, C. S., Lian, H., He, J., Duan, W., & Liu, Y. (2019). Ion intercalation engineering of electronic properties of two-dimensional crystals of 2H-$TaSe_2$. *Physical Review Materials*, *3*(10), 104003.

Xia, C., , Li, J., & . (2016). Recent advances in optoelectronic properties and applications of two-dimensional metal chalcogenides. *Journal of Semiconductors*, *37*(5), 051001. DOI: 10.1088/1674-4926/37/5/051001

Xiang, T. X., & Anderson, B. D. (2013). Molecular dynamics simulation of amorphous indomethacin. *Molecular Pharmaceutics*, *10*(1), 102–114. DOI: 10.1021/mp3000698 PMID: 23116319

Xiang, T. X., & Anderson, B. D. (2017). Molecular dynamics simulation of amorphous hydroxypropylmethylcellulose and its mixtures with felodipine and water. *Journal of Pharmaceutical Sciences*, *106*(3), 803–816. DOI: 10.1016/j.xphs.2016.10.026 PMID: 27986292

Xiao, M., & Yang, L. (2014). *Binary and ternary metal chalcogenide materials and method of making and using same* (U.S. Patent No. [insert patent number]). U.S. Patent and Trademark Office.

Xiao, D.. (2012). Coupled spin and valley physics in monolayers of MoS_2 and other group-VI dichalcogenides. *Physical Review Letters, 108*, 196802.

Xia, X., Zhou, Z., Wu, S., Wang, D., Zheng, S., & Wang, G. (2018). Adsorption removal of multiple dyes using biogenic selenium nanoparticles from an *Escherichia coli* strain overexpressed selenite reductase CsrF. *Nanomaterials (Basel, Switzerland), 8*(4), 234. DOI: 10.3390/nano8040234 PMID: 29649129

Xie, C., Yang, P., Huan, Y., Cui, F., & Zhang, Y. (2020). Roles of salts in the chemical vapor deposition synthesis of two-dimensional transition metal chalcogenides. *Dalton Transactions (Cambridge, England), 49*(30), 10319–10327. DOI: 10.1039/D0DT01561J PMID: 32648888

Xie, H., Hao, S., Bao, J., Slade, T. J., Snyder, G. J., Wolverton, C., & Kanatzidis, M. G. (2020). All-Inorganic Halide Perovskites as Potential Thermoelectric Materials: Dynamic Cation off-Centering Induces Ultralow Thermal Conductivity. *Journal of the American Chemical Society, 142*(20), 9553–9563. DOI: 10.1021/jacs.0c03427 PMID: 32320237

Xiong, B., Chen, L., & Shi, J. (2018). Anion-containing noble-metal-free bifunctional electrocatalysts for overall water splitting. *ACS Catalysis, 8*, 3688–3707. DOI: 10.1021/acscatal.7b04286

Xiong, W., Huang, K., & Yuan, S. (2019). The mechanical, electronic and optical properties of two-dimensional transition metal chalcogenides MX 2 and M 2 X 3 (M = Ni, Pd; X = S, Se, Te) with hexagonal and orthorhombic structures. *Journal of Materials Chemistry. C, Materials for Optical and Electronic Devices, 7*(43), 13518–13525. DOI: 10.1039/C9TC04933A

Xuan, R., Mo, J., Chen, J., Dou, Y., Li, X., Jiang, Z., Chai, B., Wang, C., Ding, D., Yan, J., & Wang, X. (2024). Enhanced photocatalytic CO_2 conversion of a CdS/Co-BDC nanocomposite via Co(II)/Co(III) redox cycling. *RSC Advances, 14*(35), 25247–25255. DOI: 10.1039/d4ra04842c PMID: 39139241

Xu, L., Li, D., Zhou, W., Ding, Y., Wu, Y., Xu, J., & Duan, X. (2020). High-efficiency electrodeposition of polyindole nanocomposite using MoS2 nanosheets as electrolytes and their capacitive performance. *Arabian Journal of Chemistry, 13*(7), 6061–6071. DOI: 10.1016/j.arabjc.2020.05.006

Xu, L., Xiong, Y., Mei, A., Hu, Y., Rong, Y., Zhou, Y., Hu, B., & Han, H. (2018). Efficient Perovskite Photovoltaic-Thermoelectric Hybrid Device. *Advanced Energy Materials, 8*(13), 1702937. DOI: 10.1002/aenm.201702937

Xu, X. B., Zhong, W., Zhang, X., Dou, J., Xiong, Z. G., Sun, Y., Wang, T. T., & Du, Y. W. (2019). Flexible symmetric supercapacitor with ultrahigh energy density based on NiS/MoS_2@N-rGO hybrids electrode. *Journal of Colloid and Interface Science, 543*, 147–155. DOI: 10.1016/j.jcis.2019.02.054 PMID: 30797998

Xu, Y., Li, R., Bai, S., Li, Y., Jia, Z., Yang, Y., Cui, E., Yao, F., Wang, D., Lei, C., & Lin, Q. (2023). Chalcogenide-Based Narrowband Photodetectors for Imaging and Light Communication. *Advanced Functional Materials, 33*(11), 2212523. DOI: 10.1002/adfm.202212523

Yadav, S., Yashas, S. R., & Shivaraju, H. P. (2021). Transitional metal chalcogenide nanostructures for remediation and energy: A review. *Environmental Chemistry Letters, 19*(5), 3683–3700. DOI: 10.1007/s10311-021-01269-w

Yalçın, M., Turgut, N., & Gökbulut, C., & Others. (2023). Removal of pesticide residues from apple and tomato cuticle. *Environmental Science and Pollution Research International, 30*, 15821–15829. DOI: 10.1007/s11356-022-23269-1 PMID: 36171324

Yang, G. (2024). *Composite Strategy to Enhance the Thermoelectric Performance of Chalcogenide-based Materials* [Thesis, University of Wollongong]. https://ro.uow.edu.au/articles/thesis/Composite_Strategy_to_Enhance_the_Thermoelectric_Performance_of_Chalcogenide-based_Materials/27669117/1

Yang, H., Liang, A., Chen, C., Zhang, C., Schroeter, N. B., & Chen, Y. (2018). Visualizing electronic structures of quantum materials by angle-resolved photoemission spectroscopy. *Nature Reviews. Materials, 3*(9), 341–353.

Yang, S., Jiao, S., Nie, Y., Zhao, Y., Gao, S., Wang, D., & Li, Y. (2024). Boosting photoresponse and stability of a self-powered photoelectrochemical-type Bi2Te3 imaging photodetector by a novel quasi-solid-state gel electrolyte. *Journal of Alloys and Compounds, 982*, 173718.

Yang, X., Liu, H., Li, T., Huang, B., Hu, W., Jiang, Z., Chen, J., & Niu, Q. (2020). Preparation of flower-like ZnO@ZnS core-shell structure enhances photocatalytic hydrogen production. *International Journal of Hydrogen Energy, 45*(51), 26967–26978. DOI: 10.1016/j.ijhydene.2020.07.027

Yang, X., Zhao, Z., Yu, X., & Feng, L. (2019). Electrochemical hydrogen evolution reaction boosted by constructing Ru nanoparticles assembled as a shell over semimetal Te nanorod surfaces in acid electrolyte. *Chemical Communications (Cambridge), 55*, 1490–1493. DOI: 10.1039/C8CC09993F

Yang, Y., & Lin, Q. (2025). Chalcogenide semiconductors: Charge carrier dynamics and optoelectronic applications. *Chemical Communications*. Advance online publication. DOI: 10.1039/D5CC03022F PMID: 40665834

Yan, L., Wu, Z., Li, C., & Wang, J. (2023). Sb-doped SnS_2 nanosheets enhance electrochemical reduction of carbon dioxide to formate. *Journal of Industrial and Engineering Chemistry*, *123*, 33–40. DOI: 10.1016/j.jiec.2023.03.014

Yao, Z., Yang, J., Ren, Z., Ren, X., Sun, Y., Zhao, Y., Si, J., Lei, Q., Zhang, W., Li, Z., Yin, Y., Chen, J., Wen, W., Zhu, D., Li, X., & Tai, R. (2023). Triggering Mixed Cationic-Anionic Redox in Cu_{2-x} Se Cathodes via Tailored Charge-Carrier for High Energy Density Aqueous Zn Batteries. *Advanced Energy Materials*, *13*(22), 2300236. DOI: 10.1002/aenm.202300236

Yavorskiy, D., Szoła, M., Karpierz, K., Rudniewski, R., Bożek, R., Karczewski, G., Wojtowicz, T., Wróbel, J., & Łusakowski, J. (2020). Polarization of Magnetoplasmons in Grating Metamaterials Based on CdTe/CdMgTe Quantum Wells. *Materials (Basel)*, *13*(8), 1811. DOI: 10.3390/ma13081811 PMID: 32290498

Ye, A., Fan, W., Zhang, Q., Deng, W., & Wang, Y. (2012). CdS–graphene and CdS–CNT nanocomposites as visible-light photocatalysts for hydrogen evolution and organic dye degradation. *Catalysis Science & Technology*, *2*, 969–978.

Ye, X. R., Jia, D. Z., Yu, J. Q., Xin, X. Q., & Xue, Z. (1999). One-step solid-state reactions at ambient temperatures: A novel approach to nanocrystal synthesis. *Advanced Materials*, *11*(11), 941–942. DOI: 10.1002/(SICI)1521-4095(199908)11:11<941:AID-ADMA941>3.0.CO;2-W

Ye, X., Zhu, T., & Hui, Z.. (2021). Revealing the transfer mechanisms of photogenerated charge carriers over g-C3N4/ZnIn2S4 composite: A model study for photocatalytic oxidation of aromatic alcohols with visible light. *Journal of Catalysis*, *401*, 149–159. DOI: 10.1016/j.jcat.2021.07.025

Yoshida, M., Suzuki, R., Zhang, Y., Nakano, M., & Iwasa, Y. (2015). Memristive phase switching in two-dimensional 1T-TaS2 crystals. *Science Advances*, *1*(9), e1500606.

Yousef, M. A., & Vangala, V. R. (2019). Pharmaceutical cocrystals: Molecules, crystals, formulations, medicines. *Crystal Growth & Design*, *19*(12), 7420–7438. DOI: 10.1021/acs.cgd.8b01898

Yuan, H., Kong, L., Li, T., & Zhang, Q. (2017). A review of transition metal chalcogenide/graphene nanocomposites for energy storage and conversion. *Chinese Chemical Letters*, *28*(12), 2180–2194. DOI: 10.1016/J.CCLET.2017.11.038

Yuan, S., Duan, X., Liu, J., Ye, Y., Lv, F., Liu, T., Wang, Q., & Zhang, X. (2021). Recent progress on transition metal oxides as advanced materials for energy conversion and storage. *Energy Storage Materials*, *42*, 317–369. DOI: 10.1016/j.ensm.2021.07.007

Yue, L., Wang, J., Zhang, Y., Qi, S., & Xin, B. (2016). Controllable biosynthesis of high-purity lead-sulfide (PbS) nanocrystals by regulating the concentration of polyethylene glycol in microbial system. *Bioprocess and Biosystems Engineering*, *39*(12), 1839–1846. DOI: 10.1007/s00449-016-1662-7 PMID: 27541823

Yu, J., Jin, J., Chenga, B., & Mietek Jaroniec, M. (2024). A noble metal-free reduced graphene oxide–CdS nanorod composite for the enhanced visible-light photocatalytic reduction of CO_2 to solar fuel. *Journal of Materials Chemistry. A, Materials for Energy and Sustainability*, *2*, 3407–3416. DOI: 10.1039/C3TA14493C

Yu, L., Zhu, Q., Song, S., McElhenny, B., Wang, D., Wu, C., Qin, Z., Bao, J., Yu, Y., Chen, S., & Ren, Z. (2019). Non-noble metal-nitride based electrocatalysts for high-performance alkaline seawater electrolysis. *Nature Communications*, *10*(1), 5106. DOI: 10.1038/s41467-019-13092-7 PMID: 31704926

Yu, S., Wu, X., Wang, Y., Guo, X., & Tong, L. (2017). 2D materials for optical modulation: Challenges and opportunities. *Advanced Materials*, *29*(14), 1606128.

Yuwono, W., Susanna, Sasmita, E. W., & Sihotang, W. H. (2023). Analysis of the Influence of the Role of Financial Literacy on Personal Financial Management. *European Journal of Business and Management Research*, *8*(3), 160–165.

Yu, Y., Xu, J., Zhang, J., Li, F., Fu, J., & Li, C.. (2020). Ultra-thin sns2-pt nanocatalyst for efficient hydrogen evolution reaction. *Nanomaterials (Basel, Switzerland)*, *10*, 2337–2411. DOI: 10.3390/nano10122337

Zakaria, M. B., Guo, Y., Na, J., Tahawy, R., Chikyow, T., El-Said, W. A., El-Hady, D. A., Alshitari, W., Yamauchi, Y., & Lin, J. (2020). Layer-by-Layer Motif Heteroarchitecturing of N,S-Codoped Reduced Graphene Oxide-Wrapped Ni/NiS Nanoparticles for the Electrochemical Oxidation of Water. *ChemSusChem*, *13*(12), 3269–3276. DOI: 10.1002/cssc.202000159

Zakery, A., & Elliott, S. R. (2003). Optical properties and applications of chalcogenide glasses: A review. *Journal of Non-Crystalline Solids*, *330*(1–3), 1–12. DOI: 10.1016/j.jnoncrysol.2003.08.064

Zakharchenko, K. V. (2011). Temperature effects on graphene: from flat crystal to 3D liquid. *Sl: sn.*

Zeng, R., Lian, K., Su, B., Lu, L., Lin, J., Tang, D., Lin, S., & Wang, X. (2021). Versatile Synthesis of Hollow Metal Sulfides via Reverse Cation Exchange Reactions for Photocatalytic CO_2 Reduction. *Angewandte Chemie International Edition, 60*(47), 25055–25062. DOI: 10.1002/anie.202110670 PMID: 34490697

Zeng, Y., Meng, Y., Lai, Z., Zhang, X., Yu, M., Fang, P., Wu, M., Tong, Y., & Lu, X. (2017). An Ultrastable and High-Performance Flexible Fiber-Shaped Ni–Zn Battery based on a Ni–NiO Heterostructured Nanosheet Cathode. *Advanced Materials, 29*(44). Advance online publication. DOI: 10.1002/adma.201702698

Zhai, P., Zhang, Y., Wu, Y., Gao, J., Zhang, B., Cao, S., Zhang, Y., Li, Z., Sun, L., & Hou, J. (2020). Engineering active sites on hierarchical transition bimetal oxides/sulfides heterostructure array enabling robust overall water splitting. *Nature Communications, 11*(1), 5462. DOI: 10.1038/s41467-020-19214-w PMID: 33122636

Zhang, D., Li, L., & Zhang, Y. (2021). Metal chalcogenides-based materials for high-performance metal ion capacitors. *Journal of Alloys and Compounds, 869*, 159352. DOI: 10.1016/j.jallcom.2021.159352

Zhang, F., Zhu, B., Guo, H., Qiu, J., Zheng, K., Chen, X., & Yu, J. (2021). Monolayer square-Ag_2X (X = S, Se): Excellent n type thermoelectric materials with high power factors. *Applied Surface Science, 550*, 149230.

Zhang, G., & Zhang, Y.-W. (2017). Thermoelectric properties of two-dimensional transition metal dichalcogenides. *Journal of Materials Chemistry. C, Materials for Optical and Electronic Devices, 5*(31), 7684–7698. DOI: 10.1039/C7TC01088E

Zhang, H., Xu, P., Ni, Y., Geng, H., Zheng, G., & Dong, B.. (2014a). In situ chemical synthesis of SnO2/reduced graphene oxide nanocomposites as anode materials for lithium-ion batteries. *Journal of Materials Research, 29*, 617–624. DOI: 10.1557/jmr.2014.37

Zhang, J., & Dai, L. (2016). Nitrogen, phosphorus, and fluorine tri-doped graphene as a multifunctional catalyst for self-powered electrochemical water splitting. *Angewandte Chemie International Edition, 55*, 13296–13300. DOI: 10.1002/anie.201607405

Zhang, J., Sun, E., Feng, X., Liu, H., Redfern, S. A., Kanchana, V., Liu, G., & Wang, H. (2018). Phase transition and superconductivity in ReS_2, $ReSe_2$ and $ReTe_2$. *Physical Chemistry Chemical Physics, 20*(46), 29472–29479.

Zhang, K., & Guo, L. (2013). Metal sulphide semiconductors for photocatalytic hydrogen production. *Catalysis Science & Technology, 3*, 1672–1690.

Zhang, L. D., & Mou, J. M. (1994). *Nami cailiaoxue* [Nanomaterials]. Liaoning Science and Technology Press.

Zhang, L., Kong, D., Wang, H., Jiao, L., Zhao, X., Song, J., & Lu, Y. (2021). Cocrystal of apixaban–quercetin: Improving solubility and bioavailability of drug combination of two poorly soluble drugs. *Molecules (Basel, Switzerland), 26*(9), 2677. DOI: 10.3390/molecules26092677 PMID: 34063645

Zhang, S., Zhang, X., Sun, Y., Zhou, W., Cao, W., & Cheng, D. (2023). Investigation of the annealing treatment on the performance of TiO2 photoanode. *International Journal of Hydrogen Energy, 48*(77), 29996–30005.

Zhang, T., Liu, Y., Yu, J., Ye, Q., Yang, L., Li, Y., & Fan, H. J. (2022). Biaxially Strained MoS_2 Nanoshells with Controllable Layers Boost Alkaline Hydrogen Evolution. *Advanced Materials, 34*(27), 2202195. DOI: 10.1002/adma.202202195 PMID: 35474349

Zhang, Y., Cao, N., Szunerits, S., Addad, A., Roussel, P., & Boukherroub, R. (2019). Fabrication of ZnCoS nanomaterial for high energy flexible asymmetric supercapacitors. *Chemical Engineering Journal, 374*, 347–358. DOI: 10.1016/j.cej.2019.05.181

Zhang, Y., Zhang, L., Lv, T., Chu, P. K., & Huo, K. (2020). Two-Dimensional Transition Metal Chalcogenides for Alkali Metal Ions Storage. *ChemSusChem, 13*(6), 1114–1154. DOI: 10.1002/cssc.201903245

Zhang, Y., Zhou, Q., Zhu, J., Yan, Q., Dou, S. X., & Sun, W. (2017). Nanostructured Metal Chalcogenides for Energy Storage and Electrocatalysis. *Advanced Functional Materials, 27*(35). Advance online publication. DOI: 10.1002/adfm.201702317

Zhang, Z., Liu, B., Zhou, J., & Sun, Z. (2025). Amorphous MoTex Nanomaterials Promote Visible-Light Co-Catalytic Degradation of Methylene Blue. *Materials (Basel), 18*(14), 3388. DOI: 10.3390/ma18143388 PMID: 40731598

Zhao, D., Wu, T., & Zhou, Y., Y. (2019). Dual II heterojunctions metallic phase MoS_2/ZnS/ZnO ternary composite with superior photocatalytic performance for removing contaminants. *Chemistry (Weinheim an der Bergstrasse, Germany), 25*, 9710–9720.

Zhao, D., Zhang, X., Wang, H., Zeng, H., Ma, H., Adam, J. L., & Chen, G. (2008a). Thermal properties of chalcogenide glasses in the GeSe2–As2Se3–CdSe system. *Journal of Non-Crystalline Solids, 354*(12–13), 1281–1284. DOI: 10.1016/J.JNONCRYSOL.2006.10.079

Zhao, J., Guan, B., Hu, B., Xu, Z., Wang, D., & Zhang, H. (2017). Vulcanizing time-controlled synthesis of NiS microflowers and its application in asymmetric supercapacitors. *Electrochimica Acta, 230*, 428–437. DOI: 10.1016/j.electacta.2017.02.032

Zhao, W.-W., Bothra, P., Lu, Z., Li, Y., Mei, L.-P., Liu, K., Zhao, Z., Chen, G., Back, S., Siahrostami, S., Kulkarni, A., Nørskov, J. K., Bajdich, M., & Cui, Y. (2019). Improved Oxygen Reduction Reaction Activity of Nanostructured CoS_2 through Electrochemical Tuning. *ACS Applied Energy Materials*, *2*(12), 8605–8614. DOI: 10.1021/acsaem.9b01527

Zhao, Y., Wei, S., Pan, K., Dong, Z., Zhang, B., Wu, H.-H., Zhang, Q., Lin, J., & Pang, H. (2021). Self-supporting transition metal chalcogenides on metal substrates for catalytic water splitting. *Chemical Engineering Journal*, *421*, 129645. DOI: 10.1016/j.cej.2021.129645

Zheng, D., Yang, X., Čuček, L., Wang, J., Ma, T., & Yin, C. (2024). Revolutionizing dye-sensitized solar cells with nanomaterials for enhanced photoelectric performance. *Journal of Cleaner Production*, *464*, 142717. DOI: 10.1016/j.jclepro.2024.142717

Zheng, J., Cao, Y., Cheng, C., Chen, C., Yan, R. W., Huai, H. X., Dong, Q. F., Zheng, M. S., & Wang, C. C. (2014). Facile synthesis of Fe_3S_4 hollow spheres with high-performance for lithium-ion batteries and water treatment. *Journal of Materials Chemistry. A, Materials for Energy and Sustainability*, *2*(46), 19882–19888.

Zheng, J., Ma, M., Yang, G., Wu, Y., & Mei, D. (2024). Progress in the study of binary chalcogenide-based thermoelectric compounds. *Journal of Solid State Chemistry*, *334*, 124617. DOI: 10.1016/j.jssc.2024.124617

Zheng, J., Zhang, R., Wang, X., & Yu, P. (2018). Synthesizing a flower-like NiO and ZnO composite for supercapacitor applications. *Research on Chemical Intermediates*. Advance online publication. DOI: 10.1007/s11164-018-3441-x

Zheng, Y., Wang, X., Zhao, W., Cao, X., & Liu, J. (2018). Phytic acid-assisted synthesis of ultrafine NiCo2S4 nanoparticles immobilized on reduced graphene oxide as high-performance electrode for hybrid supercapacitors. *Chemical Engineering Journal*, *333*, 603–612. DOI: 10.1016/j.cej.2017.10.008

Zheng, Z., Yu, L., Gao, M., Chen, X., Zhou, W., Ma, C., Wu, L., Zhu, J., Meng, X., Hu, J., Tu, Y., Wu, S., Mao, J., Tian, Z., & Deng, D. (2020). Boosting hydrogen evolution on MoS2 via co-confining selenium in surface and cobalt in inner layer. *Nature Communications*, *11*(1), 3315. DOI: 10.1038/s41467-020-17199-0 PMID: 32620781

Zhou, Y. M., & Xin, X. Q. (1999). Synthetic chemistry for solid state reaction at low-heating temperatures. *Wuji Huaxue Xuebao*, •••, 15.

Zhu, C., Liu, C., Fu, Y., Gao, J., Huang, H., Liu, Y., & Kang, Z. (2019). Construction of CDs/CdS photocatalysts for stable and efficient hydrogen production in water and seawater. *Applied Catalysis B: Environmental*, *242*, 178–185. DOI: 10.1016/j.apcatb.2018.09.096

Zhu, J.-J., & Wang, H. (2003). Synthesis of metal chalcogenide nanoparticles. In Nalwa, H. S. (Ed.), *Encyclopedia of nanoscience and nanotechnology* (Vol. 10, pp. 1–21).

Zhu, L., Wang, Y., Qin, C., & Cao, J. (2022). First Principles Study of Photocatalytic Reduction of CO_2 to CH_4 on WS_2-Supported Pt Clusters (2022). *The Journal of Physical Chemistry. C, Nanomaterials and Interfaces*, *126*, 16702–16709. DOI: 10.1021/acs.jpcc.2c06126

Zhu, Q., Wang, S., Wang, X., Suwardi, A., Chua, M. H., Soo, X. Y. D., & Xu, J. (2021). Bottom-Up Engineering Strategies for High-Performance Thermoelectric Materials. *Nano-Micro Letters*, *13*(1), 119. DOI: 10.1007/s40820-021-00637-z PMID: 34138379

Znidi, F., Morsy, M., & Uddin, Md. N. (2024). Navigating challenges and solutions for metal-halide and carbon-based electrodes in perovskite solar cells (NCS-MCEPSC): An environmental approach. *Heliyon*, *10*(12), e32843. DOI: 10.1016/j.heliyon.2024.e32843 PMID: 38988552

Zuo, X., Qiu, Y., Zhen, M., Liu, D., & Zhang, Y. (2025). Review on MXenes-based electrocatalysts for high-energy-density lithium–sulfur batteries. *Nano-Micro Letters*, *17*. Advance online publication. DOI: 10.1007/s40820-025-01726-z PMID: 40208556

About the Contributors

Karthik Kannan is currently serving as a Research Faculty in the Department of Mechanical Engineering at National Chung Cheng University, Taiwan, and as Adjunct Professor at Saveetha Institute of Technical and Medical Sciences, Chennai, India and Karpagam Academy of Higher Education, Coimbatore, India. Dr. Kannan is the recipient of the prestigious TEGC (Science and Technology specialization), approved by the NSTC. He was also honored with the highly competitive Brain Pool Fellowship from the NRF, which he held at Kumoh National Institute of Technology, South Korea. Dr. Karthik Kannan has an extensive publication record with over 170 peer-reviewed research articles with 7,960 citations (Google Scholar, November 2025), with an h-index of 56 and an i10-index of 122. Notably, Stanford University has listed him among the World's Top 2% Scientists for six consecutive years (2020–2025), with distinction in both single-year and career-long impact rankings (1960–2023).

Tari Vinaya Satyawan Savitri is working as a Postdoctoral fellow at Universitas Airlangga, Indonesia. She formerly worked as a 'Grants Writer - Offshore' at Envirotech Education Pty. Ltd., Australia. She has completed her Ph.D. in Environmental Science from the University of Mumbai. She did undergraduate studies in Zoology, M.Sc., in Environmental Science, and a Diploma in Biotechnology (2 years) from the University of Mumbai and a B.Ed. from S.N.D.T. University, Mumbai, India. She has qualified for the State Level Eligibility Test (SLET) for Assistant Professorship (valid in Maharashtra and Goa, India). She has been awarded many prestigious awards, including "Young Scientist Award" from the Deccan Environmental Research Organization (DERO), India "ADAMYA -The Unconquerable, Empowering Women" from the Gender Sensitization Cell, SIES College of Arts, Commerce, and Science, Nerul, Navi Mumbai, Maharashtra, India "A Friend of the Blue Planet" from the Australian Centre for Sustainable Development, Research, and Innovation (ACSDRI), Australia "Citation of Honour;

Rising Star of Viksit Bharat" from Vigyan Setu Foundation, Mumbai, India She is an 'Academic editor" at PLOS ONE, Journal (USA), an early career editorial board member at "Microbial Biotechnology" (United Kingdom), an editorial board member at "Farming and Management" (India), "Open Journal of Agricultural Research" (USA) etc. Recently, her edited book entitled "Novel Energy Storage and Conversion Technologies for Two-Dimensional MXenes and MBenes" was published by IGI Global Publishing, USA. She has attended around 40 conferences/ workshops/ webinars/ faculty development programs. She has published over 40 research items in reputed Journals, Books, Magazines/e-newsletters, and conferences. She has authored book chapters with several reputed publishers such as Elsevier, Springer, IGI Global, CRC Press, Taylor and Francis, etc. She was an Associate Researcher at the Australian Centre for Sustainable Development, Research, and Innovation (ACSDRI), Australia (2020-2023). She has successfully served as an organizing committee member, scientific committee member, conference treasurer, and hostess for many national and international conferences. She is also associated with various national and international scientific societies as a panel member, professional/ official member, life member, etc. Her current research interests are environmental economics, sustainable crop protection practices, circular economy, environmental pollution, nanotoxicology, and biomedical applications of nanomaterials. She has also been a reviewer for around 30 reputed scientific journals. She has collaborated with overseas countries such as the USA, Australia, Qatar, Indonesia, Nigeria, Thailand, Malaysia, Turkey, Taiwan, Russia, etc.

Shoeab Akhter is a passionate researcher and practicum student with a B.Sc. in the Agriculture Department of the College of Agricultural Sciences at IUBAT—International University of Business Agriculture and Technology, Dhaka 1230, Bangladesh. He is doing an internship at the Bangladesh Atomic Energy Commission. His research title is 'Green Synthesis of Nanoparticles for Agricultural Use.' Loved by nature and with a passion for the environment, his pursuits include areas of interest like agricultural sciences, food security and food science, environmental science, waste management, green energy, AI, IoT-based farming, and agritourism, among others. Along with his objective of creating an impact through writing and research, he combines new, innovative, suitable technologies and nature conservation with real-world issues. He has contributed book chapters (Emerald, CABI, Bentham Science, Springer, Cambridge Scholars) and has participated in various major projects on research. With the hope of bringing about a change in the world, he aspires to contribute to global knowledge and create a lasting impact on sustainability in society.

Hari Shankar Biswas is Assistant Professor in Chemistry Surendranath College, Kolkata. Dr. Biswas is an esteemed Assistant Professor in Chemistry at Surendranath College, Kolkata, with a multifaceted academic background and a strong dedication to research and education. He embarked on his academic journey by obtaining his B.Sc. degree in Chemistry Honours from R.K.M.V.C. College in 2005, followed by his M.Sc. degree from Presidency College, Kolkata, in 2008. Dr. Biswas further pursued his scholarly pursuits by earning a Ph.D. in Material Science from the Saha Institute of Nuclear Physics, University of Calcutta, in 2015. His professional experience spans various roles, including a stint as a School teacher at Keshub Academy, Kolkata, from 2008 to 2010, before transitioning into academia as an Assistant Professor at Surendranath College. Dr. Biswas specializes in the research domain of Carbon thin Film, Graphene, and Graphene Oxide-based nano Composites, focusing on their synthesis, characterization, and application across Environmental, Biological, and Technological fields. Dr. Biswas's scholarly contributions are significant, with authorship of 15 book chapters and over 31 papers published in esteemed International Journals. He is actively involved in academic societies, holding life membership in the Indian Science Congress Association and the Indian Chemical Society, along with an affiliated membership with the Royal Society of Chemistry. In addition to his research and academic endeavors, Dr. Biswas plays a vital role in the dissemination of scientific knowledge as a series book Editor and reviewer for reputable international publishers and journals. He serves as an Editorial board member for the American Journal of Chemical Engineering, Reseapro Journal, Lincoln University College, etc., and is affiliated with Rotary International. Dr. Biswas's editorial contributions extend to editing various books, including titles such as "Emerging Concepts in Chemical and Biological Science" and "Modern Approaches in Chemical and Biological Sciences," among others. Through his comprehensive involvement in research, education, and editorial responsibilities, Dr. Hari Shankar Biswas continues to make valuable contributions to the field of Chemistry, fostering advancements and innovations for the betterment of society.

G.Dhanalakshmi is working as Assistant Professor,Department of Physics,at Mannar Thirumalai Naicker College,(Autonomous)Madurai.She has 16 years of teaching experience.Her research area is nanomaterials,crystaloography and computational Physics.She has published 15 papers in reputed journals.

Gurushankar Krishnamoorthy is currently working as a Senior Researcher, Laboratory of Computational Modelling of Drugs, South Ural State University, Chelyabinsk in Russia. He also holds the position of Distinguished Adjunct Faculty at Saveetha Institute of Technical and Medical Sciences in India and serves as an

Assistant Professor, Department of Physics, Kalasalingam Academy of Research and Education, Krishnankoil, Tamil Nadu, India. He has started working as the guest editor for MDPI Hydrogen special issue. Prior to joining South Ural State University, Dr. Gurushankar Krishnamoorthy was a Postdoctoral Researcher (July 2018-Oct 2018) at University of Airlangga, Surabaya, Indonesia. Gurushankar Krishnamoorthy is the author and co-author of over 50 research articles in peer-reviewed scientific journals, with more than 865 citations (as of May 2024, data from Google Scholar). He has an h-index of 17 and an i10-index of 27. Gurushankar Krishnamoorthy has finished 4 students (One Ph.D., Two Undergraduate and one PG student) under his supervision. Furthermore, he has filed an Indian patent and finished three projects for the Russian Ministry of Science and Education. His research interests include the biochar phytocompounds for practical applications in supercapacitors, and biomedicine.

A.J.M. Sirajul Karim (sirajul_bsmrau@yahoo.com), International University of Business Agriculture and Technology, Dhaka-1230, Bangladesh Prof. A. J. M. Sirajul Karim was born on January 15, 1954 in Feni. He earned a B.Sc.Ag. in 1977 and an M.Sc.Ag. (Soil Science) in 1979 from Bangladesh Agricultural University (BAU). Joining Bangladesh Agriculture Research Institute (BARI) in 1979, he became Senior Scientific Officer in 1981. He received a USAID scholarship for an M.S. at Colorado State University in 1983. He joined IPSA Gazipur Agriculture University (GAU) as Assistant Professor in 1987. He earned a Ph.D. in 1995 and Postdoc in 2000 from Kyushu University, Japan. He supervised 17 Ph.D. and 37 M.S. students, published over 115 papers, served Gazipur Agriculture University (GAU) in key roles, and led a national project on Hill Soils from 2013 to 2018. At present he is Chair and Professor, College of Agricultural Sciences, International University of Business Agriculture and Technology (IUBAT).

Sudipta Koley is an Assistant Professor of Physics at Amity University, Kolkata. She received her Ph.D. degree from IIT Kharagpur, specializing in topological materials, quantum phase transitions, and strongly correlated systems in 2013. With over 30 research publications (Orcid ID: 0000-0002- 9174-4229) in reputed journals, her work spans charge-density waves, excitonic phenomena, superconductivity, and energy storage materials. She has also been a DST Woman Scientist awardee. She has over 12 years of experience in teaching and research.

Aman Kumar is a distinguished physicist with a strong academic background and a deep passion for research. He earned his B.Sc, M.Sc, and Ph.D. degrees from Chaudhary Charan Singh University, Meerut, establishing a solid foundation in the field of Physics. For the past three years, Dr. Kumar has been serving as an Assistant

Professor in the Department of Physics at Swami Vivekanand Subharti University, Meerut, India, where he has actively contributed to academia and research. His dedication to scientific exploration is evident in his extensive body of work, which includes over 33 research papers published in national and international journals and conference proceedings. In addition to his research publications, Dr. Kumar has authored one book, four patents, and ten book chapters, further demonstrating his expertise and commitment to advancing knowledge. His primary research focus lies in Theoretical Materials Science, where he continues to push the boundaries of scientific understanding, making significant contributions to the field.

Amit Kumar Kundu is Assistant Professor & HOD Department of Chemistry Sripat Singh College, Jiaganj, Murshidabad. Dr. Amit Kumar Kundu is an Assistant professor and HOD at the Department of Chemistry, Sripat Singh College, Jiaganj, Murshidabad. Dr. Kundu did B.Sc. from Krishnagar Government College, University of Kalyani in 2007 and completed MSc. Degree from Department of Chemistry, University of Kalyani in 2009. After that he received his PhD in Analytical Chemistry in 2017 from University of Kalyani. His research mainly focuses on Environmental Analytical Chemistry, carbon nano materials, Graphene, Graphene Oxide-basednano composites Synthesis, Characterization and its application. He has huge experience in Groundwater and surface water onsite sampling, sediment sampling from vadose zone and aquifer. He is also proficient in operating and managing various advanced analytical instruments. To date, he has published near about eighteen peer-reviewed journal papers and he is the author of five book chapters. He received World Environment and Livelihood Award form 'International Foundation for Environment and Ecology'.

Dilip Kumar Maiti is an eminent chemist with a distinguished academic and professional career. He completed his B.Sc. in Chemistry in 1991 from Ramakrishna Mission Vidyamandir, University of Calcutta, followed by an M.Sc. in 1993 from the University College of Science, Calcutta University. In 1998, he earned his Ph.D. from the Indian Institute of Chemical Biology, Jadavpur University. Dr. Maiti pursued postdoctoral research at Jadavpur University (2000–2002) and later at Wayne State University, Detroit, USA (2002–2004). He began his career as a Scientist in the R&D departments of RPG Life Sciences and Ciba India. Transitioning to academia, he served as a Senior Lecturer at Dumkal College, Kalyani University, before joining the University of Calcutta as a Professor in 2005. Currently, he is the Vice-Chancellor of Biswa Bangla Biswabidyalay, Bolpur. Dr. Maiti's research interests include Organic Synthesis and Material Chemistry. He has been honored with several prestigious awards, including the FRSC (London), UGC Mid-Career Award (2017), and Basudev Banerjee Memorial Award. He is actively involved with numerous scientific societies and editorial boards, contributing significantly to the global scientific community.

Prasenjit Mandal is an accomplished Assistant Professor in Chemistry at Santipur College, where he began his role on August 7, 2020. Residing in Kalyani, Nadia, Dr. Mandal has a strong research background in graphene and its composite applications, with around 15 published papers that explore various functional aspects and uses of these materials. His contributions extend to about 5 book chapters, showcasing his expertise in advanced materials and their potential in innovative applications. Through his research and academic work, Dr. Mandal actively contributes to advancing knowledge in the field of graphene-based materials.

T.C. Manjunath was born on 6th February 1967. He did his B.E. from RV College of Engg. in 1989 (B'lore Univ), M.E. in Robotics & Automation from Govt.'s LD College of Engg., (Gujarat Univ., Ahmedabad) in 1992 with FCD-First Rank & PhD in Systems & Control Engg. from IIT Bombay in the year 2007, got best research scholar award for the entire IITB. He is currently working as the Dean Research (R & D) and as Professor in Dept. of Computer Science & Engg. with specialization in IoT, Cyber Security, Blockchain Technology in Rajarajeswari College of Engineering, Bengaluru, Karnataka. He has got a teaching (academic), research & administrative experience of more than 35+ years (10 years as Principal & 8 years as HoD, 1 year as Dean Research & rest as faculty) and has published more than 1500+ research papers in various national & international journals & conferences, with 175+ patents, 100+ book-chapters, 25 awards, 100s of Keynote talks and also published 25 text books in various field of science, engineering & technology, completed projects worth Rs. 9 Crores till date & guided 10 PhD scholars in VTU. He is a member of 30 Professional Society Chapters and is a fellow of IE, IETE, AMIEE, Sr. Member of IEEE and a chartered engineer.

M Shohidullah Miah (drshohidullah@iubat.edu), International University of Business Agriculture and Technology, Dhaka-1230, Bangladesh; Dr. Md. Shohidullah Miah is a professor at the College of Agricultural Sciences, IUBAT. Dr. Miah did his B.Sc. and Master's in Fisheries from Bangladesh Agricultural University. He was awarded his Ph.D. in Aquaculture at the University of Malaya, Malaysia. Dr. Miah has also been a Post-Doctoral Fellow at the Japan Society for the Promotion of Science (JSPS) and the Center for Excellence (COE) Japan. He has professional experience of over 33 years in fisheries research and environmental water pollution control. His special interests are wastewater treatment, fish population dynamics, mass culture of animal food, biochemical analysis, and food preservation.

Dr. Miah has published more than 40 academic papers and is the author of several technical reports.

N. Suganthi is an accomplished academic and researcher serving as an Assistant Professor in the Department of Physics at Government Arts College (Autonomous), Karur, Tamil Nadu, India. Dr. Suganthi expertise lies in the synthesis of nanomaterials and studying their diverse applications, including sensors, catalysts, energy storage and conversion. Additionally, Dr. Suganthi is the author and co-author of over 20 research articles in peer-reviewed scientific journals, with more than 400 citations (as of October 2025, according to Google Scholar). Her h-index is 12, and his i10-index is 13.

Santosh Kumar Nathsharma is a distinguished faculty member in the Department of Chemistry at Stewart Science College, located in Cuttack, Odisha. Known for his dedication to academic excellence and passion for chemical sciences, Dr. Nathsharma has played a vital role in mentoring students and contributing to the growth of the department. His commitment to both teaching and research reflects his deep engagement with the subject and his desire to inspire the next generation of scientists.

P. Selvakumar is currently an Associate Professor in the Department of Science and Humanities at Nehru Institute of Technology, Coimbatore, Tamil Nadu, India. Prior to joining Nehru Institute of Technology, he held positions as an Assistant Professor in the Department of Science and Humanities various engineering colleges in Coimbatore, Tamil Nadu, India, spanning from 2011 to 2023. Dr.P. Selvakumar's academic and research interests lie primarily in the field of phytochemistry. He has made significant contributions to his field, with an impressive record that includes 30 journal publications, 12 patents, and 20 conference papers. Additionally, he has organized 5 seminars and conferences, further demonstrating his commitment to academic and scientific exchange. Throughout his career, Dr. P.Selvakumar has been recognized for his dedication and excellence in teaching, having been honored with several awards, including the Best Teacher Awards. He holds a prominent position in the academic community, contributing actively to research.

P.Periasamy obtained his PhD from Anna university in 2020. He is authored more than 20 international reputed journals.

K. Pushpanathan is an accomplished academic and researcher serving as an Associate Professor in the Department of Physics at Government Arts College (Autonomous), Karur, Tamil Nadu, India. Dr. Pushpanathan expertise lies in the synthesis of nanomaterials and studying their diverse applications, including sensors, catalysts, energy storage and conversion. Additionally, Dr. Pushpanathan is the author and co-author of over 45 research articles in peer-reviewed scientific journals, with more than 1500 citations (as of October 2025, according to Google Scholar). His h-index is 20, and his i10-index is 29.

Mohammed Ataur Rahman (marahman@iubat.edu) is the Chairman, Department of Environmental Science, Director of IUBAT Institute of SDG Studies (IISS); Chairperson, Centre for Global Environmental Culture (CGEC) and Program on Education for Sustainability at IUBAT University. Professor Rahman is a Plantation Crop, Forestry and Agriculture Specialist and also Coordinator of WWOOF Bangladesh. He teaches Ecology and Crop Climatology, Environmental Science, Sustainable and Urban Agriculture and Plant Physiology. Professor Rahman worked as Contributing Author with Late Professor M Alimullah Miyan, Former Vice Chancellor and Lead Author IPCC Working Group II and Asian Disaster Preparedness Centre (ADPC) and Disaster Management. Moreover, he is the Executive Director and Coordinator of RCE Greater Dhaka acknowledged by UNU-IAS. He obtained specialized training on Sustainability from the Association of Advancement of Sustainability in Higher Education (aashe), USA and participated in the UNESCO-UNFCCC webinar series on Climate Change Education in 2023. Professor Rahman is also working as Chief Consultant of Forestry, Environment, Plantation Crops and Permaculture Consultancy and Research (FEPPCAR), Dhaka, Bangladesh. Previously, He worked as a Senior Manager, Forestry and Plantation Crops, with 13 and a half years of experience in the Finlays' Plantation Industry of Bangladesh working with a number of the World's leading brands including Finlay Tea, Timber, Bamboo, Rubber, Canes and Black pepper, Betel Leaf etc. He possess specialised knowledge of environmental and habitat of Tropical Forest Plants, especially, Tea and Rubber and Non-Timber Forest crops. He is also experienced with plantation establishment and management of plantation crops viz. Tea, Timber, Coconut, Bamboo, Fruits and Spices, etc. and forest resource exploitation and with specialisation in wood preservation techniques. He is actively associated with multidisciplinary activities like ecoagriculture, organic farming, permaculture, multiple and multi-strata cropping culture for sustainable rural livelihood. He worked for Sustainable Agriculture and Development with the World Bank's International Assessment of Agricultural Science and Technology (IAAST). Before joining Finlays, he served Bangladesh Forest Industries Development Corporation (BFIDC) and Rural Electrification Board (REB) of Bangladesh for twelve years with significant contribution in development of forest industries and timber products. He is an Author of 12 books, booklets and book chapters. Scientific Articles are being published regularly in the English and Bangla News Papers like The Daily Star, Bangladesh Observer, News Today, Financial Express and Inquilab, Destiny etc.

Habibur Rahman is a dedicated and promising undergraduate student and research collaborator in the Department of Mechanical Engineering at the IUBAT-International University of Business Agriculture and Technology, in Dhaka-1230, Bangladesh. He is developing a strong academic and research profile with particular interests in renewable energy, agricultural materials, and composite materials. In addition to his research contributions, he actively participates in academic activities at IUBAT. Through a combination of academic excellence, research engagement, and forward-thinking vision, Habibur Rahman is making notable strides in his journey toward becoming a skilled and impactful mechanical engineer.

S M Maksudur Rahman completed his bachelor's degree in the Department of Mechanical Engineering from IUBAT-International University of Business Agriculture and Technology, Dhaka, Bangladesh. His research area is composite materials, agricultural materials, nanoparticles, and additive manufacturing, where he is specifically interested in the utilization of natural fibers and advanced materials in engineering practices. Maksudur has worked on several academic resources, and some of his works include reviews of applications of natural fiber composites and composite materials in automobiles. By conducting the research, he is trying to contribute to sustainable and innovative engineering solutions, which makes him one of the promising young researchers in the area of mechanical engineering.

Sakibul Islam Ratul is a researcher in the Department of Agricultural Science of Daffodil International University, Bangladesh. He is highly academic and committed towards research-driven innovation. His scholarly and research work is on sustainable agrifood solutions, especially in post-harvest technology, plant pathology, soil health, and climate-resilient agrifood systems. He is also pursuing more advanced research on solar-biomass hybrid drying systems, soil microplastic pollution and its effects on water retention, carbon sequestration measures, plant extract to combat fungal pathogens, UV light treatment of farm soils, and microbial interactions of soil ecologies. In addition to coursework, he has acquired hands-on experience in fieldwork, laboratory work, and academic writing, which has allowed him to relate theoretical learning to practical agricultural problems. His interests lie in supporting a science-based approach to solving the problems of smallholder and off-grid farm communities in South Asia. He would like to advance to a new stage of education and to participate in the research work of the whole world, and in the long run, he hopes to develop sustainable technologies in farming that will enable the world to be free of food security problems and environmental issues.

Julieta Salazar-Echeagaray, Full-time teaching career professor of the Faculty of Sciences Economic Administrative in Mazatlán of the Autonomous University of

Sinaloa. Degree in Accounting, Degree in Management. Master's degree in corporate finance. Doctorate in administration.

Teresa Salazar-Echeagaray, Full-time Professor and Researcher. PhD in Administration. PRODEP Profile. Candidate Level of the National System of Researchers. Member of the Sinaloa System of Researchers and Technologists with the distinction of Honorary Researcher, ANFECA Certification. Publications in refereed and indexed journals. Active member of different académica networks. Legal representative of the Lupita y Naty Integral Wellbeing Center, A.C. Peer-reviewed of journals and indexed publications.

Index

Symbols

2D materials 20, 86, 88, 90, 116, 117, 118, 119, 120, 121, 124, 131, 133, 134, 140, 174, 301, 334, 396

A

Agriculture 381, 409, 410, 411, 412, 415, 416, 417, 418, 419, 421, 424, 425, 426, 427, 428, 429, 430, 431

Analytical Techniques 40, 200, 244, 334

and Non-volatile memory devices 141

B

Bandgap Engineering 215, 288, 403

Biomedical 6, 13, 59, 60, 64, 70, 76, 88, 98, 104, 300, 361, 432

C

chalcogenide 1, 2, 3, 4, 5, 6, 7, 8, 9, 10, 11, 12, 13, 14, 15, 16, 17, 18, 19, 20, 21, 22, 23, 24, 25, 26, 27, 28, 29, 30, 31, 32, 33, 34, 35, 36, 37, 38, 39, 40, 41, 42, 46, 47, 49, 50, 52, 59, 60, 61, 63, 74, 75, 76, 78, 79, 82, 84, 91, 98, 105, 107, 108, 110, 111, 112, 135, 141, 142, 144, 146, 147, 150, 151, 152, 153, 154, 155, 156, 157, 158, 159, 161, 162, 163, 164, 168, 177, 183, 211, 212, 213, 214, 215, 216, 217, 218, 219, 220, 221, 222, 223, 225, 226, 227, 228, 229, 230, 231, 232, 233, 234, 235, 236, 237, 238, 239, 240, 241, 243, 245, 246, 247, 249, 251, 277, 279, 282, 283, 284, 285, 286, 287, 288, 289, 290, 291, 292, 293, 294, 295, 296, 297, 298, 303, 304, 305, 306, 307, 308, 311, 314, 315, 316, 317, 318, 324, 326, 329, 332, 333, 334, 335, 336, 338, 339, 340, 341, 342, 343, 344, 345, 346, 347, 348, 349, 350, 351, 352, 353, 354, 355, 359, 360, 361, 362, 366, 367, 368, 369, 370, 372, 373, 374, 375, 376, 377, 379, 381, 383, 384, 385, 386, 387, 388, 389, 390, 391, 392, 393, 394, 396, 397, 400, 401, 403, 405, 409, 410, 411, 412, 413, 414, 415, 416, 418, 419, 420, 422, 423, 424, 425, 426, 427, 428, 429, 430, 431, 432, 433, 434, 435, 436, 437

Chalcogenide-based materials 28, 141, 153, 154, 213, 235, 286, 295, 297, 324, 342, 345, 348, 350, 359, 392, 409, 412, 418, 419, 426, 430, 434, 435, 437

Chalcogenide nanocomposites 1, 2, 3, 4, 5, 6, 7, 8, 9, 11, 12, 14, 15, 17, 18, 20, 21, 22, 23, 24, 25, 26, 27, 28, 29, 30, 31, 32, 35, 39, 46, 52, 161, 162, 163, 168, 177, 211, 212, 216, 217, 218, 223, 225, 226, 227, 229, 230, 231, 232, 233, 234, 235, 241, 246, 285, 288, 316, 317, 318, 333, 334, 338, 349, 350, 359, 360, 361, 372, 373, 374, 376, 391, 392

chalcogenides 1, 2, 3, 4, 5, 6, 7, 9, 11, 15, 16, 17, 18, 20, 21, 22, 23, 24, 27, 29, 30, 31, 32, 33, 35, 36, 37, 38, 39, 40, 41, 46, 59, 60, 61, 62, 63, 64, 70, 72, 74, 81, 82, 83, 84, 85, 87, 90, 91, 92, 101, 102, 103, 105, 106, 107, 108, 110, 111, 112, 115, 116, 119, 122, 135, 136, 137, 141, 142, 143, 145, 148, 149, 150, 153, 155, 156, 157, 162, 163, 164, 165, 167, 168, 169, 170, 171, 172, 173, 177, 178, 179, 180, 181, 182, 183, 184, 213, 214, 215, 217, 218, 219, 220, 221, 223, 224, 225, 226, 227, 229, 230, 231, 232, 233, 234, 237, 238, 239, 242, 243, 244, 245, 246, 250, 282, 283, 284, 286, 287, 288, 291, 292, 296, 297, 300, 304, 307, 308, 311, 312, 313, 314, 315, 317, 318, 320, 321, 322, 323, 334, 335, 336, 337, 339, 340, 342, 343, 345, 346, 347, 348, 349, 350, 351, 352, 360, 361, 362,

363, 364, 365, 366, 367, 368, 369, 370, 371, 372, 373, 375, 377, 378, 379, 383, 384, 385, 386, 387, 389, 390, 391, 392, 393, 395, 397, 398, 402, 404, 410, 411, 412, 414, 415, 416, 417, 418, 419, 420, 422, 423, 427, 429, 432, 433, 434, 435, 437
Chalcogenides-based nanocomposites 379, 397, 398
Cheminformatics 442, 450, 456, 458, 461, 463, 465
CO2 transformation 359, 366, 369, 370, 374, 376
Cost-Effective Chalcogenide-Based Nanocomposite Electrodes for High-Performance Electrochemical and Photoelectrochemical Water Splitting 277
Counter Electrode 182, 212, 216, 219, 222, 231, 237, 239, 242, 246, 249, 250, 251, 252, 254, 265, 266, 268, 269, 270, 271, 272, 275, 323, 324, 434
Counter Electrodes 212, 213, 214, 215, 219, 220, 221, 227, 228, 229, 236, 238, 241, 243, 249, 250, 251, 252, 254, 259, 263, 265, 266, 267, 268, 269, 270

D

Density Functional Theory 115, 120, 122, 124, 129, 130, 133, 235, 299, 314, 338, 340, 347, 370
Doping 7, 46, 84, 85, 89, 90, 99, 104, 106, 118, 121, 123, 126, 130, 131, 138, 143, 144, 145, 155, 167, 170, 171, 180, 215, 220, 231, 234, 270, 271, 275, 280, 284, 286, 288, 289, 292, 295, 296, 298, 312, 313, 314, 317, 335, 338, 341, 342, 343, 344, 345, 346, 347, 349, 352, 366, 369, 399, 464, 465
Drug Discovery 440, 441, 443, 457, 467, 468, 469
Dynamical Mean Field Theory 122, 130

E

energy storage 1, 2, 3, 4, 5, 7, 9, 11, 12, 15, 20, 21, 22, 23, 24, 25, 30, 31, 32, 33, 34, 35, 38, 40, 41, 52, 53, 54, 59, 60, 61, 64, 70, 73, 85, 106, 107, 108, 117, 120, 130, 131, 134, 137, 160, 162, 164, 165, 167, 168, 169, 171, 172, 175, 176, 177, 179, 180, 183, 184, 240, 241, 242, 243, 246, 270, 271, 273, 279, 281, 295, 300, 316, 321, 322, 323, 326, 349, 355, 373, 377, 378, 384, 401, 402, 405, 410, 435
Environmental Applications 3, 37, 59, 60, 64, 300, 379, 392, 405, 433, 435

F

Financial Energy Literacy 185, 186, 190
Food Chain 395, 409, 412

G

Green Synthesis 13, 28, 59, 60, 63, 71, 72, 73, 74, 76, 77, 100, 216, 227, 233, 242, 244, 293, 297, 412, 413, 414, 416, 418, 420, 429, 430, 432
green technology 25

H

hybrid nanocomposites 103, 225, 340

I

Intercalation 42, 131, 136, 137, 140, 171, 342, 347, 348

M

Machine Learning 132, 134, 135, 153, 159, 201, 208, 235, 243, 317, 318, 439, 440, 441, 442, 446, 450, 460, 461, 464, 467
Metal Chalcogenides 23, 32, 33, 35, 38, 59, 60, 61, 62, 63, 64, 70, 72, 74, 81, 82, 83, 84, 85, 87, 90, 101, 102, 103,

106, 107, 110, 112, 115, 135, 137, 164, 165, 167, 168, 169, 170, 171, 172, 173, 178, 179, 180, 182, 183, 184, 214, 220, 237, 238, 239, 243, 244, 246, 250, 282, 291, 297, 304, 307, 308, 311, 312, 315, 317, 320, 321, 322, 323, 339, 340, 342, 343, 348, 351, 352, 364, 367, 368, 369, 371, 377, 379, 383, 386, 389, 391, 392, 402, 404, 411, 419, 434, 437
Micronutrient 410, 422

N

Nanocomposites 1, 2, 3, 4, 5, 6, 7, 8, 9, 11, 12, 13, 14, 15, 16, 17, 18, 19, 20, 21, 22, 23, 24, 25, 26, 27, 28, 29, 30, 31, 32, 33, 35, 37, 38, 39, 40, 42, 46, 47, 49, 52, 54, 81, 82, 83, 84, 85, 92, 93, 100, 101, 102, 103, 104, 107, 110, 111, 156, 161, 162, 163, 168, 177, 179, 183, 211, 212, 213, 214, 216, 217, 218, 219, 223, 225, 226, 227, 229, 230, 231, 232, 233, 234, 235, 236, 237, 241, 246, 249, 252, 253, 254, 256, 259, 261, 262, 265, 267, 268, 269, 271, 272, 274, 279, 282, 283, 284, 285, 286, 287, 288, 290, 291, 292, 294, 295, 297, 298, 299, 303, 304, 305, 306, 307, 308, 311, 312, 314, 315, 316, 317, 318, 319, 322, 326, 329, 332, 333, 334, 335, 336, 337, 338, 339, 340, 341, 342, 343, 344, 345, 347, 348, 349, 350, 351, 352, 353, 358, 359, 360, 361, 366, 367, 369, 370, 371, 372, 373, 374, 375, 376, 378, 379, 381, 382, 387, 389, 390, 391, 392, 393, 394, 395, 397, 398, 400, 401, 402, 404, 405, 407
Nanofertilizers 410, 412, 413, 414, 417, 420, 431
Nanomaterials 2, 12, 13, 20, 30, 32, 36, 37, 56, 57, 59, 60, 63, 75, 78, 105, 106, 111, 115, 137, 160, 162, 163, 164, 177, 212, 213, 216, 218, 219, 226, 227, 229, 231, 235, 237, 238, 240, 241, 243, 244, 246, 247, 248, 291, 292, 300, 304, 310, 324, 344, 346, 350, 355, 357, 378, 379, 380, 383, 384, 390, 391, 395, 396, 402, 403, 404, 406, 413, 415, 416, 422, 423, 426, 427, 428, 429, 430, 431, 432, 433, 434, 435, 436
nanotechnology 35, 59, 60, 75, 79, 110, 116, 138, 161, 163, 165, 170, 239, 243, 383, 403, 409, 410, 412, 414, 431
next-generation energy 2, 35, 117, 162, 163, 168, 173, 308
Novel Chalcogenide-Based Nanocomposites as Electrocatalysts for Nitrogen Reduction Reactions 329

O

Optical Properties 5, 6, 8, 12, 15, 17, 31, 40, 41, 49, 54, 95, 124, 126, 136, 138, 158, 165, 179, 180, 181, 182, 183, 216, 226, 295, 297, 320, 334, 415, 425
Optoelectronic 1, 6, 7, 8, 9, 11, 12, 13, 15, 19, 30, 31, 35, 38, 40, 47, 52, 55, 81, 82, 83, 84, 85, 86, 88, 91, 92, 95, 96, 97, 98, 99, 100, 101, 102, 103, 104, 110, 111, 115, 116, 117, 119, 121, 129, 130, 131, 132, 133, 134, 135, 141, 153, 156, 162, 163, 164, 165, 166, 168, 170, 172, 178, 213, 231, 243, 254, 287, 288, 334, 360, 384, 410
optoelectronics 1, 2, 3, 4, 7, 9, 12, 15, 16, 30, 40, 42, 46, 52, 59, 60, 64, 69, 70, 82, 83, 87, 91, 95, 96, 102, 110, 115, 134, 135, 142, 149, 182, 282
Oxygen Evolution 166, 170, 246, 251, 276, 282, 283, 296, 299, 303, 305, 306, 308, 309, 310, 314, 318, 320, 321, 323, 325, 326, 348, 354

P

perovskites 35, 36, 81, 82, 83, 84, 85, 89, 94, 95, 96, 97, 98, 99, 100, 101, 102, 103, 105, 108, 109, 110, 112, 225, 240, 243, 270, 318
Photoanodes 212, 214, 215, 216, 217, 218, 219, 227, 229, 231, 236, 237, 243, 244,

245, 246, 247, 254, 274, 287, 289, 299
photocatalysis 2, 3, 11, 26, 27, 32, 36, 37, 64, 67, 74, 105, 165, 166, 181, 238, 245, 246, 332, 333, 360, 361, 383, 384, 387, 388, 389, 391, 395, 397, 432
photocatalytic approach 382
Photodetectors 1, 3, 4, 6, 7, 9, 12, 13, 14, 15, 16, 17, 30, 31, 55, 82, 83, 85, 87, 88, 91, 92, 96, 97, 98, 105, 106, 117, 119, 121, 131, 132, 134, 141, 142, 148, 166, 168, 169, 334, 361, 436
Photovoltaic Materials 51, 156, 174, 231, 246

Q

Quantum Dot-Sensitized Solar Cells 36, 241, 244, 249, 268

R

Reduction Reactions 303, 306, 308, 309, 329, 330, 331, 332, 339, 349, 367

S

Solar cells 1, 2, 4, 6, 7, 9, 10, 11, 12, 14, 15, 17, 18, 29, 30, 32, 34, 35, 36, 40, 41, 48, 53, 55, 64, 69, 72, 75, 77, 83, 88, 89, 96, 97, 102, 104, 109, 110, 111, 113, 117, 132, 134, 141, 142, 143, 144, 148, 150, 154, 155, 156, 157, 160, 166, 167, 172, 173, 175, 182, 211, 212, 214, 216, 219, 221, 223, 230, 231, 232, 233, 236, 237, 238, 239, 240, 241, 242, 244, 245, 246, 247, 248, 249, 268, 270, 271, 272, 273, 274, 275, 276, 316, 323, 324, 355, 360, 362, 363, 384, 402, 404, 405, 433, 434, 436
Stability 4, 7, 8, 11, 17, 20, 22, 23, 24, 31, 40, 41, 47, 48, 57, 70, 84, 86, 89, 91, 92, 95, 96, 97, 98, 100, 101, 102, 103, 104, 120, 124, 132, 135, 138, 143, 144, 145, 148, 150, 153, 155, 159, 161, 163, 164, 166, 167, 169, 170, 171, 172, 173, 174, 175, 176, 186, 191, 194, 204, 213, 214, 217, 218, 219, 220, 221, 222, 223, 224, 225, 226, 227, 229, 230, 231, 232, 234, 235, 236, 244, 247, 268, 269, 278, 279, 280, 281, 284, 285, 289, 291, 292, 293, 294, 295, 296, 298, 304, 306, 307, 308, 310, 312, 314, 315, 316, 317, 331, 339, 341, 345, 346, 347, 348, 349, 350, 351, 352, 353, 363, 366, 369, 370, 371, 376, 397, 399, 415, 418, 424, 428, 431, 436, 440, 441, 442, 443, 448, 453, 461, 462, 468
Structure 5, 6, 8, 13, 14, 46, 48, 50, 71, 83, 85, 91, 93, 95, 97, 99, 101, 115, 116, 118, 119, 120, 121, 122, 123, 124, 125, 127, 128, 129, 130, 131, 132, 134, 136, 143, 146, 147, 148, 153, 157, 164, 168, 179, 183, 214, 215, 225, 235, 239, 249, 251, 252, 254, 283, 284, 285, 304, 311, 312, 317, 323, 325, 333, 335, 338, 340, 341, 342, 343, 344, 345, 346, 347, 349, 352, 360, 362, 363, 364, 365, 370, 373, 377, 385, 392, 401, 403, 405, 411, 435, 440, 441, 442, 444, 446, 450, 451, 452, 453, 454, 456, 458, 459, 460, 461, 462, 463, 468, 469, 470
Sulfur Fertilizer 410
Sustainability 1, 2, 3, 4, 13, 20, 25, 28, 29, 30, 31, 77, 83, 100, 107, 110, 133, 135, 152, 159, 162, 164, 167, 172, 173, 174, 175, 176, 177, 185, 186, 187, 189, 190, 191, 196, 197, 199, 201, 202, 204, 205, 207, 209, 216, 221, 232, 233, 239, 247, 250, 273, 275, 276, 280, 281, 283, 293, 296, 297, 298, 301, 305, 316, 351, 357, 373, 374, 380, 407, 413, 415, 416, 417, 419, 425, 429, 433, 435
Sustainable Development 59, 60, 61, 63, 64, 70, 185, 188, 189, 190, 201, 204, 205, 208, 209, 299, 300, 332, 356, 412, 415
Synthesis Route 253

T

Thermoelectric 7, 12, 13, 14, 20, 24, 29, 31, 36, 53, 54, 57, 81, 82, 83, 84, 85, 86, 90, 91, 92, 93, 95, 96, 99, 100, 101, 102, 103, 104, 105, 107, 108, 110, 111, 112, 113, 143, 144, 145, 148, 149, 151, 152, 154, 155, 162, 163, 164, 165, 167, 178, 181, 182, 243, 360, 361, 407, 433, 434, 435, 436, 437

Thermo electrics 141

Transition Metal Dichalcogenides 101, 107, 108, 111, 113, 115, 117, 118, 119, 120, 122, 128, 138, 139, 170, 214, 283, 320

U

UV-assisted techniques 161, 163

V

valuable chemicals and clean energy generation 359